PHYSICS FOR ENGINEERS

PHYSICS FOR ENGINEERS

(For B.E., B.Tech., AMIE and Other Degree Classes)

By

Dr. H.C. SHARMA

Professor, H.C.T.M., Kaithal

Former Professor and Chairman, Physics Department,

Kurukshetra University, Kurukshetra

Haryana

Dr. RAJESH KHARB

Lecturer, Physics Department.

Kurukshetra University, Kurukshetra

Haryana

Dr. RAJESH SHARMA

Asst. Professor

H.C.T.M., Kaithal

Haryana

UNIVERSITY SCIENCE PRESS

(An Imprint of Laxmi Publications Pvt. Ltd.)

NGALORE • CHENNAI • COCHIN • GUWAHATI • HYDERABAD
DHAR • KOLKATA • LUCKNOW • MUMBAI • PATNA
RANCHI • NEW DELHI

Published by :
UNIVERSITY SCIENCE PRESS
(*An Imprint of Laxmi Publications Pvt. Ltd.*)
113, Golden House, Daryaganj,
New Delhi-110002

Phone : 011-43 53 25 00
Fax : 011-43 53 25 28

www.laxmipublications.com
info@laxmipublications.com

Price : **Rs. 275.00** *Only.*

First Edition : 2004, *Second Edition* : 2005
Third Edition : 2006, *Fourth Edition* : 2008
Fifth Edition : 2009, *Sixth Edition* : 2010

OFFICES

© **Bangalore**	080-26 75 69 30	© **Chennai**	044-24 34 47 26
© **Cochin**	0484-237 70 04, 405 13 03	© **Guwahati**	0361-251 36 69, 251 38 81
© **Hyderabad**	040-24 65 23 33	© **Jalandhar**	0181-222 12 72
© **Kolkata**	033-22 27 43 84	© **Lucknow**	0522-220 99 16
© **Mumbai**	022-24 91 54 15, 24 92 78 69	© **Patna**	0612-230 00 97
© **Ranchi**	0651-221 47 64		

UPE-9252-275-PHYSICS FOR ENGINEERS-SHA C—2157
Typeset at : Goswami Associates, Delhi. *Printed at* : Ajit Print

CONTENTS

FOREWORD

Physics course of B. Tech. first semester and second semester in Engineering Colleges involve all branches of modern Physics like Nuclear Physics, Physical Optics, Lasers, Material Science, Modern Physics and Theory of Relativity etc. At present, there is no book, which exclusively deals with these topics according to the syllabus requirement of the Engineering Colleges, hence, it is an important step forward in this direction.

While basic physical concepts have been kept intact, mathematical intricacies have been avoided. Therefore, present book 'Physics for Engineers' is ideally suited to the needs of B. Tech. students whose level in physics is generally low. Book contains several interesting solved as well as unsolved numerical problems.

The book is compiled on the basis of classroom lectures of highly experienced teachers, and is of manageable size. It will be very useful not only for the engineering students of Haryana but also of other states.

D.K. Chaturvedi
Professor and Chairman
Department of Physics,
Kurukshetra University, Kurukshetra.

Preface to The Sixth Edition

The present book consists of both of earlier books 'Physics for Engineers' Vol. I and Vol. II which contains the syllabus of B.Tech. 1st Semester as well as of 2nd Semester. Further, topics on Interaction of Radiation with Matter, Elementary Idea of Quantum Statistics and Point Defects in crystals have been added in the last chapter of 'Special Topics'. Also, chapter on X-ray diffraction has been strengthened. Further, topic of 'Quark and Gluons' has been replaced by 'Guass law and its applications' in Chapter 29. Chapter 30 on Nanosciences has been added as per revised syllabus of M.D.U., Rohtak. Thus, the present volume now fully covers the syllabus of B.Tech. 1st year of K.U., Kurukshetra; M.D.U., Rohtak and Ch. Devi Lal University, Sirsa.

While preparing the present edition of the book, Mr. Manoj Kumar, HEC, Jagadhari had made a very critical review leading to useful alterations.

Lastly, we acknowledge with thanks the suggestions from Dr. Sanjay Sharma, Director General, E-max Institute of Engineering & Technology, Ambala; Dr. D. S. Rana, USIC K.U., Kurukshetra; Anil Singhal, NCCE Israna (Panipat); and Atul Ahuja, SDIET, Barwala(Panchkula). We will welcome constructive suggestions as well as comments from readers in future also.

—Authors

Preface to The First Edition

The present book consists of both of our earlier books 'Physics for Engineers' Vol. I and Vol. II. Part-I of the present book contains the syllabus of B. Tech. 1st semester while Part-II contains that of 2nd semester. Further, topics on Interaction of Radiation with Matter, Elementary Idea of Quantum Statistics, and Point Defects in crystals have been added in the last chapter of 'Special Topics'. Also, chapter on X-Ray diffraction has been strengthened. Thus, the present volume now fully covers the syllabus of B. Tech. 1st year of K.U., Kurukshetra, M.D.U., Rohtak and Ch. Devi Lal University, Sirsa.

While preparing the present book, Mr. Manoj Kumar, HEC, Jagadhari had made a very critical review leading to useful alterations.

Lastly, we acknowledge with thanks the suggestions from Dr. Sanjay Sharma, Ambala College of Engineering and Applied Research, Mithapur, Dr. D.S. Rana, USIC K.U., Kurukshetra, Anil Singhal, NCCE Israna (Panipat), and Atul Ahuja, SDIET, Barwala (Panchkula). We will welcome constructive suggestions as well as comments from readers in future also.

—Authors

Chapter 1

INTERFERENCE

1.1 CONCEPT OF INTERFERENCE

Intensity of light is the amount of energy across unit area perpendicular to the direction of propagation of light. This quantity is found as proportional to $<E^2>$ or $<H^2>$, where $< >$ denotes the time average over many time periods. Therefore in the following we would regard $<E^2>$ as a measure of intensity of light.

Now electric vector associated with light (which consists of electromagnetic radiation) is given by

$$\mathbf{E}(r,t) = \frac{1}{2}\,[\mathbf{A}^*(r)\, e^{i\omega t} + \mathbf{A}(r)\, e^{-i\omega t}] \qquad ...(1.1)$$

where angular frequency ω and time period T are related as $\omega = 2\pi/T$.

Then $$I = <E^2> = \frac{1}{4}\,[A^2<e^{-2i\omega t}> + A^{*2}<e^{2i\omega t}> + 2(\mathbf{A}.\mathbf{A}^*)] = \frac{1}{2}\,|\,\mathbf{A}\,|^2 \qquad ...(1.2)$$

Also it is common to write $\mathbf{A}(\mathbf{r}) = (A_1, A_2, A_3)$

such that $$A_1(r) = a_1(r)\, e^{ig1(r)}$$

$$A_2(r) = a_2(r)\, e^{ig2(r)}$$

and $$A_3(r) = a_3(r)\, e^{-ig3(r)}$$

where $a_i(r)$ and $g_i(r)$ are real quantities and are functions of position r

so, $$I = \frac{1}{2}\ (a_1^{\,2} + a_2^{\,2} + a_3^{\,3})$$

If we have two different beams of light of intensity say I_1 and I_2, then

$$I_2 = <E_2^{\,2}> = \frac{1}{2}\ |\,\mathbf{B}\,|^2$$

where $\mathbf{B}(\mathbf{r}) = (B_1, B_2, B_3)$ for the second beam such that $B_i(r) = b_i(r)\, e^{ihi(r)}$

Obviously $$I_1 = \frac{1}{2}(a_1^2 + a_2^2 + a_3^3)$$

and $$I_2 = \frac{1}{2}(b_1^2 + b_2^2 + b_3^3)$$

When these two beams superimpose, the resultant electric vector $\mathbf{E} = (\mathbf{E}_1 + \mathbf{E}_2)$ and the resultant intensity would be given by

$$I = \langle E^2 \rangle = \langle E_1^2 \rangle + \langle E_2^2 \rangle + \langle 2(\mathbf{E}_1.\mathbf{E}_2) \rangle = I_1 + I_2 + 2\langle (\mathbf{E}_1.\mathbf{E}_2) \rangle = I_1 + I_2 + J_{12}$$

where $2\langle (\mathbf{E}_1.\mathbf{E}_2) \rangle = J_{12}$ is called the interference term because when $J_{12} = 0, I = I_1 + I_2$, just the sum of the two intensity.

Calculation of J_{12}

Since $$\mathbf{E}_1 = \frac{1}{2}(\mathbf{A}e^{-i\omega t} + \mathbf{A}^* e^{i\omega t})$$

and $$\mathbf{E}_2 = \frac{1}{2}(\mathbf{B}e^{-i\omega t} + \mathbf{B}^* e^{i\omega t})$$

$$2\langle (\mathbf{E}_1.\mathbf{E}_2) \rangle = \frac{1}{2}(\mathbf{A.B}^* + \mathbf{A}^*.\mathbf{B})$$

as on time averaging other terms get cancelled. Further,

$$\mathbf{A}(r) \equiv (a_1 e^{ig1}, a_2 e^{ig2}, a_3 e^{ig3})$$

and $$\mathbf{B}(r) \equiv (b_1 e^{ih1}, b_2 e^{ih2}, b_3 e^{ih3})$$

where a_i, g_i, b_i and h_i are real functions of r.

$$\mathbf{A.B}^* = (a_1 b_1 + a_2 b_2 + a_3 b_3)e^{i\delta}$$

because usually $$g_1 - h_1 = g_2 - h_2 = g_3 - h_3 = \delta$$

Similarly $$\mathbf{A}^*.\mathbf{B} = (\Sigma_i a_i b_i)\, e^{-i\delta}$$

Thus, $$J_{12} = 2\langle \mathbf{E}_1.\mathbf{E}_2 \rangle = (\Sigma_i a_i b_i) \cos \delta$$

and $$I = I_1 + I_2 + (\Sigma_i a_i b_i) \cos \delta \qquad ...(1.3)$$

While deriving the result of eq. (1.3) we did not use any property of electromagnetic radiation *e.g.* (*i*) two beams polarised at right angle to each other do not interfere or (*ii*) the fact that electric and magnetic vectors of electromagnetic radiation, which the light is, are transverse in nature. Let us now use these facts in eq. (1.3).

Suppose the electric vector of one beam is in *X-Z* plane and that of other is in *Y-Z* plane while the direction of propagation of beam is *Z*-axis. As a result

$$a_2 = b_1 = 0$$

and $$J_{12} = a_3 b_3 \cos \delta$$

if these beams, which are polarized at right angle to each other do not interfere, then

$$J_{12} = 0 \quad i.e. \quad a_3 = b_3 = 0.$$

In other words, the electric vectors associated with the two beams are transverse to the direction of propagation an important property of electromagnetic radiation. Therefore, let us assume that electric vectors in the beams are vibrating along *X*-axis. Then

$$J_{12} = a_1 b_1 \cos \delta$$

While $$I_1 = \frac{a_1^2}{2}$$

and $$I_2 = \frac{b_1^2}{2}$$

or $$J_{12} = 2\sqrt{I_1 I_2}\ \cos\delta$$

This result shows that when $|\delta| = 0, 2\pi, 4\pi, \ldots$ the intensity $I = \left(\sqrt{I_1} + \sqrt{I_2}\right)^2$ is maximum or interference is constructive and when $|\delta| = \pi, 3\pi, 5\pi\ldots$ the intensity $I = \left(\sqrt{I_1} - \sqrt{I_2}\right)^2$ is minimum or interference is destructive.

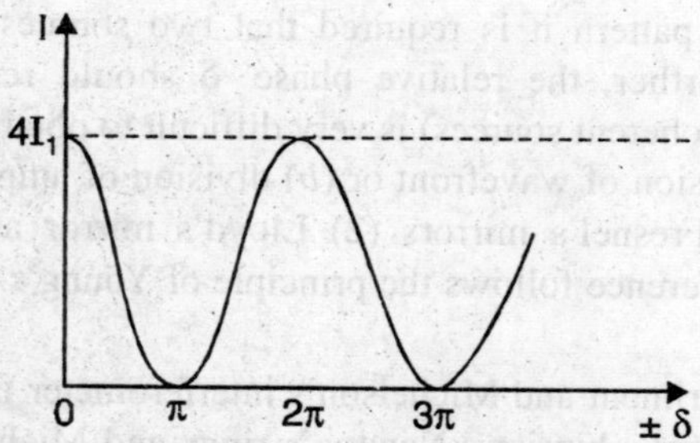

FIGURE 1.1 **The intensity variation due to interference ; $I_1 = I_2$**

Finally when the two beams are of same intensity

$$I = 4I_1 \quad \text{for} \quad |\delta| = 0, 2\pi, 4\pi, \ldots$$

and $$= 0 \quad \text{for} \quad |\delta| = \pi, 3\pi, 5\pi\ldots$$

The variation of I with δ would be as shown in Fig. 1.1.

It may also be noted that even for natural unpolarized light the above result remains valid *i.e.* when the phase difference is $0, 2\pi, 4\pi, \ldots$ the interference would be constructive and when $|\delta| = \pi, 3\pi, 5\pi\ldots$ the interference will be destructive.

Alternatively, linear polarized beams may also be represented by

$$E_1 = A_1 \cos \omega t$$

and $$E_2 = A_2 \cos(\omega t + \delta)\ ; \text{ where } \delta = \text{phase difference}$$

so that $$E = E_1 + E_2 = (A_1 + A_2 \cos\delta)\cos\omega t - A_2 \sin\delta \sin\omega t = A\cos(\omega t + \alpha)$$

where $A = (A_1^2 + A_2^2 + 2A_1A_2\cos\delta)^{1/2}$

and $$\alpha = \tan^{-1}\{A_2 \sin\delta/(A_1 + A_2\cos\delta)\}$$

Also then $$I_1 = \langle E_1^2 \rangle = \frac{A_1^2}{2}$$

$$I_2 = \langle E_2^2 \rangle = \frac{A_2^2}{2}$$

and $$I = \langle E^2 \rangle = \frac{A^2}{2}$$

$$= \frac{1}{2}(A_1^2 + A_2^2 + 2A_1A_2 \cos \delta)$$

$$= \frac{1}{2}(2I_2 + 2I_2 + 4\sqrt{I_1I_2} \cos \delta)$$

$$= (I_1 + I_2 + 2\sqrt{I_1I_2} \cos \delta)$$

(obtained by putting the value of A in terms of A_1 and A_2 from above)

Thus, we obtain the same conditions of maxima and minima as discussed earlier.

It may be noted that phase difference δ is related to path difference p_d as $\delta = \frac{2\pi}{\lambda} \cdot p_d$

$\therefore$ For maxima and minima $\quad p_d = n\lambda$ and $(2n + 1)\lambda/2$.

For a good interference pattern it is required that two sources should emit light having same wavelength and amplitude. Further, the relative phase δ should remain constant. Such a pair of monochromatic sources (called coherent sources) is very difficult to obtain directly. However, such sources can be obtained through (*a*) division of wavefront or (*b*) division of amplitude of light from same source. In the first category falls (1) Fresnel's mirrors (2) Lloyd's mirror and (3) Fresnel's biprism. These arrangements of observing interference follows the principle of Young's double slit experiment which has been discussed in Appendix-I

The Newton's rings experiment and Michelson's interferometer fall in the second category *i.e.* the division of amplitude. The Fresnel's biprism, Newton's rings and Michelson's interferometer have been discussed in detail here.

1.2 FRESNEL'S BIPRISM

It is an optical device, which is used to obtain two coherent sources of light from a single source. These two coherent sources are in turn used to study the interference of light.

Construction

Fresnel biprism consists of a combination of two acute angled prisms placed base to base. In fact, this combination is obtained from a single optically plane glass plate by proper grinding and polishing. The obtuse angle of the biprism is 179° and other two acute angles are 30′ each.

Working

The experimental arrangement to study the interference of light using Fresnel biprism is shown in Fig. 1.2.

The slit S is illuminated by a monochromatic source of light like sodium lamp. The light emerging from slit falls on the biprism. The edge B of the biprism divides the incident wave front into two parts. One of the parts passes through the upper half, ABD, of the biprism and is deviated through a small angle towards the lower half while the other part passes through the lower half, CBD, and is deviated towards the upper half. As a result, the light appears to come from the virtual sources S_1 and S_2 as shown in Fig. 1.2. Since the virtual sources S_1 and S_2 are derived from a single primary source S, therefore, these acts as two coherent sources. The interference fringes so obtained lie in overlapping region EF and can be seen through eyepiece.

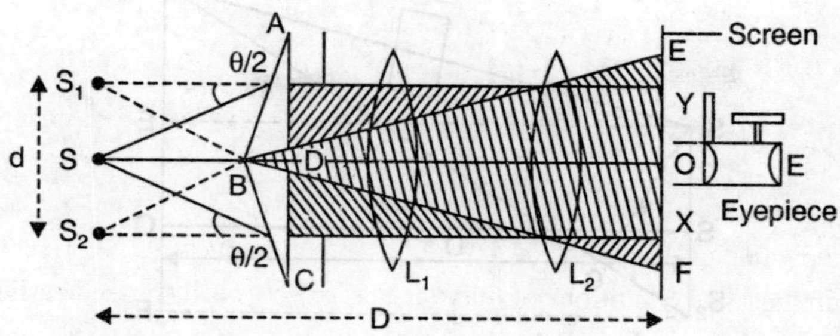

FIGURE 1.2 **Interference phenomenon by Fresnel biprism experiment ; ∠CBA = 179°.**

Theory

The wavelength λ, of the light can be determined through the following relation

$$\lambda = \frac{\beta d}{D} \qquad \text{(derived in Appendix I)}$$

where β is fringe width, d is distance between two virtual sources and D is the distance between the slit and the eyepiece, where fringes are observed. Hence knowing β, d and D, wavelength can be calculated. The procedures of the measurement of these quantities are being given in the following discussion.

Measurement of fringe width β

The vertical cross wire of the eyepiece is adjusted on any one of the bright fringes and the reading of the micrometer screw is noted. Now we set the cross wire one by one on different bright fringes and note down the reading in each case. The mean distance between two fringes, β, is then determined as

$$\beta = \text{Distance travelled / No. of fringes moved.}$$

Measurement of D

The distance D can be directly read on the optical bench.

Measurement of d

To determine d, a lens with focal length less than 1/4th of the distance between eyepiece and biprism, is mounted between eyepiece and biprism. The lens is adjusted in two different positions L_1 and L_2 such that sharp images of S_1 and S_2 are obtained in the field of view of the eyepiece. The distance d_1 and d_2 between the images in two cases are measured. Now, d is calculated through the relation $d = (d_1 d_2)^{1/2}$.

Uses of Biprism

The biprism can be used to determine the thickness of a thin sheet of transparent material such as glass, mica, etc.

When a thin transparent sheet of thickness t and refractive index, μ, is introduced in the path of one of the interfering beam, the central bright fringe is shifted to some other point due to change in path difference.

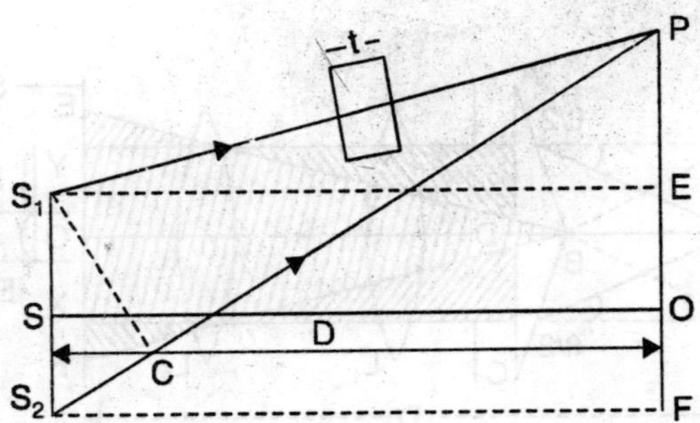

FIGURE 1.3 **Experimental determination of thickness of thin sheet of transparent material by Fresnel's biprism experiment.**

However, the fringe width remains unchanged. Let it shift to point P. Join P to S_1 and S_2. Now consider path S_1P. The length $(S_1P - t)$ of this path is travelled in air with velocity c and the length t is travelled with velocity c_m where c_m is the velocity of light in medium. Therefore, time taken to cover the distance S_1P

$$\frac{(S_1P - t)}{c} + \frac{t}{c_m} = \frac{(S_1P - t)}{c} + \frac{\mu t}{c} = \frac{[S_1P + (\mu - 1)t]}{c} \qquad \left[\because \ \mu = \frac{c}{c_m}\right]$$

Thus the path S_1 to P is equivalent to path in air equal to $S_1P + (\mu - 1)t$.

Now, path difference at P = path S_2P in air – path S_1 to P

$$= S_2P - [S_1P + (\mu - 1)t]$$

$$= \frac{xd}{D} - (\mu - 1)t \quad \text{...(1.4)} \qquad \left[\because \ S_2P - S_1P = \frac{xd}{D} \text{ See Appendix I}\right]$$

for nth maxima, path difference = $n\lambda$ ...(1.5)

from eq. (1.4) and (1.5) for nth maxima

$$\frac{x_n d}{D} - (\mu - 1)t = n\lambda \qquad \text{...(1.6)}$$

or

$$x_n = \frac{D}{d}\,[n\lambda + (\mu - 1)t]$$

In the absence of plate $t = 0$

Then

$$x_n = \frac{nD\lambda}{d}$$

If S denotes the displacement of nth maxima by introducing the transparent sheet, then

$$S = \frac{D}{d}\,[n\lambda + (\mu - 1)t] - \frac{Dn\lambda}{d}$$

or

$$S = \left(\frac{D}{d}\right)(\mu - 1)t \qquad \text{...(1.7)}$$

Eq. (1.7) is independent of n, hence the displacement of each maxima is same.

Thickness of Thin Sheet

The displacement S of any maxima by introducing a thin sheet of thickness t is given by the relation

$$S = \frac{D}{d}(\mu - 1)t \quad \text{or} \quad t = \frac{Sd}{D(\mu - 1)} \qquad ...(1.8)$$

So through the knowledge of S, d, D and μ, t can be determined or if t is known μ can be determined.

1.3 INTERFERENCE IN THIN FILMS

Every one is quite familiar with the beautiful colours produced by a thin film of oil floating over the surface of water and also by the thin film of soap bubble. These interesting phenomenon could be explained on the basis of interference between light reflected from the top and bottom surface of a thin film. The interference in case of thin film takes place due to reflected light as well as transmitted light.

Interference Due to Reflected Light

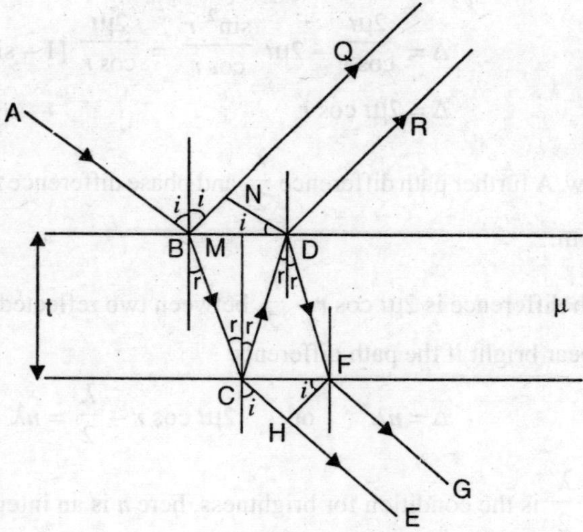

FIGURE 1.4 **Colour formation in thin films.**

Consider an incident ray AB of monochromatic light of wavelength λ, falling at an angle of incidence i on the film of thickness t and refractive index μ as shown in Fig. 1.4.

The ray is partly reflected along BQ and partly refracted along BC. Let angle of refraction be r. The ray BC is again partly reflected along CD from lower surface of the film and partly refracted along CE parallel to AB at C it is again partly reflected along CD inside the film and partly refracted along DR. Similar reflections and refraction take place at D and F as shown in Fig. 1.4. The rays BQ and DR interfere with each other due to reflected light and rays CE and FG interfere due to transmitted light.

Now we calculate path difference between two reflected rays BQ and DR. The rays BQ and DR have a common path beyond the plane DN, where DN is perpendicular to BQ. Also, these rays have a common path up to B. The path difference is therefore given by

$$\Delta = (BC + CD) \text{ in film} - BN \text{ in air}$$

$$\Delta = (BC + CD)\,\mu - BN \quad ...(1.9)$$

here μ is refractive index of the film.

Now from Fig. 1.4, $BC = CD = t/\cos r$ ...(1.10)

From ΔBND $\sin i = \frac{BN}{BD}$ or $BN = BD \sin i = 2BM \sin i$ ...(1.11)

From ΔBCM $\tan r = \frac{BM}{MC} = \frac{BM}{t}$ or $BM = t \tan r$ ...(1.12)

And as we know $\mu = \frac{\sin i}{\sin r}$ or $\sin i = \mu \sin r$ ...(1.13)

Substituting eqns. (1.12) and (1.13) from eq. (1.11), we get

$$BN = 2t \tan r \,.\, \mu \sin r$$

$$BN = \frac{2\mu t \sin^2 r}{\cos r} \quad ...(1.14)$$

Inserting the values of BC and BN in eq. (1.9), we obtain

Path difference $\Delta = \frac{2\mu t}{\cos r} - 2\mu t \frac{\sin^2 r}{\cos r} = \frac{2\mu t}{\cos r}[1 - \sin^2 r]$

or $\Delta = 2\mu t \cos r$ ...(1.15)

which is known as cosine law. A further path difference $\frac{\lambda}{2}$ and phase difference π is produced when reflection takes place at denser medium.

So the effective path difference is $2\mu t \cos r - \frac{\lambda}{2}$ between two reflected waves.

1. The film will appear bright if the path difference

$$\Delta = n\lambda \quad \text{or} \quad 2\mu t \cos r - \frac{\lambda}{2} = n\lambda$$

or $2\mu t \cos r = (2n + 1)\frac{\lambda}{2}$ is the condition for brightness, here n is an integer including 0.

2. The film will appear dark if the path difference

$$\Delta = \frac{(2n+1)\lambda}{2} \quad \text{or} \quad 2\mu t \cos r - \frac{\lambda}{2} = \frac{(2n+1)\lambda}{2}$$

or $2\mu t \cos r = n\lambda$ is the condition for darkness.

Interference Due to Transmitted Light

The path difference between the two transmitted waves CE and FG is also given by

$$(CD + DF)\mu - CH = 2\mu t \cos r$$

there is no additional path difference of $\frac{\lambda}{2}$ as reflection takes place from rarer medium.

1. So, the film will appear bright if

$$2\mu t \cos r = n\lambda$$

2. The film will appear dark if

$$2\mu t \cos r = (2n + 1)\frac{\lambda}{2} \quad \text{where } n = 0, 1, 2, 3\ldots$$

It means that the reflected and transmitted systems are complimentary to each other. In other words, if a film appear bright for reflected light, it will appear dark for transmitted light and *vice-versa*.

Colours of Thin Films

When white light is incident on a thin film the light which comes from any point from it will not include the colour whose wavelength satisfies the eq. $2\mu t \cos r = n\lambda$, in the reflected system. Therefore the film will appear coloured and the colour will depend upon the thickness and the angle of inclination. If r and t are constant, the colour will be uniform. In the case of oil on water different colours are seen because r and t vary.

1.4 NEWTON'S RINGS

When the convex surface of a plano-convex lens is placed on a plane glass plate, a thin air film is produced between the two surfaces. The thickness of the air film so produced increases as we go outwards from the point of contact. Also, the loci of all the points corresponding to same thickness are circles concentric with the point of contact. Hence the interference due to such a film gives circular fringes concentric with the point of contact. The circular fringes so produced are known as Newton's rings. Newton's rings can be formed due to reflected as well as transmitted light.

Newton's Rings by Reflected Light

To produce Newton's rings by reflected monochromatic light, the light coming from a point source is made parallel by a convex lens L. The light then falls on a plane glass plate G which is inclined at an angle 45° to

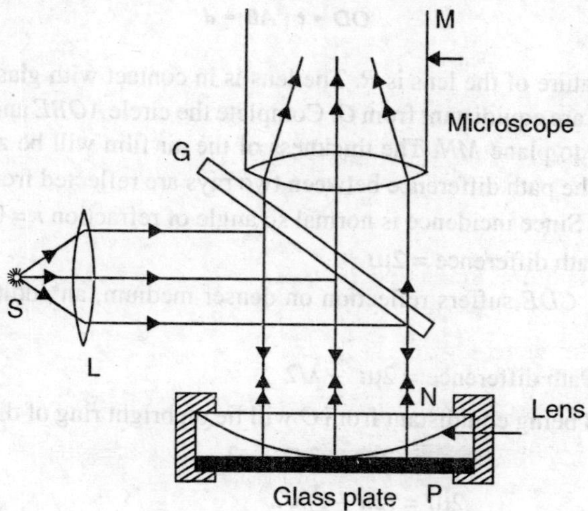

FIGURE 1.5 **Experimental arrangement of Newton's ring experiment.**

the incident beam and is reflected normally on the surface of plano-convex lens of large radius of curvature placed with its plane surface upside on a glass plate P in a holder shown in Fig. 1.5.

Light is reflected upwards from the upper and lower face of the wedge-shaped air film formed between the curved surface of the lens and the glass plate P. These two reflected light superimpose upon each other with a path difference depending on the thickness of the air film at the point of incidence and so produce alternate dark and bright circular fringes. The circularity of the fringes is due to the sharp contact between the glass plate and the plano-convex lens. The circular fringes are observed by means of a low power travelling microscope and their diameter can be measured.

Theory

Let a monochromatic light ray AB fall normally on air film at the point B after passing through the lens. Part of it is reflected from point B (rarer medium) on glass-air boundary and goes upward along BF without phase reversal. Other part refracts into air film along BC. At point C again reflection takes place on glass plate (denser medium). The reflected light goes along CDE with a phase change of π or path difference $\lambda/2$. The two rays BF and CDE produce interference fringes depending upon their path difference.

To obtain an expression for the diameter of a ring consider the vertical section AOB of plano-convex lens through the center of curvature as shown in Fig. 1.6.

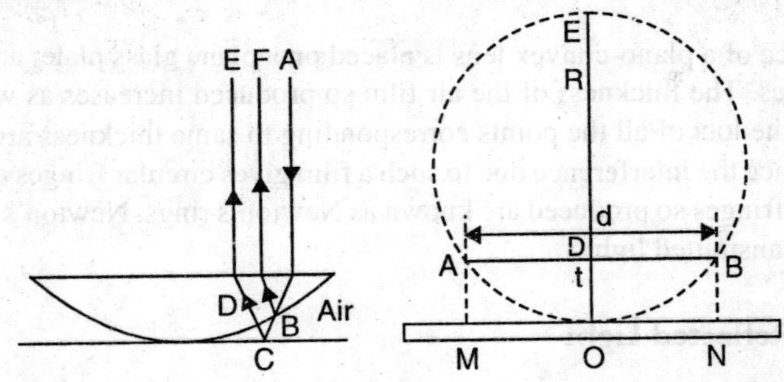

FIGURE 1.6 **Ring's formation in reflected system and mathematical determination of Newton's ring diameter.**

$OD = t$; $AB = d$

The radius of curvature of the lens is R. The lens is in contact with glass plate MON at O in such a manner that points A and B are equidistant from O. Complete the circle $AOBE$ and draw diameter ODE. Draw BN and AM perpendicular to plane MN. The thickness of the air film will be zero at O, around O circular fringes will be obtained. The path difference between two rays are reflected from $B(BF)$ and other reflected from $C(CDE)$ is $2\mu t \cos r$. Since incidence is normal so angle of refraction $r = 0$.

$$\therefore \quad \text{Path difference} = 2\mu t$$

As one of the rays CDE suffers reflection on denser medium, an additional path difference $\lambda/2$ is introduced.

$$\therefore \quad \text{Path difference} = 2\mu t + \lambda/2$$

The points A and B being equidistant from O will lie on bright ring of diameter AB if path difference

$$= 2\mu t + \lambda/2 = n\lambda$$

or

$$2\mu t = (2n - 1)\lambda/2 \qquad \text{...(1.16)}$$

Similarly, the points A and B will lie on the dark ring when path difference

$$2\mu t + \lambda/2 = (2n + 1)\lambda/2 \qquad \text{...(1.17)}$$

or

$$2\mu t = n\lambda$$

Now from the geometry of the circle in Fig. 1.6

$$AD \times DB = OD \times DE = OD(OE - OD) = t(2R - t) = 2Rt - t^2 \qquad \text{...(1.18)}$$

where

$$OD = t$$

If t is very small so t^2 can be neglected.

$$\therefore \qquad AD \times DB = 2Rt$$

If d is the diameter of ring then $AD = DB = \frac{d}{2}$

Hence,

$$\frac{d}{2} \times \frac{d}{2} = 2Rt \quad \text{or} \quad \frac{d^2}{4} = 2Rt \quad \text{or} \quad 2t = \frac{d^2}{4R} \qquad \text{...(1.19)}$$

If A and B lie on nth bright ring of diameter d_n then putting the value of $2t$ in eq. (1.16)

$$\frac{\mu d_n^2}{4R} = (2n - 1)\frac{\lambda}{2} \qquad \text{...(1.20)}$$

and for the points to lie on nth dark ring, we have from eq. (1.17)

$$\frac{\mu d_n^2}{4R} = n\lambda \qquad \text{...(1.21)}$$

or

$$d_n^2 = \frac{4Rn\lambda}{\mu} \quad \text{or} \quad d_n \propto \sqrt{n}$$

The diameters of dark rings are proportional to the square root of natural numbers. From eq. (1.20), we have diameter of nth bright ring

$$d_n^2 = 4R(2n - 1)\frac{\lambda}{2\mu} \quad \text{or} \quad d_n \propto \sqrt{2n - 1}$$

Thus the diameter of bright rings are proportional to square root of odd natural number.

Determination of Wavelength of Sodium Light

Let d_n and d_m be the diameter of nth and mth bright rings. The medium enclosed between convex lens and glass plate is air *i.e.* $\mu = 1$, from eq. (1.20), we have

$$\frac{d_n^2}{4R} = (2n - 1)\frac{\lambda}{2} \quad \text{and} \quad \frac{d_m^2}{4R} = (2m - 1)\frac{\lambda}{2},$$

Subtracting these eqns., we have

$$\frac{d_n^2 - d_m^2}{4R} = 2(n - m)\frac{\lambda}{2} \quad \text{or} \quad \lambda = \frac{d_n^2 - d_m^2}{4R(n - m)}$$

Determination of Refractive Index of Liquids

A drop of liquid is placed in the space between plano-convex lens and glass plate then Newton's rings are observed. If d_n and d_m are the diameter of nth and mth bright rings, then applying eq. (1.20) we have

$$\frac{\mu d_n^2}{4R} = (2n-1)\frac{\lambda}{2} \quad \text{and} \quad \frac{\mu d_m^2}{4R} = (2m-1)\frac{\lambda}{2}$$

Subtracting these eqns., we have $\mu \dfrac{\left(d_n^2 - d_m^2\right)}{4R} = (n-m)\lambda$

or
$$\mu = (n-m)\lambda \frac{4R}{\left(d_n^2 - d_m^2\right)}$$

Knowing the wavelength of light λ and by measuring d_n, d_m and R in the usual manner μ can be calculated.

If λ is not known than knowing the d_{air} and d_{liq}, μ can be calculated as

$$d_{air}^2 = 4Rn\lambda$$

$$d_{liq}^2 = \frac{4Rn\lambda}{\mu}$$

or
$$\mu = \frac{d_{air}^2}{d_{liq}^2}$$

Newton's Rings by Transmitted Light

When Newton's rings are obtained by reflected light, the central spot was dark due to additional path difference $\frac{\lambda}{2}$ in one of the interfering rays. The interference can be obtained by transmitted light rays BE and DF as in Fig. 1.7. The ray BE originates from AB by single reflection at B while the ray DF originate from AB by two

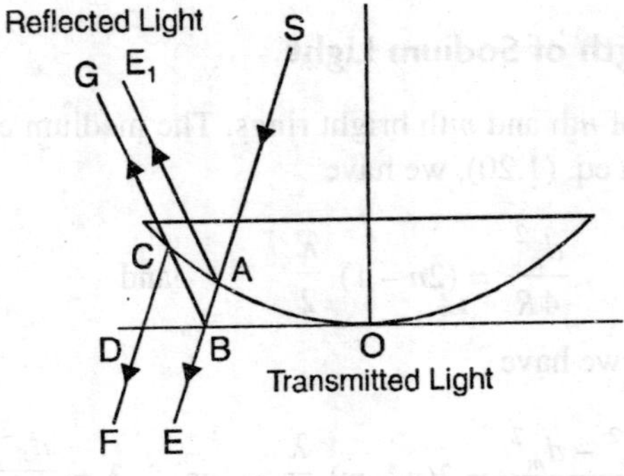

FIGURE 1.7 **Ring's formation in transmitted system.**

reflection one at B and other at C on denser medium. At each at reflection phase change of π or path difference $\frac{\lambda}{2}$ is introduced.

$\therefore$ Total phase change is $= \pi + \pi = 2\pi$

or path difference $= \frac{\lambda}{2} + \frac{\lambda}{2} = \lambda$

which means no phase change or path difference. So only path difference $= 2\mu t \cos r$ in two interfering beams BE and DF.

If $\mu = 1(air)$ or $r = 0$

then for bright rings $2t = n\lambda$

for dark rings $2t = (2n-1)\frac{\lambda}{2}$

Proceeding in the similar manner as in reflected system, we can express diameter of rings in terms of R and λ.

$\therefore$ For nth bright ring $d_n^2 = 4Rn\lambda$

and for nth dark ring $d_n^2 = 2(2n-1)R\lambda$

On comparing corresponding equations in two systems it will be clear that dark ring in reflected system is bright in transmitted system and *vice-versa.* Therefore, the central spot, which is dark in reflected system, will be bright in transmitted system. Thus two systems are complimentary to each other.

1.5 MICHELSON INTERFEROMETER

Principle

The amplitude of light beam from a source is divided in two parts of equal intensities by partial reflection and transmission. These beams are then sent in two directions at right angles and are brought together after they suffer reflection from plane mirrors to produce interference fringes.

Construction

It consists of two highly polished plane mirrors M_1 and M_2 and two plane parallel glass plates G_1 and G_2 of exactly the same thickness. The plate G_1 is half silvered at the back so that the incident beam is divided into a reflected and transmitted beam of equal intensity. The plates G_1 and G_2 are parallel to each other and are placed 45° to the mirrors M_1 and M_2, which in turn are mutually perpendicular to each other. The mirrors M_1 and M_2 are provided with screws on their backside to adjust them exactly perpendicular to each other. The interference bands are observed with the help of telescope T.

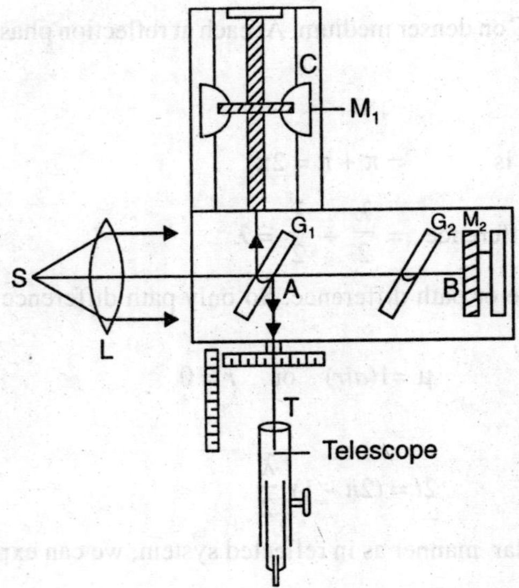

FIGURE 1.8 **Experimental arrangement of Michelson's interferometer.**

Working

Light from the source *S* is made parallel by a collimating lens *L* and then it is allowed to fall on the glass plate G_1. It is partly reflected at the back surface of G_1 along *AC* and partly transmitted along *AB*. The ray *AC* travels normally towards plane mirror M_1 and so it is reflected along the same path and comes out along *AT*. The refracted or transmitted light is received by the mirror M_2 normally, and is reflected along the same path and then moves along *AT* after reflection from back surface of G_1. So the two beams received along *AT* are produced from a single source by the division of amplitude and we may get interference patterns depending upon the path difference between them.

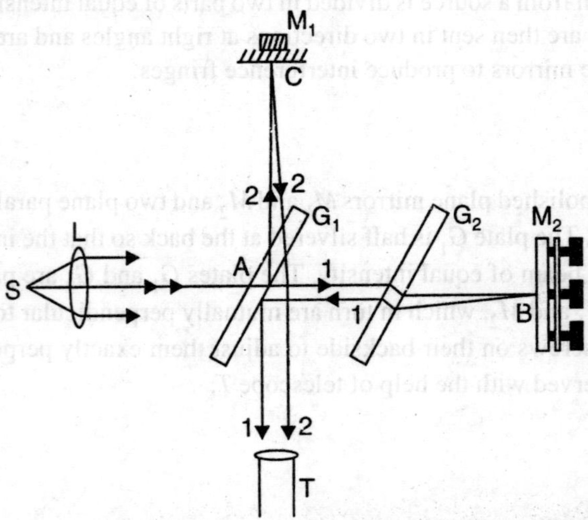

FIGURE 1.9 **Different path travelled by two waves which are reflected from two mirrors M_1 and M_2.**

It is clear from Fig. 1.9 that a ray starting from source S and suffering reflection at the mirror M_2 travels through the glass plate G_1 only once. To compensate it a second glass plate G_2 is introduced exactly parallel to plate G_1 in the path AB.

Formation of Circular Fringes

If eye is placed to see the mirror M_1 directly then in addition it will see virtual image of M_2 formed by reflection in the glass plate G_1. It means that one of the interfering beam comes from M_1 and other appears to come from the virtual image M_2' of the mirror M_2. This system is similar to an air film enclosed between mirrors M_1 and M_2', however, with the difference that in the case of real film between two surfaces, multiple reflections are observed, while in the present case only two reflections take place.

The air film will have uniform thickness if M_2' is exactly parallel to M_1 and G_1 is exactly at an angle of 45° to both of them. The formation of air film shown in Fig. 1.10 corresponds to the case when M_2 is nearer to plate G_1 than M_1 by a distance d.

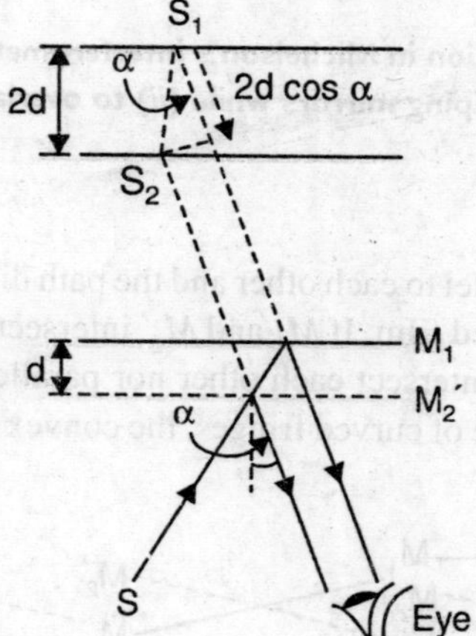

FIGURE 1.10 **Ring formation in Michelson's interferometer.**

Therefore, the two interfering beams appears to come from two virtual images S_1 and S_2 of the original source S, the distance between two virtual sources being $2d$. Now since the ray AB coming after reflection from M_2' suffers reflection at the silvered surface of the glass plate G_1, so an additional path difference of $\frac{\lambda}{2}$ is introduced.

$$\therefore \quad \text{Total path difference} = 2d + \frac{\lambda}{2}$$

If the eye is placed at an angle α w.r.t. the normal to the mirrors.

$$\text{Total path difference} = 2d \cos \alpha + \frac{\lambda}{2}$$

The condition for constructive interference is, therefore,

$$2d \cos \alpha + \frac{\lambda}{2} = n\lambda$$

where $n = 0, 1, 2, 3, \ldots$ etc.

When M_2' coincides with M_1 then the air paths AC and AB are exactly equal, and hence the total path difference will be $\frac{\lambda}{2}$ *i.e.* the case of destructive interference and we will get total darkness as shown in Fig. 1.11 (*ii*). When M_2 is moved either forward or backward parallel to itself then the circular fringes are observed [see Fig. 1.11. (*i*) and (*iii*)] for an extended monochromatic source of light because of the fact that for a given n, the value of λ is constant and so the locus of the fringes is circle.

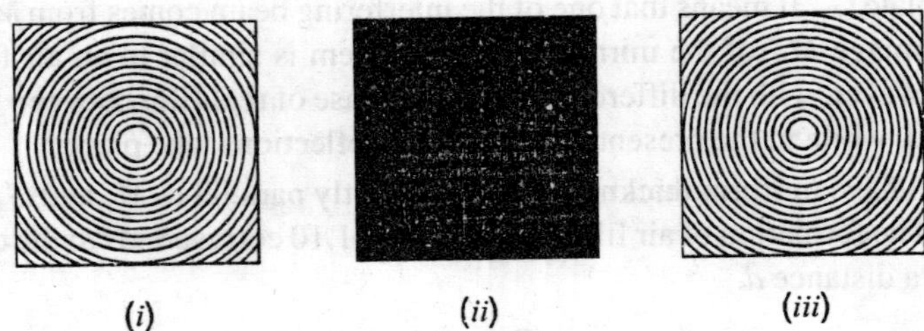

FIGURE 1.11 **Ring formation in Michelson's interferometer. (*i*) and (*iii*) correspond to non-overlaping mirrors while (*ii*) to overlaping mirrors.**

Localized Fringes

When M_1 and M_2' are not exactly parallel to each other and the path difference is very small then one observes fringes as in the case of a wedge shaped film. If M_1 and M_2' intersect each other then the fringes are straight lines. But, when M_1 and M_2' neither intersect each other nor parallel to each other then curved fringes are observed as shown in Fig. 1.12. In case of curved fringes, the convex surface is towards the narrow wedge of the air film.

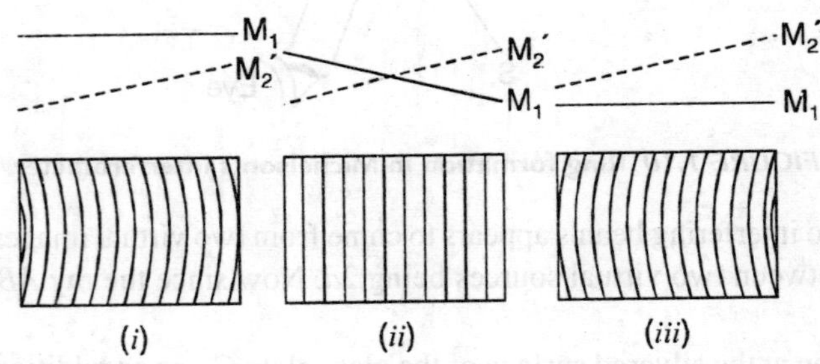

FIGURE 1.12 **Fringes formation by different position of mirrors. (*i*) and (*iii*) correspond to when M_1 and M'_2 neither intersect nor parallel and (*ii*) corresponds to the case when these intersect.**

1.6 APPLICATIONS OF MICHELSON'S INTERFEROMETER

1. Determination of Wavelength

Mirrors M_1 and M_2 are adjusted exactly perpendicular to each other. The position of mirror M_1 is adjusted till some bright fringes appear in the field of view of telescope. Now the cross wire of the telescope is adjusted on a particular bright fringe. The mirror M_1 is displaced slightly and with the help of a micrometer screw the distance travelled is measured for displacing some finite number of bright fringes, say n. Let this distance be d, then we have

$$2d = n\lambda \qquad \text{or} \qquad \lambda = 2\frac{d}{n}$$

Hence through the knowledge of d and n one may now easily determine the wavelength of the light used.

2. Determination of Refractive Index or the Thickness of a Transparent Medium

To find either t or μ, the two path AB and AC are made exactly equal by setting the apparatus with white light so that a straight dark fringe coincide with the cross wire of the telescope. The thin transparent material is now introduced between glass plate G_2 and mirror M_2 and the adjustment are made so that light passes through it normally. In this position the path difference will be $(\mu - 1)t$. The position of the mirror M_1 is adjusted till again the dark fringe of zero path difference coincides with the cross wire of the telescope. The distance d through which mirror M_1 has moved can be read on the micrometer and we have $(\mu - 1)\, t = d$. Hence, if t is known μ can be calculated and *vice-versa.*

SOLVED NUMERICAL PROBLEMS

Problem 1. *Interference fringes are produced by a biprism whose acute angle is 20′ and refractive index 1.5. The slit is illuminated with a light of wavelength 6000 Å and is kept at a distance of 10 cm from biprism. Calculate the fringe width at a distance of 100 cm from the biprism.* (Jan.1999, K.U.K.)

Solution. Here $$A = 20' = \frac{1}{3}\left(\frac{\pi}{180}\right) \text{ radians}$$

$$\mu = 1.5, \qquad \lambda = 6000 \text{ Å} = 6 \times 10^{-5} \text{ cm.}$$

Distance between slit and biprism, $a = 10$ cm.

Distance between biprism and screen, $b = 100$ cm

$$\therefore \qquad D = a + b = 10 + 100 = 110 \text{ cm.}$$

Now $$\beta = \frac{\lambda D}{2d} = \frac{\lambda D}{2a(\mu - 1)A} = \frac{(6 \times 10^{-5} \times 110 \times 3 \times 180)}{[2 \times 10 \times (1.5 - 1) \times \pi]}$$

$$= \frac{11350.31847}{100000} = 0.0011350 \text{ cm.}$$

Problem 2. *In a Fresnel biprism experiment we obtain 75 fringes in the field of view with wavelength 5893 Å. What will be the number of fringes in the field of view with green light of wavelength 5550 Å.* (June 2000, K.U.K.)

Solution. The thickness is given by

$$2\mu\Delta t = n\lambda$$

for $$\lambda_1 = 5893 \text{ Å},\ n_1 = 75$$

let for $$\lambda_2 = 5550 \text{ Å},\ n_2 = ?$$

$$\therefore \qquad 2\mu\,\Delta t = n_1\lambda_1 = n_2\lambda_2$$

$$n_2 = \frac{n_1\lambda_1}{\lambda_2} = \frac{75 \times 5893}{5550} = 79.6 \text{ fringes.}$$

Problem 3. *Interference fringes are produced by monochromatic light of wavelength 5460 Å. When a thin sheet of transparent material of thickness 6.3 × 10⁻⁴ cm is introduced in the path of one of the interfering beams, the central fringe shifts to a position occupied by sixth bright fringe. Compute the refractive index of the sheet.* (May 1999, K.U.K.)

Solution. Here $\lambda = 5460 \times 10^{-8}$ cm

$t = 6.3 \times 10^{-4}$ cm

Central fringe shift $x_0 = 6\beta$

Now, $$x_0 = \left(\frac{\beta}{\lambda}\right)(\mu - 1)t$$

$\therefore$ $$6\beta = \left(\frac{\beta}{\lambda}\right)(\mu - 1)t$$

$$\mu = \frac{6\lambda}{t} + 1 = \frac{6 \times 5460 \times 10^{-8}}{6.3 \times 10^{-4}} + 1$$

$$\mu = 0.52 + 1 = 1.52 \quad \text{or} \quad \mu = 1.52$$

Problem 4. *The distance between the slit and biprism (angle of biprism and refractive index are respectively 179° and 1.5) and between biprism and screen are 50 cm each. Calculate the wavelength of the light if the distance between successive fringes is 0.0135 cm.* (June 2001, K.U.K.)

Solution. Here $\beta = 0.0135$ cm

$D = 50 + 50 = 100$ cm

$\mu = 1.5$

As we know that, $2d = 2a(\mu - 1)A$

When $$A = \frac{(180° - 179°)}{2} = \frac{1}{2°} = \frac{1}{2} \times \frac{\pi}{180} = \frac{\pi}{360} \text{ radians}$$

and $a = 50$

$\therefore$ $$2d = 2 \times 50 \times (1.5 - 1) \times \frac{\pi}{360} \text{ cm}$$

Now, $$\lambda = \frac{2\beta d}{D} = 0.0135 \times 2 \times 50 \times (1.5 - 1) \times \frac{\pi}{100} \times 360$$

$$= 5893 \times 10^{-8} \text{ cm} = 5893 \text{ Å}.$$

Problem 5. *A soap film 500 nm thick is viewed at an angle of 35° to the normal. Find the wavelength of light in the visible spectrum which will be absent from the reflected light (μ = 1.33).* (Dec. 1999, K.U.K.)

Solution. Here angle of incidence $i = 35°$

$\mu = 1.33$, $\lambda = ?$ $n = 1$, $t = 500 \times 10^{-7}$ cm $= 5 \times 10^{-5}$ cm

Now $$\mu = \frac{\sin i}{\sin r} = \frac{\sin 35°}{\sin r} \quad \text{or} \quad \sin r = \frac{0.5735}{1.33}$$

$\therefore \qquad r = 25.54° \qquad \text{or} \qquad \cos r = 0.9022$

for a bright band in reflected system

$$2\mu t \cos r = (2n+1)\frac{\lambda}{2}$$

$$\lambda = \frac{(4\mu t \cos r)}{(2n+1)}$$

for $\quad n = 0, \quad \lambda = 4 \times 1.33 \times 5 \times 10^{-5} \times 0.9022 = 24 \times 10^{-5}$ cm

This wavelength does not lie in the visible region.

For $\quad n = 1, \quad \lambda = \dfrac{24 \times 10^{-5}}{3} = 8000$ Å

For $\quad n = 2, \quad \lambda = \dfrac{24 \times 10^{-5}}{5} = 4800$ Å

This wavelength is in the visible region.

EXERCISE

1. Explain how the wavelength of a source of monochromatic light may be measured using a biprism. *(May, 1998, K.U.K.)*
2. Explain the formation of Newton's rings and prove that the areas enclosed between successive dark rings are equal. *(May, 1998, 2002, K.U.K.)*
3. Draw a neat diagram to observe Newton's rings interference fringes and show that the square of the diameters of successive rings have equal difference. *(May, 1995, K.U.K.)*
4. What are coherent sources ? Explain the interference phenomenon with the help of Michelson's interferometer. *(Jan., 1997, K.U.K.)*
5. Describe with necessary theory the Fresnel's biprism method of determining the wavelength of light. What will happen if the width of the slit is increased gradually ? *(Jan., 1999, K.U.K.)*
6. (*a*) Explain the formation of Newton's rings in reflected light.
   (*b*) Discuss the conditions for obtaining stationary interference pattern. *(Dec. 2000, K.U.K.)*
7. Describe Michelson interferometer and explain its working. How it can be used to find the thickness of a thin transparent film ? *(Dec. 1999, May 2002, K.U.K.)*
8. A biprism is placed 5 cm from a slit illuminated by sodium light ($\lambda = 5890$ Å). The width of the fringes obtained on a screen 75 cm from the biprism is 9.424×10^{-2} cm. What is the distance between the two coherent sources ? **[Ans.** 0.05 cm**]**
9. A thin sheet of glass ($\mu = 1.5$) of 6 microns thickness introduced in the path of interfering beams in a biprism arrangement shift the central fringe to a position normally occupied by the fifth. Find the wavelength of light used. (1 micron = 10^{-4} cm) **[Ans.** 6000 Å**]**
10. In Newton's ring experiment, the diameter of the 4th and 12th dark rings are 0.400 cm and 0.700 cm respectively. Find the diameter of the 20th dark ring. **[Ans.** $D_{20} = 0.906$ cm**]**
11. In a Michelson's interferometer 200 fringes cross the field of view when the movable mirror is displaced through 0.0589 mm. Calculate the wavelength of monochromatic light used. **[Ans.** 5890×10^{-8} cm**]**
12. A this film ($\mu = 1.6$) is introduced in one of the beams in biprism experiment and central fringe is found to be displaced to the position of 20th dark fringe for a light of $\lambda = 6000$ Å. Calculate the thickness of the film. **[Ans.** 0.0195 mm**]** *(Jan. 2003, K.U.K.)*

Chapter 2

DIFFRACTION

2.1 HUYGEN-FRESNEL EXPLANATION OF DIFFRACTION

The bending of light around the sharp edged obstacle and spreading of light into the geometric shadow are regarded as diffraction. Its explanation could be possible only by assuming the wave nature of the light along with the following postulates now known as Huygen-Fresnel principle.

Postulates

(*a*) Every point source of light emits spherical waves which spread out radially with a velocity $c = 3 \times 10^8$ m/sec.

(*b*) Each element of the wavefront at a later instant becomes a center of spherical waves called secondary wavelets.

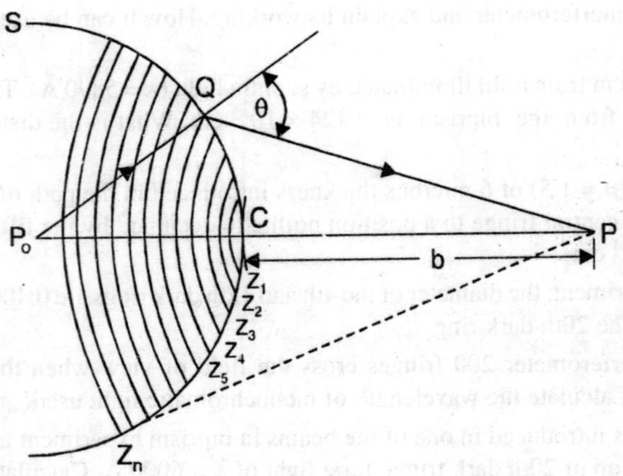

FIGURE 2.1 **The propagation of light on the basis of Huygen-Fresnel principle.**

(*c*) The diffraction is the result of the interference of secondary wavelets which reach the given point of observation.

Let us suppose P_0 as the point source of light and P the point of observation [see Fig. 2.1]. At any instant of time let S be the position of spherical wavefront between P_0 and P such that $P_0C = a$ and $PC = b$ where C is the point where P_0P line intersects the spherical wavefront S. As per Huygen's construction, spheres are drawn with P as center and $b, b + \frac{\lambda}{2}, b + 2\frac{\lambda}{2}, b + 3\frac{\lambda}{2}, \ldots\ldots, b + j\frac{\lambda}{2}$ as radii. The intersections of these spheres with spherical wavefront divide the latter into regions $Z_1, Z_2, \ldots Z_j, \ldots$ called Fresnel zones. Obviously the wavelets after emerging from two adjoining zones will reach P in out of phase (*i.e.* phase difference $= \pi$). By second postulate every element of these zones becomes the center of secondary wavelets some of which may reach point P. The sum of all these contributions therefore will determine the intensity of light at P. The calculations of Huygen and Fresnel show that the resultant amplitude of light disturbance at P is given by

$$U(P) = \frac{2iA\lambda e^{ik(a+b)}}{(a+b)} [k_1 - k_2 + k_3 \ldots\ldots k_n]$$

where $\lambda =$ wavelength

$A =$ amplitude at unit distance from P_0.

$k = \frac{2\pi}{\lambda}$; the wave number

and $k_j =$ inclination factor for jth zone since these depend on the angle θ. These may be written as $k_j(\theta)$. Also, n is the no. of the last zone from which secondary wavelets could reach the point P. These factors, k_j, decrease with increasing θ [see Fig. 2.1 for the definition of angle θ] such that $k_j(\theta) = k_{n+1} = 0$ and that k_j has same value for all wavelets emerging from jth zone and reaching P.

As the factors outside the square bracket [ ] is same and constant for a particular pair of points P_0 and P, it is the value of [ ] $= k_1 - k_2 + k_3 \ldots\ldots k_n$, which would be a measure of the resultant amplitude of disturbance of light at P.

Simple calculations show that

$$[\ ] = \frac{1}{2}[k_1 + k_n] \text{ when } n \text{ is odd}$$

and

$$= \frac{1}{2}[k_1 - k_n] \text{ when } n \text{ is even.}$$

In other words,

$$U(P) = \frac{iA\lambda e^{ik(a+b)}}{(a+b)} [k_1 \pm k_n]$$

That is, only the contributions from the 1st and the last zone would determine the resultant amplitude of light at P. Intensity, on the other hand, is given by

$$I(P) = |U(P)|^2 = \frac{(A\lambda)^2}{(a+b)^2} [k_1 \pm k_n]^2$$

Further, since we shall be calculating only the intensity of light at P, factor $i.e^{ik(a+b)}$ will be contributing nothing and will be ignored here after. In short, we shall be writing

$$U(P) = [U_1 - U_2 + U_3 \ldots\ldots\ldots\ldots \pm U_n]$$

with $$U_i(P) = \frac{2\lambda A}{(a+b)} k_i$$

Hence, $$U(P) = \frac{1}{2}[U_1(P) \pm U_n(P)]$$

$$= \frac{1}{2} \text{ [sum of contributions from 1st and last zone]}$$

$$\approx U_1(P)/2 \text{ as } U_n(P) \to \text{zero with increasing } n.$$

and $$I(P) = \frac{1}{4} U_1^2$$

So far we have assumed that there was no obstruction present between P_0 and P. There could be obstructions between P_0 and P in the form of small aperture in an opaque screen or small circular opaque obstacle or a semi-infinite screen with sharp edge which would block certain number of zones. As a result, the wavelets from these zones would not reach P. For example, let us take the case of a screen having an opening such that it blocks all zones except the central half zone. Then

$$U(P) = \frac{1}{2} U_1(P)$$

and the intensity at P will be same as if there were no screen. If the size of the opening is increased such that it exposes the full central zone to P, then intensity will become $I(P) = U_1^2$ *i.e.* four times the intensity without screen. On increasing the size further such that 1st two zones get exposed to P, the intensity of light will be equal to $(U_1 - U_2)^2$ which will be zero because U_1 is nearly equal to U_2. The contributions from neighbouring zones do not differ much. Thus, we find that the intensity of light at P fluctuates rapidly as the size of the opening (or aperture) is increased. The same will be the pattern when P approaches the screen keeping the position of source P_0 and size of the aperture in the screen fixed as more and more zones get exposed in doing so.

Next, take the case of a circular obstacle (say disc) such that for some point P it blocks the central zone Z_1 completely (see Fig 2.2). As a result, the intensity at P

$$= [U_2 - U_3 + U_4 \ldots\ldots\ldots\ldots U_n]^2 = \left(\frac{U_2}{2}\right)^2 \approx \left(\frac{U_1}{2}\right)^2 \text{ same as if there were no disc.}$$

This was most astonishing conclusion of Huygen-Fresnel theory of propagation of light because this means that there is light at P which lies in the geometric shadow of the circular disc. However, this result as well as the earlier ones discussed above were experimentally verified thereby establishing the wave nature of light beyond any doubt whatsoever.

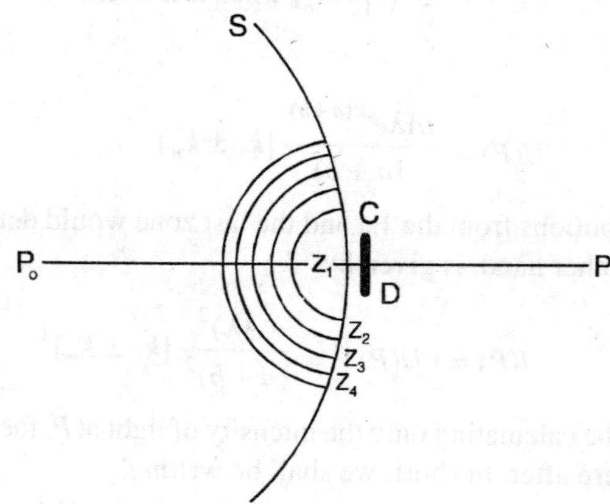

FIGURE 2.2 **The circular obstacle, blocking the effect of central zones.**

As a last example, let us consider semi-infinite screen with sharp edge [Fig. 2.3]. The semi-infinite screen CD will obstruct the lower parts of all the zones w.r.t. point P. Hence amplitude of light at P would be given as

$$U(P) = \frac{1}{2}[U_1 - U_2 + U_3 - \ldots\ldots\ldots\ldots] = \frac{U_1(P)}{4} \text{ as } n \to \infty$$

and
$$I(P) = \frac{1}{4}\left(\frac{U_1^2}{4}\right) = \frac{1}{4} \text{ times the intensity without screen.}$$

As the point P moves upwards say to the positions P_1, P_2 etc. more and more of lower half of the zones get exposed w.r.t. the new positions. Suppose, P_1 is such that the central zone get fully exposed, then

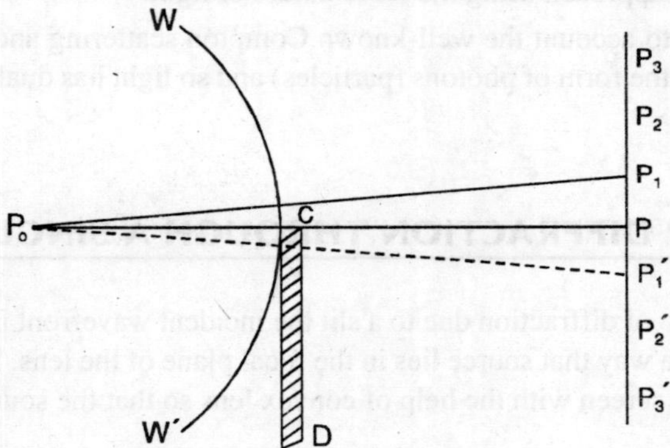

FIGURE 2.3 **Diffraction by semi-infinite plane obstacle.**

$$U(P_1) = \frac{1}{2}[2U_1 - U_2 + U_3 - \ldots\ldots\ldots\ldots] \approx \frac{1}{2}\left(\frac{3}{2}U_1\right)$$

and $I(P_1) = 9/4$ times the intensity without screen.

When we go to P_2 such that zones Z_1 and Z_2 are fully exposed to P_2 while other only half. Then $U(P_2)$

$$U(P_2) = \frac{1}{2}[2U_1 - 2U_2 + U_3 - U_4 \ldots\ldots\ldots\ldots] = \frac{1}{2}[U_1 - U_2] + \frac{1}{2}[U_1 - U_2 + U_3 \ldots\ldots\ldots\ldots]$$

$$\approx \frac{1}{2}\left[\frac{U_1}{2}\right] \text{ as } U_1 \approx U_2$$

and
$$I(P_2) = \frac{1}{4}\left(\frac{U_1}{2}\right)^2 \approx I(P)$$

but much smaller than $I(P_1)$.

Thus, when we move into the illuminated part of the space, the intensity first show fluctuation and then becomes constant at a value as if there were no screen. On the other hand when we move into the geometric shadow say to P_1' such that Z_1 is completely covered. Then

$$U(P_1') = \frac{1}{2}[-U_2 + U_3 - U_4 + \ldots\ldots\ldots] \approx \frac{1}{2}\left[\frac{-U_2}{2}\right]$$

and
$$I(P_1') = \frac{1}{4}\left(\frac{U_2}{2}\right)^2 \approx \frac{1}{4}\left(\frac{U_1}{2}\right)^2 \text{ as } U_1 \text{ and } U_2 \text{ are nearly equal.}$$

As we go further down to P'_2 such that 1st two zones Z_1 and Z_2 get covered completely, the amplitude of light would be

$$U(P'_2) = \frac{1}{2}[+U_3 - U_4 + U_5 - \ldots\ldots] \approx \frac{1}{2}\left[\frac{+U_3}{2}\right]$$

and
$$I(P'_2) = \frac{1}{4}\left(\frac{U_3}{2}\right)^2 < I(P) < I(P_1')$$

Thus, we conclude that in the geometric shadow, light spreads with decreasing intensity.

All these examples, therefore, convincingly explain the phenomenon of diffraction of light thereby confirming Huygen-Fresnel approach using the wave nature of light.

In fact, if we take into account the well-known Compton scattering and the photoelectric effect we find that light propagates in the form of photons (particles) and so light has dual character ; wave nature and particle nature.

2.2 FRAUNHOFER DIFFRACTION THROUGH A SINGLE SLIT

To obtain a Fraunhofer class of diffraction due to a slit the incident wavefront is plane which is obtained by using a convex lens in such a way that source lies in the focal plane of the lens. The diffracted light from the slit is again collected on the screen with the help of convex lens so that the source and screen are at infinite distance from the obstacle.

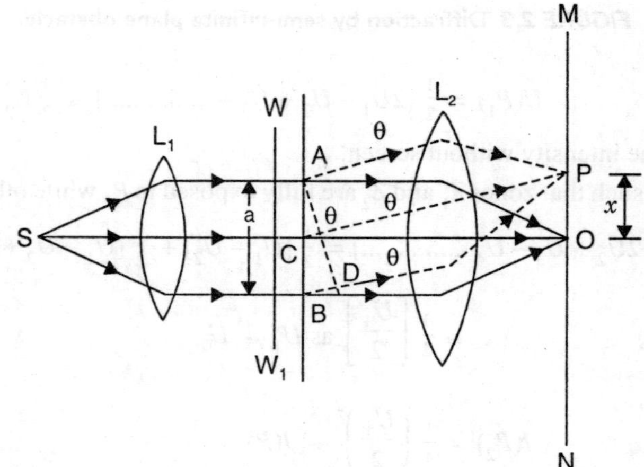

FIGURE 2.4 **Experimental arrangement of Fraunhofer diffraction through a single slit.** ***AB* = *a*; ∠*BAD* = θ; *OP* = *x*.**

The experimental arrangement is shown in Fig. 2.4. S is a source of monochromatic light of wavelength λ, L_1 is a collimating lens and AB is a narrow slit of width a. WW_1 is a plane wavefront incident on the slit AB and each point on this wavefront is a source of secondary disturbance. The diffracted secondary waves are focussed on the screen MN with the help of a convex lens L_2.

The waves reaching at point O, just opposite the middle of the slit form a bright central image. This is because the secondary waves from points equidistant from C and situated in the upper and lower halves CA and CB of the wavefront travel the same distance in reaching O and hence the path difference is zero. The secondary waves thus reinforce each other and O will be a point of maximum intensity.

Let us now consider the intensity at a point P above O produced by the rays diffracted at an angle θ as shown in Fig. 2.4. The secondary waves coming from all points from the slit start in the same phase but they travel different distances in reaching the point P in different phases. To calculate the path difference between the extreme rays reaching the point P draw AD perpendicular to the diffracted rays coming from B. The path difference between the extreme rays travelling from A and B is given by BD.

Let us suppose that $BD = \lambda$ and C is the middle point of the slit AB. Hence path difference between CP and AP is equal to $\frac{\lambda}{2}$. Therefore, the waves travelling from A and C shall reach P in opposite phases and cancel each other. Similarly any wave coming from a point below A and corresponding point below C from CB will cancel each other's effect and thus for path difference $BD = \lambda$, the point P will be having the minimum intensity.

Now from ΔABD, the path difference

$$BD = AB \sin\theta = a \sin\theta, \quad \text{where } a \text{ is the width of the slit.}$$

Therefore, $a \sin\theta = \lambda$, the point P will have a minimum intensity or a dark fringe.

Similarly, we can explain for path difference 2λ, we get a point of minimum intensity. Hence if the path difference $a \sin\theta = 3\lambda$, 4λ etc. additional dark fringes are obtained.

In general $\quad a \sin\theta_n = n\lambda \quad$ ...(2.1)

is the condition of nth dark fringes. The value of θ_n becomes

$$\theta_1, \theta_2 \ldots\ldots \text{ for } n = 1, 2, \ldots\ldots$$

It should be noted that when $n = 0$, $\theta_n = 0$ which corresponds to point O on the screen which is central maxima.

If however the path difference is odd multiple of $\frac{\lambda}{2}$ the directions of secondary maxima is obtained. In this case,

$$a \sin\theta_n = (2n+1)\frac{\lambda}{2} \quad \text{...(2.2)}$$

where $n = 1, 2, 3 \ldots$ etc.

Thus the diffraction pattern due to single slit consists of a central bright maxima at O followed by the secondary maxima and minima on both sides of central maxima. The intensity distribution on the screen is given in Fig. 2.4. The point O corresponds to the position of central maxima and the points on the screen for which the path difference between the points A and B is λ, 2λ, 3λ etc. correspond to the position of secondary minima. The secondary maxima are obtained at $3\frac{\lambda}{2}$, $5\frac{\lambda}{2}$, $7\frac{\lambda}{2}$ etc. and are of much less intensity. The intensity falls off rapidly as we move outwards from the point O.

If the lens L_2 is very near to the slit or the screen is far away from the lens then

$$\sin\theta = \frac{x}{f} \quad \text{...(2.3)}$$

where f is the focal length of the lens and x is the distance between O and P on the screen.

But $\quad \sin\theta = \frac{\lambda}{a}, \quad$ ...(2.4)

From eqs. (2.3) and (2.4), we have $\quad x/f = \frac{\lambda}{a} \quad$ or $\quad x = f\frac{\lambda}{a}$

Hence width of central maxima $= 2x$

$$\therefore \quad 2x = 2f\frac{\lambda}{a}$$

The width of central maxima is proportional to λ, the wavelength of light. With red light the width of central maxima is more and with violet the width is less. The width of central maxima is also inversely proportional to the width of slit, therefore, the width of central maxima is more for a narrow slit.

Analytical Treatment of Intensity Distribution Due to Single Slit

It is clear from Fig. 2.4 that after AD all the waves travel the same distance upto screen. Therefore, the path difference between the extreme waves from the slit will be $BD = a \sin \theta$. It means that the path difference of different waves vary uniformly from zero to $a \sin \theta$ as we move from the point A of the slit to the point B of the slit.

The corresponding phase difference δ between the first and the last wave is given by

$$\delta = \frac{2\pi}{\lambda} \times \text{path difference}$$

$$\delta = \left(\frac{2\pi}{\lambda}\right) a \sin \theta.$$

The resultant amplitude A at point P can be found out by the vector addition theorem and is given by

$$A = A_0 \frac{\sin \alpha}{\alpha} \quad \text{...(2.5)}$$

where A_0 is the incident amplitude on AB and α is the half of the phase difference between the first and the last wave *i.e.*

$$\alpha = \frac{\delta}{2} = \left(\frac{\pi}{\lambda}\right) a \sin \theta \quad \text{...(2.6)}$$

Also the intensity can be written as $I = A^2$ and using eq. (2.5)

$$I = A_0^2 \frac{\sin^2 \alpha}{\alpha^2} \quad \text{...(2.7)}$$

It is clear from these equations that as θ changes, α, A and I also change. It means different sets of parallel rays *i.e.* having different θ will focus at different points on the screen. Now it is very difficult to find the value of A or I at every point of the screen. However at points of some special interest it may be obtained as follows.

1. Position of Central Maxima

From eq. (2.7) we find that A will be maximum when $\sin \alpha/\alpha$ will be maximum. Now $\sin \alpha/\alpha = 1$ where $\alpha \to 0$. Therefore A or I will be maximum where $\alpha = 0$ or $\theta = 0$ *i.e.* in the direction of incident light. This maximum is known as principal or central maximum shown in Fig. 2.4 by point O. Therefore in this direction

$$A_{max} = A_0 \quad I_{max} = A_0^2 = I_0$$

2. Position of Secondary Minima

The resultant amplitude will be minimum *i.e.* zero if $\sin\alpha = 0$ but $\alpha \neq 0$ (because if $\alpha = 0$, $\frac{\sin\alpha}{\alpha} = 1$ and it correspond to central maximum)

Thus the condition for minima is $\sin\alpha = 0$

or
$$\alpha = \pm\pi, \pm 2\pi, \pm 3\pi, \ldots\ldots \pm n\pi$$

or
$$\alpha = n\pi \text{ where } n = \pm 1, \pm 2, \pm 3, \ldots\ldots\ldots$$

From eq. (2.6)
$$(\pi/\lambda)\, a\sin\theta = n\pi$$
$$a\sin\theta = n\lambda$$

This equation gives the direction of minimum intensity. For different values of n we have different values of θ *i.e.* different directions of minimum intensity. So it is clear that several minima on both sides of the principal maxima will be obtained corresponding to $n = 1, 2, 3 \ldots$ and $n = -1, -2, -3\ldots$. These are known as secondary minima.

3. Position of Secondary Maxima

Apart from the principal maximum or central maximum, we obtain secondary maxima approximately at

$$\alpha = \pm\frac{3\pi}{2}, \pm\frac{5\pi}{2}, \pm\frac{7\pi}{2} \text{ etc. in between of which lie secondary minima.}$$

Also substituting these values of $\alpha = \frac{3\pi}{2}, \frac{5\pi}{2}$ in eq. (2.7) we get

$$(I_1)_{\text{at } \alpha = \frac{3\pi}{2}} = \frac{A_0^2}{22} = \frac{I_0}{22}$$

and
$$(I_2)_{\text{at } \alpha = \frac{5\pi}{2}} = \frac{A_0^2}{62} = \frac{I_0}{62}$$

i.e. intensity in secondary maxima is very small as compared to principal maximum and also goes on decreasing towards higher order secondary maxima.

Now keeping the results of cases 1, 2, 3 in view if we plot a graph between the intensity I and α we get a curve as shown in Fig. 2.5. This curve is known as intensity distribution curve due to diffraction at a single slit.

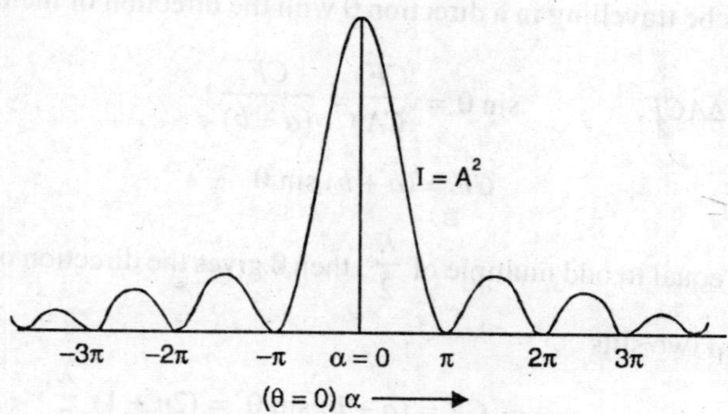

FIGURE 2.5 **Intensity distribution on the screen.**

Fraunhofer Diffraction at Double Slit

Consider two rectangular slits AB and CD parallel to each other and perpendicular to the plane of the paper. Let the width of slit be a and width of the opacity be b. Let a plane wavefront be incident on the surface of a double slit. L is a collecting lens and MN is screen perpendicular to the plane of the paper. All the secondary waves travelling undiffracted come to focus at O. Therefore O corresponds to the position of central maxima as shown in Fig. 2.6. The diffraction pattern due to two slits consists of two parts.

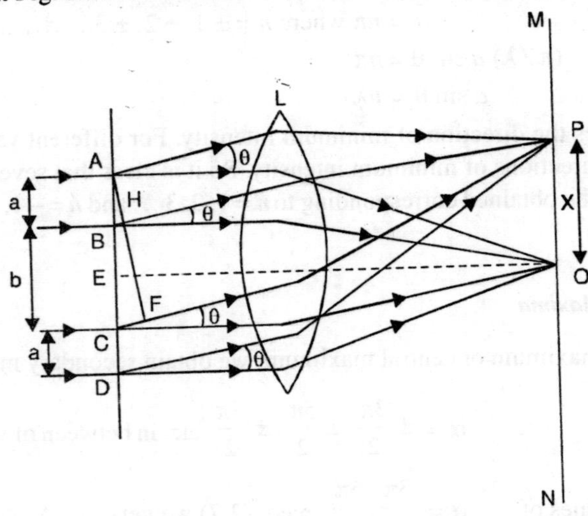

FIGURE 2.6 **Experimental arrangement of Fraunhofer diffraction at double slit. θ is the angle of diffraction.**

(*i*) The interference phenomenon due to secondary waves coming from the corresponding points of two slits.

(*ii*) The diffraction pattern due to the secondary waves from the two slits individually.

For determining the position of interference maxima and minima the diffracting angle is denoted by θ and for diffraction maxima and minima the diffracting angle is denoted by φ. Both the angles θ and φ refer to the angle between the directions of incident waves and the secondary waves.

(i) Interference Maxima and Minima

Let the secondary waves be travelling in a direction θ with the direction of incident wave.

From Fig. 2.6 in ΔACF,
$$\sin\theta = \frac{CF}{CA} = \frac{CF}{(a+b)}$$

or
$$CF = (a+b)\sin\theta$$

If this path difference is equal to odd multiple of $\frac{\lambda}{2}$, then θ gives the direction of minima due to interference of secondary waves from two slits

$$\therefore \quad CF = (a+b)\sin\theta_n = (2n+1)\,\frac{\lambda}{2} \qquad \text{...(2.8)}$$

Putting $n = 0, 1, 2, 3,\ldots$ the values of $\theta_1, \theta_2, \theta_3$ etc. corresponding to the directions of minima can be obtained.

From eq. (2.8) we have $$\sin\theta_n = \frac{(2n+1)\lambda}{2(a+b)} \qquad ...(2.9)$$

On the other hand, if the secondary waves travel in a direction θ′ such that the path difference is even multiple of $\frac{\lambda}{2}$, then θ′ gives the direction of maxima due to interference of secondary waves from two slits.

$$\therefore \qquad CF = (a+b)\sin\theta' = n\lambda$$

or $$\sin\theta' = \frac{n\lambda}{(a+b)} \qquad ...(2.10)$$

Putting $n = 0, 1, 2, 3$ etc. the values of $\theta_1', \theta_2', \theta_3'$, etc. corresponding to the direction of maxima can be obtained from eq. (2.10).

Putting $n = 0, 1, 2$ etc. in eq. (2.9), we have

$$\sin\theta_0 = \frac{\lambda}{2(a+b)},\ \sin\theta_1 = \frac{3\lambda}{2(a+b)},\ \sin\theta_2 = \frac{5\lambda}{2(a+b)}$$

and $$\sin\theta_2 - \sin\theta_1 = \frac{\lambda}{(a+b)} \text{ and so on} \qquad ...(2.11)$$

Same results are obtained from eq. (2.10). Hence the angular separation between any two consecutive minima or maxima is equal to $\frac{\lambda}{(a+b)}$. The angular separation is directly proportional to the wavelength and inversely proportional to $(a + b)$, the distance between the two slits.

(ii) Diffraction Maxima and Minima

Let the secondary waves be travelling in a direction inclined at an angle φ (not shown in Fig. 2.6) with the initial direction of incident wave. If the path difference BH is equal to λ, then φ gives the direction of diffraction minima (see Fig. 2.6). That is the path difference between the secondary waves coming from the extreme points of a slit *i.e.* from A and B is equal to λ. Considering the slit to be divided into two equal halves such that the path difference between the corresponding points of the upper and lower half is equal to $\frac{\lambda}{2}$. The effect at P due to the wavefront on AB is, therefore, zero. Similarly, for the same direction of secondary waves the effect at P due to wavefront incident on CD is also zero. In general

$$a\sin\varphi_n = n\lambda \qquad ...(2.12)$$

$$\sin\varphi_n = \frac{n\lambda}{a} \qquad ...(2.13)$$

Putting $n = 1, 2, 3$, etc. the values of $\varphi_1, \varphi_2, \varphi_3$ etc. corresponding to the directions of diffraction minima can be obtained. The value of the angle φ corresponding to $n = 0$ is zero which corresponds to a central maxima.

Comparing eqs. (2.9) and (2.13), we find that θ is much smaller than φ which shows that a number of interference maxima and minima will be within each diffraction maxima.

The intensity distribution due to Fraunhofer diffraction at two slits is shown in Fig. 2.7. The full line represents equally spaced interference maxima and minima and the dotted curves represents the diffraction maxima and minima. In the region of central maxima of single slit diffraction pattern equally spaced interference maxima and minima are observed. Hence the pattern due to diffraction at a double slit consists of a diffraction pattern due to the individual slits of width a each and interference maxima and minima depends upon the value of a and b.

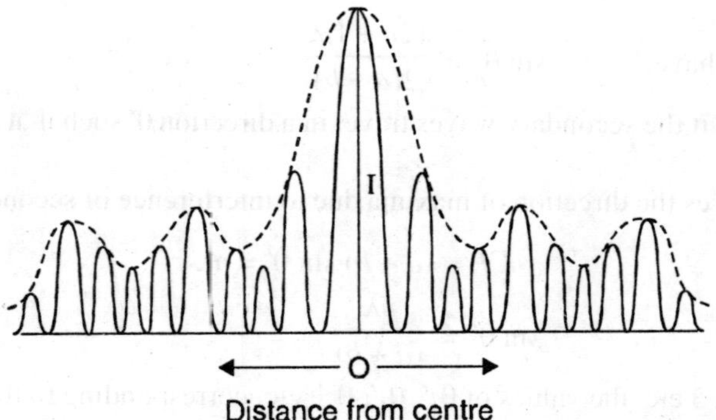

FIGURE 2.7 **Intensity distribution curve due to diffraction at double slit.**

2.3 PLANE TRANSMISSION DIFFRACTION GRATING

An arrangement, which consists of a large number of equidistant parallel slits of same widths and shape, is called a diffraction grating. No. of slits are of the order of 20000 slits per inch. These slits are obtained by ruling equidistant parallel lines either on a plane transparent plane glass surface or on a plane or concave silvered surface. These lines are drawn with a fine diamond point. If the lines are drawn on plane transparent surfaces, then the grating is called plane transmission grating and if the lines are drawn on silvered surface, then the grating is called reflection grating.

Theory of Plane Transmission Grating

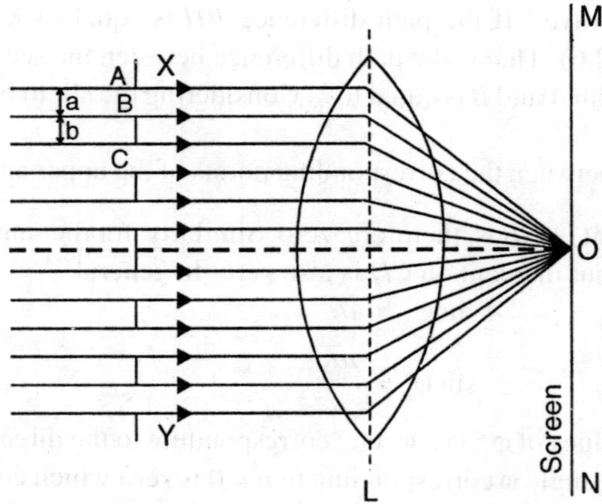

FIGURE 2.8 **Diffraction through plane transmission grating.**

Let XY be the grating surface and MN be the screen placed perpendicular to the plane of the paper [see Fig. 2.8]. Let $AB = a$ and $BC = b$, be the widths of each slits and opaque portion between two consecutive slits, respectively. The distance $a + b$ is called grating element. Now consider a parallel beam of light of wavelength λ is allowed to fall normally on the grating surface. Most of the light passing from the spacing of the slits goes undeflected and is focussed by the collecting lens on the screen at the point O. The point O is, therefore, called central maxima or principal maxima.

Principal Maxima

When the width of the spacing is comparable to the wavelength of light, then the part of the light is diffracted on both sides of the grating leaving the slits. Let us now consider the secondary waves travelling in a direction θ with the direction of incident light and are focussed at the point O_1 on the screen [see Fig. 2.9]. The intensity at O_1 will depend upon the path difference between the secondary waves coming from the corresponding points A and C of two neighbouring slits. The path difference between the secondary waves coming from A and C is equal to CD which is given by

$$CD = (a + b) \sin \theta$$

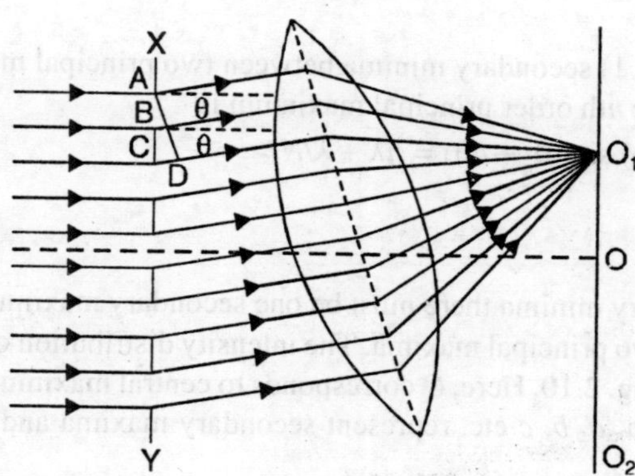

FIGURE 2.9 **Diffraction at an angle θ. *CD* is the path difference.**

The point O_1 will be of the maximum intensity if this path difference is equal to integral multiple of wavelength λ of light.

For nth principal maxima

$$(a + b) \sin \theta_n = n\lambda \qquad ...(2.14)$$

where θ_n is the direction of nth principal maxima and n is any integer $n = 0, 1, 2, 3, 4....$

For $n = 0$, we get central bright maxima at O on the screen.

For $n = 1$, $\quad (a + b) \sin \theta_1 = \lambda \qquad ...(2.15)$

This gives the condition of 1st principal maxima on either side of O on the screen. The intensity at this point is much less than that at O.

For $n = 2$ $\quad (a + b) \sin \theta_2 = 2\lambda \qquad ...(2.16)$

This gives the condition for second principal maxima on either side of the screen and so on. In general the direction θ_n of nth principal maximum is given by

$$(a + b) \sin \theta_n = n\lambda$$

Secondary Minima

The direction θ_n for the nth principal maximum is given by

$$(a + b) \sin \theta_n = n\lambda$$

from this direction θ_n let the angle of diffraction change by a small value $d\theta$ such that the path difference between the secondary wavelets from A and C changes by $\frac{\lambda}{N}$ where N is the total number of lines on the grating.

Then the path difference between the secondary wavelets from the extreme ends of the grating will be $\frac{\lambda}{N} \times N = \lambda$. Now the grating can be divided in two equal parts. Therefore, the path difference between the secondary wavelets from the corresponding points of two halves will be $\frac{\lambda}{2}$. Hence these waves interfere destructively in this direction and cancel out the effects of each other. Thus $(\theta_n \pm d\theta)$ will give the direction of the first secondary minimum after nth primary maximum. Similarly for path difference $2\lambda/N$ and so on. Hence we get second secondary minima on either side of the nth primary maxima. Same will happen for

$$\frac{3\lambda}{N}, \frac{4\lambda}{N}, \frac{5\lambda}{N} \cdots \frac{(N-1)}{N}\lambda.$$

Thus there will be $(N-1)$ secondary minima between two principal maxima. Thus the condition for 1st secondary minima after the nth order principal maximum is

$$(a+b)\sin(\theta_n + d\theta) = n\lambda + \lambda/N.$$

Secondary Maxima

Since in between two secondary minima there must be one secondary maximum hence there will be $(N-2)$ secondary maxima between two principal maxima. The intensity distribution on the screen due to diffraction from the grating is shown in Fig. 2.10. Here, O corresponds to central maximum. O_1, O_2 etc. on either side of O represent principal maxima, a, b, c etc. represent secondary maxima and d, e etc. represent secondary minima.

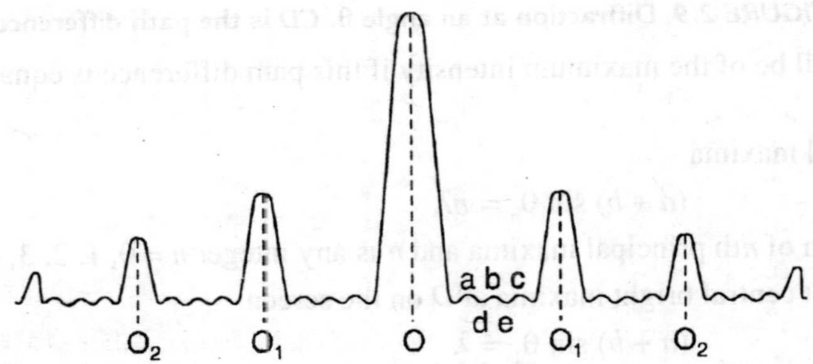

FIGURE 2.10 **Intensity distribution on the screen due to diffraction from grating.**

Width of Principal Maxima

The angular width of principal maximum is the angle between the first secondary minima lying on either side of it. The direction of nth principal maximum is given by

$$(a+b)\sin\theta = n\lambda \quad \text{...(2.17)}$$

Let $(\theta + d\theta)$ and $(\theta - d\theta)$ be the directions of the first secondary minima on either side of it [Fig. 2.11].

Then $$(a+b)\sin(\theta \pm d\theta) = n\lambda \pm \lambda/N. \quad \text{...(2.18)}$$

Dividing eq. (2.18) by eq. (2.17), we obtain

$$\frac{\sin(\theta \pm d\theta)}{\sin\theta} = \frac{\left(n\lambda \pm \frac{\lambda}{N}\right)}{n\lambda}$$

$$\text{or} \quad \frac{[\sin\theta \cos d\theta \pm \cos\theta \sin d\theta]}{\sin\theta} = 1 \pm \frac{1}{nN}$$

Now for small value of $d\theta$, $\cos d\theta = 1$ and $\sin d\theta = d\theta$

Therefore, we have $$1 \pm \frac{\cos\theta}{\sin\theta}\, d\theta = 1 \pm \frac{1}{nN}$$

$$\cot\theta\, d\theta = \frac{1}{nN} \quad \text{or} \quad d\theta = \frac{1}{[nN \cot\theta]} \qquad ...(2.19)$$

where $d\theta$ refer to half the angular width of the nth principal maximum.

It is clear from eq. (2.19) that

1. $d\theta$ **is inversely proportional to** N**, the total number of lines on the grating.**

2. It is inversely proportional to $n \cot\theta$**. The value of** $n \cot\theta$ **is more for higher orders. Thus, the half angular width is less for higher orders.**

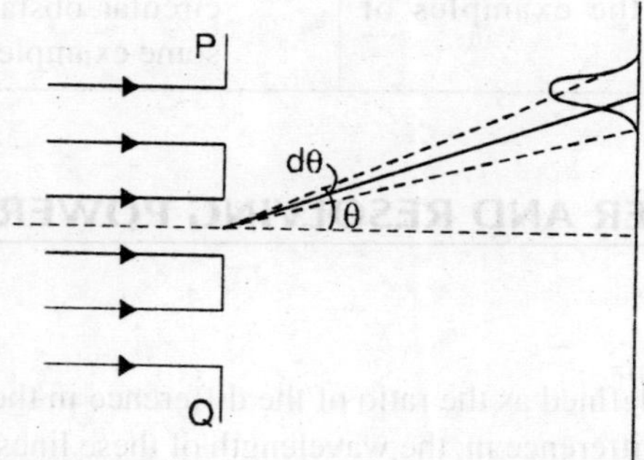

FIGURE 2.11 **The width of central maxima.**

2.4 DIFFERENCE BETWEEN FRAUNHOFER AND FRESNEL DIFFRACTIONS

The diffraction phenomenon is generally divided into two classes :

(*a*) Fresnel diffraction

(*b*) Fraunhofer diffraction

The difference between two classes of diffraction is studied as follows.

	Fraunhofer Diffraction		*Fresnel Diffraction*
1.	In Fraunhofer diffraction, the source and screen are at infinite distance from aperture or the obstacle.	1.	In Fresnel diffraction either the source or the screen or both are at finite distance from the diffracting element *i.e.* aperture or obstacle.
2.	A convex lens is used to make the light from the source parallel before it falls on the aperture and another convex lens is used to focus the light after diffraction on the screen. This arrangement makes the source and the screen to infinity from the obstacle.	2.	No lenses are used to make the rays parallel or convergent.

3. The incident wavefront is plane.	3. The incident wavefront is not plane but is either cylindrical or spherical.
4. Due to plane wavefront the secondary waves originating from the exposed part of wavefront are in same phase at every point in the plane of the aperture.	4. The phase of secondary waves is not same at all the points in the plane of the aperture or the obstacle.
5. The diffraction is produced by interference between parallel rays which are brought to focus with a convex lens.	5. The resultant amplitude at any point on the screen is obtained by mutual interference of secondary waves from different elements of exposed part of wavefront.
6. In this class of diffraction angular inclinations are important and not the distance.	6. In this case of diffraction distances are important not the angular inclination.
7. This class of diffraction can be discussed with more accuracy in mathematical terms.	7. This class of diffraction is studied with approximation only.
8. Diffraction through single slit, double slit and transmission grating are the examples of Fraunhofer diffraction.	8. Diffraction through circular aperture, by circular obstacle and by straight edge are some examples of Freshel diffraction.

2.5 DISPERSIVE POWER AND RESOLVING POWER

Dispersive Power

Dispersive power of a grating is defined as the ratio of the difference in the angle of diffraction of any two neighboring spectral lines to the difference in the wavelength of these lines. Mathematically

$$\text{Dispersive power} = \frac{d\theta}{d\lambda}$$

It can also be defined as the rate of change of angle of diffraction with respect to the wavelength. The principal maxima is given by

$$(a + b)\sin\theta = n\lambda$$

Differentiating this eq. on both sides, we get

$$(a + b)\cos\theta\, d\theta = n d\lambda$$

or

$$\frac{d\theta}{d\lambda} = \frac{n}{(a + b)\cos\theta}$$

or

$$\frac{d\theta}{d\lambda} = \frac{N'n}{\cos\theta}$$

where n is the order of the spectrum and $N' = \frac{1}{(a+b)}$ is the number of lines per cm of the grating surface and θ is the angle of diffraction for nth order principal maximum of wavelength λ. One can conclude from this equation that

1. The dispersive power is directly proportional to the order n, of the spectrum. It means that the angular separation of any two spectral lines is double in second order than that in the Ist order.

2. The dispersive power is directly proportional to the no. of lines per cm N', on the grating surface. It means that angular separation of spectral lines is more for a grating having larger no. of lines per cm.

3. For small value of θ, $\cos\theta = 1$ and the factor $\cos\theta$ can be neglected. If we neglect $\cos\theta$ then it is clear that in a particular order with a given value of N', the angular dispersion $d\theta$ of any two spectral lines is directly proportional to the difference in wavelength of these lines *i.e.* equal increment of θ correspond to equal increment of wavelength. Such a spectrum is called normal spectrum.

Resolving Power

The resolving power of a grating is its ability to show two neighbouring lines in a spectrum as separate. It may also be defined as the ratio of the wavelength of any spectral line to the difference in wavelengths between this line and a neighbouring line such that the two spectral lines can be just seen as separate. Consider two wavelets of wavelengths λ and $\lambda + d\lambda$ are incident normally on the surface of grating. According to Reyleigh's criteria, the two patterns would be just resolved if the principal maximum of one fall on the 1st secondary minimum of the other in any order.

The condition for first secondary minimum after the nth order maximum is given by

$$(a + b)\sin(\theta + d\theta) = n\lambda + \frac{\lambda}{N} \qquad ...(2.20)$$

Now, let the nth order principal maximum corresponding to wavelength $\lambda + d\lambda$ has the same direction as that of the first secondary minimum of the nth order principal maximum corresponding to wavelength λ (see Fig. 2.12), then

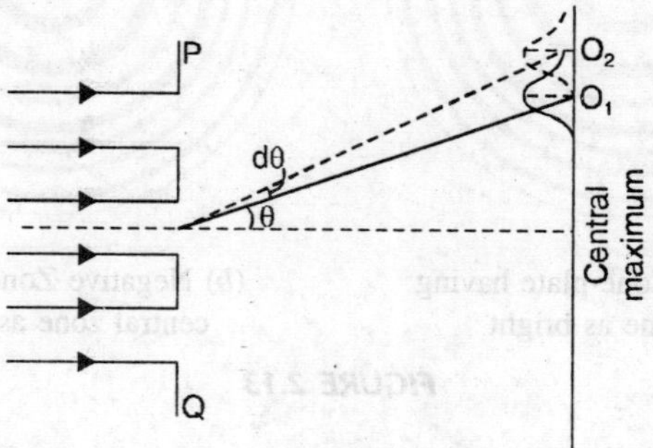

FIGURE 2.12 **Resolving power of a grating.**

$$(a + b)\sin(\theta + d\theta) = n(\lambda + d\lambda) \qquad ...(2.21)$$

From eqns. (2.20) and (2.21), we have

$$n\lambda + \frac{\lambda}{N} = n\lambda + nd\lambda$$

or

$$\frac{\lambda}{N} = nd\lambda \quad \text{or} \quad \frac{\lambda}{d\lambda} = nN$$

which shows that resolving power depends on the order of spectrum and the no. of lines on the grating but does not depend on the grating element.

2.6 ZONE PLATE

Zone plate was designed to explain the rectilinear propagation of light using Huygen-Fresnel principle. This principle consists in dividing the wavefront into half wave zones and assuming that each element of the zone becomes the center of secondary wavelets. These wavelets on reaching the given point of observation would then determine the intensity of light there.

To prepare a zone plate take a white paper and draw concentric circles with radii in the ratio of square root of natural number *i.e.*

$$r_1 : r_2 : \ldots\ldots\ldots r_j \ldots\ldots\ldots : r_n = \sqrt{1} : \sqrt{2} : \ldots\ldots\ldots : \sqrt{n}$$

Then alternate regions between circles are blackened starting with the central (1st) circle as black or white. Thus we can have two such drawings ; one having the central circle as white and second with central circle as blackened. Now take the photograph of these drawings on the thin glass plate which are called as the negatives of the above mentioned drawings. The one with central part as transparent is called the positive zone plate while the other having the central part opaque is called the negative zone plate. The remaining regions would be opaque and transparent alternatively. These zone plates are shown in Fig. 2.13.

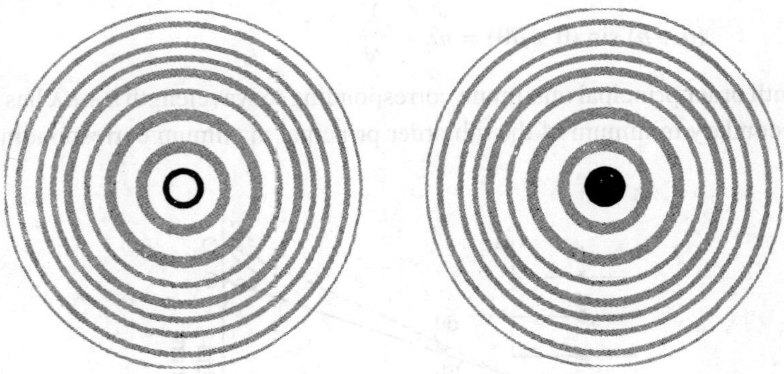

(*a*) Positive Zone-plate having central zone as bright

(*b*) Negative Zone-plate having central zone as dark

FIGURE 2.13

Working of the Zone Plate

Let *X-Y* be the section of such a zone plate (Fig. 2.14) held perpendicular to the plane of the paper and OQ_1, OQ_2, OQ_3.....OQ_n be the radii of the half wave zones denoted by r_1, r_2,.........r_n respectively. Suppose, *S* is the source of light and *P* the position of bright image of *S*. Then, if $SO = a$, and $OP = b$,

$$SQ_1 + Q_1P = a + b + \frac{\lambda}{2}$$

$$SQ_2 + Q_2P = a + b + \frac{2\lambda}{2}$$

$$\vdots \quad \vdots \quad \vdots \quad \vdots$$

$$SQ_n + Q_nP = a + b + \frac{n\lambda}{2} \quad \ldots(2.22)$$

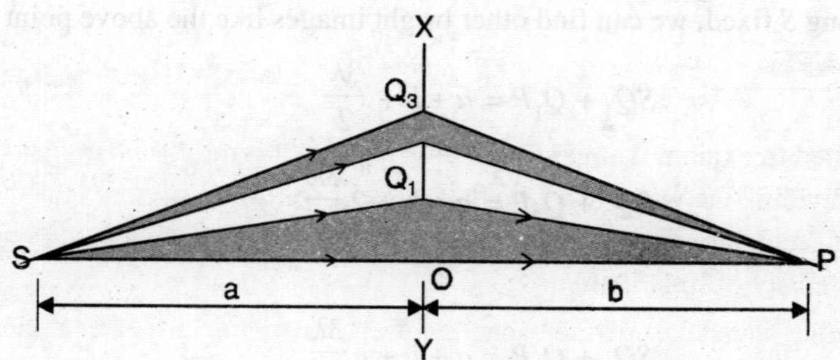

FIGURE 2.14 **Action of the zone plate forming image at *P* of the source at *S*.**

Here λ denotes the wavelength of light from S.

Also, $$(SQ_1)^2 = a^2 + r_1^2$$

or $$SQ_1 = a + \frac{1}{2}\frac{r_1^2}{a} \text{ when } \frac{r_1^2}{a} << 1$$

Similarly $$Q_1P = b + \frac{1}{2}\frac{r_1^2}{b}$$

Finally for the nth half wave zone

$$SQ_n = a + \frac{1}{2}\frac{r_n^2}{a}$$

and $$Q_nP = b + \frac{1}{2}\frac{r_n^2}{b} \quad ...(2.23)$$

Now using eqs. (2.22) and (2.23), we get

$$a + \frac{1}{2}\frac{r_n^2}{a} + b + \frac{1}{2}\frac{r_n^2}{b} = a + b + \frac{n\lambda}{2}$$

or $$\left(\frac{r_n^2}{2}\right)\left(\frac{1}{a} + \frac{1}{b}\right) = \frac{n\lambda}{2}$$

By using the sign convention one gets

$$\frac{1}{b} - \frac{1}{a} = \frac{n\lambda}{r_n^2} \quad ...(2.24)$$

This result resembles lens formula

$$\frac{1}{b} - \frac{1}{a} = \frac{1}{f_n} \quad ...(2.25)$$

such that $f_n = \frac{r_n^2}{n\lambda}$, the equivalent focal length of the zone plate.

Further, keeping S fixed, we can find other bright images like the above point P such that

$$SQ_1 + Q_1P = a + b + \frac{3\lambda}{2}$$

$$SQ_2 + Q_2P = a + b + 2\frac{3\lambda}{2}$$

$$\vdots \quad \vdots \quad \vdots \quad \vdots$$

and $$SQ_n + Q_nP = a + b + n\frac{3\lambda}{2}$$

Following the same procedure as above for the new observation point we would get

$$f_n = \frac{r_n^2}{3n\lambda}$$

Similarly for other bright point we can obtain

$$f_n = \left[\frac{1}{(2m+1)}\right]\frac{r_n^2}{n\lambda} \; ; m = 0, 1, 2, \ldots\ldots$$

Since, the wavelets from the alternate transparent zones would reach the point P in same phase, P is bright. If $U_1(P)$, $U_3(P)$, ... $U_n(P)$ represent the amplitudes of the light disturbance which reach P from these half wave zones, the resultant amplitude would be

$$U(P) = U_1(P) + U_3(P) + \ldots\ldots\ldots\ldots\ldots\ldots U_n(P) \; ; n \text{ is odd (central zone is bright)}$$

That is $U(P) > \frac{U_1(P)}{2}$; (for the case when there is no zone plate in between S and P)

Hence, intensity at P with the zone plate in position will be more than when there were no obstruction. This behaviour of zone plate is termed as focussing action.

Lastly, we can obviously compare the action of a lens with that of zone plate in the following manner :

1. In case of a lens the image is brighter than that of zone plate.

2. Lens produces image only at one place while zone plate produces at more than one place.

3. Image produced by a lens is due to refraction while by a zone plate it is due to diffraction.

4. Variation of focal length with colour of the light in case of zone plate has a reverse pattern to that in case of a lens. For example, in case of lens $f_r > f_b$ while in case of zone plate $f_r < f_b$ where r means red colour and b means blue.

Apart from the above mentioned points, the action of the lens and of zone plate are similar for example, with regard to the formation of real image and chromatic aberration.

SOLVED NUMERICAL PROBLEMS

Problem 1. *How many orders of diffraction will be seen when light of wavelength 6000 Å passes through a grating ruled with 6000 lines/cm.* (May 1998, K.U.K.)

Solution. Here, $(a + b) \sin \theta_n = n\lambda$

The maximum possible value of $\sin \theta_n = 1$

$\therefore$ $(a + b) \times 1 = n\lambda$

Again $(a+b) = \frac{1}{6000}, \qquad \lambda = 6000 \times 10^{-8}$ cm

$$\therefore \qquad n = \frac{(a+b)}{\lambda} = \frac{1}{6000 \times 6000 \times 10^{-8}} = \frac{100}{36} = 2.77.$$

Problem 2. *What should be the minimum number of lines per inch in a half inch width grating to resolve the D_1(5896 Å) and D_2(5890 Å) lines of sodium?* (Jan 1999, K.U.K.)

Solution. As the order is not mentioned in this problem,
Let us consider the case of first order
Here $n = 1$

$$\lambda_1 = 5896 \times 10^{-8} \text{ cm}, \lambda_2 = 5890 \times 10^{-8} \text{ cm}$$
$$d\lambda = (5896 - 5890) \times 10^{-8} = 6 \times 10^{-8} \text{ cm.}$$

λ can be taken as

$$\frac{\lambda_1 + \lambda_2}{2} = 5893 \times 10^{-8}$$

Now $N = \frac{1}{n} \times \frac{\lambda}{d\lambda} = \frac{5893 \times 10^{-8}}{(1 \times 6 \times 10^{-8})} = 983$ (Appro.)

Since the grating is ½ inch wide, we should have the number of lines per inch as

$$983 \times 2 = 1966$$

Hence number of lines per cm = 1966/2.54 = 775.

Problem 3. *Light is incident normally on a grating 0.5 cm wide with 2500 lines. Find the angular separation of the two sodium lines in the first order spectrum. Can they be seen distinctly, wavelength of the lines are 5890 Å and 5896 Å.* (May 1995, K.U.K.)

Solution. Width of grating = 0.5 cm
Number of lines = N = 2500

Grating element $= (a+b) = \frac{0.5}{2500} = \frac{1}{5000}$

Mean wavelength of two lines $\lambda = \frac{(5890 + 5896)}{2} = 5893$ Å

Also difference $d\lambda$ in wavelengths = 6 Å = 6×10^{-8} cm

If θ be the mean angle of diffraction for the first order spectrum, then

$$\sin\theta = \frac{\lambda}{(a+b)} = 500 \times 5893 \times 10^{-8} = 0.2946$$

$$\therefore \qquad \theta = 17°8'$$

We know dispersive power $\frac{d\theta}{d\lambda} = \frac{n}{(a+b)\cos\theta}$

or $d\theta = \frac{n\,d\lambda}{(a+b)} \cdot \frac{1}{\cos\theta} = \left(\frac{1 \times 6 \times 10^{-8} \times 5000}{\cos 17° 08'}\right) \times \frac{180}{\pi}$ radians

or $d\theta = 1° 47.5'$

The two lines will be seen distinctly if they are resolved by grating.

Now resolving power $\dfrac{\lambda}{d\lambda} = \dfrac{5893 \times 10^{-8}}{6 \times 10^{-8}} = 982\ N$

Actual resolving power of diffraction grating $= n \times N = 1 \times 2500 = 2500$

The actual resolving power is greater than the required resolving power, so the lines will be distinctly visible.

Problem 4. *A parallel beam of monochromatic light is allowed to be incident normally on a plane grating having 1250 lines per cm and a second order spectral line is observed to be deviated through 30°. Calculate the wavelength of the spectral line.* (June 2000, K.U.K.)

Solution. Here number of lines per cm = 1250

$\therefore$ Grating element $(a + b) = \dfrac{1}{1250}$

and $\theta = 30°, n = 2, \lambda = ?$

Now $(a + b)\sin\theta = n\lambda$

$\therefore$ $\lambda = (a + b)\dfrac{\sin\theta}{n} = \dfrac{1 \times \sin 30}{1250 \times 2} = \dfrac{1}{1250 \times 2 \times 2} = 2 \times 10^{-8}$ cm

$\therefore$ $\lambda = 2 \times 10^{-4}$ cm.

Problem 5. *A diffraction grating used at normal incidence gives a line (5400 Å) in a certain order superposed on the violet line (4050 Å) of the next higher order. How many lines per cm are there in the grating if the angle of diffraction is 30° ?* (Dec. 2000, K.U.K.)

Solution. Here
$$\lambda_1 = 5400\ \text{Å} = 5400 \times 10^{-8}\ \text{cm}$$
$$\lambda_2 = 4050\ \text{Å} = 4050 \times 10^{-8}\ \text{cm}$$
$$\theta_1 = \theta_2 = 30°$$

$\therefore$ $\sin\theta_1 = \sin\theta_2 = \dfrac{1}{2}$

$$(a + b)\sin\theta_1 = n_1\lambda_1 \quad ...(1)$$

and $$(a + b)\sin\theta_2 = (n_1 + 1)\lambda_2 \quad ...(2)$$

Dividing the above equations, we get

$$1 = (n_1 + 1)\frac{\lambda_2}{n_1\lambda_1} = \frac{(n_1 + 1)}{n_1} \times \left(\frac{40.5 \times 10^{-6}}{54 \times 10^{-6}}\right)$$

$$1 = \frac{(n_1 + 1)}{n_1} \times \left(\frac{40.5}{54}\right)$$

or $n_1 = (n_1 + 1) \times 0.75$ or $n_1 = 3$

From eq. (1)

$(a + b)\sin 30 = 3 \times 54 \times 10^{-6}$ or $(a + b) = 324 \times 10^{-6}$

The number of lines per cm $= \dfrac{1}{(a + b)} = \dfrac{1}{324 \times 10^{-6}} = 3086$ lines/cm.

Problem 6. *What should be the minimum number of lines on a grating which will just resolve sodium lines of wavelength 5890 Å and 5896 Å in the second order.* (Dec. 1999, K.U.K.)

Solution. The resolving power of grating is given by

$$\frac{\lambda}{d\lambda} = nN$$

Here $n = 2$,

$$d\lambda = 5896 - 5890 = 6 \text{ Å}$$

$$\lambda = \frac{(5890 + 5896)}{2} = 5893 \text{ Å}$$

$$\therefore \quad N = \frac{5893 \times 10^{-8}}{6 \times 10^{-8} \times 2} = 491 \text{ (app.)}$$

Problem 7. *The separation of sodium line ($\lambda = 5893$ Å) in the second order spectrum of a transmission grating containing 5000 lines per cm is 2.5 minutes for normal incidence. What is the difference in wavelength of two yellow lines.*

(June 2001, K.U.K.)

Solution. We know that,

$$\frac{d\theta}{d\lambda} = \frac{n}{(a+b)\cos\theta} = \frac{n}{(a+b)\sqrt{1-\sin^2\theta}}$$

$$\therefore \quad d\lambda = \frac{(a+b)}{n}[1-\sin^2\theta]^{1/2} \times d\theta = \frac{(a+b)}{n}\left[1 - \frac{n^2\lambda^2}{(a+b)^2}\right]^{1/2} d\theta$$

$$(a+b)\ \frac{1}{5000} \text{ cm}, \ n = 2, \ \lambda = 5893 \times 10^{-8} \text{ cm}$$

and $\quad d\theta = 2.5 \text{ minutes} = 2.5 \times 1 \times \frac{\pi}{60} \times 180 \text{ rad.}$

$$\therefore \quad d\lambda = \left\{\left(\frac{1}{5000}\right)\left[1 - 2^2 \times (5893 \times 10^{-8})^2 \times (5000)^2\right]^{\frac{1}{2}}\right\} \times 2.5 \times 1 \times \frac{\pi}{60} \times 180$$

$$= 5.81 \times 10^{-8} \text{ cm} = 6 \text{ Å (app.)}$$

Problem 8. *The diameter of the first ring of a zone plate is 1 mm if $\lambda = 5000$ Å, find the distance of the brightest spot when a plane wave falls on the zone plate.*

Solution. $\quad f_n = \frac{r_n^2}{n\lambda}; \quad n = 1, \lambda = 5 \times 10^{-7} \text{ m and } = r_1 = \frac{10^{-3}}{2} \text{ m}$

$$\therefore \quad f = \frac{10^{-6}}{4 \times 5 \times 10^{-7}} = \frac{1}{2} \text{ m} = 50 \text{ cm.}$$

Problem 9. *Calculate the radii of first three transparent zones of a zone plate which is to bring a parallel beam of light with $\lambda = 6 \times 10^{-7}$m to its principal focus at a distance of 2m.*

Solution. As, $\quad f_n = \frac{r_n^2}{n\lambda}; \quad f_1 = 2 \text{ m} = b$

$$r_1^2 = f_1\lambda = 2 \times 6 \times 10^{-7}$$

$\therefore \qquad r_1 = \sqrt{1.2 \times 10^{-3}}$ m = 0.109 cm

Further, $\qquad r_n^2 = nb\lambda$

$$r_3^2 = 3 \times 2 \times 6 \times 10^{-7}$$

$$r_3 = \sqrt{3.6 \times 10^{-3}} \text{ m} = 0.19 \text{ cm}$$

Similarly, $\qquad r_5^2 = 5 \times 2 \times 6 \times 10^{-7} = 6 \times 10^{-6}$

or the third radius $\qquad = \sqrt{6 \times 10^{-3}}$ m = 0.245 cm.

EXERCISE

1. Give the construction and theory of plane transmission grating and explain the formation of spectra by it. What will be the effect of increasing the number of lines per inch in the grating on the spectra ? *(Jan.1999 & June 2003, K.U.K.)*
2. Explain the action of a plane transmission grating and discuss the factors which control its dispersive and resolving powers. *(May 1998, K.U.K.)*
3. Distinguish between Fraunhofer and Fresnel diffraction and discuss Fraunhofer diffraction at a single slit. *(May 1995 & June 2000, Jan. 2003, K.U.K.)*
4. Derive the expressions for resolving power and dispersive power of a diffraction grating. *(Jan. 1997, K.U.K.)*
5. Describe and explain the Fraunhofer diffraction pattern obtained with a narrow slit illuminated by a parallel beam of monochromatic light. *(Jan. 2003, June 2000, K.U.K.)*
6. What is the plane transmission grating ? Describe how would you employ it for determining the wavelength of light ? *(Jan. 2003, Dec. 2000, K.U.K.)*
7. Explain the distinction between interference and diffraction of light and discuss the diffraction through a slit. *(June 2001, K.U.K.)*
8. Calculate the angles at which the first dark band and the next bright band are formed in the Fraunhofer diffraction pattern of a slit 0.3 mm wide. (λ = 5890 Å) **[Ans.** $\theta = 6.7'$, $\theta' = 10'$ **]**
9. A grating with 15000 ruling per inch is illuminated normally with white light extending from 4000 Å – 7000 Å. Show that only the first order spectrum is isolated but the second and third order overlap. **[Ans.** 1st order 13", 24", 2nd order 28", 56", 3rd order 45"**]**
10. How many orders will be visible if the wavelength of incident radiation is 5000 Å and the number of lines on the grating is 2620 in one inch. **[Ans.** 19**]**
11. A diffraction grating which has 4000 lines to a cm is used at normal incidence. Calculate the dispersive power of the grating in the 3rd order spectrum in the wavelength region 5000 Å. **[Ans.** 15000**]**
12. What is the highest order spectrum, which may be seen with monochromatic light of wavelength 6000 Å by means of diffraction grating with 5000 lines per cm ? **[Ans.** 3**]**
13. The radius of the first ring of the zone plate is 1 mm. If the plane wave of $\lambda = 50 \times 10^{-8}$ m falls on the plate find where will be the brightest spot. **[Ans.** 2m**]**
14. If the diameter of the central zone is 2.5 mm and a point source of light of $\lambda = 7.5 \times 10^{-7}$ m is placed 5m away from the zone plate, find the position of primary and secondary images. **[Ans.** 357.1 cm, 80.64 cm**]**

Chapter 3 POLARISATION

3.1 POLARISED AND UNPOLARISED LIGHT

Introduction

The successful explanation of the phenomenon of interference and diffraction has established that light is a form of wave motion. Now the wave motion is of two types : (*i*) Longitudinal wave motion, (*ii*) Transverse wave motion. However, these phenomenon do not decide whether the light waves are longitudinal or transverse. An important phenomenon known as polarisation of light confirms that light is a transverse wave motion.

Longitudinal Wave Motion

In a longitudinal wave motion the particles of the medium vibrate about their mean position parallel to the direction of propagation of the wave. It means that longitudinal waves are symmetrical about their direction of propagation *i.e.* any two such waves are alike in their properties.

Transverse Wave Motion

In transverse wave motion the particles of the medium vibrate about their mean position perpendicular to the direction of propagation of the wave. It means that transverse waves are asymmetrical about their direction of propagation, because in the case of transverse waves there may be any number of directions in which the particle vibrate all perpendiculars to the direction of propagation of the wave. So two otherwise similar transverse waves differ from each other in the sense that their directions of vibration may be different. If by some arrangement, the vibrations are restricted to one particular direction or plane only, the light is said to be polarised and the phenomenon is known as polarisation of light.

The phenomenon in which the vibrations of the particles of the medium produced by light waves are restricted to one particular direction or plane is called polarisation of light. The light wave having one particular direction of vibration is called polarised light. Ordinary light (from any source) in which the particles of the medium can vibrate in any number of directions all perpendicular to the direction of propagation is called unpolarised light.

Transverse Nature of Light Waves i.e. Polarised and Unpolarised Light

The phenomenon of interference and diffraction firmly confirm that light is a form of wave motion. Further, it can be proved from the following simple experiment that light is a transverse wave motion.

Take a stretched rubber string *AB* passing through two narrow slits S_1 and S_2 parallel to each other in the vertical plane as shown in Fig. 3.1. Set up a longitudinal wave by moving the end *A* to and fro parallel to the length of the string. Rotate any of the slits about *AB* as axis. We find that this rotation does not effect the passage of the wave. Now, if we set up a transverse wave in the string by moving the end *A* up and down in the vertical plane it will be seen that the wave passes through the slits undisturbed as shown in Fig. 3.1 (*a*). Now rotate the slit S_2 through 90° about *AB* as axis we find that the wave passes through the slit S_1 undisturbed as before but it is not able to pass through S_2 so that beyond S_2 the string remains practically undisturbed as shown in Fig. 3.1 (*b*).

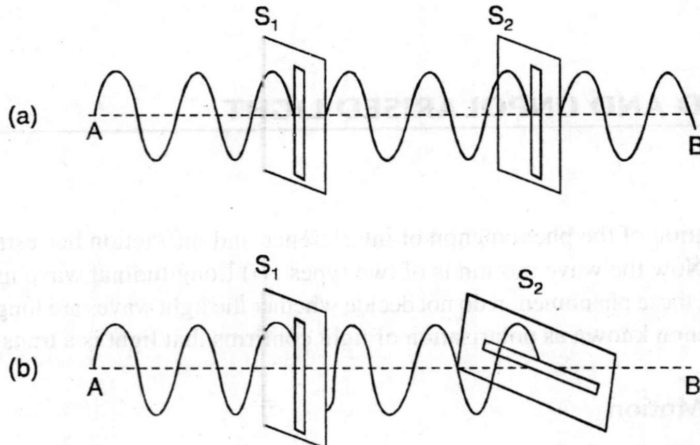

FIGURE 3.1 **Transverse nature of light.**

This shows that the transverse waves can pass through the slits only when they are parallel to each other and lie in the same plane in which the wave lies. Thus transverse waves have different properties w.r.t. vertical plane as compared with other planes. Such waves are said to be polarised after passing through the first slit. The phenomenon in which the wave oscillations are restricted to a particular direction in a plane is called polarisation. This analogy helps us to distinguish between the longitudinal and transverse waves. The tourmaline and several other transparent crystalline substances behave w.r.t. light wave in exactly the same manner as the slit do w.r.t. transverse waves in cord.

Take a tourmaline crystal *A* cut parallel to its crystallographic axis and placed normally to the path of a beam of light coming from a source *S* as shown in Fig. 3.2 (*a*). The emergent beam will be slightly coloured due to natural colour of crystal. Rotate the crystal *A* about *SO* as axis, it will be seen that there is no change in the intensity or the colour of the emergent beam. Take a second crystal *B* cut similarly and hold it in the path of the beam so that its axis is parallel to the axis of the crystal *A*.

The beam of light will be seen to pass through both the crystal and will be still coloured [Fig. 3.2 (*b*)]. Now rotate the crystal *B* about the beam as axis. It will be seen that the intensity of emergent beam decreases and when the axis of both the crystals are at right angles to each other, no light comes out of the crystal *B* [Fig.3.2 (*c*)]. If the crystal *B* is further rotated, light reappears and its intensity becomes maximum again when their axes are parallel. In one complete rotation of the crystal *B*, intensity of light become maximum twice when the axis of *A* and *B* are parallel and the intensity becomes zero twice when the axis of *A* and *B* are at right angles to each other. Comparing the observations in this experiment with the mechanical analogue discussed earlier, we can firmly conclude that the light waves are transverse and the vibrations in them are at right angles to the direction in which the wave is travelling.

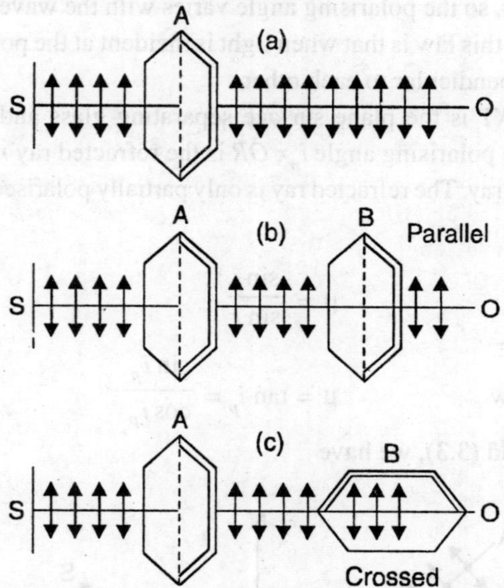

FIGURE 3.2 **(*a*) Tourmaline crystal *A* cut parallel to its cryatallographic axis.**
(*b*) Axis of crystal *B* is parallel to the axis of crystal *A*.
(*c*) Axis of crystal *B* perpendicular to the axis of crystal *A*.

The first crystal, which polarises the light incident on it, is called polariser. The second crystal, which analyses whether the light is polarised or not, is called analyser.

Plane of Vibration

The plane in which the vibration occurs is called the plane of vibrations.

Plane of Polarisation

The plane perpendicular to the plane of vibration is called the plane of polarisation. The plane of polarisation is the plane in which no vibration occur as shown in Fig. 3.3.

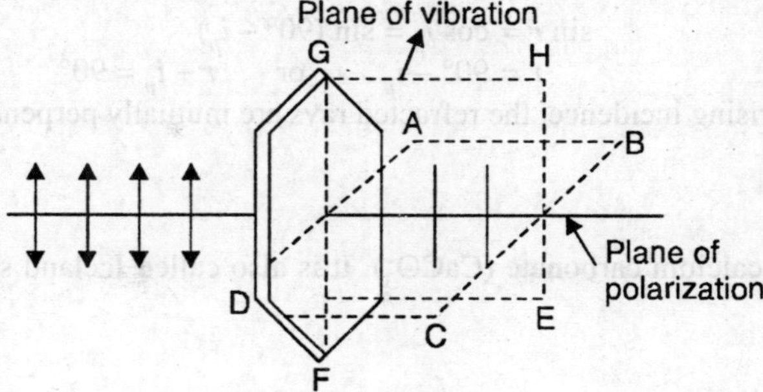

FIGURE 3.3 **Plane of vibration (*GHEF*) and plane of polarisation (*ABCD*).**

3.2 BREWSTER'S LAW

It states that the tangent of the polarising angle is numerically equal to the refractive index of the medium *i.e.*

$$\mu = \tan i_p \qquad ...(3.1)$$

where μ is refractive index and i_p is the polarising angle. As the refractive index of the medium depends on the wavelength of incident light, so the polarising angle varies with the wavelength of light.

A direct consequence of this law is that when light is incident at the polarising angle, the reflected and refracted rays are mutually perpendicular to each other.

As shown in Fig. 3.4, *XY* is the plane surface separating glass and air. *AO* is the incident ray of unpolarised light incident at the polarising angle i_p ; *OR* is the refracted ray *r* being the corresponding angle of refraction ; *OS* is the reflected ray. The refracted ray is only partially polarised and reflected ray is completely polarised.

From Snell's Law
$$\mu = \frac{\sin i}{\sin r} \qquad ...(3.2)$$

And from Brewster's law
$$\mu = \tan i_p = \frac{\sin i_p}{\cos i_p} \qquad ...(3.3)$$

Comparing eqs. (3.2) and (3.3), we have

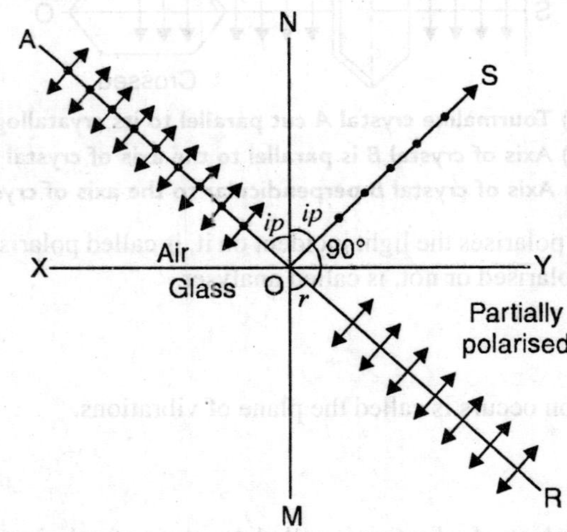

FIGURE 3.4 **Reflected and refracted rays are perpendicular to each other.**

$$\sin r = \cos i_p = \sin (90° - i_p)$$

Therefore, $r = 90° - i_p$ or $r + i_p = 90°$

That is, at the polarising incidence, the refracted rays are mutually perpendicular to each other.

Calcite Crystal

It is a hydrated crystal of calcium carbonate ($CaCO_3$). It is also called Iceland spar. The important points about it are as follows.

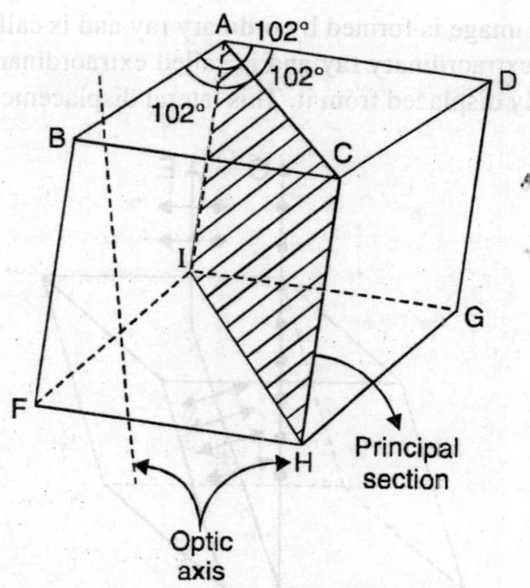

FIGURE 3.5 **Rhombus shape calcite crystal.**

(*i*) It occurs in nature in several forms. On cleavage or breakage, it is always reduced to the form of a rhombus as shown in Fig. 3.5. The six faces of the rhombus are parallelogram of angles 78° and 102° approximately. At the two opposite corners such as *A* and *H* the three angles of the faces meeting there are obtuse (each of 102°). These corners are known as blunt corners. At each of the remaining four corners, one of the angles is obtuse while the other two are acute.

(*ii*) The line through a blunt corner *A* or *H* making equal angles with the three edges meeting there is called an optic axis. In fact it is just a direction and not a fixed line in the crystal. The substances like calcite and quartz contains only one optic axis and is called uniaxial crystals. Substances like mica and borax contain two optic axes and are called biaxial crystals.

(*iii*) A plane containing an optic axis and perpendicular to the two opposite faces of the crystal is called principal plane or principal section.

3.3 DOUBLE REFRACTION

The phenomenon in which a ray of light incident on a calcite or quartz crystal is split up into two refracted rays is called double refraction. One refracted ray obeys the ordinary laws of refraction *i.e.* the ratio $\frac{\sin i}{\sin r}$ is always constant for all values of angle of incidence *i* for this ray. This ray is called the ordinary ray (*O*-ray).

The second refracted ray does not obey the ordinary laws of refraction *i.e.* ratio $\frac{\sin i}{\sin r}$ does not remain constant for all values of the angle of incidence *i*. This suffers refraction even at normal incidence. This ray is called extraordinary ray (*E*-ray). The crystal like calcite or quartz, which show this phenomenon, are called double refracting or birefringent crystals.

The phenomenon of double refraction can be demonstrated with a simple experiment. Mark an ink dot on a piece of paper and place a calcite crystal on it. The two images will be seen as shown in Fig. 3.6. On rotating the crystal about the incident ray as axis, one image remains stationary while the other revolves around the first. The stationary image is formed by ordinary ray and is called ordinary image 'O' while the revolving image is formed by extraordinary ray and is called extraordinary image E. The emergent ray is parallel to O-ray though laterally displaced from it. This lateral displacement depends upon the thickness of the crystal.

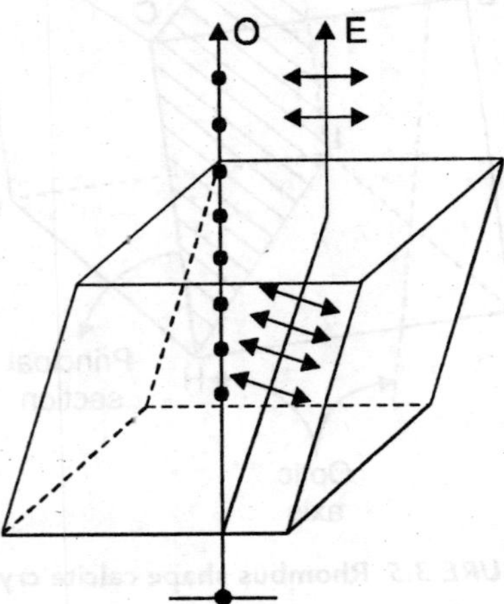

FIGURE 3.6 **Double refraction.**

Following are some important points about double refraction :

1. The ratio $\frac{\sin i}{\sin r}$ for ordinary ray remains constant and is called refractive index μ_0 of the crystal for ordinary ray. As $\frac{\sin i}{\sin r}$ is constant, therefore μ_0 is same in all direction and hence velocity of ordinary ray is same in all directions in the crystal.

2. For extraordinary ray, $\frac{\sin i}{\sin r}$ is not constant but varies with the angle of incidence showing that the refractive index is different in different directions for E-ray. Therefore E-ray travels with different velocities in different directions. However, along the optic axis both O-ray and E-ray travel with the same velocity *i.e.* no double refraction occur along this axis.

3. It is observed that both the ordinary and extraordinary rays are plane polarised but in perpendicular planes, the O-ray has vibrations perpendicular to the plane of paper and E-ray has vibrations in the plane of paper.

4. Malus law is obeyed in double refraction *i.e.* sum of the intensities of the two refracted rays is equal to the intensity of the incident ray.

Huygen's Theory of Double Refraction in Uniaxial Crystals

Huygen explained the phenomenon of double refraction by applying his principle of secondary wavelets to uniaxial crystals. Accordingly, he put forward the following postulates.

1. When an incident wave falls on the surface of a crystal then each disturbed point inside the crystal becomes the origin of two wave fronts or wave surfaces, one corresponding to ordinary ray and the other to extraordinary ray.

2. For the ordinary ray, for which the velocity of light is same in all directions, the wave surface is spherical and is called ordinary wave surface.

3. For the extraordinary ray, for which the velocity of light varies with direction, the wave surface was supposed to be of the next order in simplicity to the sphere *i.e.* an ellipsoid.

4. Since the two rays (*O*-ray and *E*-ray) travel with the same velocity along the optic axis therefore the two wave surfaces touch each other at the extremities of the optic axis.

Now as regards double refraction, the uniaxial crystals can be divided into two different categories.

1. Negative Crystal

In such crystals, the diameter of the ordinary spherical wave surface is equal to the minor axis of the extraordinary ellipsoid wave surface as shown in Fig. 3.7 (*a*). It is clear from the figure that

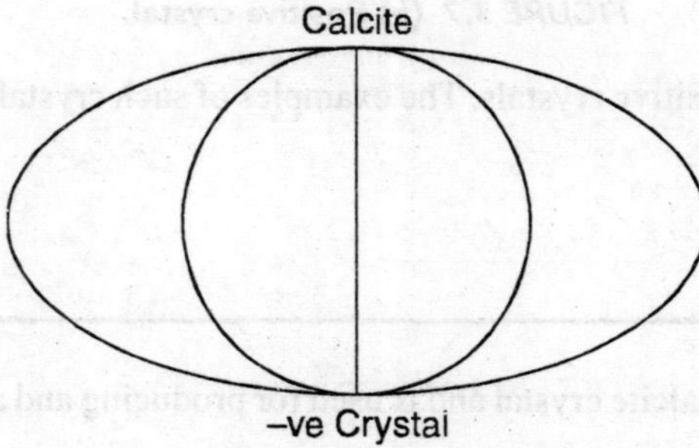

FIGURE 3.7 **(*a*) Negative crystal.**

(*a*) The wave surface due to ordinary ray *i.e.* spheres lies inside the extraordinary wave surface *i.e.* ellipsoid.

(*b*) The velocity of extraordinary ray is different in different directions. It is minimum along the optic axis *XY* and maximum in a direction perpendicular to optic axis.

(*c*) The velocity of ordinary ray is same in all directions.

(*d*) In this case velocity of *E*-ray is greater than that of *O*-ray in all directions except along optic axis. Therefore the refractive index for *E*-ray is less than that of *O*-ray *i.e.* $\mu_e < \mu_o$.

Such crystals are called negative crystals. The examples of such crystals are calcite, tourmaline and nitrate of soda.

2. Positive Crystal

In such crystals, the diameter of the ordinary spherical wave surface is equal to the major axis of the extraordinary ellipsoid wave surface, as shown in Fig. 3.7 (*b*). It is clear from the figure that

(*a*) The wave surface due to extraordinary ray *i.e.* ellipsoid lies inside the wave surface due to ordinary ray *i.e.* spheres.

(*b*) Velocity of extraordinary ray is different in different direction. It is maximum along the optic axis *XY* and is minimum in a direction perpendicular to optic axis.

(*c*) The velocity of ordinary ray is same in all directions.

(*d*) In this case velocity of *O*-ray is greater than that of *E*-ray in all directions except along optic axis. Therefore the refractive index for *O*-ray is less than that of *E*-ray *i.e.* $\mu_o < \mu_e$.

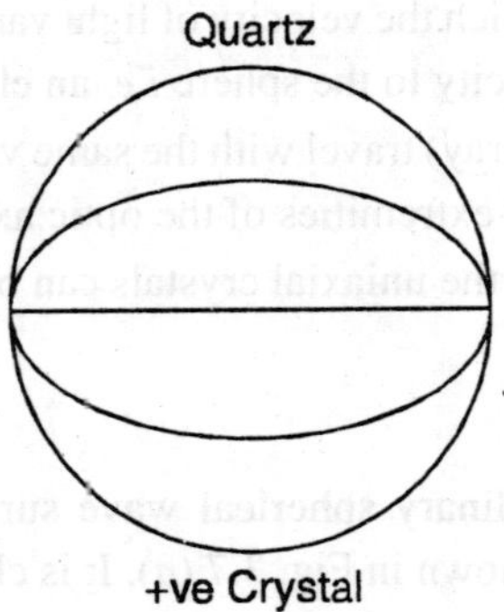

FIGURE 3.7 **(*b*) Positive crystal.**

Such crystals are called positive crystals. The examples of such crystals are quartz, sulphate of potash ice and zicron.

3.4 NICOL PRISM

It is an optical device made from calcite crystal and is used for producing and analysing plane polarised light.

Principle

It is based upon the principle of double refraction in calcite crystal. When light is incident on the calcite crystal it is split up into two rays *i.e.* ordinary ray and extraordinary ray. The ordinary ray is then generally eliminated by total internal reflection and the extraordinary ray is transmitted through the prism. The prism gives an intense beam of plane polarised light. Willium Nicol invented this prism in 1828.

Construction

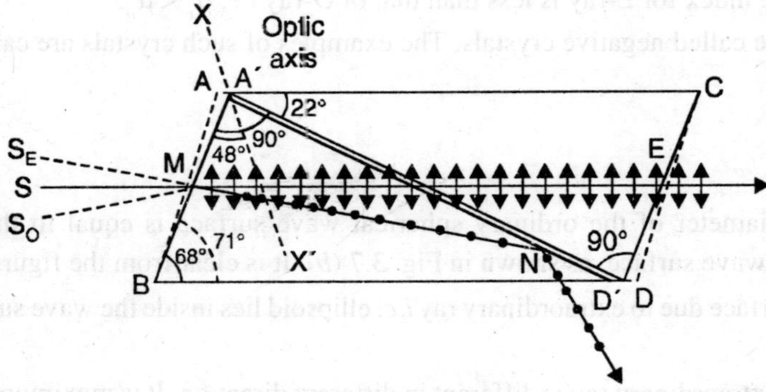

FIGURE 3.8 **Nicol Prism.**

A calcite crystal whose length is three times its breadth is taken. *ABCD* is the principal section of the crystal as shown in Fig. 3.8. Smaller end faces *AB* and *CD* of its principal section are grounded until the angles *B* and *D* are reduced from 71° to 68°. Then the crystal is cut diagonally into two halves by a vertical plane. The cut surfaces of the two halves of the crystal are properly polished and then cemented together in the original position with a thin layer of Canada balsam. The combination so obtained is known as Nicol prism.

Action

The canada balsam is a transparent substance whose refractive index lies in between the refractive indices of *O*-ray and *E*-ray.

$$\text{Refractive index of } O\text{-ray } (\mu_O) = 1.658$$

$$\text{Refractive index of canada balsam } (\mu_{CB}) = 1.55$$

$$\text{Refractive index of } E\text{-ray } (\mu_e) = 1.486$$

It is clear from above that the canada balsam acts as a rarer medium for *O*-ray and acts as a denser medium for *E*-ray. Let us consider a ray *SM* incident on the Nicol prism. It suffers double refraction and is then split up into two rays *i.e.* *O*-ray and *E*-ray. As mentioned above, *O*-ray travel from denser to rarer medium and the *E*-ray travels from rarer to denser medium. The *O*-ray can suffer total internal reflection at canada balsam layer if it falls on this layer at an angle greater than the critical angle which is given below

$$\text{Critical angle} = \sin^{-1}\left(\frac{\mu_{CB}}{\mu_O}\right) = 69°$$

Now, the incident ray *SM* is made to fall on the Nicol prism in such a way that *O*-ray inside the prism travelling from calcite (denser medium) to canada balsam (rarer medium) falls on canada balsam at angle greater than 69° and hence it is totally internally reflected and is absorbed by the blackened surface of the Nicol prism and is thus eliminated. On the other hand the *E*-ray does not suffer total internal reflection and is simply transmitted through the Nicol prism. In this way we get a plane polarised beam (*E*-ray) with a Nicol prism. It may be noted that the Nicol prism can not be used for highly convergent and divergent beams. The angle of incidence of the incident ray on the face *AB* should be limited to a narrow angle of 14°. For any angle, greater than this, the *O*-ray will strike the balsam layer at angle less than the critical angle and so will not be totally internally reflected but rather transmitted. The dimensions of the Nicol prism are such that the angle of incidence on *AB* for *E*-ray should not exceed 14°. Thus to avoid the total internal reflection of *E*-ray the angle between the extreme rays of light incident on the Nicol prism is limited to 28°.

Uses of Nicol Prism

1. It acts as a polariser *i.e.* when ordinary unpolarised light is incident on it, the light emerging out of it is plane polarised one.
2. It also acts as an analyser used to detect the plane polarised light.
3. It is oftenly used with quite a few other optical devices to detect circularly or elliptically polarised light.
4. It is also used to measure the change in the plane of polarisation of light.

Drawbacks of Nicol Prism

1. It cannot be used for ultraviolet light because the canada balsam layer absorbs ultraviolet radiation.
2. The light emerging out of it is not uniformly plane polarised.
3. As *E*-ray passes through it obliquely, the emergent ray is always displaced a little to one side.

3.5 QUARTER AND HALF WAVE PLATES

Quarter Wave Plate (Q.W.P.)

A crystal plate which can introduce a path difference of $\frac{\lambda}{4}$ or a phase difference of $\frac{\pi}{2}$ between the *O*-ray and *E*-ray is called a quarter wave plate. Such a plate is prepared from a doubly refracting uniaxial crystal of calcite or quartz of suitable thickness whose refracting faces are cut parallel to the direction of optic axis.

When a beam of plane polarised monochromatic light of wavelength λ is allowed to fall normally on such a crystal, it is split up into *O* and *E*-ray, which travel in the same direction but with different velocities as shown in Fig. 3.9.

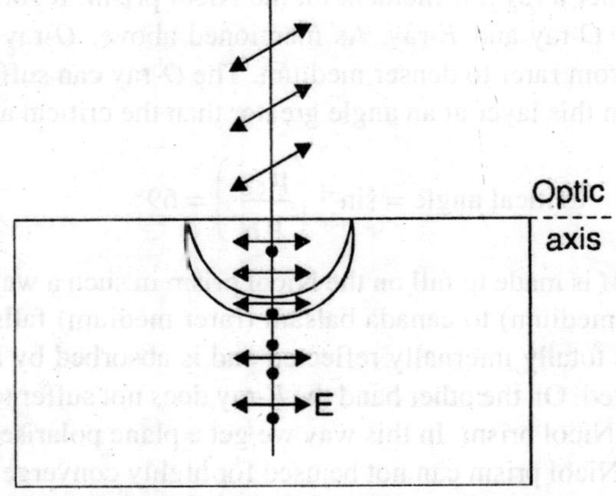

FIGURE 3.9 **Quarter wave plate.**

In negative crystals like calcite the *E*-ray travels faster than *O*-ray so that $\mu_o > \mu_e$.

Within the plate, the optic path for *E*-ray $= \mu_e t$

The optic path for *O*-ray $= \mu_o t$

with *t* as the thickness of the plate.

Therefore, the path difference $= \mu_o t - \mu_e t = t(\mu_o - \mu_e)$.

Now in case of quarter wave plate this path difference must be equal to $\frac{\lambda}{4}$.

Therefore, $t(\mu_o - \mu_e) = \frac{\lambda}{4}$, and hence the thickness *t* of Q.W.P. is given by

$$t = \frac{\lambda}{[4(\mu_e - \mu_o)]}$$

For a positive crystal, like quartz, $\mu_e > \mu_o$ therefore the thickness of positive crystal is given by

$$t = \frac{\lambda}{[4(\mu_e - \mu_o)]}.$$

Uses of Q.W.P.

1. It is used for producing and detecting circularly polarised and elliptically polarised light.
2. It is used for analyzing all kinds of polarised light.

Half Wave Plate (H.W.P.)

A crystal plate which can introduce a path difference of $\frac{\lambda}{2}$ or a phase difference of π between the O-ray and E-ray is called a half wave plate. Such a plate is prepared from a doubly refracting uniaxial crystal of calcite or quartz of suitable thickness whose refracting faces are cut parallel to the direction of optic axis.

When a beam of plane polarised monochromatic light of wavelength λ is allowed to fall normally on such a crystal, it is split up into O and E-ray, which travel in the same direction but with different velocities. In the case of a negative crystals like calcite the E-ray travels faster than O-ray so that $\mu_o > \mu_e$.

If t be the thickness of the plate, then

The optical path for E-ray $= \mu_e t$

The optical path for O-ray $= \mu_o t$

Therefore, the net path difference between the two rays is given by

Path difference $= \mu_o t - \mu_e t = t(\mu_o - \mu_e)$.

Now in case of half wave plate, this path difference must be equal to $\frac{\lambda}{2}$.

Therefore, $t(\mu_o - \mu_e) = \frac{\lambda}{2}$, and hence the thickness t of H.W.P. is given by

$$t = \frac{\lambda}{2(\mu_o - \mu_e)}$$

For a positive crystal, like quartz, $\mu_e > \mu_o$ therefore the thickness of H.W.P. of a positive crystal is given by

$$t = \frac{\lambda}{2(\mu_e - \mu_o)}$$

Uses of H.W.P.

1. A half wave plate is used in the construction of Laurent's half shade device, which in turn is used in polarimeter.
2. It is used for detecting and analyzing plane-polarised light.

Specific Rotation

The angle of rotation produced by any optically active substance depends upon

(*i*) the thickness of the substance

(*ii*) concentration of the solution or the density of the active substance

(*iii*) temperature, and

(*iv*) wavelength of light.

For a fixed thickness of the substance or solution the angle of rotation increases as the wavelength of light decreases roughly inversely as the square of the wavelength. The optical activity of a substance is measured by its specific rotation or specific rotatory power. Specific rotation at a given temperature and for a given wavelength of light used is defined as the rotation (in degree) produced by one-decimeter length of solution containing 1 gm of optically active material per c.c. of solution.

If θ is the rotation produced by l decimeter of a solution and c, the concentration in grams per c.c., then the specific rotation S at a given temperature t and wavelength is given by

$$[s]^t_\lambda = \frac{\theta}{l \times c} = \text{rotation in degrees/length in decimeter} \times \text{concentration in gm/c.c.}$$

Molecular Rotation

It is defined as the rotation produced by one decimeter length of solution containing one gram molecules of the active substance per c.c. of the solution.

Molecular rotation is obtained by the product of the specific rotation and the molecular weight of the substance.

3.6 LAURENT'S HALF SHADE AND BIQUARTZ POLARIMETER

Polarimeter

It is an instrument used to measure the rotation produced by a substance. When it is used for finding the optical rotation of sugar, it is called saccharimeter. If the specific rotation of the sugar is known, the concentration of the sugar solution can be found out.

The simplest form of a polarimeter consists of two Nicol prisms N_1 and N_2 placed at a little distance apart and capable of rotation about a common axis *i.e.* about the incident beam as axis [Fig. 3.10]. The prisms are adjusted in the crossed position so that field of view is dark. Then a tube containing an optically active solution is introduced between two Nicol prisms. Now the field of view is found to be no longer dark because the solution has rotated the plane of polarisation. The Nicol N_2 is rotated till the field of view again become dark. The angle through which the Nicol N_2 has been rotated to restore extinction gives the angle of optical rotation θ. In such a simple polarimeter, it is very difficult to find the angle of rotation θ accurately because it is not possible to ascertain one position of the Nicol N_2 for complete extinction. Therefore, improved form of polarimeter are used in actual practice which are discussed below.

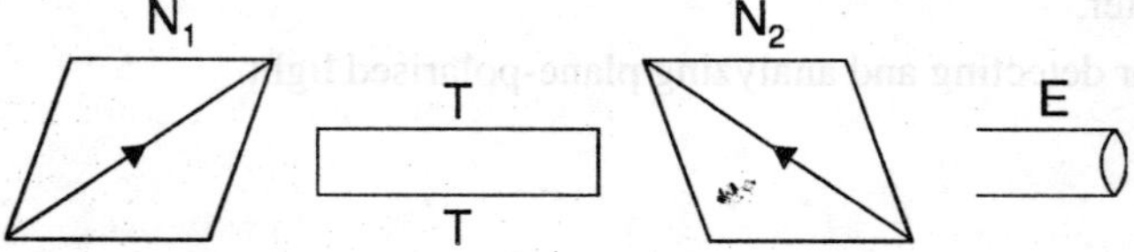

FIGURE 3.10 **Simplest from of po' meter.**

Laurent's Half Shade Polarimeter

The arrangement of this polarimeter is shown in Fig. 3.11. It consists of Nicol prisms N_1 and N_2, used as polariser and analyzer respectively. Behind the polariser N_1 a spherical device known as half shade device is placed. The half shade device consists of a semicircular plate of glass G cemented to a semicircular plate of quartz Q. The thickness of the quartz plate is such that it introduces a phase difference of π between the O-ray and E-ray. So it is half wave plate. The thickness of the glass plate is such that it absorbs the same amount of light as is done by the quartz plate Q while other half is covered by glass plate G. The optically active solution can be taken in the hollow glass tube T. The light emerging out of the analyzer N_2 can be viewed through a telescope E. The analyzer N_2 can be rotated about the axis of the tube and its rotation can be measured with the help of a vernier scale on the graduated circular scale divided in degrees.

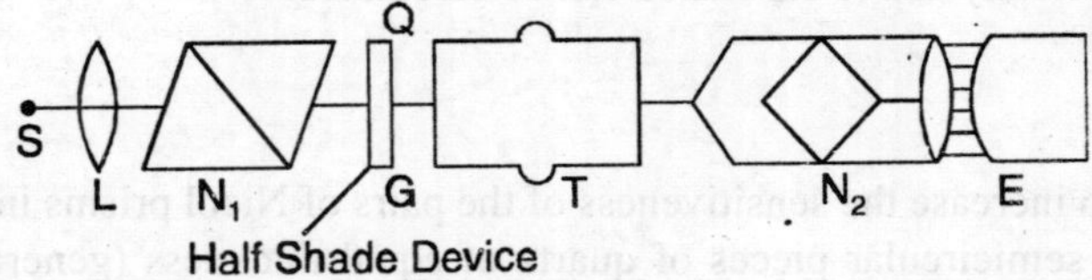

FIGURE 3.11 **Laurent's half shade polarimeter.**

Monochromatic light from a source S is rendered parallel by the lens L. This parallel beam falls on the polariser N_1. After passing through N_1 the beam is plane polarised. The plane polarised beam so obtained falls normally on the half shade device and it has vibrations along OD, [Fig. 3.12]. On passing through the glass half, the vibrations will remain along OD but on passing through the quartz half there will be split up into E and O components. These components travel along the same direction but with different velocities and therefore on emergence a phase difference of π is introduced between them. Thus the vibrations of the beam emerging out of quartz plate will be along OB whereas the vibrations of the beam emerging out of glass plate will be along OD. If the analyzing Nicol N_2 is fixed with the principal plane parallel to OD, the plane polarised light through glass half will pass and hence it will appear brighter than the quartz half from which light partially obstructed.

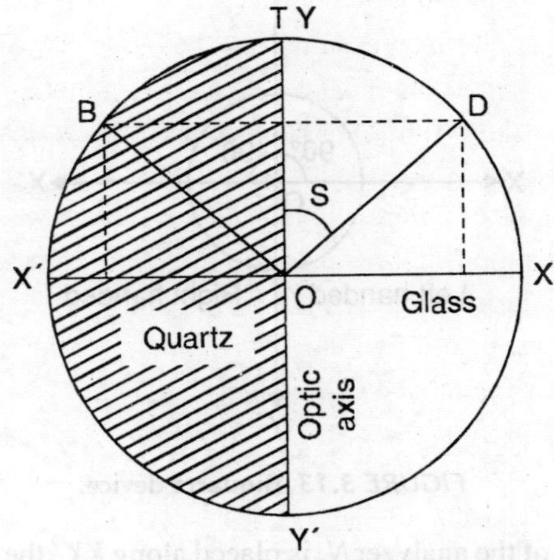

FIGURE 3.12 **Laurent's half shade device.**

If the Nicol N_2 is placed with its principal plane parallel to OB, the quartz half will appear brighter than the glass half. When the analyzer N_2 has its principal plane parallel to YOY' *i.e.* along the direction which bisects the angle BOD, the amplitudes of light incident on the analyzing Nicol N_2 from both the halves (quartz half and glass half) will be equal. Therefore the field of view will be equally bright. When the principal plane of N_2 is at right angle to YOY' again the components of OD and OB are equal. The two halves are again equally illuminated but as the intensity of the component passing through is small as compared to that in the previous case, the two halves are said to be equally dark.

To find specific rotation of an optically active substance (say sugar solution) the analyzer N_2 is set in the position for equal darkness of the field of view, first without the solution in the tube T. The reading on the circular scale is noted. Now the tube containing the solution of known concentration is placed. This will produce optical rotation and therefore the field of view will no longer be equally dark. The analyzer is again rotated to the position so that the field of view is equally dark. The new reading on the circular scale is noted. The difference in the two readings gives the angle of rotation θ. Then from the relation $[S]_\lambda{}^t = \theta/l \times c$ the specific rotation of the optically active substance can be calculated.

Biquartz Polarimeter

Biquartz device is used to increase the sensitiveness of the pairs of Nicol prisms in the polarimeter work. A biquartz consists of two semicircular pieces of quartz of equal thickness (generally 3.75 mm) cemented together to form a circular plate. Out of these two one consists of right handed and the other of left handed quartz *i.e.* one of them rotates the plane of vibration of polarized light incident on it towards left while the other towards right. The refracting faces are cut perpendicular to the optic axis. Each half produces equal rotation of the plane polarized light incident normally upon them but in opposite directions. Now like half shade device, it is also placed in a polarimeter just after the polarising Nicol N_1. Such a polarimeter is called biquartz polarimeter. If monochromatic light (sodium light) is used and thickness of each half is 3.75 mm, then rotation produced by each half is 90°. The plane polarised light emerging out of the Nicol N_1 falls on biquartz device with vibrations say along YY' (see Fig. 3.13). Now each half produces rotation through 90° but in opposite direction, therefore light of right half will have vibrations along OX while that of left along OX.

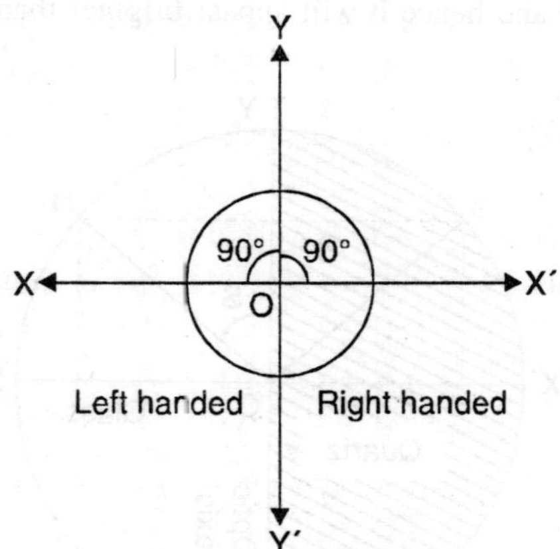

FIGURE 3.13 **Biquartz device.**

When the principal plane of the analyzer N_2 is placed along XX' the two halves will appear with equal brightness in the field of view. And when the principal planes of N_2 is placed along YY' the two halves will appear completely dark. At all other positions of N_2 two halves will appear as unequally bright. Hence Nicol N_2 may be placed in desired position accurately.

If white light is used, the different colours will be rotated through different angles. As the thickness of each half is 3.75 mm so the yellow colour suffers a rotation different from 90°. Thus it is clear that after this device the direction of vibrations in light of various colours will be obtained as shown in Fig. 3.14. The biquartz thus furnishes a highly sensitive device for measuring the optical rotation and it is used exactly in the same manner as Laurent's half shade polarimeter. So first of all, the analyzer is set for tint of passage without solution in the tube T and the reading on the circular scale is noted. Then the tube is filled with the solution of known concentration and again the analyzer is set for tint of passage and the reading is noted. The difference in the two readings gives the angle of rotation θ. Hence the specific rotation can be measured by the following relation $[S]_\lambda{}^t = \dfrac{\theta}{lc}$

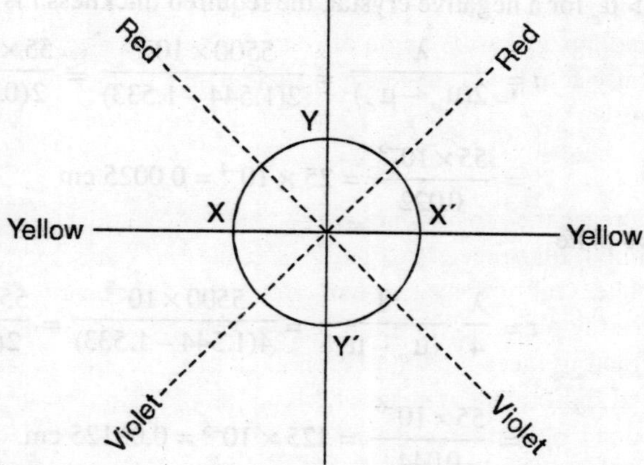

FIGURE 3.14 **Direction of vibration of various colours.**

3.7 PHOTOELASTICITY

The photoelasticity deals with the analysis of stress distribution in structural and mechanical components with the help of polarised light. It has been observed that certain transparent substances like glass celluloid and Bakelite etc. are not doubly refracting but they show the phenomenon of double refraction when subjected to mechanical stress.

When the specimen of these substances are placed between two crossed nicols they do not change the field of view. But when stress is applied to these specimens the emergent light in general is elliptically polarized and hence partly transmitted the analyzing Nicol. Thus the field of view is changed. The stress distribution in complex engineering structures such as bridges, girders, boiler plates, gear teeth etc. can be analyzed by photoelastic studies of transparent models. The scale models of these structures are constructed by materials like zylonite or Bakelite. They are subjected to loads and stresses to be investigated when placed between crossed Nicols. From the study of interference patterns the position and the extent of stresses in various parts are calculated. This helps to determine the stresses to be expected in actual practice.

SOLVED NUMERICAL PROBLEMS

Problem 1. *Calculate the thickness of a quarter wave plate from the following data : $\mu_o = 1.553$, $\mu_e = 1.544$, $\lambda = 5000$ Å* (May 1999, K.U.K.)

Solution. Since $\mu_o > \mu_e$ for a negative crystal, the required thickness t is given by

$$t = \frac{\lambda}{4(\mu_o - \mu_e)} = \frac{5000 \times 10^{-8}}{4(1.553 - 1.544)} = \frac{5000 \times 10^{-8}}{4(0.009)}$$

$$= \frac{5000 \times 10^{-8}}{0.036} = \frac{5 \times 10^{-5}}{0.036} = 0.001388 \text{ cm.}$$

Problem 2. *Calculate the thickness of (i) quarter wave plate, (ii) half wave plate. Given that $\mu_o = 1.544$, $\mu_e = 1.533$ and $\lambda = 5500$ Å.* (Jan. 1999, K.U.K.)

Solution. Since $\mu_o > \mu_e$ for a negative crystal, the required thickness t is given by

$$t = \frac{\lambda}{2(\mu_o - \mu_e)} = \frac{5500 \times 10^{-8}}{2(1.544 - 1.533)} = \frac{55 \times 10^{-6}}{2(0.011)}$$

$$= \frac{55 \times 10^{-6}}{0.022} = 25 \times 10^{-4} = 0.0025 \text{ cm}$$

(*ii*) For quarter wave plate

$$t = \frac{\lambda}{4} \cdot \frac{1}{(\mu_o - \mu_e)} = \frac{5500 \times 10^{-8}}{4(1.544 - 1.533)} = \frac{55 \times 10^{-6}}{2(0.011)}$$

$$= \frac{55 \times 10^{-6}}{0.044} = 125 \times 10^{-5} = 0.00125 \text{ cm.}$$

Problem 3. *A ray of light is incident on the surface of a glass plate (μ = 1.53) at the polarising angle. Calculate the angle of refraction.*

Solution. Here μ = 1.55, from Brewster's Law

$$\tan i = \mu = 1.55 \text{ or } i = 57°11'$$

also
$$i + r = 90°$$

$$\therefore \quad r = 90° - 57°11' = 32°49'.$$

EXERCISE

1. What are quarter wave and half wave plates ? Give one application of each. *(May 1998, K.U.K.)*
2. Explain the phenomenon of double refraction in a calcite crystal. *(May 1995 & June 2000, K.U.K.)*
3. (*a*) Give the construction and working of Laurent's half shade polarimeter.
   (*b*) Write short note on photoelasticity. *(May 1995, 99 & June 2000, K.U.K.)*
4. Give the various methods for obtaining polarised light. Describe in detail the construction and working of Nicol's prism. *(Jan. 1997, 2002, K.U.K.)*
5. What do you mean by optical activity ? Give with full details the construction and working of Biquartz polarimeter. *(Jan. 2003, 1999, June 2001, K.U.K.)*
6. What is polarisation of light ? Describe the construction of Nicol prism and show how it can be used as a polariser and analyser ? *(June 2000, K.U.K.)*
7. Give the construction and working of Laurent's half-shade polarimeter. What is its main drawback ? *(Dec. 1999, May 2002, K.U.K.)*
8. What is the plane polarised light ? Explain the phenomenon of double refraction and describe the construction and working of Nicol prism. *(Jan. 2003, May 1999, K.U.K.)*
9. Calculate the specific rotation if the plane of polarisation is turned through 26.4°, traversing 20 cm length of 20% suger solution. **[Ans. 66°]**
10. A 15 cm tube containing cane sugar solution (specific rotation = 66°) shows optical rotation 7°. Calculate the strength of the solution. **[Ans. 0.07 gm/cc]**
11. The critical angle of clear crystal for green light is 24.4°. What will be its angle of polarisation ?
12. Length of 25 cm of a solution, containing 50 gm of solute per litre causes a rotation of the plane of polarisation of light by 5°. Find the rotation of plane of polarisation by a length of 75 cm of a solution containing 100 gm of solute per litre. **[Ans. 30°]**

Chapter 4 FIBER OPTICS

4.1 OPTICAL FIBERS

With the advent of fiber optics, data can be very easily and rapidly transmitted over long distances, through glass fibers, of the size of a human hair. It is expected that the fiber optics technology would be as common as electrical wiring in the house in near future. This technology no doubt will take its own time to develop fully. However, it is increasingly replacing wire transmission lines in communication systems.

What are Optical Fibers

Optical fibers are the light equivalent of microwave wave-guide with the additional advantage of very wide bandwidth. Physically an optical fiber is a very thin and flexible medium having a cylindrical shape consisting of three sections :

1. Core 2. Cladding 3. Jacket

1. *The Core.* The core is the innermost section and is made of glass or plastic. This is the actual fiber and has the remarkable property of conducting an optical beam.

2. *Cladding.* The core is surrounded by its own cladding, a glass or plastic coating, that has optical properties which are very different from those of the core.

3. *Jacket.* The outer section is called the jacket made of plastic or polymer and other materials that is provided for the protection against moisture, absorption crushing and other environmental dangers.

The core acts like a continuous layer of two parallel mirrors. A signal is first encoded into a light beam which is then passed in between the two boundaries and propagated as a result of multiple internal reflections as shown in Fig. 4.1.

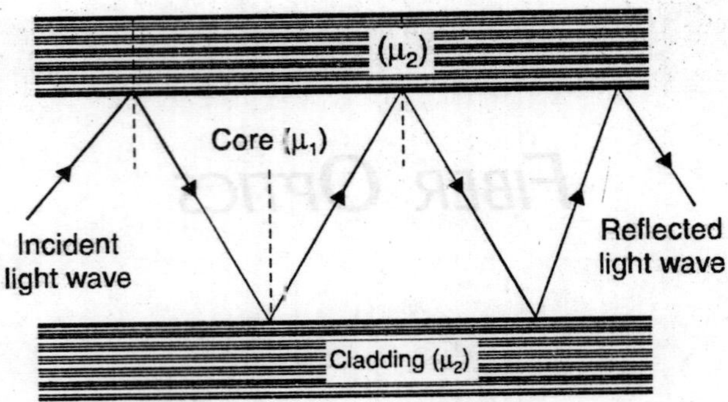

FIGURE 4.1 **Light wave propagation along a glass fiber core.**

4.2 NUMERICAL APERTURE

Another important term associated with a fiber is the numerical aperture sometimes called the figure of merit for optical fibers.

The numerical aperture of an optical fiber is defined as

$$NA = \sin\theta_0\,(\max)\sqrt{\frac{(\mu_1^2 - \mu_2^2)}{\mu_0}} \quad \text{where } \theta_0 \text{ is acceptance angle} \qquad ...(4.1)$$

If the fiber is surrounded by air ($\mu_0 = 1$) then

$$NA = \sqrt{(\mu_1^2 - \mu_2^2)} \qquad ...(4.2)$$

Generally μ_1 is only few percent greater than μ_2

$$NA = \sqrt{(\mu_1 + \mu_2)(\mu_1 - \mu_2)} \approx \sqrt{2\mu_1(\mu_1 - \mu_2)}$$

$$= \sqrt{\frac{2\mu_1^2(\mu_1 - \mu_2)}{\mu_1}} = \mu_1\sqrt{2\Delta} \qquad ...(4.3)$$

where $\qquad (\mu_1 + \mu_2) = 2\mu_1$

and $\qquad \Delta = \dfrac{(\mu_1 - \mu_2)}{\mu_1}$

Δ is the fractional difference between the core and cladding refractive indices.

Fig. 4.2 shows the variation of *NA* with the acceptance angle. It has been observed that numerical aperture for the fibers used in short distance communication is in the range of 0.1 to 0.3.

Now one question arises how much *NA* is expected for better performance. It has been observed that smaller *NA* is not an advantage because it makes it harder to launch power into the fiber.

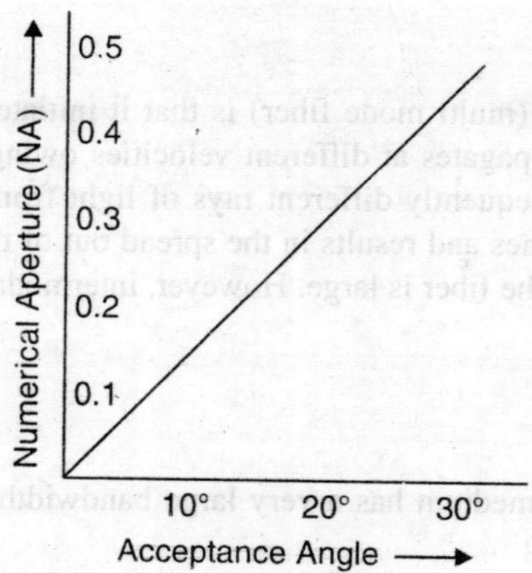

FIGURE 4.2 **Variation of numerical aperture with acceptance angle.**

4.3 SIMPLE IDEA OF SINGLE MODE AND MULTIMODE FIBERS ; APPLICATIONS

Number of Modes and Cut-off Parameters of Fibers

The number of modes supported by a fiber is determined by an important parameter called cut-off parameter. Its various aliases are "normalized frequency of cut-off" or "*V*-number".

The mathematical expression for the *V*-number is

$$V = \frac{\pi d}{\lambda_0}\sqrt{\mu_1^{\,2} - \mu_2^{\,2}} \approx \frac{\pi d}{\lambda_0} NA$$

where d denotes the diameter of the core and λ_0 is the wavelength of incident light.

For $NA = \sqrt{\mu_1^{\,2} - \mu_2^{\,2}}$ provided that the environment of the fiber is air. If it is in a medium of refractive index μ_0 then $V = \frac{\pi d}{\lambda_0}\mu_0 NA$. The approximate number of modes, which the fiber will support, is expressed as. Number of modes $\approx \frac{1}{2}V^2$, provided the V number is considerably larger than unity.

Single Mode Propagation

In the study of the number of modes supported by fiber we have seen that for each mode there is a particular V number. If V number is less than that value the mode will be absent. It is also evident that the lowest order mode, which the cylindrical fiber will support, is the HE_{11} mode having zero V number. The next higher modes are TE_{01} and TM_{01} having V number of 2.405. Hence if a fiber is fabricated so that its V number is made slightly less than 2.405 all the modes except HE_{11} mode will only be present and it is called single mode fiber.

Now let us see how a single mode fiber condition is achieved.

V number is expressed as $\quad V = \frac{\pi d}{\lambda_0} NA$

V number is reduced either by reducing the diameter of the fiber or by reducing the NA.

Multi Mode Fibers

The greatest disadvantage of *MMF* (multi mode fiber) is that it initiates the phenomenon of intermodal dispersion. Each mode in *MMF* propagates at different velocities owing to different angles of incidence at the core-cladding interface. Consequently different rays of light from the source of light reach at the outer end of the fiber at different times and results in the spread out of the pulse in the time domain. This becomes prominent when length of the fiber is large. However, intermodal dispersion can be reduced using GRIN fibers.

Applications of Fibers

1. When the transmmission medium has a very large bandwidth, a single mode fibre is used *e.g.* underseas cable systems.

2. When the system bandwidth requirements are between 2 Hz and 200 MHz a graded index multimode fibers would be best choice *e.g.* in intracity trunks between telephone central office.

When the system bandwidth requirements are lower, a step index multimode fiber would be better *e.g.* data links.

4.4 IMPORTANCE OF OPTICAL FIBERS

The optical fibers have a lot of advantages over the wireless or radio systems and this is why communication industries have introduced the fiber optic systems significantly.

The following are the main advantages of optical fibers :

1. *Low Attenuation.* Attenuation in a fiber is markedly lower than that of coaxial cable or twisted pair and is constant over a very wide range. So transmission within wide range of distance is possible without repeaters etc.

2. *Smaller Size and Light Weight.* Optical fibers are considerably thinner than coaxial cable or bundled twisted pair cable. So they occupy much less space.

3. *Electromagnetic Isolation.* Electromagnetic waves generated from electrical disturbances or electrical noise do not interfere with light signals. As a result the system is not vulnerable to interference impulse noise or cross talk.

4. *Physical Connections are not Needed.* No physical electrical connection is required between the senders and receiver.

5. *More Reliability.* The fiber is much more reliable because it can better withstand environmental conditions such as pollution radiation and salts produce no corrosion. Moreover, it is nominally affected by the nuclear radiation. Its life is longer in comparison to copper wire.

6. *Security and Privacy.* Almost there is no cross talk in optical fibers and hence transmission is more secure and private, as it is very difficult to tap into a fiber.

7. *Greater Bandwidth.* Bandwidth of the optical fibers is higher than that of an equivalent wire transmission line.

8. *No Need of Isolation Coating.* As fibers are very good dielectrics, isolation coating is not required.

9. *Higher Data Rate.* Data rate is much higher in a fiber and hence much more information can be carried by each fiber than by equivalent copper cables.

10. *Lower Cost.* The cost per channel is lower than that of an equivalent wire cable system.

SOLVED NUMERICAL PROBLEMS

Problem 1. *Compute the NA and the acceptance angle of an optical fibre from the following data : μ_1 (core) = 1.55 and μ_2 (cladding) = 1.50.* (June, 2003 K.U.K.)

Solution.

$$NA = \mu_1 \sqrt{(2\Delta)}$$

where

$$\Delta = \left(\frac{\mu_1 - \mu_2}{\mu_1}\right) \quad \text{or} \quad \Delta = \left(\frac{1.55 - 1.5}{1.55}\right) = 0.03226$$

$$NA = 1.55 \sqrt{(2 \times 0.03226)} = 0.3094$$

Acceptance angle $\theta_0 = \sin^{-1} NA = 23.2°$.

Problem 2. *Compute the NA, acceptance angle and the critical angle of the fiber having μ_1 = 1.5 and the refractive index of the cladding = 1.45.*

Solution.

$$\Delta = \frac{\mu_1 - \mu_2}{\mu_1} = \left(\frac{1.5 - 1.45}{1.5}\right) = 0.033$$

$$NA = \mu_1 \sqrt{(2\Delta)} = 1.5 \sqrt{(2 \times 0.033)} = 0.387$$

$$\theta_o = \text{acceptance angle} = \sin^{-1} NA = 22.78°$$

$$\theta_c = \sin^{-1} \frac{\mu_2}{\mu_1} = \sin^{-1} \frac{1.45}{1.5} = 75.2°$$

EXERCISE

1. Explain numerical aperture and give simple idea about single mode and multimode fibers.
2. Write a brief note on fibers optics and its application.
   (*Jan. 1997, May 1998, May 1999, Dec. 1999, June 2001, May 2002, K.U.K.*)
3. Calculate the refractive indices of the core and cladding material of a fiber from following data :
   NA = 0.22 $\quad \Delta = 0.012$. **[Ans.** $\mu_1 = 1.42, \mu_2 = 1.40$**]**
4. Calculate the wavelength of a light wave in free space for a frequency of 600 GHz. velocity = 3×10^8 m/s. **[Ans.** 0.5 μm**]**
5. Distinguish between fibre optics and optical fibres. (*Jan. 2004, K.U.*)

Chapter 5

DIELECTRICS

5.1 DIELECTRICS—MOLECULAR THEORY

Dielectrics are the insulators and an insulator is a material, which contains no free electrons or number of such electrons is too low to constitute the electric current. In such substances the electrons are tightly bound to the nucleus of the atom.

When we place a dielectric in an electric field say between the plates of a charged capacitor, the positive charge gets displaced in the direction of applied electric field whereas the negative charges are displaced in the opposite direction of the field. The applied electric field separates positive charge from the negative charge. This effect is known as the polarization of dielectric.

NON-POLAR AND POLAR DIELECTRICS

Non-Polar Dielectrics

The dielectrics which are polarized only when they are placed in an electric field are called non-polar dielectric. Thus a non-polar dielectric does not possess any permanent electric dipole moment. They acquire dipole moment by induction when placed in an electric field as shown in Fig. 5.1. The non-polar dielectrics have symmetrical structure *e.g.* H_2, N_2, O_2, CO_2, CCl_4 etc.

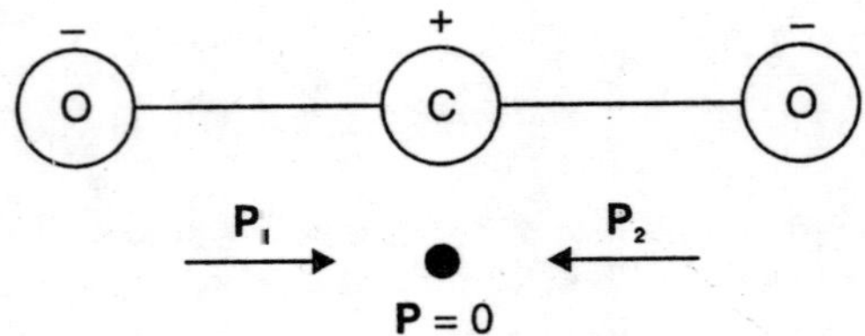

FIGURE 5.1 CO_2 molecule has zero electric dipole moment hence is non-polar.

Polar Dielectrics

A polar molecule is one in which the center of gravity of positive charge does not coincide with the center of gravity of the negative charges. Such dielectric have permanent electric dipole moments with their random orientation in the absence of external electric field as shown in Fig. 5.2 (*a, b*) *e.g.* HCl, H_2O etc.

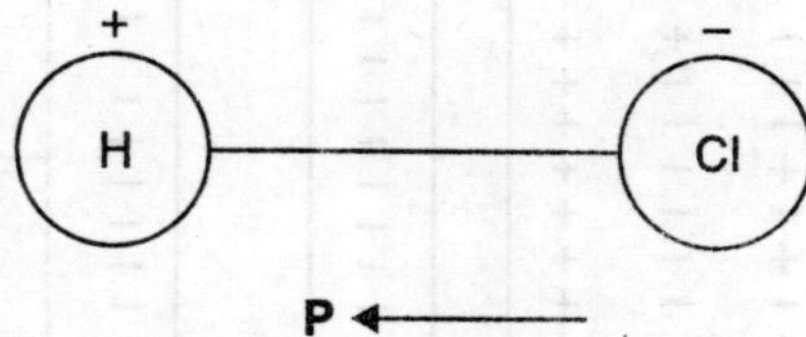

FIGURE 5.2 **(*a*) HCl molecule is polar and has net electric dipole moment.**

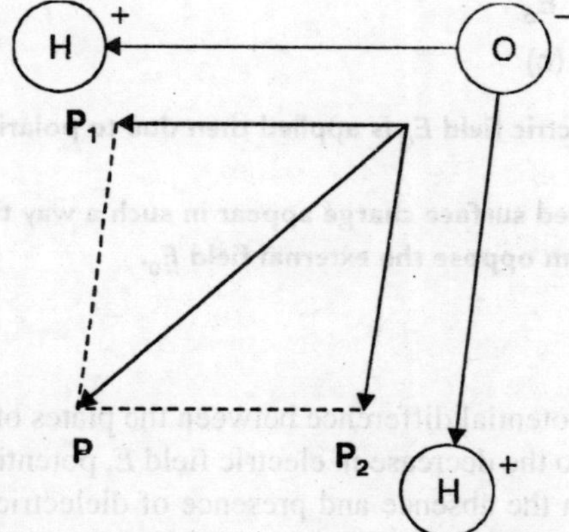

FIGURE 5.2 **(*b*) H_2O molecule has some electric dipole moment hence H_2O is polar molecule.**

Non-Polar Dielectric in an Electric Field

FIGURE 5.3 **(*a*) The random distribution of positive and negative charges in non-polar dielectric.**

Fig. 5.3 (*a*) shows the random distribution of positive and negative charges in a non-polar dielectric. When it is placed in an electric field $\mathbf{E}_0$ (say between the plates of a parallel plate capacitor), then due to polarization surface charges appear *i.e.* positive charge is induced on one surface while negative charge on other surface [Fig. 5.3 *b, c*]. The induced surface charge appear in such a way that the electric field $\mathbf{E}'$ set up by them opposes the external field $\mathbf{E}_0$. The resultant field $\mathbf{E}$ in the dielectric is the vector sum of $\mathbf{E}'$ and $\mathbf{E}_0$ *i.e.* $\mathbf{E} = \mathbf{E}_0 + \mathbf{E}'$ or $E = E_0 - E'$.

Hence, the external electric field weakens the original electric field within the dielectric. Thus we conclude that if we place a dielectric in an electric field, induced surface charges appear which leads to weakening the original field within the dielectric.

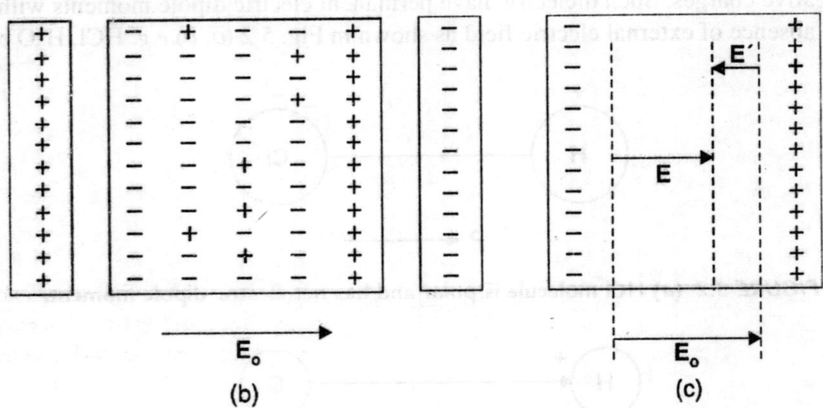

FIGURE. 5.3 **(*b*) When electric field E_0 is applied then due to polarization, surface charge appears.**
(*c*) The induced surface charge appear in such a way that the electric field E' set up by them oppose the external field E_0.

Dielectric Constant (K)

When we introduce dielectric, the potential difference between the plates of a capacitor also falls. From the relation $V = Ed$, one finds that due to the decrease in electric field E, potential difference, V, should also fall if E_0 and E are the electric fields in the absence and presence of dielectric and V_0 and V are the potential difference between the plates without the dielectric and with dielectric respectively, then

$$\frac{E_0}{E} = \frac{V_0}{V} = K \text{ where } K \text{ is dielectric constant}$$

As $E_0 > E,\ K > 1.$

The dielectric constant may also be defined as the ratio of the capacitance of a capacitor with dielectric (C) to the capacitance of the same capacitor without dielectric (C_0) *i.e.* $K = \dfrac{C}{C_0}$

Effect of Dielectric on Capacitance

When we place a dielectric material in between the plates of a charged capacitor, surface charges appear which reduce the potential difference between the plates, thereby increasing the capacitance.

5.2 GAUSS LAW IN DIELECTRICS

The dielectrics when placed in the electric field get polarized. So, because of this polarization also causes electric field as well as the existance of bound charge distribution.

Thus, while applying Gauss Law we are to consider two types of charges *i.e.* free as well as bound charges. In other wards the surface integral of normal component of electric field, $\vec{E}$ will satisfy

$$\oint_S \vec{E} \cdot \hat{n}\, ds = \frac{1}{\varepsilon_0}(Q + Q_P) \quad ...\text{[SI]} \qquad ...(5.1)$$

where Q = Free charge

Q_P = bound charge due to polarization of the dielectric

and ε_o = permittivity of the vacuum.

Here, S is the gaussian surface drawn in the dielectric. S_1, S_2 and S_3 are the conductors carrying free charges q_1, q_2 and q_3 respectively [Fig. 5.4].

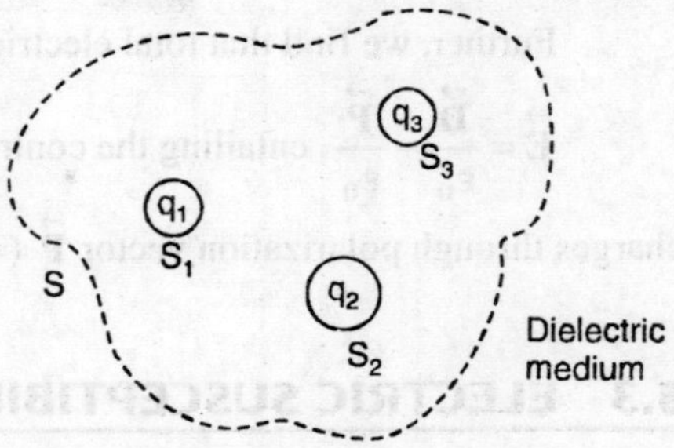

$$Q = \sum_{i=1}^{3} q_i$$

FIGURE 5.4 Gaussian surface S and location of free charges. V is the volume enclosed by S.

Now, if $\vec{P}$ is the polarization vector, the bound charge distribution will be given as

$$Q_P = \oint_{S_1+S_2+S_3} \sigma_p\, ds + \int_V \rho_P\, dV$$

$$= \oint_{S_1+S_2+S_3} \vec{P} \cdot \hat{n}\, ds + \int_V (-\vec{\nabla} \cdot \vec{P})\, dV$$

because $\sigma_p = \vec{P} \cdot \hat{n}$, (surface density of bound charges)

$\rho_P = -\vec{\nabla} \cdot \vec{P}$ (volume density of bound charges)

$\hat{n}$ being the unit vector normal to the surface under consideration. Now,

$$\int_V \rho_P\, dV = -\int_V \vec{\nabla} \cdot \vec{\mathbf{P}}\, dV = -\oint_{S+S_1+S_2+S_3} \vec{P} \cdot \hat{n}\, ds \qquad \text{(Divergence Theorem)}$$

Therefore, $Q_P = -\oint_S \vec{\mathbf{P}} \cdot \hat{n}\, ds$

with this result Gauss Law becomes (Eq. (5.1))

$$\oint_S \vec{\mathbf{E}} \cdot \hat{n}\, ds = \frac{Q}{\varepsilon_0} - \frac{1}{\varepsilon_0}\oint_S \vec{\mathbf{P}} \cdot \hat{n}\, ds$$

Hence, $\oint_S (\varepsilon_0 \vec{\mathbf{E}} + \vec{\mathbf{P}}) \cdot \hat{n}\, ds = Q$

$$\oint_S \vec{\mathbf{D}} \cdot \hat{n}\, ds = Q\,;\ \vec{\mathbf{D}} = \varepsilon_0 \vec{\mathbf{E}} + \vec{\mathbf{P}} = \textit{electric displacement vector.}$$

and by using divergence theorem

$$\int_V \vec{\nabla} \cdot \vec{\mathbf{D}}\, dV = Q = \int_V \rho dV$$

leading to

$\vec{\nabla} \cdot \vec{\mathbf{D}} = \rho$, where ρ denotes the free charge density within the dielectric

The surface integral

$$\oint_S \vec{\mathbf{D}} \cdot \hat{n}\, ds = Q \qquad ...(5.2)$$

is called Gauss Law in a Dielectric while

$$\vec{\nabla} \cdot \vec{D} = \rho, \text{ its differential form.}$$

Further, we find that total electric field $\vec{E}$ has two terms as

$\vec{E} = \frac{\vec{D}}{\varepsilon_0} - \frac{\vec{P}}{\varepsilon_0}$, entailing the contribution of free charge density through $\vec{\nabla} \cdot \vec{D} = \rho$ and also of bound charges through polarization vector $\vec{P}$ (= Dipole moment per unit volume of the dielectric).

5.3 ELECTRIC SUSCEPTIBILITY AND DIELECTRIC CONSTANT

For most of the dielectrics one finds that

$$\vec{P} \propto \vec{E} \quad \text{or} \quad \vec{P} = \chi\,\vec{E}$$

where χ is the constant of proportionality called *electric susceptibility* of the material.

Therefore

$$\vec{D} = \varepsilon_0\,\vec{E} + \vec{P} \qquad ...(5.3)$$

$$= \varepsilon_0\,\vec{E} + \chi\,\vec{E} = (\varepsilon_0 + \chi)\,\vec{E} = \varepsilon\vec{E}$$

where ε is called the permittivity of the material such that

$$\varepsilon = \varepsilon_0 + \chi \qquad ...(5.4)$$

Such materials are called linear dielectrics. Further, it is more convenient to write

$$K = \frac{\varepsilon}{\varepsilon_0} = 1 + \frac{\chi}{\varepsilon_0} \qquad ...(5.5)$$

K is called *dielectric constant.*

Thus, we have following relations for linear dielectrics

$$\vec{P} = \chi\,\vec{E}, \qquad ...(5.6)$$

$$\vec{D} = \varepsilon\,\vec{E}\,; \qquad ...(5.7)$$

$$\varepsilon = (\varepsilon_0 + \chi) \qquad ...(5.4')$$

and

$$K = \frac{\varepsilon}{\varepsilon_0} = 1 + \frac{\chi}{\varepsilon_0} \qquad ...(5.5')$$

We list below value of *K* for some materials.

Material	*K*	E_{max} *(Volts/m)*
Glass	5 – 10	9×10^6
Mica	6.0	5.2×10^6
Nylon	3.5	16×10^6
Rubber	2 – 3.5	$16 - 40 \times 10^6$
Sulfur	4.0	
Petroleum oil	2.1	12×10^6
Water (20°C)	80	
Air (1 atm)	1.00	3×10^6

[Note : E_{max} is the maximum electric field which the dielectric can withstand without breakdown].

5.4 ATOMIC POLARIZABILITY AND PERMITTIVITY

Atomic Polarizability

When we place non-polar molecule in electric field, the center of gravity of negative charges is displaced relative to that of positive charges. In this way the molecule acquires an induced electric dipole moment. The induced dipole moment $\mathbf{p}_m$ is directly proportional to electric field $\mathbf{E}_m$.

Hence $\mathbf{p}_m \propto \mathbf{E}_m$ or $\mathbf{p}_m = \alpha \mathbf{E}_m$ where α is known as polarizability. Here m means molocular and

$\mathbf{E}_m = \mathrm{E} + \dfrac{\mathbf{p}}{3\,\varepsilon_0}$, **E** being external field.

5.5 ENERGY STORED IN ELECTRIC FIELD

The electrostatic field in a medium is established due to the free charge distribution. The medium consists of dielectrics and conductors. The volume charge distribution will therefore exist within the dielectric while surface charge on the conductors. Let ρ be the volume charge density and σ the surface charge density. We further assume that the charge system is bounded *i.e.* it has a finite volume. Then by definition of electric displacement vector, $\vec{\mathbf{D}}$, these distributions are given by

$$\rho = \nabla \cdot \mathbf{D} \quad \text{and} \quad \sigma = \mathbf{D} \cdot \hat{n}$$

where $\hat{n}$ is a unit vector normal to the surface and directed outwardly. Also, if U denotes the potential at a point the energy of the system of charges would be given by

$$W = \frac{1}{2}\int_V U\rho\, dV + \frac{1}{2}\int_S U\sigma\, dS$$

where V is the entire volume and S the entire surface of the system of conductors having surface charge distribution. Putting the values of ρ and σ we obtain

$$W = \frac{1}{2}\int_V U\nabla \cdot \mathbf{D}\, dV + \frac{1}{2}\int_S U\mathbf{D} \cdot \hat{n}\, dS$$

Now use the vector identity

$$\nabla \cdot (U\mathbf{D}) = U\,\nabla \cdot \mathbf{D} + \mathbf{D} \cdot \nabla U \text{ to get}$$

$$W = \frac{1}{2}\int_V \nabla \cdot (U\mathbf{D}) dV - \frac{1}{2}\int_V \mathbf{D} \cdot \nabla U\, dV + \frac{1}{2}\int_S U\mathbf{D} \cdot \hat{n}\, dS$$

Taking the outer boundary of the system as at infinity and using divergence theorem as well as

$$\mathbf{E} = -\nabla U, \text{ we get}$$

$$W = \frac{1}{2}\int_{S+S'} U\mathbf{D} \cdot \hat{n}'\, dS + \frac{1}{2}\int_V \mathbf{D} \cdot \mathbf{E}\, dV + \frac{1}{2}\int_S U\mathbf{D} \cdot \hat{n}\, dS$$

where $S + S'$ denotes the surface of conductors as well as the boundary at infinity while $\hat{n}'$ is the unit normal to the bounding surfaces and directed outwardly. In fact S' is the surface of the boundary at infinity. Since U varies as r^{-1}, **D** as r^{-2} and S as r^2 the integration over S' will be zero. As a result,

$$W = \frac{1}{2}\int_S U\mathbf{D} \cdot \hat{n}'\, dS + \frac{1}{2}\int_V \mathbf{D} \cdot \mathbf{E}\, dV + \frac{1}{2}\int_S U\mathbf{D} \cdot \hat{n}\, dS$$

Now, it is obvious that $\hat{n}' = -\hat{n}$ everywhere over S as $\hat{n}'$ is directed into the conductors while $\hat{n}$ outwardly. So, these surface integrals will cancel each other.

Hence, $$W = \frac{1}{2}\int_V \mathbf{D}\,.\,\mathbf{E}\,dV \qquad ...(5.8)$$

In this volume integral the volume of the conductors may be excluded as **E** inside the conductors is always zero. Thus, above expression (5.8) gives the electrostatic energy of the field. Further,

$$w = \frac{\mathbf{D}\,.\,\mathbf{E}}{2},\ \mathbf{D} = \varepsilon\mathbf{E} \qquad ...(5.9)$$

$$= \frac{\varepsilon E^2}{2} \quad \text{(for a linear dielectric medium)}$$

is called the energy density of the electrostatic field. ε is the permittivity of the medium.

[Note : There could be surface charge distribution on the conductors embeded within the dielectric.]

EXERCISE

1. Define (*i*) Electric polarization, (*ii*) Displacement, (*iii*) Susceptibility, (*iv*) Permittivity. (*May 1998, K.U.K.*)
2. State and prove Gauss's theorem. How is it modified in a dielectric medium ? (*Jan. 1997, May 1998, May 1999 and Jan. 2001, K.U.K.*)
3. What are dielectrics ? Define dielectric constant of a material and explain the physical significance. (*May 1995, May 2002, K.U.K.*)
4. Obtain the general form of Gauss's law valid in the presence of dielectric media. Obtain the expression relating electric field vector **E**, electric displacement vector **D** and polarization vector **P**. (*May 1995, K.U.K.*)
5. Find an expression for the energy stored in an electric field. (*Dec. 1999, K.U.K.*)
6. Define dielectric constant and electric susceptibility and obtain the relation between them. (*Dec. 1999, K.U.K.*)
7. State and prove Gauss's theorem in electrostatics. How this theorem will be amended in the presence of a dielectric ? (*May 1999 & June 2000, K.U.K.*)

Chapter 6 MAGNETIC MATERIALS

6.1 MAGNETIZATION AND MAGNETIC INTENSITY

Introduction

The materials, which are capable of becoming magnetized in an external magnetic field, are known as magnetic materials. On the basis of their magnetic properties, these substances are classified into the following three categories :

1. Diamagnetic 2. Paramagnetic 3. Ferromagnetic

To study the various properties of these materials it is essential to know the following important definitions.

Important Definitions

Magnetic Dipole

Two opposite poles separated by a small distance from each other are said to form a magnetic dipole. On placing a magnetic dipole in an external magnetic field, a torque acts upon it which rotates it to set in a particular dirction.

Let τ be the torque acting on the dipole in an external magnetic field **B**, then its magnetic moment, $\mathbf{p}_m$, is defined by

$$\tau = \mathbf{p}_m \times \mathbf{B}$$

or in magnitude $\tau = p_m B \sin\theta$

where θ is the angle between magnetic dipole moment $\mathbf{p}_m$ and magnetic induction **B**.

Magnetic Permeability

The degree to which the magnetic field can penetrate or permeate a material is called relative permeability (μ_r) of the material.

Relative permeability of a medium may also be defined as the ratio of the number of lines of magnetic induction per unit area (*i.e.* flux density **B**) in that medium to the number of lines per unit area that would be present if the medium is replaced by vacuum (*i.e.* flux density $\mathbf{B}_0$).

$$\therefore \qquad \mu_r = \frac{B}{B_0}$$

μ_r is the dimensionless quantity and its value is unity for vacuum.

Since, $\mathbf{B} = \mu\mathbf{H}$ and $\mathbf{B}_0 = \mu_0\mathbf{H}$

Therefore, relative permeability may also be defined as the ratio of the permeability of the medium (μ) and the permeability of free space (μ_0).

$$\therefore \qquad \mu_r = \frac{\mu}{\mu_0} \quad \text{or} \quad \mu = \mu_r\mu_0$$

SI unit of permeability is henry/meter. For free space $\mu_0 = 4\pi \times 10^{-7}$ henry/meter.

Magnetic Induction (or Flux Density) B

If **F** is the force experienced by a positive charge q moving with velocity **v** through a point in a magnetic field then the magnetic induction **B** at that point is defined by

$$\mathbf{F} = q(\mathbf{v} \times \mathbf{B}) \quad \text{or} \quad F = qvB \sin\theta$$

Or magnitude of **B** is given by $B = \dfrac{F}{[qv \sin\theta]}$

where θ is the angle between **v** and **B**. SI unit of **B** is weber/m^2 or Tesla and CGS unit is Gauss (= maxwell/cm^2). 1 Tesla = 10^4 Gauss.

Magnetic Intensity or Magnetizing Force H

The degree to which a magnetic field can magnetize a magnetic material is expressed in terms of **H**. The magnitude of magnetizing force may be defined as the number of ampere turns flowing round the unit length of a solenoid required to produce that magnetic field.

Magnetic intensity due to a current carrying coil of n turns per meter is

$$H = ni \quad \text{ampere turns/meter} \qquad \text{...(6.1)}$$

where i is the current in the coil. Magnitude of magnetic induction at a point well inside a long solenoid having a material core of permeability μ is

$$B = \mu ni \qquad \text{...(6.2)}$$

Therefore, from eqs. (6.1) and (6.2), we have

$$B = \mu H \qquad \text{...(6.3)}$$

If inside the solenoid there is free space, then magnetic induction

$$B_0 = \mu_0 H \qquad \text{...(6.4)}$$

where μ_0 is the permeability of free space.

Intensity of Magnetization

Intensity of magnetization '*I*' is defined as the magnetic moment per unit volume of the material. If **M** = Magnetic moment, V = Volume of the material then

$$\mathbf{I} = \mathbf{M}/V$$

If a is the area of cross section and l the length of the material, then

$|\mathbf{M}| = ml$ where m is pole strength and $V = al$

$$\therefore \quad |\mathbf{I}| = \frac{ml}{al} = \frac{m}{a}$$

Therefore, the intensity of magnetization of a material may also be defined as the pole strength per unit area of cross-section of the material.

SI unit of intensity of magnetization is weber/meter^2.

Relation Between B, H and I

We can regard flux density **B** in a material due to a magnetizing force **H**, as the sum of the flux density $\mathbf{B}_0$ in vacuum produced by the magnetizing force and flux density $\mathbf{B}_m$ due to magnetization of material.

Therefore $\quad \mathbf{B} = \mathbf{B}_0 + \mathbf{B}_m$

Also, we know that $\quad \mathbf{B}_0 = \mu_0\mathbf{H}$ and $\mathbf{B}_m \propto \mathbf{I}$ or $\mathbf{B}_m = \mu_0\mathbf{I}$

Therefore, $\quad \mathbf{B} = \mu_0\mathbf{H} + \mu_0\mathbf{I}$

or $\quad \mathbf{B} = \mu_0(\mathbf{H} + \mathbf{I})$. It is the relation in SI units.

And $\quad \mathbf{B} = \mathbf{H} + 4\pi\mathbf{I}$ in CGS system.

Magnetic Susceptibility

It is defined as the ratio of intensity of magnetization I produced in the given material to the magnetic intensity H.

Magnetic susceptibility $\quad \chi = I/H$.

Relation between Permeability and Susceptibility

We know that, $\quad \mathbf{B} = \mu_0\mathbf{H} + \mu_0\mathbf{I}$

or $$\frac{B}{H} = \mu_0 + \mu_0\frac{I}{H}$$

Also, $$\frac{B}{H} = \mu \quad \text{(Permeability of the material)}$$

and $$\frac{I}{H} = \chi \quad \text{(susceptibility)}$$

$$\therefore \quad \mu = \mu_0 + \mu_0\chi \quad \text{or} \quad \mu = \mu_0(1 + \chi)$$

or $$\frac{\mu}{\mu_0} = 1 + \chi \quad \text{or} \quad \mu_r = 1 + \chi,$$ where μ_r is relative permeability.

6.2 PARA ; DIA AND FERROMAGNETISM

As mentioned earlier, depending on the behaviour of different magnetic materials in an external magnetic field, these can be classified into three categories *viz.* Diamagnetic, Paramagnetic and Ferromagnetic.

A. Diamagnetic Substances

The substances which when placed in a magnetic field acquire feeble magnetization in a direction opposite to that of the applied field are called diamagnetic substances.

Common examples of diamagnetic substances are Bismuth, Antimony, Water, Alcohol and Hydrogen etc.

Properties of Diamagnetic Substances

1. These substances set themselves at right angles to the direction of the applied magnetic field.
2. Diamagnetic substances when subjected to a non-uniform magnetic field move from stronger to the weaker part of the field.
3. The permeability μ for diamagnetic substance is less than 1 but positive *i.e.* $\mu < 1$.
4. The susceptibility (χ) for diamagnetic substances have a small negative value. Its value is almost constant and does not vary with field or temperature.

B. Paramagnetic Substances

The substances which when placed in a magnetic field are feebly magnetized in the same direction as that of the field.

Common examples of paramagnetic substances are Aluminium, Chromium, Platinum, Manganese, Oxygen etc.

Properties of Paramagnetic Substances

1. In a uniform magnetic field they rotate until their longest axis is parallel to the field.
2. These substances when subjected to a non-uniform magnetic field experience an attractive force towards the stronger part of the field.
3. As for paramagnetic substances, when placed in a magnetic field, the magnetic induction B becomes more than the magnetizing field H. The permeability μ for a paramagnetic substance is slightly greater than one.
4. The magnetic susceptibility (χ) has small positive value.
5. The susceptibility does not change with H for a given temperature.
6. The susceptibility varies inversely as the absolute temperature. Its value decreases with increase in temperature and at higher temperature its value becomes negative and the substance becomes diamagnetic.

C. Ferromagnetic Substances

The magnetization of dia and paramagnetic substances exists only in the presence of an external magnetic field. The magnetization of ferromagnetic is retained even in the absence of external magnetic field and the field generated by this magnetization exists independently. Such substances acquire high degree of magnetization in the same sense as the applied magnetic field.

Common examples of these materials are Iron, Steel, Nickel and Cobalt etc.

Properties of Ferromagnetic Substances

1. They show the properties of Paramagnetic to much higher degree and have permeability of the order of hundreds and thousands.
2. The value of susceptibility (χ) is also very large and positive.

3. As temperature increases the value of χ decreases. Above a certain temperature ferromagnetic become ordinary paramagnetic and this temperature is called Curie temperature.

4. The susceptibility, $\chi = \frac{I}{H}$, remains constant for small values of H and for moderate values of H the value of I increases rapidly with H and for larger value of H, I attains a constant value. Hence the value of χ increases for moderate value of H and for larger value of H it decreases with increasing H.

Origin of Permanent Magnetic Moment

Atoms or molecules of a substance possess permanent magnetic moment. These permanent moments arise from the orbital and spin motion of the electrons.

First, we consider the orbital moment. An orbital electron revolving round the nucleus in an atom is equivalent to small circulating current giving to a magnetic moment

$$\mu_m = i\,A$$

i = Current in the circular loop

A = Area enclosed by the circulating current.

Then current around the orbit is the amount of charge which passes any point in the orbit per unit time *i.e.*

$$i = \frac{ev}{2\pi r}$$

v is the velocity of the electron in the orbit. The magnetic moment is then

$$\mu_m = \frac{evA}{2\pi r} = \frac{ev\pi r^2}{2\pi r} = \frac{evr}{2} \quad [\because \quad A = \pi r^2] \qquad ...(6.5)$$

The angular momentum of the electron in the circular orbit of radius r is equal to

$$L = mvr \quad \text{or} \quad r = \frac{L}{mv}, \; m \text{ being the electron mass}$$

Putting in eq. (6.5), we get $\mu_m = e\mathbf{L}/2m$

Since the charge on electron is negative, μ_m and $\mathbf{L}$ will be oppositely directed vectors

Thus $$\mu_m = -|e|\frac{\mathbf{L}}{2m} \qquad ...(6.6)$$

The constant co-efficient $e/2m$ is called gyromagnetic ratio. The Bohr magneton is defined as the magnetic moment of an electron with an angular momentum $h/2\pi$.

Then $$\mu_B = \left(\frac{e}{2m}\right)\left(\frac{h}{2\pi}\right) = 9.27 \times 10^{-24}\ \text{A} - \text{m}^2$$

We know that quantum numbers l and m_l refer to the orbital angular momentum of the electron in its orbit and that of its z-component. The number l is connected to the total angular momentum by the relation

$$L = \sqrt{l(l+1)}\,\frac{h}{2\pi} \qquad l = 0, 1, 2, 3, ... \qquad ...(6.7)$$

Whereas the number m_l is given as

$$L_z = m_l \frac{h}{2\pi} \qquad ...(6.8)$$

with $-l \le m_l \le l$; $2l+1$ values in all

Therefore putting eqs. (6.8) and (6.7) in eq. (6.6), we get

$$(\mu_m)_{orbit} = \left(\frac{e}{2m}\right)\left(\frac{h}{2\pi}\right)\sqrt{[l(l+1)]} = \frac{e}{2m}\cdot\frac{h}{2\pi} m_l \qquad ...(6.9)$$

The magnitude of spin magnetic moment $(\mu_m)_{spin}$ is found to be twice as large as eq. (6.6). Therefore

$$(\mu_m)_{spin} = 2\left(\frac{e}{2m}\right)\left(\frac{h}{2\pi}\right)m_s$$

where s is the spin quantum number. Since $s = ½$ and $m_s = -½$ and ½, the magnitude of the magnetic moment due to spin is always $eh/4\pi m$ and is equal to the Bohr Magneton.

Classical Theory of Paramagnetism

1. Langevin Theory

Langevin considered a paramagnetic gas in which each atom or molecule possess a permanent magnetic moment. The mutual interaction between the different gas particles is neglected. When gas is subjected to an external magnetic field, the state of magnetism will be determined by the two factors.

(a) The applied magnetic field. Which tends to align the magnetic axis and molecules in its own direction.

(b) The thermal agitation. That works to disorganize the orderly state so produced. The possibility that a dipole is inclined at an angle θ to the field direction in thermal equillibrium is proportional to

$$e^{\frac{\mu H \cos\theta}{k_B T}}$$

if n is total number of dipoles and dn is number of dipoles inclined at an angle θ with field direction, then

$$dn = ne^{\frac{\mu H \cos\theta}{k_B T}} \sin\theta \, d\theta$$

Therefore, average magnetic moment $\bar{\mu}$ in the field direction will be obtained by dividing the sum of the resolved components of the magnetic moments of all the dipoles *i.e.*

$$\bar{\mu} = \frac{\int \mu\cos\theta\, dn}{\int dn} = \frac{\int_0^\pi \mu\cos\theta\, e^{\frac{\mu H\cos\theta}{k_B T}} \sin\theta\, d\theta}{\int_0^\pi e^{\frac{\mu H\cos\theta}{k_B T}} \sin\theta\, d\theta}$$

putting $\alpha = \dfrac{\mu H}{k_B T}$, we can write the above expression as

$$\bar{\mu} = \mu\frac{d}{d\alpha}\log \int_0^\pi e^{\alpha\cos\theta}\sin\theta\, d\theta = \mu\frac{d}{d\alpha}\log\left[\frac{e^{\alpha}-e^{-\alpha}}{\alpha}\right]$$

$$= \frac{\mu\alpha}{(e^{\alpha} - e^{-\alpha})}\left[\frac{(e^{\alpha} + e^{-\alpha})}{\alpha} - \frac{(e^{\alpha} - e^{-\alpha})}{\alpha^2}\right]$$

$$= \mu\left[\coth(\alpha) - \frac{1}{\alpha}\right] = \mu\left[\coth(\alpha) - \frac{1}{\alpha}\right] = \mu L(\alpha) \qquad ...(6.10)$$

where $L(\alpha)$ is called Langevin function.

Case I. When $\mu H >> k_B T$, this means $\alpha = \frac{\mu H}{k_B T}$ will be quite large. For such a case temperature, T is small or H is large we find that

$$|L(\alpha)|_{\alpha \to \infty} \Rightarrow \left|\frac{(1 + e^{-2\alpha})}{(1 - e^{-\alpha})} - \frac{1}{\alpha}\right|_{\alpha \to \infty} = 1$$

i.e. Langevin function tends to unity. Then from eq. (6.10), $\bar{\mu} = \mu$

If N is the number of dipoles per unit volume, the magnetization, M per unit volume will be

$$M = N\mu = M_s$$

where M_s is the saturation value of magnetization.

Case II. When $\mu H << k_B T$, this means $\alpha = \frac{\mu H}{k_B T}$ will be quite small. For such a case temperature, T is high or H is small we find that

$$|L(\alpha)|_{\alpha \to 0} \Rightarrow \left|\frac{(1 + e^{-2\alpha})}{(1 - e^{-2\alpha})} - \frac{1}{\alpha}\right|_{\alpha \to 0} = \left|\frac{(e^{\alpha} + e^{-\alpha})}{(e^{\alpha} - e^{-\alpha})} - \frac{1}{\alpha}\right|_{\alpha \to 0}$$

$$\left|\frac{\left(2 + \frac{2\alpha^2}{2!}\right)}{\left(2\alpha + \frac{2\alpha^3}{3!}\right)} - \frac{1}{\alpha}\right|_{\alpha \to 0} = \left|\left(\frac{1}{\alpha} + \frac{\alpha}{3}\right) - \frac{1}{\alpha}\right|_{\alpha \to 0} = \frac{\alpha}{3} = \frac{\mu H}{3k_B T}$$

so that

$$M = \frac{N\mu^2 H}{3k_B T} \qquad ...(6.11)$$

this gives the magnetic susceptibility

$$\chi = \frac{M}{H} = \frac{N\mu^2 H}{3k_B T}$$

relation for susceptibility can also be written as χ = constant/T and is called Curie Law.

2. Weiss Theory

Langevin theory failed to explain the complicated type of dependence of susceptibility upon temperature exhibited by several paramagnetic substances. Furthermore, theory does not throw any light on the intimate relationship between para and ferromagnetism. Weiss in 1907, modified the Langevin theory by introducing a concept of internal molecular field. He assumed that in a real gas, the molecules are mutually influenced by their magnetic moments and consequently there should exist with in the gas a molecular field. This field, produced at any point by all the neighbouring molecules, is proportional to and acting in the same sense as the intensity of magnetization M. Let this internal molecular field be represented by H_i, then

$$H_i \propto M \quad \text{or} \quad H_i = \lambda M,$$

where λ is molecular field co-efficient.

Therefore, net effective field should be $\qquad H_e = H + H_i \qquad$ ...(6.12)

i.e. a vector sum of external applied H and internal molecular field H_i.

Then eq. (6.11) becomes

$$M = \frac{N\mu^2 H_e}{3k_B T} = \frac{N\mu^2 (H + H_i)}{3k_B T} = \frac{N\mu^2 (H + \lambda M)}{3k_B T}$$

or

$$M\left(1 - \frac{N\mu^2\lambda}{3k_B T}\right) = \frac{N\mu^2 H}{3k_B T} \quad \text{or} \quad M = \frac{N\mu^2 H}{3k_B\left(T - \dfrac{N\mu^2\lambda}{3k_B}\right)}$$

hence the magnetic susceptibility will be

$$\chi = \frac{M}{H} = \frac{N\mu^2}{2k_B}\left(\frac{1}{T - \dfrac{N\mu^2\lambda}{3k_B}}\right)$$

or

$$\chi = \frac{c}{(T - \theta)} \qquad \text{...(6.13)}$$

where

$$c = \frac{N\mu^2}{3k_B} \quad \text{and} \quad \theta = \frac{N\mu^2\lambda}{3k_B}$$

θ is called paramagnetic Curie point and eq. (5.13) is called Curie Weiss Law.

6.3 HYSTERESIS ; B-H AND I-H CURVE

The lagging of magnetization or magnetic induction behind the magnetizing field is called Hysteresis. If a soft iron bar is placed in a magnetizing field H. It acquires induced magnetism. Fig. 6.1 shows the variation of magnetization I with the magnetizing field H. As the magnetizing field H is increased, the magnetization also increases as represented by part OA. At A it reaches a saturation value, where any further increase in the value of H does not increase I. If now the magnetizing field is gradually reduced to zero, the magnetization I also decreases but does not become zero when $H = 0$ but has a certain value represented by OB. This value of magnetization left in the bar even when H is made zero is called Residual Magnetism or Retentivity or Remanance.

If now the direction of H is reversed, the part of the curve BCD is obtained. The residual magnetism reduced to zero for a value OC of H in the reverse direction. This value of magnetizing field required to reduce the residual magnetism or retentivity zero is called Coercivity of the material. On further increasing H in the opposite direction the saturation point D, symmetrical to A is reached.

If the field is increased again in the forward direction (*i.e.* less negative) in steps, we reach back to point *A* along the path *DEFA*. Thus a closed curve *ABCDEFA*, called Hysteresis loop. A similar curve is obtained on plotting magnetic induction *B* against *H*. It is clear from above discussion, that whenever a magnetic material is taken round a cycle of magnetization, the value of magnetization *I* or magnetic induction *B* always legs behind *H*.

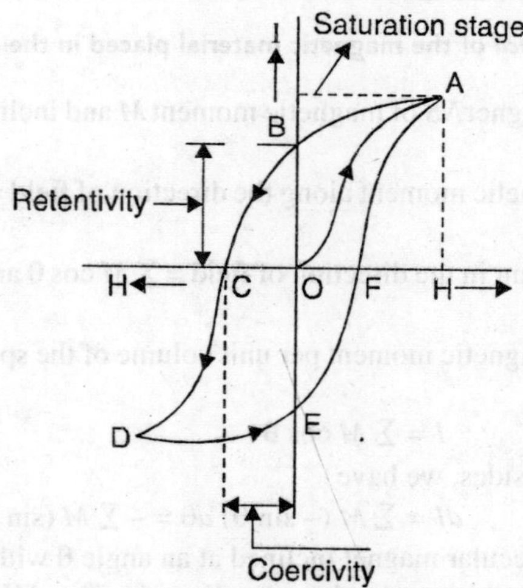

FIGURE 6.1 **The variation of magnetization *I* with the magnetizing field *H*.**

Energy Loss due to Hysteresis

When a magnetic material is taken through a complete cycle of magnetisation, some energy is always spent during the said process. When a specimen of magnetic material is taken through a cycle of magnetisation, this energy is spent in orienting and re-orienting the molecular magnets time and again. The molecular motion within the specimen results in production of heat. This energy cannot be recovered completely on reversing the magnetic field. The loss of energy per unit volume per cycle due to hysteresis is equal to the area of the *I-H* loop of the specimen or $1/4\pi$ times the area of its *B-H* loop in C.G.S. units. This energy is dissipated in the form of heat.

The study of the hysteresis loops of different magnetic materials provides useful information about their properties like retentivity, coercivity, energy loss, maximum magnetisation or induction etc. which determine their use in industry. The above study helps in selecting magnets, electromagnets, transformer cores, telephone diaphragms, cores of dynamos and motors etc.

Mathematical Treatment of Energy Loss in Hysteresis

When a ferromagnetic material is made to pass through a cycle of magnetization, there is expenditure of energy. This energy is generally dissipated in the form of heat in the specimen.

When a specimen is placed in a magnetic field, induced magnetism is acquired by the specimen the magnetic dipoles or the molecular magnets are oriented and reoriented when the material is passed through a cycle of magnetization and this molecular motion results in the production of heat.

In order to find the energy loss per cycle, let us consider a unit volume *abcd* of the magnetic material which is placed in a magnetising field having an intensity *H* as shown in Fig. 6.2.

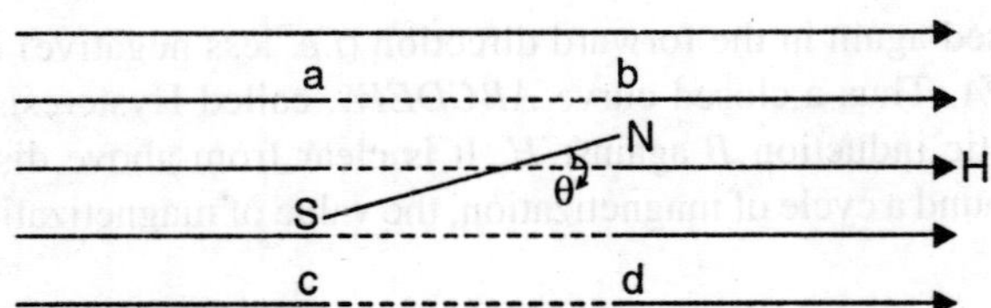

FIGURE 6.2 **Volume *abcd* of the magnetic material placed in the magnetising field *H*.**

Let there be a molecular magnet *NS* of magnetic moment *M* and inclined at an angle θ to the direction of magnetizing field.

The component of the magnetic moment along the direction of field = $M \cos \theta$ and that perpendicular to the field = $M \sin \theta$.

So the total magnetic moment in the direction of field = $\sum M \cos \theta$ and that perpendicular to the field = $\sum M \sin \theta = 0$.

Now from definition the magnetic moment per unit volume of the specimen $\sum M \cos \theta$ is equal to the intensity of magnetization *I*.

$$\therefore \qquad I = \sum M \cos \theta$$

Taking differential on both sides, we have

$$dI = \sum M(-\sin \theta)\, d\theta = -\sum M (\sin \theta)\, d\theta \qquad ...(6.14)$$

The couple acting on a molecular magnet inclined at an angle θ with the direction of the field = $MH \sin \theta$ and the work done in rotating this magnet by a small angle $d\theta = MH \sin \theta(-d\theta)$. The negative sign indicates that $d\theta$ is decreased in angle.

Total work done in rotating all the molecular magnets in unit volume

$$= -\sum MH \sin \theta\, d\theta$$

$$= -H\{\sum M \sin \theta\, d\theta\} \qquad ...(6.15)$$

From eqns. (6.14) and (6.15), the work done in increasing the intensity of magnetization by

$$dI = H\, dI$$

If intensity of magnetization changes from I_1 to I_2, we have

$$\text{Work done} \qquad \int_{I_1}^{I_2} H dI \qquad ...(6.16)$$

Let us consider the *I-H* curve of the material, and let it be having the shape *ABCDEFA* as shown in Fig. 6.3 (*a*). Consider two points *r* and *s* very near to each other on the curve. Since these points are very near

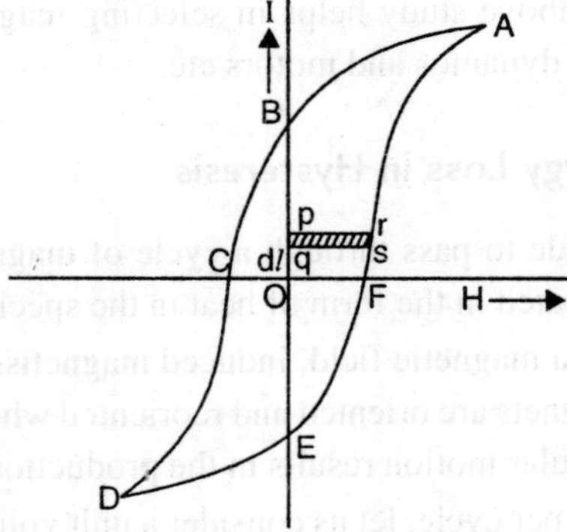

FIGURE 6.3 **(*a*) *I-H* curve of material with small area *pqsr*.**

so the value of H is constant and let the increase in value of I be dI *i.e.* the difference between values of I at p and q.

From eq. (6.16), the amount of work done on the specimen to increase the intensity of magnetization by $dI = \int H\,dI$ which is corresponding to the area of the strips $pqsr$. Similarly we can imagine the whole cycle to be made up of small rectangular strips and the total work done on the specimen when H increases or by the specimen when H decreases to zero can be calculated.

Let us consider the complete cycle as shown in Fig. 6.3(b).

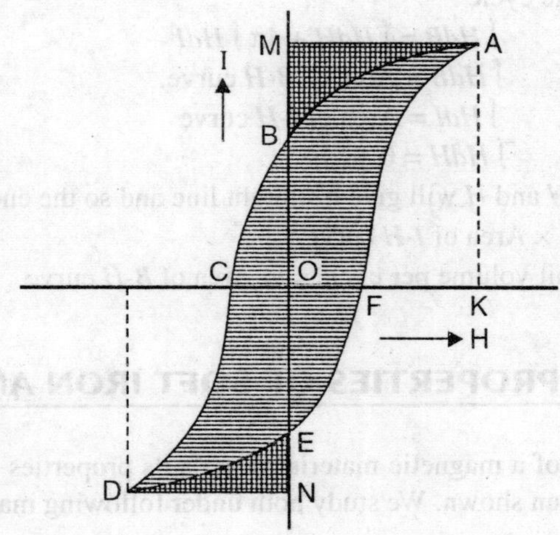

FIGURE 6.3 (*b*) Complete cycle of *I-H* curve.

(i) work done on the specimen when the value of H increases from zero to K

$$\int_E^M H\,dI = \text{Area of } EFAMBOE$$

(ii) work done by the specimen when the value of H decreases to zero

$$\int_M^B H\,dI = \text{Area of } AMBA$$

(iii) work done on the specimen when the value of H increases from zero to L in opposite direction

$$\int_B^N H\,dI = \text{Area of } BCDNEOB$$

(iv) work done by the specimen when the value of H decreases to zero

$$= \int_N^E H\,dI = \text{Area of } DNED$$

So the total work done on the specimen per unit volume per cycle

$$= \text{Area } EFAMBOE - \text{Area } AMBA + \text{Area } BCDNEOB - \text{Area } DNED$$

$$= \text{Area } ABCDEFA = \text{Area of } I\text{-}H \text{ curve.}$$

So the work done on the specimen per unit volume when taken through a complete cycle of magnetization is equal to the area enclosed by I-H curve. The larger is the area of the curve the larger is the loss of energy per unit volume per cycle.

Area of B-H Curve

We have the relation for magnetic induction as

$$B = H + 4\pi I$$

Taking differential on both sides,

$$dB = dH + 4\pi dI$$

Multiplying both sides by H,

$$HdB = HdH + 4\pi HdI$$

Integraing for a complete cycle

$$\int HdB = \int HdH + 4\pi \int HdI$$

Now $\int HdB$ = Area of B-H curve.

$\int HdI$ = Area of I-H curve

and $\int HdH = 0$

Since a curve between H and H will give a straight line and so the enclosed area = 0

Area of B-H curve = $4\pi \times$ Area of I-H curve

So loss of energy per unit volume per cycle = ¼ Area of B-H curve.

6.4 DIFFERENCE IN PROPERTIES OF SOFT IRON AND STEEL

A study of I-H and B-H curve of a magnetic material reveals its properties and behaviour. In Fig. 6.4 I-H curve for iron and steel have been shown. We study both under following magnetic properties.

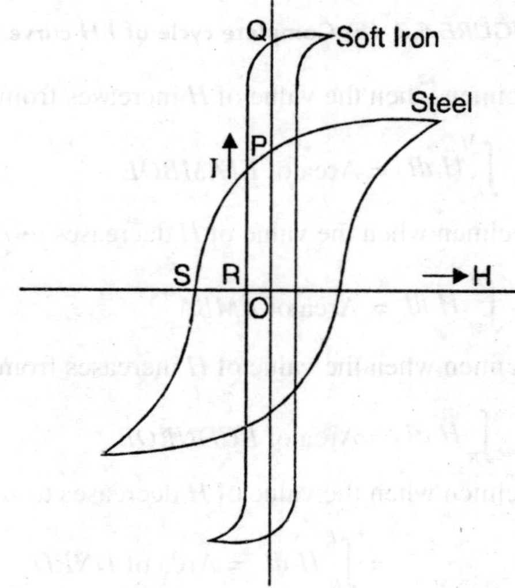

FIGURE 6.4 ***I-H* curves of soft iron and steel.**

Susceptibility. The intensity of magnetisation of soft iron, I, is greater than that of steel for the same value of H and hence susceptibility is greater for soft iron than for steel.

Permeability. We plot B-H curve for both iron and steel. It is seen that the value of magnetic induction B for same value of H is greater for soft iron than for steel. Hence permeability (B/H) of soft iron is greater than that of steel.

Retentivity. When the magnetising field H is reduced to zero, soft iron retains greater intensity of magnetization OQ than that of steel which is OP and hence the retentivity of soft iron is greater than that of steel.

Coercivity. To demagnetise steel a greater force OS is required than that for soft iron which is OR. Hence the coercivity of soft iron is less than that of steel.

Hysteresis Loss. The area of I-H curve for soft iron is smaller than that of steel. Hence the hysteresis loss per c.c. per cycle is less for soft iron than that for steel.

6.5 DETERMINATION OF MAGNETIC SUSCEPTIBILITY AND PERMEABILITY

Magnetometer Method for B-H and I-H Curves

This method is most suitable for long rod or wire as the demagnetising effect is minimised.

As shown in Fig. 6.5 the specimen PQ, free from magnetism is placed in the solenoid S, carrying a current. The solenoid is connected through a commutator K to a battery E_1 along with an ammeter A and a rheostat R_1. A second coil is wound over S and its ends a and b are connected to battery E_2 along with a resistance box in series. The current through this coil is so adjusted in magnitude and direction that the effect of magnetic field produced in it neutralizes the magnetism produced in it by the vertical component of the earth's field.

A deflection magnetometer M in the 'end on' position is placed in line with upper end of the specimen as shown in Fig. 6.5.

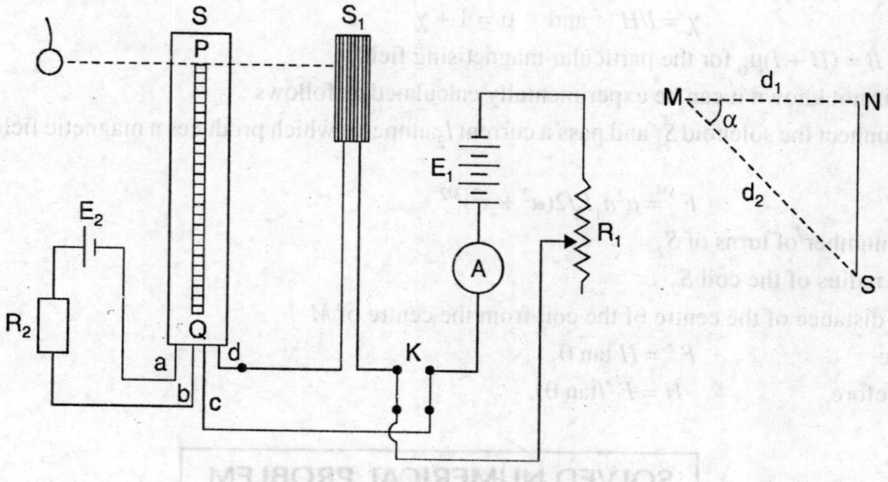

FIGURE 6.5 **Magnetometer Method.**

A compensating coil S_1 is fixed vertically with the axis in line with the centre of the magnetometer needle is included in series with the solenoid to neutralise the magnetising effect of the solenoid S itself on the magnetometer. The distance from the centre of M and the direction of the winding of this coil are so adjusted that in the absence of the rod or specimen the needle is not effected by the current by the current through the coil S and S_1 *i.e.* the field due to S is equal and opposite to that due to S_2 at M.

After making these adjustments the current is switched on with the help of key K in one direction. The specimen gets magnetised and causes deflection in the magnetometer and current in the ammeter which are recorded.

Let the radius of specimen $PQ = r$

Intensity of magnetisation due to current in the solenoide = I

Horizontal component of the earth's magnetic field = H

If the poles produced are N and S as shown in Fig. 6.5 we have

Pole strength = $\pi r^2 I$

Field at M due to pole $N = \pi r^2 I/d_1^{\,2} \times 4\pi\mu_0$

And field at M due to pole $S = [\pi r^2 I \; 4\pi\mu_0 \cos(\alpha)/d_2^{\,2}] \times [\pi r^2 d_1/d_2^{\,2} d_2]$

(along the same horizontal line as due to pole N)

$\therefore$ Total intensity of magnetic field at M due to poles N and S will be

$$F = 4\pi^2 r^2 I\mu_0/d_1^{\,2} - 4\pi^2 r^2 I\mu_0 d_1/d_2^{\,3} \qquad \text{...(6.17)}$$

or

$$F = I\mu_0\,(1 - d_1^{\,3}/d_2^{\,3})\,\pi r^2/d_1^{\,2}$$

As the magnetic needle is subjected to two mutually perpendicular field F and H, the needle is deflected through an angle θ

Therefore, $F' = H \tan\theta$ ...(6.18)

From eqs. (6.17) and (6.18), we have

$$I = 4\pi\mu_0 d_1^{\,2}\, H \tan\theta/\pi r^2\,(1 - d_1^{\,3}/d_2^{\,3}) \text{ Weber/m}^2$$

The above equation gives relation for I in terms of measurable quantities thus its value can be determined. The intensity of magnetising field $H = nI_1$ where I_1 is the current in amperes and n is the number of turns per cm length of S. The value of χ and μ is then given by

$$\chi = I/H \quad \text{and} \quad \mu = 1 + \chi$$

Also $B = (H + I)\mu_0$ for the particular magnetising field.

If H is not known it can be experimentally calculated as follows :

Disconnect the solenoid S_1 and pass a current I_2 amperes which produces a magnetic field at M which is given by

$$F' = a^2 n_1 I_2/2(a^2 + x^2)^{3/2}$$

where n_1 = number of turns of S_2

a = radius of the coil S_2

x = distance of the centre of the coil from the centre of M

Since $F' = H \tan\theta_1$

Therefore, $H = F'/\tan\theta_1$.

SOLVED NUMERICAL PROBLEM

Problem 1. *An iron ring of 1 m mean circumference is made from round iron of the cross-section 24 cm². Its permeability is 500. It is wound with 600 turns. What current will be required to produce a flux of 2.4 × 10⁵ lines ?*

Solution. Let I be the current required.

Here flux $f = 2.4 \times 10^5$ lines $= 2.4 \times 10^{-3}$ Wb [as 10^3 lines = 1 Wb]

Permeability $m_r = 500$

Area of cross-section $A = 24 \text{ cm}^2 = 2.4 \times 10^{-3} \text{ m}^2$

Length of the path $l = 1$ m

$$\therefore \quad \text{Flux density} = \text{Flux/area} = \frac{2.4 \times 10^{-3}}{2.4 \times 10^{-3}} = 1 \text{ wb/m}^2$$

$$\therefore \quad \text{Magnetic force} \quad H = \text{Flux density } (B)/\mu_0\mu_r$$

$$\therefore \quad H = \frac{1}{4\pi \times 10^{-7} \times 500} = 1591 \text{ AT/m}$$

Ampere turns required, $\text{AT} = H \times l = 1591 \times 1 = 1591$

Since ampere turns = Number of turns × Current

$\therefore$ Current = $\text{AT}/N = 1591/600 = 2.65$ A.

EXERCISE

1. Explain what is meant by hysteresis. *(Jan. 1997. K.U.K.)*
2. Show that the area of a *B-H* curve denoted the energy dissipated per unit volume of material during each magnetizing cycle. *(June 2000, K.U.K.)*
3. What are the characteristics of para, dia and ferromagnetic substances ? Explain their behaviour with the examples. *(May 1999, June 1999, K.U.K.)*
4. Explain magnetic hysteresis. Show that the area of the *B-H* cycle represents the energy dissipated per cubic meter of the material per cycle of magnetization. *(Jan. 1999, K.U.K.)*
5. Explain the following terms :
   (*i*) Magnetization and magnetic intensity
   (*ii*) Magnetic permeability and susceptibility. *(May 1998, K.U.K.)*
6. Explain in details the magnetometer method for *B-H* curve. *(Dec. 2000)*
7. Describe *B-H* hysteresis loop and its significance. *(Dec. 2000, K.U.K.)*
8. A solenoid consisting of 500 turns and carrying 4 A current is 0.5 m long. Calculate (*i*) magnetomotive force (*ii*) Total flux if area of cross-section is 0.0008 m^2. **[Ans.** 251.3 Wb]
9. A circular ring of iron is wound uniformly with 1000 turns of wire. Its mean diameter is 0.2 m and its cross-section is 4 cm^2. Find the current which will produce a flux of 2000 lines. Given permeability = 800. **[Ans.** 31.25 mA]
10. What useful informations can be derived from hystersis curve of a given material ? *(Jan. 2002, K.U.K.)*
11. (*a*) How atomic magnetic moments arise ? Explain.
    (*b*) Discuss classical theory of paramagnetism and discuss the effect of temperature.
    (*c*) Write briefly about the theory of domains. *(Jan. 2003, K.U.K.)*

Chapter 7

MODERN PHYSICS

7.1 EXCITATION AND IONIZATION POTENTIALS ; THEIR MEASUREMENT

Excitation

The process of raising or transferring the electron from lower state to a higher energy state is called excitation. And when the atom is in a state of high energy it is said to be excited.

Consider the case of a simplest atom *i.e.* hydrogen atom. This has one electron in the inner most orbit *i.e.* $n = 1$ and is said to be in the normal state. If by some means, sufficient energy is supplied to the electron it moves to higher energy states and the atom is said to be excited. The process of excitation is absorption phenomenon. The excited state is not stationary state. It has life time of the order of 10^{-8} sec because the electron under the attractive force of the nucleus jumps to the lower permitted level. This is accompanied by the emission of radiation according to Bohr's frequency condition.

The energy necessary to excite an atom can be supplied in a number of ways. The most commonly used method is by means of electron bombardment. In this process a filament is heated and the electrons are emitted due to thermionic emission. The electrons are accelerated through a potential V so that they acquire enough kinetic energy. Now the electrons are allowed to collide the atom. The kinetic energy (wholly or partially) of the electrons is transferred to the atom. The atom is now in the excited state. The minimum potential V required to accelerate the bombarding electrons to cause excitation from the ground state is called the resonance potential. The various values of potential to cause excitation of higher states are called excitation potential. We have seen that the energy required to excite the electron from first to second state is $13.6 - 3.4 = 10.2$ eV, and first to third state is $13.6 - 1.5 = 12.1$ eV and so on. Hence the successive excitation potentials of hydrogen atom are 10.2, 12.1 .....Volts.

Ionization

When the electron is completely removed from the nucleus, the atom is said to be ionized. The process of raising the atom from the normal state to the ionized state is called ionization. It is also an absorption process. The potential necessary to accelerate the bombarding electrons causing ionization is called the ionization potential. The energy required to ionize hydrogen atom is 0 – (– 13.6) = 13.6 eV. Hence ionization of hydrogen is 13.6 Volts.

7.2 ABSORPTION AND DIFFRACTION OF X-RAYS ; CHARACTERISTICS OF X-RAYS SPECTRA ; MOSLEY'S LAW

X-Rays

Actually X-Rays are electromagnetic waves of very short wavelength. When a beam of fast moving electrons strike a solid target, an invisible high penetrating radiation is produced called X-rays because of their then unknown nature. X-rays are found to travel in straight lines from the source and could not be deflected by electric and magnetic field which shows that they are not charged particles.

Production of X-rays

The X-rays are produced when fast moving electrons strike a target of suitable material. The basic requirement for their production are

(*i*) A source of electrons

(*ii*) Effective means of accelerating the electrons

(*iii*) A target of suitable material of high atomic weight.

There are two types of X-ray tubes, namely,

(*a*) The gas filled or Roentgen X-ray tube

(*b*) The modern coolidge tube or hot filament tube.

Roentgen Tube

It consists of an evacuated glass bulb containing a concave cathode of aluminium *C* [see Fig. 7.1]. Concave cathode focusses a beam of cathode rays on the tungsten anti-cathode. This anti-cathode is inclined at an angle of 45° to the path of cathode rays. Anti-cathode is sometimes called as target. The anode *A* is an aluminium rod connected externally to the anti-cathode. The anode helps to make the emission of *X*-rays steady. This tube worked by an induction coil producing a voltage ranging from 10^3 to 10^5 volts. When the voltage applied is small the X-rays produced are soft and when the voltage applied is high the X-rays produced

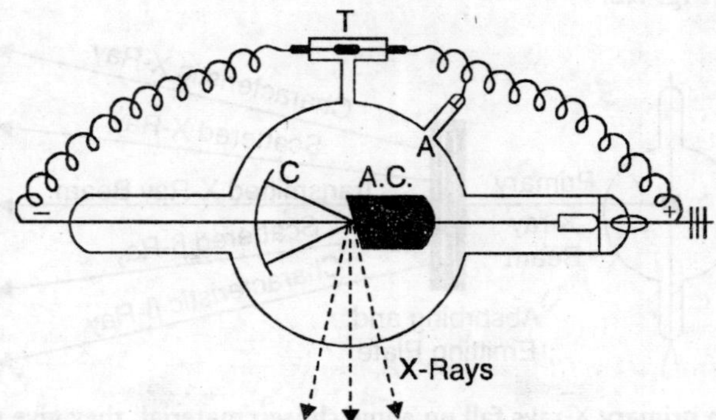

FIGURE 7.1 **Roentgen's X-ray Tube.**

are hard. The cathode rays fall on the anti-cathode and hence X-rays are produced. The functioning of the tube depends on the pressure of the gas inside the tube and the potential difference applied. During the working of the X-ray tube, the positive ions of the gas are produced. These positive ions should be removed, and hence the pressure inside the tube is lowered, therefore, a greater potential is needed to run the discharge, and the emitted X-rays are harder.

The anti-cathode should possess the following properties :

(*i*) It should have high atomic number.

(*ii*) It should have high melting point.

(*iii*) It should have high conductivity towards heat and electricity.

(*iv*) It should have low vapour pressure at high temperature.

Tantalum, platinum and tungsten are the best examples.

The main defects of the roentgen tube are :

(*i*) The intensity of the X-rays can not be controlled.

(*ii*) The softening device is not perfect.

(*iii*) The quality of X-rays can not be controlled.

(*iv*) The applied voltage required is to be too high.

Properties of X-rays

1. X-rays are not deflected by electric and magnetic field, which shows that they have no charge.
2. They are highly penetrating and can pass through many solids which are opaque towards light *e.g.* wood, flesh etc.
3. They cause flourescence in many substances *i.e.* barium, plato-cyanide, cadmium, tungsten, zinc sulphide etc.
4. They effect a photographic plate and are even more effective than light.
5. They ionize a gas and also knock out electrons from certain metals on which they fall. This effect is known as photoelectric effect.
6. They travel in straight line with velocity of light *i.e.* 3×10^8 m/sec.
7. Like light, X-rays consists of electromagnetic waves of very short wavelength and show reflection, refraction, interference, diffraction and polarization in a similar way.
8. When X-rays (primary) coming from the anti-cathode of an X-ray tube fall on a plate of some chosen material, they give rise to a complex phenomenon of secondary radiations which consists of three different types as shown in Fig. 7.2.

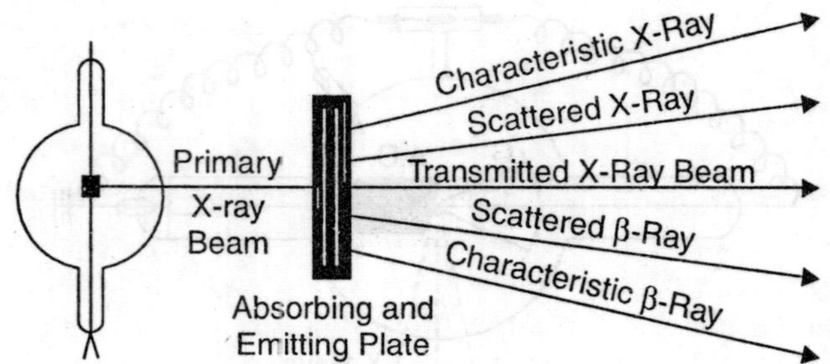

FIGURE 7.2 **When primary X-rays fall on some chosen material, they give rise to a complex phenomenon of secondary radiation as characteristics X-ray, scattered X-ray, Transmitted X-rays, Scattered β-rays and characteristics β-rays.**

(*i*) ***Scattered X-rays.*** They are practically of same nature and wavelength as the primary X-rays. Their properties do not depend upon the nature of the scattering substance.

(*ii*) ***Characteristics X-rays.*** The wavelength of the characteristics X-rays is equal to or less than the primary X-rays but it is characteristics of the substance upon which the primary X-rays are incident and do not depend upon the wavelength of primary X-rays. The characteristic X-rays from substances of high atomic weights are of shorter wavelength than those from substances of low atomic weight and hence are more penetrating. Every substance give rise to a particular set of characteristic X-rays consisting of one or more series known as *K, L, M* etc. The X-rays in *K* series of element have smaller wavelength than in *L, M* or *N* series ; whereas the wavelength of the *X*-rays in same series, say *K* or *L*, produced from an element of higher atomic weight is less than that produced from an element of low atomic weight.

(*iii*) ***β-rays.*** They consist of fast moving electrons produced by the photoelectric effect. Their properties are independent of the nature of the scattering substance but depends upon the quality of primary X-rays.

The Origin of X-rays

X-rays, like visible light, originate from the jumping of an electron from one orbit to another. When high speed electrons from the cathode of an electron X-ray tube strike the target, they ionize many of the atoms composing the surface layers of the metal.

Owing to their very high speeds (about 10% of the velocity of light), the electron penetrate the atom and remove an electron from the inner shell by collision. This is illustrated in Fig. 7.3 where an electron is knocked out of the *K*-shell.

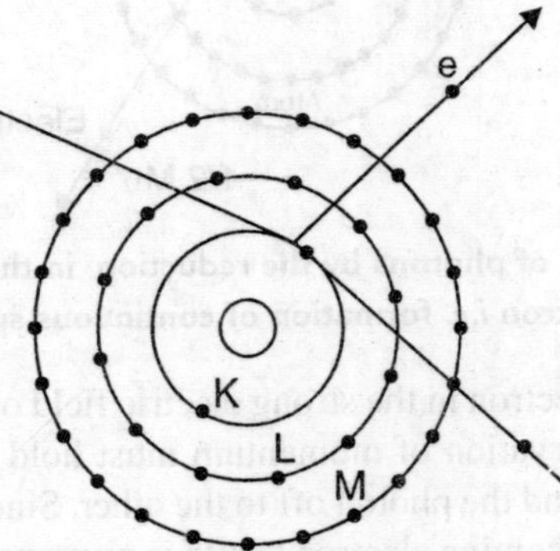

FIGURE 7.3 **When electron strike, the electron is knocked out from the *K*-shell after the collision.**

The designations *K, L, M, N, O, P* etc. for the various electron shells originated with the X-ray spectroscopist and are identified with quantum numbers $n = 1, 2, 3, 4, 5, 6$ etc. When an electron is missing in the innermost *K* shell, a nearby electron from the next shell beyond jumps into vacant space, simultaneously emitting a photon of energy $h\nu$. Such X-rays arising from millions of atoms produce the *K*-lines shown in Fig. 7.4. The jumping process continues until the outer most shell is reached, where an electron jumping gives rise to visible light. Thus we see how it is possible for a single atom to emit X-rays of different wavelengths.

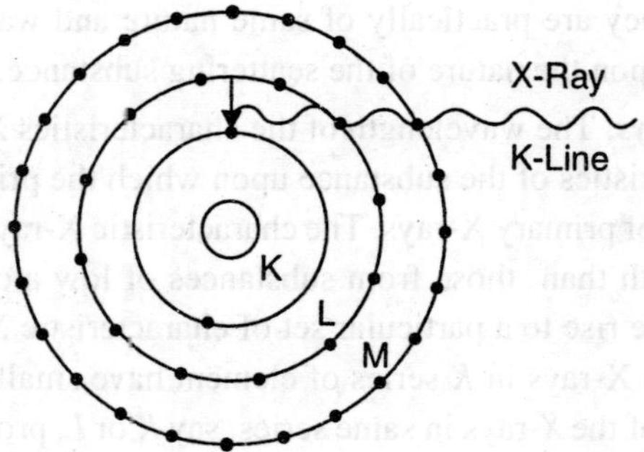

FIGURE 7.4 **Formation of *K*-lines when electron jumps from *L* shell to *K* shell.**

The continuous X-rays spectrum is due to another phenomenon often referred to as "Bremsstrahlung". These radiations are due to slowing down of high speed electrons as they pass close to nuclei of atoms within the target of the X-ray tube. The process is illustrated in Fig. 7.5. As the electron passes through the atom, it is attracted by the positive charge of the nucleus and is deflected in its path.

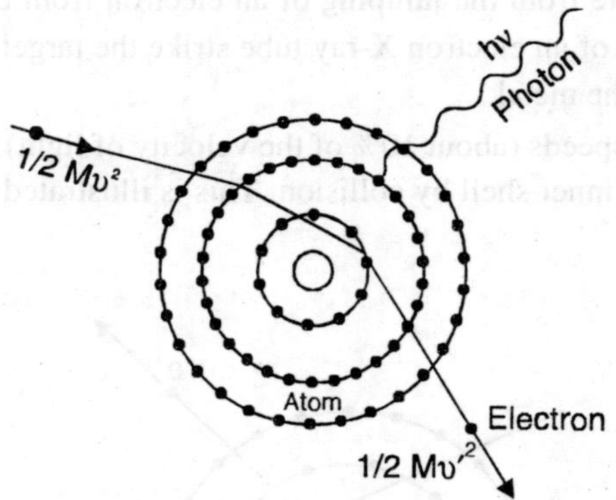

FIGURE 7.5 **Emission of photons by the reduction in the energy of the high speed electron *i.e.* formation of continuous spectrum.**

During the deflection of the electron in the strong electric field of the nucleus, a light wave of energy $h\nu$ is emitted. Since the law of conservation of momentum must hold for such a collision, the electron is deflected off to one side of the atom, and the photon off to the other. Since the law of conservation of energy must hold, some of the energy of the incoming electron $\frac{1}{2}m\upsilon^2$ is given up to the newly created photon $h\nu$, and the remainder $\frac{1}{2}m\upsilon'^2$ is retained by the electron. Thus the electron is slowed down to a velocity υ' by the encounter. The closer the electron comes to the nucleus, the greater is its loss in velocity and energy, and the greater is the energy of the radiated photon. By the law of conservation of energy

$$\frac{1}{2}m\upsilon^2 - \frac{1}{2}m\upsilon'^2 = h\nu \quad ...(7.1)$$

The highest frequency that is possible is one, in which the electron is completely stopped by the atom. In this special case,

$$\frac{1}{2}m\upsilon^2 = h\nu_{max} \quad ...(7.2)$$

Since the kinetic energy of the electrons in the beam striking the target is given by the voltage V applied to the tube, we may use equation

$$eV = \frac{1}{2}m\upsilon^2 \quad ...(7.3)$$

and obtain

$$eV = h\nu_{max} \quad ...(7.4)$$

While eq. (7.3) must be replaced by the relativistic equation., when voltage are greater than 20000 Volts eq. (7.4) holds true for all voltages.

So the production of continuous spectrum can be considered to be inverse of photoelectric effect because in this case electron K.E. = $\frac{1}{2}$ mv^2 is converted into photon energy $h\nu_{max}$.

Mosley's Law

The frequency of a spectral line in characteristics *X*-rays spectrum varies directly as the square of atomic number *Z* of the element emitting it or mathematically we can write as

$$\nu \propto (Z - a)^2 \quad \text{or} \quad \nu = b(Z - 1)^2 \qquad \text{...(7.5)}$$

where ν = frequency of characteristic radiation

b = constant, which is different for different series

a = constant, known as screening constant and is different for each series.

The characteristic X-rays spectra of different elements are similar to each other in the sense that each consists of *K, L* and *M* series.

The frequency of lines (in each series) produced from an element of higher atomic number is greater than that produced by an element of lower atomic number. The reason for this is that the binding energy of electron increases as we go from one element to another of higher atomic number *i.e.* in case of an element of higher atomic number, large amount of energy is required to liberate an electron from *K, L,* and *M* shells respectively of that element. Thus in the case of K_α line of characteristic X-ray spectrum , higher is atomic number of the target material, higher is the frequency of K_α line.

For K_α line, *b* was found to be equal to ¾ *R*, where *R* is Rydberg constant and constant, *a,* was found to be 1. So for K_α line

$$\nu_{K\alpha} = \frac{3}{4} R(Z - 1)^2 \quad \text{using eq. (7.5)} \qquad \text{...(7.6)}$$

Fig. 7.6 shows Mosley diagram for K_α and K_β lines obtained by plotting $\sqrt{\nu}$ versus atomic number of different element of periodic table. The graph is linear one as expected.

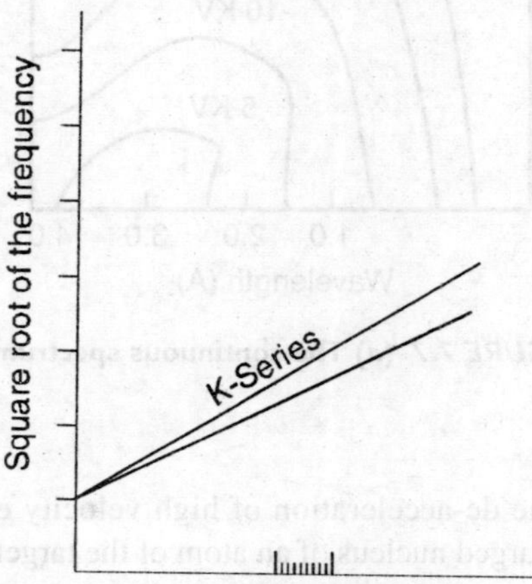

FIGURE 7.6 Mosley's diagram for K_α and K_β lines by plotting on frequency versus atomic number.

The exact form of Mosley's law is $\frac{1}{\lambda} = R(Z-\sigma)^2\left[\frac{1}{n^2} - \frac{1}{p^2}\right]$ ...(7.7)

where σ is correction factor, and n and p are the principal quantum numbers of the energy levels between which the transition takes place.

Importance of Mosley's law

1. According to Mosley's law, it is the atomic number of an element and not the atomic weight which accounts for the physical and chemical properties of an element.

2. This law provides the guidance that elements must be arranged in periodic table according to their atomic number not according to their atomic weights.

3. By considering the X-ray spectra of rare earth elements, this law has been helpful in determining their atomic number and also fixing their position in periodic table at appropriate positions.

4. This law also led to the discovery of some new elements like hafnium (72), promethium (61) etc.

X-ray Spectrum

As explained above X-ray spectrum consists of (*a*) continuous spectrum and (*b*) line spectrum. These are shown in Fig. 7.7 (*a*), (*b*).

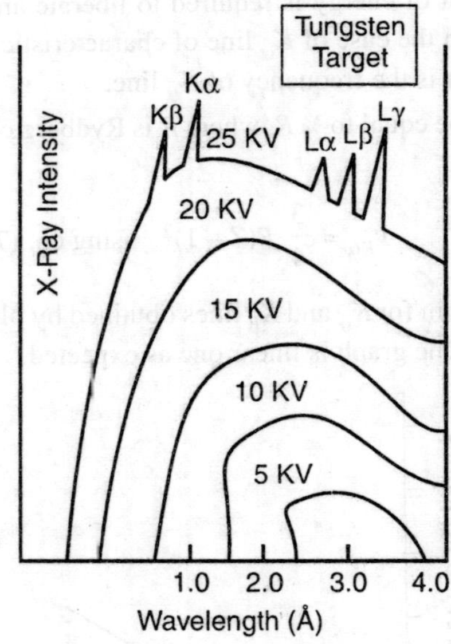

FIGURE 7.7 **(*a*) The continuous spectrum.**

(a) Continuous Spectrum

(*i*) It is produced due to the de-acceleration of high velocity electrons when they are deflected while passing near the positively charged nucleus of an atom of the target material.

(*ii*) It has sharply defined short wavelength limit given by

$$\lambda_{min} = \left(\frac{12400}{V}\right) \times 10^{-10}\text{ m} \quad \text{or} \quad \frac{12400}{V}\text{ Å}$$

(*iii*) The cut-off wavelength λ_{min} is independent of the nature of the target material but is inversely proportional to the potential difference between the cathode and anode of an X-ray tube. The value of λ_{min} decreases as this potential difference is increased.

(*iv*) The intensity of the continuous spectrum is found very nearly proportional to the square of the applied voltage for a given target and the atomic number of the target material when a constant potential difference is applied.

(*v*) There is a shift of maximum intensity position towards the short wavelength side as voltage is increased.

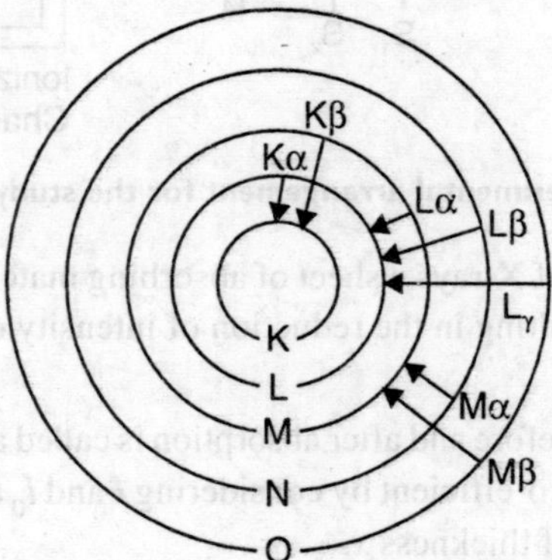

FIGURE 7.7 **(*b*) The jumping of electron from higher orbit to lower orbit.**

(b) Line Spectrum

(*i*) It is produced when electrons are dislodged from the innermost orbit of the atoms of the target material followed by electron jumps from outer orbits.

(*ii*) It consists of discrete spectral lines which constitute *K*-series, *L*-series and *M*-series etc. *K*-series consists of those lines for which electron's jump end at *K*-level.

(*iii*) *K*-series being most energetic constitute the hard X-rays whereas *L* and *M* series form the soft X-rays.

(*iv*) Line spectrum is the characteristics of the target material used. The number of lines present in the spectrum depends both on the nature of the target material and the excitation voltage.

(*v*) There is a regular shift towards shorter wavelength in the *K* spectrum as the atomic number of the target is increased. The exact relationship, as found by Mosley, is given as

$$\frac{\lambda_2}{\lambda_1} = \frac{\nu_1}{\nu_2} = \frac{(Z_1 - 1)^2}{(Z_2 - 1)^2}$$

where ν_1 = frequency of K_α line for a target material having atomic number Z_1 and ν_2 and Z_2 are similar quantities for some different target material.

Absorption of X-rays

We can study the absorption of X-rays with an apparatus shown in Fig. 7.8. A narrow beam of monochromatic X-rays when passes through matter, some part of it is absorbed while the rest is transmitted. X-rays are made to pass through slits S_1 and S_2 of lead plates. So that a narrow beam of it is allowed to fall on a crystal and due to Bragg reflection, we get a monochromatic beam and then this beam is made to enter in the ionization chamber in which the current of ionization produced will measure the intensity of X-rays.

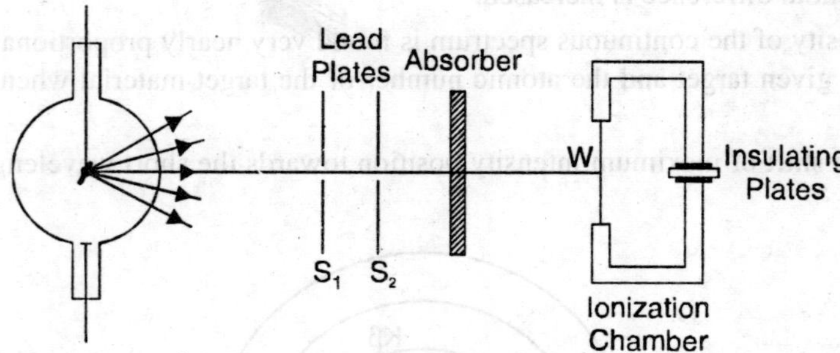

FIGURE 7.8 **An experimental arrangement for the study of X-ray absorption.**

After knowing the intensity of X-rays a sheet of absorbing material is put in the path of X-rays before it enters the ionization chamber resulting in the reduction of intensity of X-rays due to absorption of X-rays by the absorber.

The ratio I/I_0 *i.e.* the current before and after absorption is called absorption co-efficient of the material. We can get a relation for absorption co-efficient by considering I and I_0 as the intensities of X-rays before and after it passes through an absorber of thickness x.

Let dI be the further decrease in intensity when X-rays travel through an additional thickness dx of the absorber and this will be proportional to the original intensity of X-rays *i.e.*

$$-\frac{dI}{dx} \propto I \quad \text{or} \quad \frac{dI}{dx} = -\mu I \qquad ...(7.8)$$

where μ is the constant of proportionality and is called linear absorption co-efficient of the absorber or linear attenuation co-efficient of the absorber.

From eq. (7.8) we have $\qquad dI/I = -\mu dx$

integrating both sides within the limits from I_0 to I and from 0 to x (thickness)

$$\int_{I_0}^{I} \frac{dI}{I} = -\mu \int_0^x dx$$

$$\therefore \qquad \log\left(\frac{I}{I_0}\right) = -\mu x \quad \text{or} \quad I = I_0 e^{-\mu x} \qquad ...(7.9)$$

which shows that the intensity of X-rays decrease exponentially with thickness of the material [see Fig. 7.9].

We can define absorption co-efficient from eq. (7.8) as

$$\mu = \frac{\left(\frac{-dI}{I}\right)}{dx}$$

i.e. it is equal to the fractional decrease in the intensity of X-rays per unit thickness of the absorber. It is found that μ depends on the wavelength of X-rays nearly by the relation $\mu \propto \lambda^3$.

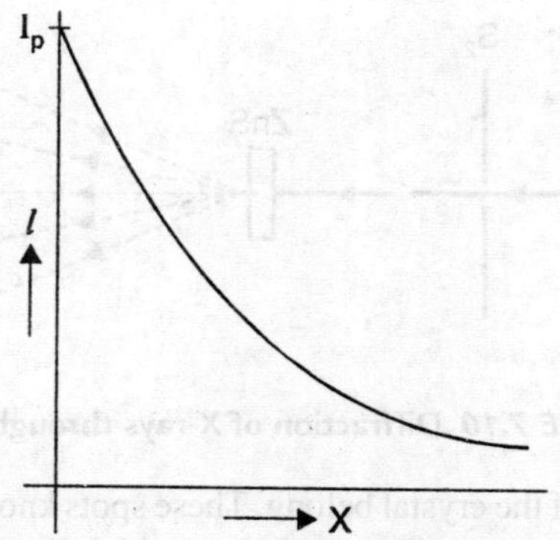

FIGURE 7.9 **Exponential decrease in intensity of X-rays with thickness.**

The value of μ can be found by finding the thickness of the absorber sheet which reduces the intensity of X-rays beam to ½ of its initial value such, thickness is called half value layer (H.V.L.) and is denoted by *T*.

If we put $\frac{I}{I_0} = \frac{1}{2}$ and $x = T$ in eq. (7.9), we get

$$e^{-\mu T} = \frac{1}{2} \quad \text{or} \quad e^{\mu T} = 2$$

$$\mu T = \log 2 = 0.693 \quad \text{or} \quad \mu = \frac{0.693}{T}$$

which gives another definition of absorption co-efficient in terms of H.V.L.

In practice, however, it is μ/σ called mass absorption co-efficient μ_m is more important than μ and may be defined as equal to the fractional decrease in the intensity of a beam of unit cross-section in passing through a unit mass of the absorbing material. μ_m is found to depend on the wavelength λ of X-rays and also on atomic number *Z* of absorber. A relation of it is given by

$$\mu_m = D\lambda^3 Z^3$$

X-ray Diffraction

Due to short wavelength of X-ray, which is of the order of 10^{-8} cm ordinary grating which have 6000 lines per cm cannot produce any appreciable diffraction with X-rays. Therefore, a crystal grating in which the atoms are arranged in a regular pattern is used for X-ray diffraction. The arranged atoms in the crystal grating corresponds to the grating lines and distance between two atoms which is of the order of 10^{-8} cm form the grating element. The difference between crystal grating and optical grating is that the atomic centers of diffraction in the crystal grating are not all in one plane but distributed in space, while in case of optical grating they are in one plane.

The crystal is thus a three dimensional space grating rather than a two dimensional plane grating. When a narrow pencil of X-rays is made to pass through a thin crystal the diffracted rays emerged in various well defined directions to form a symmetrical pattern of spots on a photographic plate placed perpendicular to the beam. A narrow beam of X-rays is allowed to pass through a thin crystal of zinc blend (ZnS) and transmitted beam is received on a photographic plate *P* as shown in Fig. 7.10. After an exposure of many hours, it can be found on developing the plate that in addition to a central black spot at *O* where the direct beam struck the plate, there are many other fainter spots regularly arranged, indicating that the X-rays have

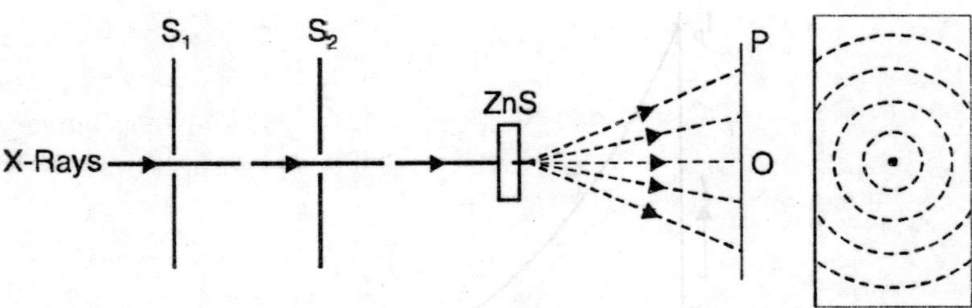

FIGURE 7.10 **Diffraction of X-rays through ZnS.**

been diffracted from various planes of the crystal belong. These spots known as Laue spots form a geometrical pattern depending upon the particuiar crystalline system to which the crystal belongs since different Laue pattern are obtained with different crystalline configuration.

Bragg's Law

Since the crystal constitute a three dimensional grating, therefore, the theory of Laue spots is very complex. Prof. W.L. Bragg has given a very simple interpretation of the different pattern obtained by supposing that reflection takes place from the various planes in the crystal. The atoms in a crystal are arranged in regular fashion having a constant space between them. When a beam of X-rays falls on the crystal, each atom acts as a scattering center and emits secondary wavelets. Since X-rays are much more penetrating than ordinary light, therefore, only partial reflection takes place at each layer of atoms. The condition for the measurement of wavelength of the X-rays from a set of parallel layers in given plane is obtained as follows.

Let a narrow beam of monochromatic X-ray of wavelength λ be incident on cleavage planes *A, B, C* etc. in a crystal and let *XY* be an advancing wavefront which meets the cleavage plane of the crystal at a glancing angle θ at *X* [see Fig. 7.11]. The ray *XL* is reflected at the cleavage plane *B* whereas the ray *YN* is reflected at the cleavage plane A. If the waves reflected from these planes of atoms are in phase with each other , at *N* they will reinforce each other, but if they are in opposite phase they will cancel each others effect and there will be no reflected beam.

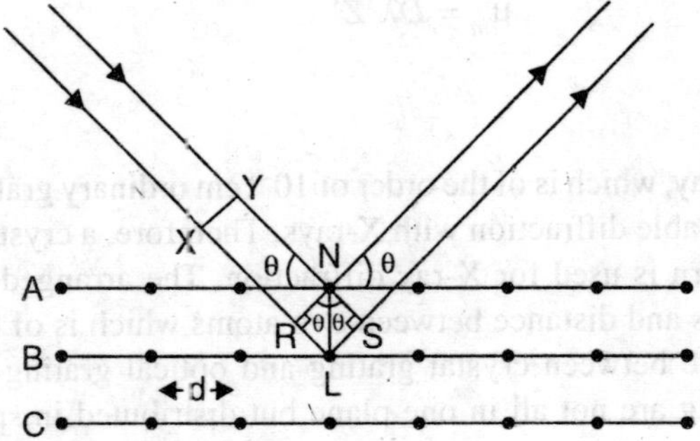

FIGURE 7.11 **Diffraction on various planes *A, B, C*.**

For reinforcement the path difference between the two coming waves from any two crystal planes should be an integral multiple of λ *i.e.* wavelength of X-rays.

The path difference between them is given by (*RL* + *LS*).

Therefore, for reinforce condition $(RL + LS) = n\lambda$ ...(7.10)

In right angled ΔNRL $\dfrac{RL}{NL} = \sin\theta$

$$\therefore \qquad \frac{LS}{NL} = \sin\theta$$

If d is the distance between two consecutive cleavage planes and is known as lattice constant

$$NL = d \quad \text{or} \quad RL = d\sin\theta$$

Similarly, in right angled ΔNSL $\qquad LS/NL = \sin\theta$

or $\qquad LS = NL\sin\theta \quad \text{or} \quad LS = d\sin\theta$

Substituting these values of RL and LS in eq. (7.10), we get

$$d\sin\theta + d\sin\theta = n\lambda \quad \text{or} \quad 2d\sin\theta = n\lambda \qquad \ldots(7.11)$$

This relation is known as Bragg's law, which is similar to the relation obtained for the ordinary plane diffraction grating having $2d$ as the grating element. It indicates that for given values of n, λ and d there is reflection only in a particular direction. In other direction there will be destructive interference. For $n = 1$ we get first order spectrum, for $n = 2$ we get second order spectrum and so on.

Powder-photograph Method for X-ray Diffraction

In the powder photograph method, the incident monochromatic radiation strikes a finely powdered specimen or a fine gained polycrystalline specimen contained in a thin walled capillary tube. The distribution of crystalline orientation will be nearly continuous. The powder method is convenient because single crystal is not required. Diffracted rays go out from individual crystals which happens to be oriented with planes making an incident angle θ with the beam satisfying Bragg's equation. Diffracted rays leaves the specimen along the generator of cones concentric with the original beam. The generator makes an angle 2θ with the direction of the original beam, where θ is the Bragg's angle. The cones intercept the film in a series of concentric rings as shown in Fig. 7.12.

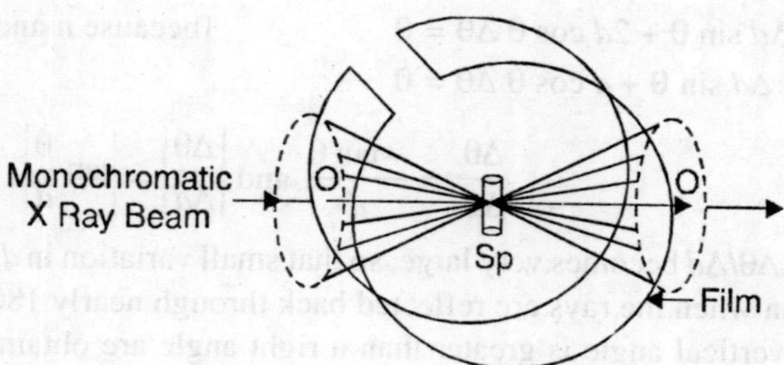

FIGURE 7.12 **X-ray diffraction camera.**

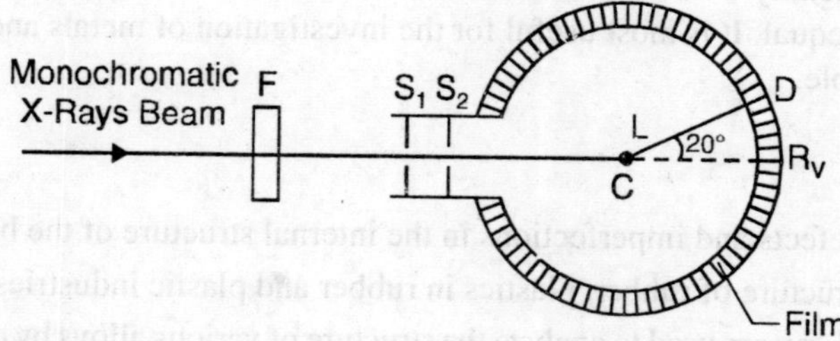

FIGURE 7.13 **An experimental arrangement of Powder-Photographic method.**

The arrangement of experiment is shown in Fig. 7.13. The X-rays are made monochromatic by passing them through Zirconium oxide filter F. A narrow beam of monochromatic X-rays limited by slits S_1 and S_2

falls on the powdered specimen C. The photographic film fits round the inner surface of the camera covering nearly the whole in circumference in order to record beams diffracted upto nearly 180°. The strip of the film is usually long in its horizontal dimensions, so that only arcs of circles appear on the film on developing. The X-rays after falling on the powder passes out of the camera through a hole cut in the film, in order to minimise the fogging produced due to direct scattering. The traces obtained on the photographic plate are shown in Fig. 7.14. The curvature of the trace changes as the angle of diffraction passes through 90°. The radius of curvature increases with the increase of angle of diffraction and at 90° straight lines are obtained.

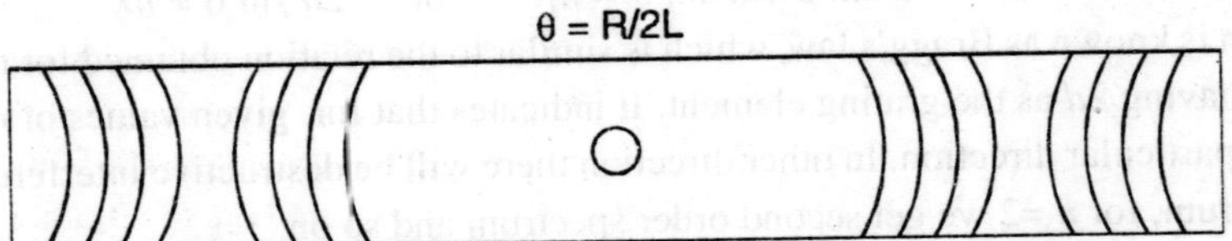

FIGURE 7.14 **A pattern record on photographic plate.**

The crystal structure can be obtained from the arrangement of the traces and their relative intensities. Consider a spectrum with glancing angle θ at a point D which is at a distance R from O where the direct incident beam strikes the film and if L be the radius of the cylinderical camera, then

$$\theta = \frac{R}{2L}$$

This value of θ will give us the spacing of the planes by using Bragg's relation,

$$2d \sin\theta = n\lambda$$

On differentiating we get

$$2\Delta d \sin\theta + 2d\cos\theta\,\Delta\theta = 0 \qquad \text{[because } n \text{ and } \lambda \text{ are constants]}$$

or

$$\Delta d \sin\theta + d\cos\theta\,\Delta\theta = 0$$

or

$$\frac{\Delta\theta}{\Delta d} = \frac{-\tan\theta}{d} \quad \text{and} \quad \left|\frac{\Delta\theta}{\Delta d}\right| = \left|\tan\frac{\theta}{d}\right|$$

When θ tends to 90°, $\Delta\theta/\Delta d$ becomes very large, so that small variation in d produce large variation in θ. Therefore, high resolution when the rays are reflected back through nearly 180°. This is the reason why spectra for which the semi-vertical angle is greater than a right angle are obtained by the use of the long cylinderical film nearly surrounding the crystal powder.

This method is applicable for all kinds of crystalline matter, however, small the individual crystal may be. This method is specially used to determine the structure of the simple cubic crystals for which the spacing of the unit cell are equal. It is most useful for the investigation of metals and alloys structure, while single crystal are not possible.

Applications of X-rays

(*i*) To locate the defects and imperfections in the internal structure of the body.

(*ii*) To study the structure of rubber, plastics in rubber and plastic industries.

(*iii*) In metallurgy X-ray are used to analyze the structure of various alloys by determining the structure of in gots with the help of X-rays diffraction.

(*iv*) X-rays are used to study the crystal structure in crystallography.

(*v*) X-rays are widely used in radiography.

(*vi*) Many types of skin diseases, malignant sores, cancers, tumors etc. have been cured by controlled exposure to X-rays of suitable quality.

(*vii*) X-rays are used to find the atomic number of the elements and also for the identification of the elements.

7.3 COMPTON SCATTERING

It can be shown that the phenomenon of Compton scattering could be explained by assuming the particle character of light. In other words, the light radiation is propagated by photons each carrying an energy $h\nu$ and momentum $h\nu/c$ where all the letters have their usual meanings. These photons are mass less entities. The photons collide with electrons conserving total energy and linear momenta. The collision process, therefore, proceeds in a manner as shown in Fig. 7.15. As shown $h\nu/c$ and $h\nu'/c$ represent the momentum of the incident photon, before and after the scattering while **p** represents the momentum of the scattered electron. ϕ and θ are the angles of scattering of photon and electron respectively.

Conservation of energy requires $$h\nu + mc^2 = h\nu' + \frac{mc^2}{\left(1 - \frac{\upsilon^2}{c^2}\right)^{1/2}} \quad ...(7.12)$$

Conservation of momenta along forward direction and transverse direction lead to

$$h\frac{\nu}{c} = \frac{m\upsilon}{\sqrt{\left(1 - \frac{\upsilon^2}{c^2}\right)}} \cos\theta + h\frac{\nu'}{c}\cos\phi \quad ...(7.13)$$

$$0 = -\frac{m\upsilon}{\sqrt{\left(1 - \frac{\upsilon^2}{c^2}\right)}} \sin\theta + h\frac{\nu'}{c}\sin\phi \quad ...(7.14)$$

From eqns. (7.13) and (7.14) one gets

$$\left(h\frac{\nu}{c} - h\frac{\nu'}{c}\cos\phi\right)^2 + \left(h\frac{\nu'}{c}\sin\phi\right)^2 = \frac{m^2\upsilon^2}{\left(1 - \frac{\upsilon^2}{c^2}\right)}$$

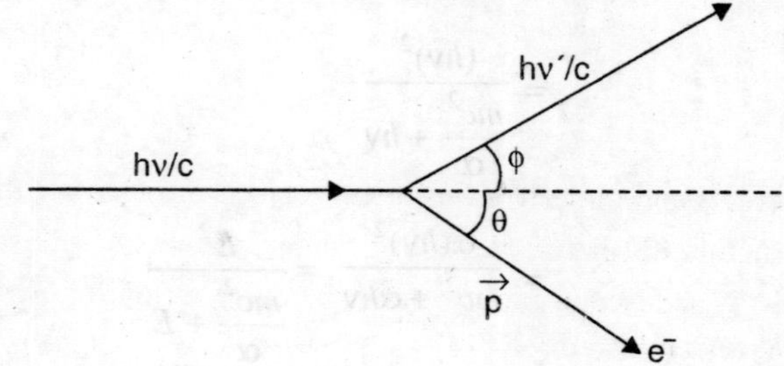

FIGURE 7.15 **Compton scattering of photon by electron. ϕ denotes the direction of scattered photon of energy $h\nu'$ and θ the direction of recoil electron.**

$$\left(h\frac{\nu}{c}\right)^2 + \left(h\frac{\nu'}{c}\right)^2 - 2h^2\frac{\nu\nu'}{c^2}\cos\phi = \frac{m^2\upsilon^2}{\left(1-\frac{\upsilon^2}{c^2}\right)} \quad ...(7.15)$$

From eq. (7.12) we get,

$$\left[h\frac{(\nu-\nu')}{c} + mc\right]^2 = \frac{m^2c^2}{\left(1-\frac{\upsilon^2}{c^2}\right)} \quad ...(7.16)$$

Subtract eq. (7.15) from eq. (7.16)

$$2h^2\,\nu\frac{\nu'}{c^2}(1-\cos\phi) = 2h(\nu-\nu')m$$

or

$$\frac{h}{mc^2}(1-\cos\phi) = \frac{1}{\nu'} - \frac{1}{\nu} = \frac{1}{c}(\lambda' - \lambda)$$

$$\frac{h}{mc}(1-\cos\phi) = (\lambda' - \lambda) \quad ...(7.17)$$

Hence,

$$\frac{2h}{(mc)}\sin^2\left(\frac{\phi}{2}\right) = \Delta\lambda \quad ...(7.18)$$

$$\Delta\lambda = 2 \times \frac{h}{(mc)} \quad (\phi = 180°)$$

$$= 0 \ (\phi = 0°),$$

Thus as ϕ varies from 0 to 180°, the extent of change in wavelength varies from 0 to twice the compton wavelength ($= h/mc$).

It is obvious that

Recoil energy of electron = Energy lost by photon

$$T_{re} = h\nu - h\nu' = h(\nu - \nu')$$

Now,

$$1/\nu' - 1/\nu = (h/mc^2)(1 - \cos\phi) = (h/mc^2)\,.\,\alpha, \text{ where } \alpha = 1 - \cos\phi$$

So,

$$T_{re} = h\left[\nu - \frac{1}{\frac{1}{\nu} + \frac{h\alpha}{mc^2}}\right] = \left[h\nu - \frac{h\nu mc^2}{mc^2 + \alpha h\nu}\right]$$

$$= \frac{(h\nu)^2}{\frac{mc^2}{\alpha} + h\nu}$$

$$= \frac{\alpha(h\nu)^2}{mc^2 + \alpha h\nu} = \frac{E^2}{\frac{mc^2}{\alpha} + E}$$

where $E = h\nu$, the energy of the incident photon.

= Maximum when $\alpha = 2$ *i.e.* $\phi = 180°$, a head on collision.

Hence, $$T_{re}^{\ max} = \frac{E^2}{\frac{mc^2}{2} + E}$$

In case we measure $T_{re}^{\ max}$ experimentally, we can obtain E, the energy of incident photon, by solving the above quadratic equation in E. In fact, we get

$$E^2 - ET_{re}^{\ max} - mc^2T_{re}^{\ max}/2 = 0$$

which gives

$$E = \left[T_{re}^{\ max} \pm \sqrt{\{(T_{re}^{\ max})^2 + (2mc^2T_{re}^{\ max})\}}\right]\Big/2$$

$$= \left(T_{re}^{\ max}/2\right)\left[1 \pm \sqrt{\{1 + (2mc^2)/(T_{re}^{\ max})\}}\right]$$

$$= \left(T_{re}^{\ max}/2\right)\left[1 + \sqrt{\{1 + (2mc^2)/(T_{re}^{\ max})\}}\right]$$

$\because$ With –ve sign E becomes –ve.

Also note that for $\phi = 90°$, the change in wavelength of the incident light or photon is given by $\Delta\lambda = \lambda' - \lambda = h/(mc)$

This is called the Compton wavelength which comes out as $6.62 \times 10^{-34}/(9 \times 10^{-31} \times 3 \times 10^8) = (2.62/27) \times 10^{-11}$m = 0.0245 Å.

SOLVED NUMERICAL PROBLEMS

Problem 1. *Calculate the cut-off wavelength when the potential difference applied to the X-ray tube is 100 kV.*

Solution. We have the relation $\lambda_{min} = \frac{12400}{10^5} = 0.124$ Å ; $V = 10^5$ Volts.

Problem 2. *Find the quantum energy of X-ray of frequency 10^{28} cps. Given $h = 6.62 \times 10^{-27}$ erg sec.*

Solution. Energy is given by $E = h\nu = 6.62 \times 10^{-27} \times 10^{28} = 66.2$ ergs.

Problem 3. *Deduce the wavelength of K_α line for an atom of $Z = 92$, using Mosley's law. Given Rydberg constant $R = 1.1 \times 10^7$ m^{-1}.* (Jan. 1999, K.U.K.)

Solution. The spectral line K_α is emitted when an electron jumps from an orbit of principal quantum number 2 to 1. Its wavelength λ is given by

$$\frac{1}{\lambda} = \frac{\nu}{c} = R(Z-a)^2\left[\frac{1}{1^2} - \frac{1}{2^2}\right]$$

but for K series $a = 1$, therefore, $$\frac{1}{\lambda} = \tfrac{3}{4}R(Z-1)^2$$

$$\therefore \quad \frac{1}{\lambda} = \tfrac{3}{4}(1.1 \times 10^7)(92-1)^2 = 6.83 \times 10^{-10}\ \text{m}^{-1}$$

or $$\lambda = \frac{1}{6.83 \times 10^{10}}\ \text{m} = 0.146 \times 10^{-10}\ \text{m} = 0.146\ \text{Å}$$

Problem 4. *Calculate the glancing angle on the cube face of a rock salt (d = 2.814 Å) corresponding to second order diffraction maxima for X-rays of wavelength 0.710 Å.*

Solution. Here $d = 2.814$ Å, $\lambda = 0.710$ Å, $n = 2$

From Bragg's law $2d \sin \theta = n\lambda$ or $\sin \theta = \dfrac{n\lambda}{2d} = \dfrac{2 \times 0.71}{2 \times 2.814} = 0.252$

or $$\theta = 14°30'.$$

Problem 5. *Calculate the wavelength of X-ray beam incident at 12° for 1st order reflection from calcite crystal if grating constant for crystal is 3.035 Å.*

Solution. Given $d = 3.035$ Å, $n = 1$, $\theta = 12°$

As from Bragg's law $2d \sin \theta = n\lambda$

or $$\lambda = \frac{2d \sin \theta}{n} = \frac{2 \times 3.035 \times \sin 12°}{1} = 1.26 \text{ Å}.$$

Problem 6. *Calculate the longest wavelength that can be analyzed by a rock salt crystal of spacing d = 2.82 Å in the first order.* (Dec. 2000, K.U.K.)

Solution. For longest wavelength, $\sin \theta = 1$.

Given $d = 2.82$ Å, $n = 1$

From Bragg's law $2d \sin \theta = n\lambda$ or $2d = n\lambda$

or $$\lambda = \frac{2d}{n} = \frac{2 \times 2.82}{1} = 5.64 \text{ Å}.$$

Problem 7. *A photon of energy 1.02 MeV is scattered through 90° by a free electron. Calculate the energy of the photon and electron after interaction.*

Solution. We have the relation

$$\Delta\lambda = \lambda' - \lambda = \frac{h}{mc}(1 - \cos \phi) = \frac{6.6 \times 10^{-34}}{9.1 \times 10^{-31} \times 3 \times 10^{8}}(1 - 0)$$

$$\Delta\lambda = 2.42 \times 10^{-12} \text{ m}$$

Change in frequency of the photon

$$\Delta\nu = \frac{c}{\Delta\lambda} = \frac{3 \times 10^{8}}{2.42 \times 10^{-12}}$$

Change in energy of photon $$\Delta E = h\Delta\nu = \frac{(6.6 \times 10^{-34} \times 3 \times 10^{8})}{(2.42 \times 10^{-12})} = 0.51 \text{ MeV}.$$

This energy is transferred to the free electron which in turn will be the K.E. of the recoil electron.

$\therefore$ Energy of photon after interaction = 1.02 – 0.51 = 0.51 MeV.

Problem 8. *Calculate the maximum energy of recoil of electron in a Compton scattering with a photon of energy E.*

Solution. The recoil energy of electron is maximum when $\phi = 180°$, ahead on collision as photon loses maximum energy then. In other words

$$\Delta\lambda = \frac{2h}{mc} \quad \text{or} \quad \lambda' - \lambda = \frac{2h}{mc}$$

$$\frac{1}{\nu'} - \frac{1}{\nu} = \frac{2h}{mc^2}$$

i.e.
$$\frac{1}{\nu'} = \frac{1}{\nu} + \frac{2h}{mc^2}$$

Now maximum recoil energy of electron = Maximum loss of energy by photon = $h\nu - h\nu'$

$$= h\nu - \frac{h}{\left[\frac{1}{\nu} + \frac{2h}{mc^2}\right]} \text{ obtained by putting the value of } \nu'$$

from above
$$= h\nu - \frac{h\nu mc^2}{[mc^2 + 2h\nu]}$$

$$= E - \frac{Emc^2}{[mc^2 + 2E]} \text{ as } E = h\nu$$

$$= \frac{E^2}{\left[E + \frac{mc^2}{2}\right]}$$

[**Note:** Similarly it can be shown that recoil energy of the electron is equal to $E^2/[E + mc^2/\alpha]$ where α is equal to $(1 - \cos\phi)$ for a general case.]

EXERCISE

1. What are X-rays ? How do we get the continuous and sharp line spectrum in X-rays ? (*Dec. 2000, K.U.K.*)
2. Explain Bragg's law for X-ray diffraction. (*Dec. 2000, K.U.K.*)
3. What is X-ray diffraction ? Deduce Bragg's law for the diffraction of X-rays and discuss Bragg's condition for X-ray diffraction. (*Jan. 1999, K.U.K.*)
4. State and explain Mosley's law. (*June, 2000, K.U.K.*)
5. Derive Bragg's law for X-ray diffraction. How it can be used in the determination of crystal structure ? (*Dec. 1999, K.U.K.*)
6. What is Compton effect ? Derive the expression for the change in wavelength of the scattered ray and the energy of the recoil electron. (*June 2000, Jan. 2003, K.U.K.*)
7. Write notes on the following :
   (*a*) Excitation and Ionization potential
   (*b*) Absorption of X-rays
   (*c*) Origin of characteristics X-rays. (*May. 1998, Dec. 2000, K.U.K.*)
8. What is Compton scattering ? Derive expression for the frequency of scattered photon for different scattering angle. How are scattering and recoil angles related to each other ? (*June 2001, 2002, K.U.K.*)
9. Calculate the minimum applied potential required to produce X-rays of 1 Å wavelength. [**Ans.** 12.4 kV]
10. An X-ray tube operated at 40 kV emits a continuous X-ray spectrum with a short wavelength limit λ_{min} = 0.310 Å. Calculate the Planck's constant. [**Ans.** 6.61×10^{-34} J-S]

11. Monochromatic X-rays of wavelength 1.4 Å are incident on a crystal having 1.5 Å as interatomic spacing. Find the various orders in which the diffraction takes place. **[Ans.** $n \leq 2$**]**
12. Electron bombarding the anode of a coolidge tube, produce X-rays of wavelength 1 Å. Find the energy of each electron at the moment of impact, given $h = 6.62 \times 10^{-34}$ J-S. **[Ans.** 12.41 keV**]**
13. In Compton experiment, the wavelength of X-ray radiation scattered at an angle of 45° is 0.022 Å. Calculate the wavelength of the incident rays. **[Ans.** 0.015 Å**]**
14. In Compton scattering show that the angle of scattering, θ, of electron is given by

$$\cot\theta = \frac{(1-\cos\phi)\left(1+\frac{h\nu}{mc^2}\right)}{\sin\phi}$$

15. Discuss X-ray powder method for investigating crystalline structure. *(K.U. June, 2007)*
16. What do you understand by X-ray spectrum ? Discuss different types of X-ray spectra. *(K.U. June, 2007)*

Chapter **8**

SPECIAL THEORY OF RELATIVITY

8.1 MICHELSON-MORLEY EXPERIMENT

Before the formulation of special theory of relativity by Albert Einstein it was being considered that velocity of light like the motion of other material bodies should depend on the relative motion between the observer and source of light. Also, it was thought that the velocity of light might depend on the motion of the transmitting medium. However, experiments designed to investigate these expectations failed to confirm them. It was then accepted that velocity of light is a constant quantity *i.e.* independent of

(*i*) velocity of source

(*ii*) state of motion of trasmitting medium

(*iii*) the velocity of the observer or the frame of observations.

Michelson and Morley experiment is also one of those experiments and require discussion in some detail.

This was designed to investigate whether the velocity of earth and hence the frame of reference of observation had any effect on the velocity of light. This experiment is based on the phenomenon of interference of monochromatic light. The various parts of the experiment as well as the paths of the light leading to interference fringes have been shown in Fig. 8.1.

It is quite obvious from the diagram that the two rays 1 and 2 travelled at right angles before interfering with each other. So, if these have equal distances say l, the time taken by them before overlapping each other to cause circular fringes would depend on the motion of the apparatus. For example, if the earth moves towards M_1 mirror with a speed υ, the time to travel for the ray 1 would be

$$t_1 = \frac{l}{(c-\upsilon)} + \frac{l}{(c+\upsilon)} = \frac{2l}{c\left(1-\frac{\upsilon^2}{c^2}\right)}$$

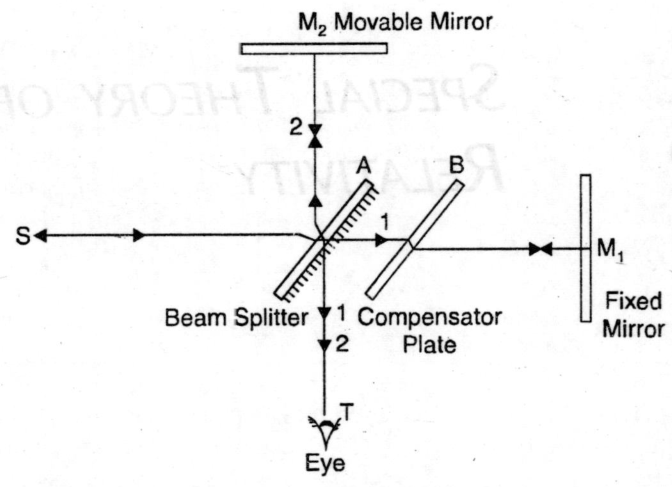

FIGURE 8.1 **Block diagram of Michelson-Morley experiment. M_1, M_2 are plane mirrors, while S and T denote positions of source and telescope respectively.**

and for the ray 2 which went towards the mirror M_2 would be* $t_2 = \dfrac{2l}{c\left(1-\dfrac{\upsilon^2}{c^2}\right)^{\frac{1}{2}}}$

So,
$$\Delta t = t_1 - t_2 = \frac{2l}{c}\left[\frac{1}{1-\dfrac{\upsilon^2}{c^2}} - \frac{1}{\left(1-\dfrac{\upsilon^2}{c^2}\right)^{1/2}}\right] = \frac{l.\upsilon^2}{c^3}$$

(obtained by assuming $\upsilon/c << 1$)

This time difference will cause a shift in the number of fringes as equal to $\nu \,.\, \Delta t = \dfrac{\nu}{c}\,.\, l \,.\, \dfrac{\upsilon^2}{c^2}$ where ν = frequency of the monochromatic light. If now the apparatus is rotated by 90° and above observations are repeated in which mirrors M_1 and M_2 exchange their roles, the total shift will be twice the above result

i.e.
$$\frac{2\nu}{c}\,.\, l \,.\, \frac{\upsilon^2}{c^2} = 4 \times 10^4 \times 10^3 \times 10^{-8} = 0.4$$

---

* From this figure it is obvious that $t_2 = 2t'$ such that

$c^2t'^2 = \upsilon^2t'^2 + l^2$ which gives $t' = \dfrac{l}{c\left(1-\dfrac{\upsilon^2}{c^2}\right)^{1/2}}$

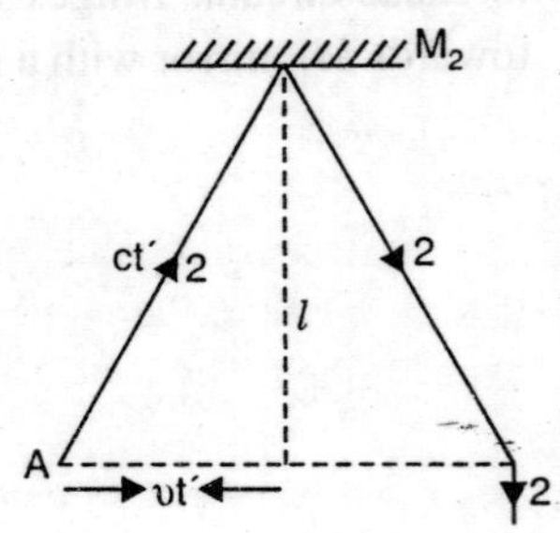

when $$\left[\frac{v}{c} = 2 \times 10^4 \text{ cm}^{-1}, l = 10^3 \text{cm } \& \frac{\upsilon}{c} = 10^{-4}\right]$$

This much fringe shift is well within experimental measurements. However, nothing of this sort was observed by Michelson and Morley leading to the conclusions that motion of the apparatus relative to the light pulse did not change the velocity of light. Hence velocity of light is same in all inertial frames whatsoever. Also, the concept of ether as the all prevailing medium has to be discarded after the negative results of Michelson-Morley experiment.

8.2 LORENTZ TRANSFORMATIONS

The transformation equation for a point in space as observed in two different inertial frames having uniform relative motion are called Lorentz transformations. We suppose that frame S' is moving with a velocity υ w.r.t. frame of reference S along their common X-axis (see Fig. 8.2). We further assume that at $t = 0$ these frames were coinciding. As it has been experimentally established that the velocity of light is same in every inertial frame, the position of the light wave (spherical) in two frames S and S' will satisfy

$$x^2 + y^2 + z^2 - c^2t^2 = x'^2 + y'^2 + z'^2 - c^2t'^2 = 0 \qquad \text{...(8.1)}$$

where (x, y, z, t) and (x', y', z', t') denote the space and time coordinates of some point P in S and S' respectively. For convenience let us denote

$$x = x_1, y = x_2, z = x_3 \text{ and } ict = x_4 \qquad (i = \sqrt{-1})$$

As a result, eq. (8.1) may be rewritten as

$$\Sigma_i\, x_i^{\,2} = \Sigma_k\, x_k'^{\,2} \;;\; i, k = 1, 2, 3, 4 \qquad \text{...(8.2)}$$

As the motion of S' is along X-axis,

$$x_2' = x_2 \quad \text{and} \quad x_3' = x_3$$

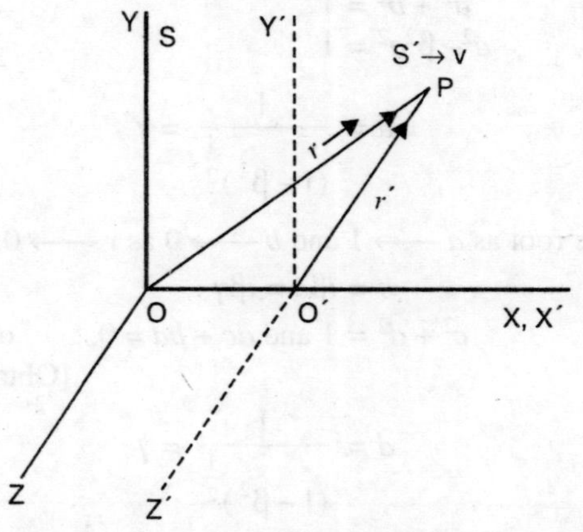

FIGURE 8.2 **Two inertial frames S and S', with S' moving w.r.t. S along common X-axis with the velocity v, $OO' = vt$.**

Because of same reasons the transformation equations of x_1' and x_4' will not involve y- and z-coordinates. If they involve, the choice of origin in y-z plane will affect the results of any event, which is against the expectations. Thus, we assume

$$x_1' = ax_1 + bx_4 \qquad \text{...(8.3)}$$

and $$x_4' = cx_1 + dx_4$$

where the coefficients a, b, c and d are to be determined on the basis of eq. (8.2). Thus, eq. (8.2) now reads as

$$x_1'^2 + x_4'^2 = x_1^2 + x_4^2 \qquad (\because\ x_2' = x_2,\ x_3' = x_3) \qquad \text{...(8.4)}$$

By using eq. (8.3) we obtain by comparing the coefficients of x_1^2 and x_4^2 on both sides that

$$a^2 + c^2 = b^2 + d^2 = 1 \qquad \text{and} \qquad ab + cd = 0 \qquad \text{...(8.5)}$$

Further, using the results of eq. (8.5) we also obtain from eqs. (8.3)

$$x_1 = ax_1' + cx_4' \qquad \text{...(8.6)}$$

and

$$x_4 = bx_1' + dx_4'$$

These are inverse transformations to eqs. (8.3)

In matrix form, eqns. (8.6) and (8.3) become

$$\begin{pmatrix} x_1 \\ x_4 \end{pmatrix} = \begin{pmatrix} a & c \\ b & d \end{pmatrix} \begin{pmatrix} x_1' \\ x_4' \end{pmatrix} \qquad \text{...(I)}$$

and

$$\begin{pmatrix} x_1' \\ x_4' \end{pmatrix} = \begin{pmatrix} a & b \\ c & d \end{pmatrix} \begin{pmatrix} x_1 \\ x_4 \end{pmatrix} \qquad \text{...(D)}$$

Now, if we use eqn. (8.6) in eqn. (8.4) and equate the coefficients of $x_1'^2$ and $x_4'^2$ we will get

$$a^2 + b^2 = c^2 + d^2 = 1 \qquad \text{and} \qquad ac + bd = 0 \qquad \text{...(8.7)}$$

Next, let us study the motion of the origin of S'. Obviously $x_1' = 0$ and $x_1 = vt$ and hence from eqs. (8.3) $0 = avt + bict$, c being the velocity of light

or

$$b = i\left(\frac{v}{c}\right) a = i\beta a \qquad \text{...(8.8)}$$

where

$$\beta = \frac{v}{c}$$

From eqns. (8.7) and (8.8) we now get

$$a^2 + b^2 = 1$$

$$a^2 - \beta^2 a^2 = 1$$

or

$$a = \frac{1}{(1-\beta^2)^{\frac{1}{2}}} = \gamma \qquad \text{...(8.9)}$$

We took positive square root as $a \longrightarrow 1$ and $b \longrightarrow 0$ as $v \longrightarrow 0$.

Also

$$b = i\beta a = i\beta\gamma$$

Again from

$$c^2 + d^2 = 1 \text{ and } ac + bd = 0, \qquad \text{or} \qquad c = -i\beta d$$

[Obtained using the values of a and b]

We obtain

$$d = \frac{1}{(1-\beta^2)^{\frac{1}{2}}} = \gamma$$

thus,

$$c = -i\beta\gamma \qquad \text{and} \qquad d = \gamma$$

or transformation matrix $\begin{pmatrix} \gamma & i\beta\gamma \\ -i\beta\gamma & \gamma \end{pmatrix}$

i.e.

$$\begin{pmatrix} x_1' \\ x_4' \end{pmatrix} = \begin{pmatrix} \gamma & i\beta\gamma \\ -i\beta\gamma & \gamma \end{pmatrix} \begin{pmatrix} x_1 \\ x_4 \end{pmatrix}$$

or

$$x_1' = \gamma(x_1 + i\beta x_4) \qquad \text{and} \qquad x_4' = \gamma(-i\beta x_1 + x_4)$$

Now by introducing original coordinates and the values of β and γ, we will have

$$x' = \frac{x - \upsilon t}{\sqrt{1 - \upsilon^2/c^2}}; \quad y' = y; \quad z' = z \quad \text{and} \quad t' = \frac{t - \frac{\upsilon}{c^2}x}{\sqrt{1 - \upsilon^2/c^2}}$$

These are called Lorentz transformations.

Proper Time

The time interval shown by a clock when at rest in a given inertial frame is called proper interval of time, $d\tau$. For example, from $dx_1^2 - c^2dt^2 = dx_1'^2 - c^2dt'^2 = -c^2d\tau^2$ as in S' where clock is at rest $dx_1' = 0$ and $dt' = d\tau$.

That is
$$d\tau = dt.\sqrt{\left(1 - \frac{\upsilon^2}{c^2}\right)}$$

$$\because \quad \frac{dx_1}{dt} = \upsilon$$

8.3 VELOCITY FOUR-VECTOR

$$u_\mu = \frac{dx_\mu}{d\tau} = \frac{dx_\mu / dt}{\sqrt{1 - \beta^2}} \quad \text{where } \mu = 1, 2, 3, 4$$

i.e.
$$u_1 = \frac{\upsilon_1}{\sqrt{(1 - \beta^2)}}$$

$$u_2 = \frac{\upsilon_2}{\sqrt{(1 - \beta^2)}}$$

$$u_3 = \frac{\upsilon_3}{\sqrt{(1 - \beta^2)}} \quad \text{and} \quad u_4 = \frac{ic}{\sqrt{(1 - \beta^2)}} \qquad \text{...(8.10)}$$

These are the components of velocity-four vector, **u**.

Lorentz Transformations of Velocity

Now, $u_{x'} = \dfrac{dx'}{dt'}$ and υ is the relative velocity of S' w.r.t. S.

As $x' = (x - \upsilon t)\gamma$, $y' = y$, $z' = z$ and $t' = [t - (\upsilon x/c^2)]\gamma$,

$$u_{x'} = \frac{(u_x - \upsilon)}{\left(1 - \frac{u_x \upsilon}{c^2}\right)}, \quad u_{y'} = \frac{u_y}{\left(1 - \frac{u_x \upsilon}{c^2}\right)}\gamma$$

$$u_{z'} = \frac{u_z}{(1 - u_x\,\upsilon / c^2)}\gamma \qquad \text{...(8.11)}$$

8.4 ADDITION OF VELOCITIES

The law of addition of velocities in special theory of relativity is quite different. However, it reduces to the same as in non-relativistic mechanics when velocities are small in comparison with that of light.

Let us assume three inertial frames S. S' and S'' such that S' moves relative to S with uniform velocity $\mathbf{v}$ and S'' with $\mathbf{w}$ relative to S'. Both $\mathbf{v}$ and $\mathbf{w}$ are parallel to common x-axis. Then by Lorentz transformations we have

$$x' = \frac{(x - vt)}{\sqrt{\left(1 - \frac{v^2}{c^2}\right)}} \qquad t' = \frac{t - \frac{vx}{c^2}}{\sqrt{1 - \frac{v^2}{c^2}}}$$

and

$$x'' = \frac{x' - wt'}{\sqrt{1 - \frac{w^2}{c^2}}} \qquad t'' = \frac{\left(t' - \frac{wx'}{c^2}\right)}{\sqrt{1 - \frac{w^2}{c^2}}}$$

Eliminating x' and t' we obtain from above

$$x'' = \frac{\left[x\left(1 + \frac{vw}{c^2}\right) - t(v + w)\right]}{\sqrt{\left(1 - \frac{w^2}{c^2}\right)\left(1 - \frac{v^2}{c^2}\right)}}$$

and

$$t'' = \frac{\left[t\left(1 + \frac{vw}{c^2}\right) - x\left(\frac{v + w}{c^2}\right)\right]}{\sqrt{\left(1 - \frac{w^2}{c^2}\right)}\sqrt{\left(1 - \frac{v^2}{c^2}\right)}}$$

By rearranging the terms appropriately, we get

$$x'' = \frac{(x - ut)}{\sqrt{\left(1 - \frac{u^2}{c^2}\right)}} \quad \text{and} \quad t'' = \frac{\left(t - \frac{u}{c^2}x\right)}{\sqrt{\left(1 - \frac{u^2}{c^2}\right)}} \qquad \text{...(8.12)}$$

where

$$u = \frac{(v + w)}{\left(1 + \frac{vw}{c^2}\right)} \qquad \text{...(8.13)}$$

or

$$\beta = \frac{(\beta_1 + \beta_2)}{(1 + \beta_1\beta_2)} \text{ with } \beta = \frac{u}{c},\ \beta_1 = \frac{v}{c} \quad \text{and} \quad \beta_2 = \frac{w}{c}$$

Thus, these results, eqs. (8.12) and (8.13), show that S'' is moving with a velocity $\mathbf{u}$ w.r.t. S, along common x-axis. This $\mathbf{u}$ is not equal to $\mathbf{v} + \mathbf{w}$ as is the case in non-relativistic mechanics. However, when v/c

and w/c are small $\mathbf{u} \longrightarrow \mathbf{v} + \mathbf{w}$. The other point which is to be noted is that β always remains less than unity so long as β_1 and β_2 are less than one.

8.5 VARIATION OF MASS WITH VELOCITY ; MASS-ENERGY EQUIVALENCE

Variation of Mass with Velocity

Let us consider two observers, O' sitting in the railway train having velocity υ along x-axis and O which is at rest in the ground. They are given two basketballs perfectly elastic and of equal masses. They are asked to carry out head on collision of the basketballs in their respective co-ordinate systems S' and S by throwing these balls perpendicular to the direction of the motion of the train say along common Y-axis. Obviously, system S' attached with observer O' will be moving with velocity v relative to S where observer O is at rest. The head on collisions between two balls will therefore, appear in S in a way as shown in Fig. 8.3.

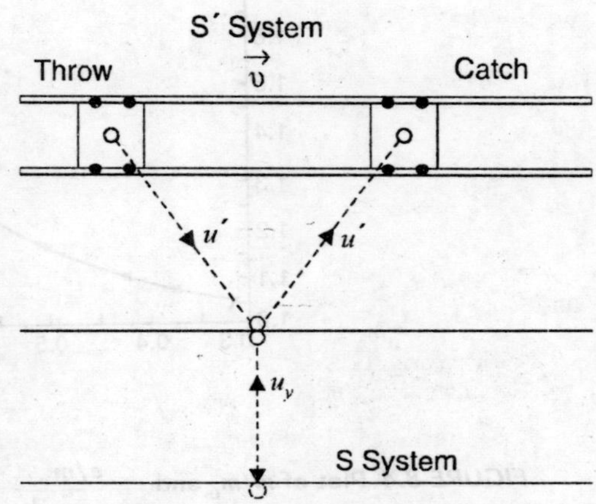

FIGURE 8.3 **The collision between two basket-balls as appears to *O* at rest in *S*.**

Let us now apply conservation law of linear momenta in the frame S assuming the velocities of the balls equal to u_y and u_y' in their respective frames of reference and these are equal. However, u_y' in S will have a value equal to $\dfrac{u_y'}{\left[\left(1 + u_x' \cdot \dfrac{\upsilon}{c^2}\right) \cdot \gamma\right]}$ as per Lorentz transformations. Now, as $u_x' = 0$ and $u_y' = u_y$, this will be finally given by u_y/γ as the velocity of the ball (of O') along y-axis as observed in the frame of reference S. Hence, by conservation law of momenta along the y-axis one would get

$$2mu_y = 2m' \frac{u_y}{\gamma}$$

where m' is the mass of the basketball as observed in S. Thus, we get

$$m' = m.\gamma = \frac{m}{\sqrt{\left(1 - \dfrac{\upsilon^2}{c^2}\right)}}$$

In other words the mass will increase as its velocity υ increases and in the non-relativistic limits there will be no change in the mass with velocity.

Experimental Verification

The variation of mass with velocity has been verified by measuring m as well as υ of β-rays (electrons) emitted by a β-source. Such a source emits electrons at various velocities. As a result, one can have m of electrons having variety of velocities and hence can plot m/m_0 and $\dfrac{e/m}{\left(1 - \dfrac{\upsilon^2}{c^2}\right)^{1/2}}$ as a function υ/c. Here,

m_0, e and c are known. One such plot has been shown in Fig. 8.4. It is clear that theoretical curves fit very well the experimental values shown as dots.

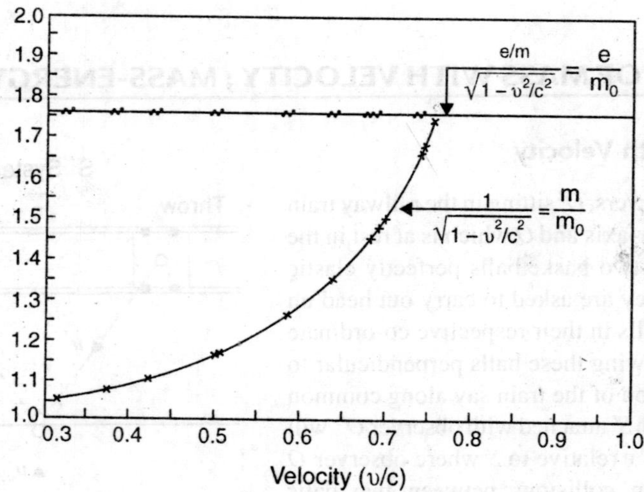

FIGURE 8.4 Plot of m/m_0 and $\dfrac{e/m}{\left(1-\dfrac{v^2}{c^2}\right)^{1/2}}$ vs v/c ; e/m has been expressed in the unit 10^7 emu/gm.

Dots represent experimental value while full line curve the theoretical values.

Table of variation of mass with velocity

$$\frac{m}{m_0} = \frac{1}{\sqrt{1-\dfrac{v^2}{c^2}}}$$

v/c	m/m_0	v/c	m/m_0
0	1.000	0.8	1.667
0.1	1.005	0.9	2.294
0.3	1.048	0.95	3.202
0.4	1.091	0.98	5.025
0.5	1.155	0.99	7.089
0.7	1.40	0.999	22.366

Mass Energy Equivalence

It is well known that $\dfrac{dE}{dt} = \vec{F} \cdot \vec{v}$ where E, F, v are energy, external force and velocity respectively.

Since $\mathbf{F} = \frac{d}{dt}\mathbf{p}$, or $\mathbf{F} = \frac{d}{dt}\left(\frac{m_0 v}{\left(1-\frac{v^2}{c^2}\right)^{1/2}}\right)$ $\left(\because \mathbf{p} = \frac{m_0 v}{\sqrt{1-v^2/c^2}}\right)$

where m_0 is the rest mass of the particle,

$$\mathbf{v} \cdot \mathbf{F} = \mathbf{v}\frac{d}{dt}\left(\frac{m_0 v}{\left(1-\frac{v^2}{c^2}\right)^{1/2}}\right) = \frac{m_0 v \cdot \frac{dv}{dt}}{\sqrt{1-\frac{v^2}{c^2}}} + \frac{m_0 v^2 \frac{v}{c^2} \cdot \frac{dv}{dt}}{\left(1-\frac{v^2}{c^2}\right)^{3/2}} = \frac{m_0 v \cdot \frac{dv}{dt}}{\left(1-\frac{v^2}{c^2}\right)^{3/2}}$$

Also $m = \frac{m_0}{\sqrt{\left(1-\frac{v^2}{c^2}\right)}}$ or $\frac{dm}{dt} = m_0 \frac{\left(v \cdot \frac{dv}{dt}\right)/c^2}{\left(1-\frac{v^2}{c^2}\right)^{3/2}}$

Hence, $\frac{dE}{dt} = \mathbf{v} \cdot \mathbf{F} = c^2 \frac{dm}{dt}$

i.e. $\frac{dE}{dt} = c^2 \frac{dm}{dt}$ (**v.F** is always equal to the rate of change of energy)

which leads to $\Delta E = c^2 \Delta m$

This is so-called mass energy equivalence. In other words, disappearance of mass will lead to the appearance of energy as per the above relation. On integration

$$\int_{v=0}^{v} dE = C^2 \int_{v=0}^{v} dm$$

$$E_v - E_0 = E_k = (m - m_0)c^2 \qquad \text{where } E_k \text{ is the K.E.}$$

Further, it can be proved that

$$E^2 = m_0^2 c^4 + p^2 c^2$$

In fact, $E = mc^2 = \frac{m_0 c^2}{\sqrt{1-\frac{v^2}{c^2}}}$

On squaring $E^2 = \frac{m_0^2 c^4}{\left(1-\frac{v^2}{c^2}\right)} = \frac{m_0^2 c^4}{1-\frac{v^2}{c^2}} \times \left(1 - \frac{v^2}{c^2} + \frac{v^2}{c^2}\right) = m_0^2 c^4 + \frac{m_0^2 v^2 c^2}{\left(1-\frac{v^2}{c^2}\right)}$

But $p = mv = \frac{m_0 v}{\sqrt{1-\frac{v^2}{c^2}}}$

Therefore, $E^2 = m_0^2 c^4 + p^2 c^2$

For mass less pacticles, $m_0 = 0$

$$E = pc \quad \text{or} \quad p = E/c$$

In case of photons, $E = h\nu$, ν being the frequency of light. The momentum of the photons would be

$$p = \frac{h\nu}{c}.$$

Note. The special theory of relativity, therefore involves, calculation in four dimensional space-time coordinates. For example, position four coordinates (x, y, z, ict), velocity four components $(\gamma V_x, \gamma V_y, \gamma V_z, i\gamma c)$ and the four momentum components $(\gamma mv_x, \gamma mv_y, \gamma mv_z, i\gamma mc)$ etc. Here, $\gamma = 1/\sqrt{(1-(v^2/c^2))}$, called the relativitics factor. The fourth dimension is imaginary. The other point of difference is that space length, linear momentum, mass are no longer invariant under Lorentz transformations. For instance, space length, density and mass change with the relative motion of the observer. Instead, it is the space-time length (four dimensional) and the momentum-energy which are invariants.

SOLVED NUMERICAL PROBLEMS

Problem 1. *Calculate the equivalent energy of 1amu of mass. Given 1amu = 1.66×10^{-27} kg*

(Jan. 2001, K.U.K.)

Solution. $E = mc^2 = 1.66 \times 10^{-27} \times (3 \times 10^8)^2 \text{ J} = 9 \times 1.66 \times 10^{-11} \text{ J} = (9 \times 1.66/1.6) \times 10^{-11}/10^{-13} \text{ MeV}$ $= 934$ MeV (1 MeV $= 1.6 \times 10^{-13}$ J)

Problem 2. *A photon of $\lambda = 3310$ Å falls on a photo cathode and an electron of energy 3×10^{-19} J is ejected. If the wavelength of the incident photon is changed to 5000 Å, the energy of the ejected electron is 0.79×10^{-19} J. Calculate the value of h, and the threshold wavelength of photo cathode.*

(Jan. 2000, K.U.K.)

Solution. By Einstein equation we have $h\nu = \varphi_0 + T$; φ_0 = work function of the photo cathode and T = K.E. of the ejected electron.

Then for two photon energies $\quad h\nu_1 = \varphi_0 + T_1$

and $\quad h\nu_2 = \varphi_0 + T_2$

or $\quad h(\nu_1 - \nu_2) = T_1 - T_2 = (3 - 0.79)\, 10^{-19} = 2.21 \times 10^{-19}.$

Hence,

$$h = \frac{2.21}{(\nu_1 - \nu_2)} \times 10^{-19} \quad \text{...(1)}$$

Now,

$$\nu_1 = \frac{c}{\lambda_1} \quad \text{and} \quad \nu_2 = \frac{c}{\lambda_2}$$

$$(\nu_1 - \nu_2) = c\left(\frac{1}{\lambda_1} - \frac{1}{\lambda_2}\right) = 3 \times 10^8 \left(\frac{1}{33.1} - \frac{1}{50}\right) \times 10^8 = 3 \times 10^{14}$$

Substituting this in eq. (1), we get

$$h = \left(\frac{2.21}{3}\right) \times 10^{-33} = 7.37 \times 10^{-34} \text{ J sec.}$$

Problem 3. *Calculate the mass equivalent to an energy 200 MeV released in a fission.*

(1998, K.U.K.)

Solution. By Einstein equation $\quad c^2 \Delta m = E$

$$\therefore \quad \Delta m = \frac{E}{c^2}\ ; \ E = 200 \times 1.6 \times 10^{-6} \text{ ergs}$$

$$\therefore \quad \Delta m = \frac{200 \times 1.6 \times 10^{-6}}{9 \times 10^{20}} \text{ gm} = 0.36 \times 10^{-24} \text{ gm} = \frac{0.36}{1.66} \text{ amu} = 0.22 \text{ amu.}$$

Problem 4. *Calculate the change in mass and speed of the particle if its total energy is twice the rest energy.* (Jan. 1999, K.U.K.)

Solution. According to the variation of mass with velocity

$$m = \frac{m_0}{\sqrt{\left(1 - \frac{\upsilon^2}{c^2}\right)}} \quad i.e. \quad mc^2 = \frac{m_0 c^2}{\sqrt{\left(1 - \frac{\upsilon^2}{c^2}\right)}}$$

As $\quad mc^2 = 2m_0c^2, \; 2 = \dfrac{1}{\sqrt{\left(1 - \frac{\upsilon^2}{c^2}\right)}} \quad$ or $\quad \dfrac{\upsilon}{c} = \dfrac{\sqrt{3}}{2} = 0.87$

$$\therefore \quad \upsilon = 0.87c \qquad \text{while} \quad \Delta m = 2m_0 - m_0 = m_0$$

Problem 5. *Calculate m/m_0, υ/c and total energy of 2.5 BeV protons. Given that rest energy of proton = 939 MeV.*

Solution.

Kinetic energy K.E. $\quad E_k = 2.5 \text{ BeV} = 2500 \text{ MeV}$

Rest energy of proton $\quad m_0c^2 = 939 \text{ MeV}$

So, total energy $\quad mc^2 = 2500 + 939 = 3439 \text{ MeV}$

Now, $\quad \dfrac{m}{m_0} = \dfrac{mc^2}{m_0c^2} = \dfrac{3439}{939} = 3.662$

Also, $\quad \dfrac{mc^2}{m_0c^2} = \dfrac{1}{\sqrt{1 - \frac{v^2}{c^2}}}$

which gives

$$\frac{v}{c} = \sqrt{1 - \left(\frac{m_0c^2}{mc^2}\right)^2} = \sqrt{1 - \left(\frac{939}{3439}\right)^2} = 0.962$$

EXERCISE

1. Show that when β_1 and β_2 are less than unity, β will also be less than unity where $\beta = \dfrac{(\beta_1 + \beta_2)}{(1 + \beta_1\beta_2)}$.

   [**Hint.** $(1 - \beta_1)(1 - \beta_2) = 1 + \beta_1\beta_2 - (\beta_1 + \beta_2) > 0$ or $1 + \beta_1\beta_2 > \beta_1 + \beta_2$]

2. Calculate m/m_0, υ/c and total energy of 400 MeV protons. Given that the rest energy of proton is 939 MeV. [**Ans.** 1.43, 0.71 and 1339 MeV]

3. Discuss mass energy equivalence relation.

4. Drive Lorentz transformation equations. *(May 2002, Jan. 2003 K.U.K.)*
5. Discuss Michelson-Morley experiment with a particular emphasis on the importance of it's negative result.
6. Describe how the variation of mass with velocity was experimentally verified.
7. Find the velocity at which mass becomes double the rest mass. $\left[\textbf{Ans. } \frac{v}{c} = 0.866\right]$
8. Prove that $E^2 = p^2c^2 + m_0^2c^4$ and for photons $p = \frac{E}{c}$.
9. Show that Lorentz Transformation matrix is

$$\begin{pmatrix} \gamma & 0 & 0 & i\beta\gamma \\ 0 & 1 & 0 & 0 \\ 0 & 0 & 1 & 0 \\ -i\beta\gamma & 0 & 0 & \gamma \end{pmatrix}$$

10. At what velocity the mass of a particle becomes 10 times. $\left[\textbf{Ans. } \frac{v}{c} = 0.995\right]$
11. Find the mass equivalent to one Photon of Na-light. [**Ans.** 3.75×10^{-34} gm]

Chapter 9 NUCLEAR ENERGY

9.1 NUCLEUS ; NUCLEAR FORCES ; BINDING ENERGY

The central (main) part of the atom is termed as nucleus. It contains neutrons, *N*, and protons, *Z*, only. The sum of two is called the mass number *A i.e.* $A = N + Z$. Every nucleus is designated by an appropriate symbol (say *X*) such as He for helium, C for carbon, O for oxygen etc. The atomic number, *Z*, and the mass number, *A*, are shown as indices like

$${}_ZX^A \text{ (general case)}$$

$${}_2He^4, {}_6C^{12}, {}_7N^{14}, {}_8O^{16} \text{ etc.}$$

The number, *Z*, is also called the atomic number of the nucleus. Further the nuclei with same *Z* are called isotopes while with same *A* as isobars and with same neutron number $N = (A - Z)$ as isotones.

Nuclear Forces

In order to understand the concept of binding energy of the nucleus one has to know the nature of nuclear forces, which binds the neutrons and protons (both called as nucleons) together within the nucleus. Despite the presence of coulomb repulsion among the protons, these remain bound in the small space of the nucleus. Therefore, the nuclear forces must be short range, attractive and stronger forces in comparison with coulomb and hence gravitational forces. Besides this, it has been found that these forces are same irrespective of whether the nucleons are neutron-proton, neutron-neutron or proton-proton. In other words nuclear forces are charge independent.

Nuclear Potentials

In the literature following forms are used for the nuclear potentials :

(*a*) Square Well $\quad V(r) = -V_0 \quad \text{for} \quad r \leq \alpha$

$\quad = 0 \quad \text{for} \quad r > \alpha$

(*b*) Exponential $V(r) = -V_0 e^{-r/\alpha} = -V_0 f(r)$

(*c*) Gaussian $V(r) = -V_0 \exp\left[\frac{-r^2}{\alpha^2}\right] = -V_0 g(r)$

(*d*) Yukawa $V(r) = -V_0 \frac{\left(e^{-r/\alpha}\right)}{\left(\frac{r}{\alpha}\right)} = -V_0 h(r)$

All these are central and two parametric. The parameters are V_0, the strength of the potential and α, the range. These are about 50 MeV and 1 fm, respectively (1 fm = 10^{-13} cm).

Binding Energy

It has been found experimentally that the mass of the nucleus is never equal to the sum of the masses of all protons and neutrons which it contains as one would like to expect. In fact, it is always smaller than this sum because some mass is lost in binding these nucleons together. The energy equivalent to this loss of mass is defined as the binding energy, *B*, of the nucleus.

The binding energy is therefore, defined as $c^2\Delta M$ where ΔM is known as mass defect.

Thus $$B = [ZM_p + (A - Z)M_n - M(A, Z)]c^2 \quad ...(9.1)$$

$$= c^2\Delta M \quad ...(9.2)$$

where M_p and M_n denote the proton mass and neutron mass respectively. Usually the masses are expressed in terms of the unit called atomic mass unit (a.m.u.). This unit is conventionally defined as that in which mass of Oxygen is 16 a.m.u. Thus, 1 a.m.u. = experimentally measured mass of O^{16} divided by 16, which comes out as 1.66×10^{-24} gm or about 931 MeV. In this system of units

$$M_p = 1.00814 \text{ a.m.u.} \quad \text{and} \quad M_n = 1.00898 \text{ a.m.u.}$$

Measured mass of the nucleus is generally shown as a function of *A* and *Z* as *M*(*A*, *Z*). There is another system of units for the masses in which C^{12} = 12 a.m.u.

Packing Fraction

Another important quantity is the packing fraction, *f*. It is defined as

$$f = \frac{[M(A, Z) - A]}{A} \quad ...(9.3)$$

Hence, $$M(A, Z) = A(1 + f) \quad ...(9.4)$$

Binding Energy Curve

Binding energy, *B*, of all the nuclei have been calculated by eq. (9.1). Then binding energy per nucleon obtained by dividing *B* by the mass number, *A*, of the corresponding nucleus is plotted versus *A*. The resulting smooth curve obtained is shown in Fig. 9.1. This curve shows that *B*/*A* first increases then stays nearly constant at around 8 MeV before it again decreases though slowly as *A* increases. However, there is an exception. While *B*/*A* of all nuclei falls on the smooth curve, that of $_2He^4$, $_6C^{12}$ and $_8O^{16}$ do not. For these three nuclei *B*/*A* has been found as significantly, different from their neighbours.

One further notes from this graph that if the heavier nucleus breaks into pair of smaller nuclei of comparable masses in some reaction, large amount of energy will always be liberated. So such a reaction has got a potential of being a source of nuclear energy.

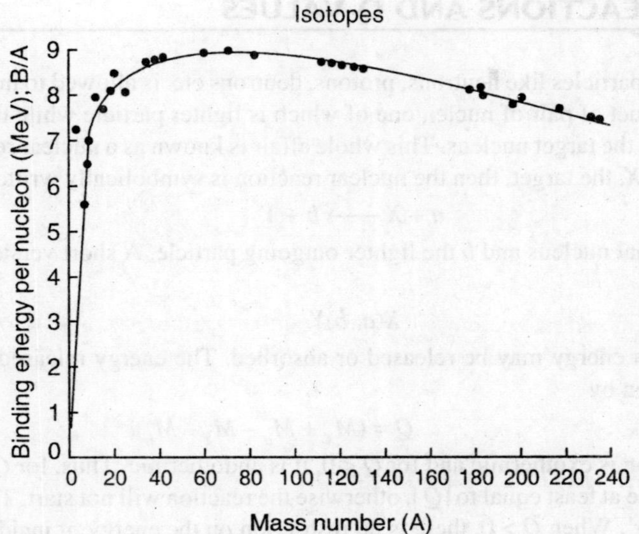

FIGURE 9.1 **A plot of binding energy per nucleon vs. mass number.**

Stable Nuclides

Lastly, we show in Fig. 9.2 the plot of known stable nuclides. It shows that though for small *A*, neutrons and protons number in stable nuclides are same, for large *A*, the neutrons number starts exceeding the proton number. Thus for the stability of the nucleus the proportional of the neutrons should be more than that of protons perhaps, because of the increasing effects of coulomb repulsion among several protons with increasing *A*.

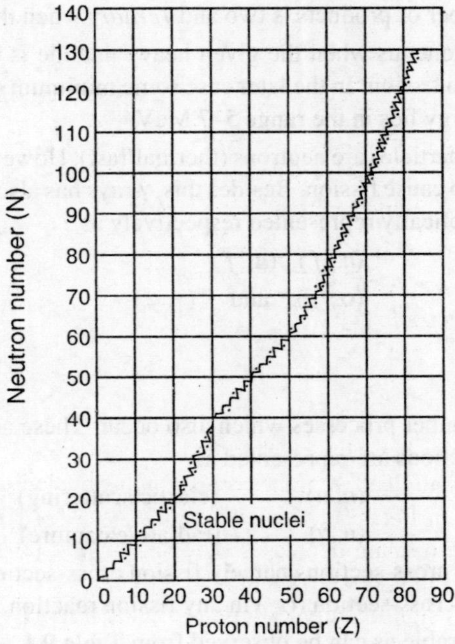

FIGURE 9.2 **A plot between neutron number and proton number of known stable nuclides.**

9.2 NUCLEAR REACTIONS AND Q-VALUES

When a beam of nuclear particles like neutrons, protons, deutrons etc. is allowed to interact with some target nucleus one finds a product of pair of nuclei, one of which is lighter particle while the other is heavier just same or comparable with the target nucleus. This whole affair is known as a nuclear reaction. If '*a*' represent the incident particle and *X*, the target, then the nuclear reaction is symbolically written as

$$a + X \longrightarrow b + Y \quad \text{...(9.5)}$$

where *Y* is heavier residual nucleus and *b* the lighter outgoing particle. A short version of such a scheme of reaction is written as

$$X(a, b)Y \quad \text{...(9.6)}$$

In such a reaction energy may be released or absorbed. The energy released/absorbed is called *Q*-value of the reaction given by

$$Q = (M_X + M_a - M_Y - M_b)c^2 \quad \text{...(9.7)}$$

For $Q > 0$, reaction is exothermic and for $Q < 0$, it is endothermic. Thus, for $Q < 0$, the energy of the incident particle should be at least equal to $|Q|$, otherwise the reaction will not start. This minimum energy is called 'Threshold Energy'. When $Q > 0$, there is no restriction on the energy of incident particles to initiate the reaction. For any energy the reaction will definitely take place.

9.3 NUCLEAR FISSION

Fission is a nuclear reaction in which a heavier nucleus, *X*, by absorbing energy splits up into two lighter nuclides of comparable masses. In such a reaction an enormous amount of energy is released. Such a reaction, now known as fission, was first observed by O. Hahn and F. Strassmann in 1939. It was also observed that in a rare event the number of product nuclei was three. This case does occur but very seldom. So, fission is of two types : *Binary* when the number of products is two and *Ternary* when the number of products is three.

Fission is said to be spontaneous when the given heavy nuclide is unstable to fission ($Q > 0$). For $Q < 0$ the given nucleus is stable to fission. In the later case some minimum (threshold) energy is required to initiate fission. Normally this energy lies in the range 5–7 MeV.

In most cases the incident particles are neutrons (thermal/fast). However, the particles like protons, α, and deutron have also been used to cause fission. Besides this, γ-rays has also been used for fission reactions. These fission processes are symbolically represented respectively as

$$(n, f), \ (d, f) \quad \text{...(9.8)}$$
$$(\alpha, f) \ \text{and} \ (\gamma, f)$$

Other Competing Reactions

Along with fission there are also other processes which also occur. These are elastic scattering of neutrons and emission of γ-rays. These reactions are represented as

$$(n, n) \quad \text{(elastic scattering)}$$

and

$$(n, \gamma) \quad \text{(radiative capture)} \quad \text{...(9.9)}$$

So, we have three types of cross-sections namely fission cross-section (σ_f), elastic scattering cross-section (σ_n) and radiative capture cross-section (σ_γ) in any fission reaction.

σ_γ and σ_f are quite comparable as can be observed from Table 9.1.

TABLE 9.1

Cross-section	$_{92}U^{235}$	$_{94}Pu^{239}$
σ_a	683*b*	1028*b*
σ_f	577*b*	742*b*
σ_γ	101*b*	286*b*

Here, σ_a denotes the sum of σ_f and σ_γ and is called the absorption cross-section (1 $b = 10^{-24}$ cm^2). Since, the cross-section is the ratio of number of events of particular type occurring per second to the intensity of incident particles, its unit is that of area. The unit, used frequently is barn and 1 barn = 10^{-24} cm^2.

Fissionable Nuclides

The commonly known heavy nuclides, which show fission, are the isotopes of Th, U and Pu. Some of these show fission with even thermal neutrons (incident energy E_n = fraction of eV) *e.g.* U^{233}, U^{235}, Pu^{239} while others require fast neutrons. The neutrons having velocity, v = 2200 m/sec were used to cause fission. These are termed as thermal neutrons (E_n = 0.025 eV). Along with these, 14 MeV neutrons (fast) were also used to study fission.

Energy Released In Fission

The energy available per fission can be calculated in terms of binding energy per nucleon of the fissionable and product nuclides. For this, one has to consult the *B*/*A* vs. A curve (Fig 9.1) where fissionable nuclide (heavier) has a value of about 7.5 MeV and the product (lighter) nuclides have a value 8.5 MeV each.

The energy released in a fission reaction $_0n^1 + {}_zX^A = {}_{z1}Y^{A1} + {}_{z2}Z^{A2} + 2{}_0n^1$ is given by

$$E_f = M_x(A, Z) - M_y(A_1, Z_1) - M_z(A_2, Z_2) - M_n \quad ...(9.10)$$

where all masses have been expressed in MeV.

Now use the relation of masses in terms of binding energy. As a result we get

$$E_f = ZM_p + (A - Z)M_n - B_X - Z_1M_p - (A_1 - Z_1)M_n + B_Y - Z_2M_p - (A_2 - Z_2)M_n + B_z - M_n \quad ...(9.11)$$

$$= B_Y + B_Z - B_X$$

(using the conditions $A = A_1 + A_2 + 1$; $Z = Z_1 + Z_2$)

If **B** denote the binding energy per nucleon the above eq. becomes

$$E_f = A_1\mathbf{B}_Y + A_2\mathbf{B}_Z - A\mathbf{B}_X = A(\mathbf{B}_Y - \mathbf{B}_X) - \mathbf{B}_Y \qquad (\because \ \mathbf{B}_y \approx \mathbf{B}_z)$$

$$\therefore \quad \approx A(8.5 - 7.5) - 8.5 = A - 8.5 \text{ MeV}$$

With A as the mass number of the fissionable nucleus, X, the amount of energy available in a single fission is more than 200 MeV. This order shows that huge amount of energy will be available for every gram of the fissionable material. In fact, the principle of reactor has a bearing on this very fact.

Production of Neutrons

Neutrons required to initiate fission are obtained through other nuclear reactions involving the incident beam of charged particles. This is because the charged particles like proton and deutron can easily be accelerated.

Therefore, the neutron produced in such reactions can have any desired energy. We mention some of these reactions here.

(*a*) $_1H^3(p, n)\,_2He^3$ (*b*) $_3Li^7(p, n)\,_4Be^7$

(*c*) $_6C^{13}(d, n)\,_7N^{14}$ (*d*) $_1H^3(d, n)\,_2He^4$

(*e*) $_4Be^9(\alpha, n)\,_6C^{12}$

The range of the energy of the neutrons obtained through these reactions is about 5 keV – 20 MeV. Besides this, the neutrons produced above are approximately monoenergetic.

Production of γ-Rays

The fission in the heavy nuclides can also be induced by γ-rays. For this reason, one requires reactions in which such γ-rays ($E_\gamma \longrightarrow 5 — 6$ MeV) are available. Such reactions are being given below :

(*i*) $_9F^{19}(p, \gamma)_{10}Ne^{20}$ ($E_\gamma = 6.3$ MeV)

(*ii*) $_3Li^7(p, \gamma)_4Be^8$ ($E_\gamma = 17.5$ MeV)

Prompt and Delayed Neutrons

In a binary fission neutrons are also produced along with two product nuclides. These neutrons are termed as Prompt Neutrons. The product nuclides on the other hand are generally neutron rich and are unstable to neutron emission. So, these product nuclides also emit neutrons as well as β-(electrons) particles through series of steps. Such neutrons are termed as Delayed Neutrons.

Mass Distribution

In a binary fission of a particular nuclide variety of products are emitted. So, there will be variation in the mass numbers of the product nuclides after each fission. All these products have been identified in the case of

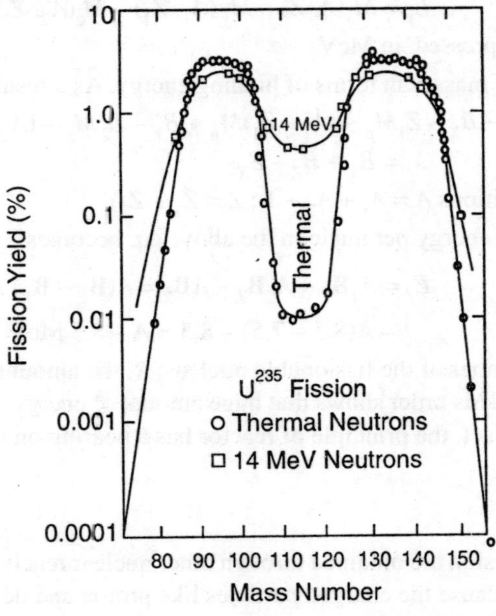

FIGURE 9.3 **Yield-Mass curve for fission of U ^{235} induced by neutrons. Curves plotted from "best" values.**

U^{235} and Pu^{239}. The mass distribution in the former case are available for these fission by thermal neutrons (v = 2200 m/sec) as well as fast neutrons (E_n = 14 MeV). This is being shown in Fig. 9.3 in the form of % yield *vs.* mass number A. One finds from this figure that the curve has two maxima separated by a minimum (~ 0.01%) *i.e.* the distribution is asymmetric. Secondly, the asymmetry starts disappearing as the energy of neutrons increases *e.g.* from thermal to 14 MeV.

Explanation of Binary Fission

People have made attempts to explain binary fission by assuming liquid drop model for the nucleus. Under this approach (first given by N. Bohr and J.A.Wheeler in 1939) one takes into account the two competing energies in the nucleus. One is surface energy and the other is coulomb energy. It is assumed that as the heavy nuclides absorbs say a neutron, its shape starts deforming with the above mentioned energies coming into play in opposite direction. In fact, the surface energy increases, while the coulomb energy decreases with increasing deformation. In this process the nuclear shapes becomes critical and then there is no coming back to original shape. As a result, the nucleus splits into two smaller nuclides with comparable masses. The way the fission process proceeds has been shown in Fig. 9.4.

The theory of the fission process

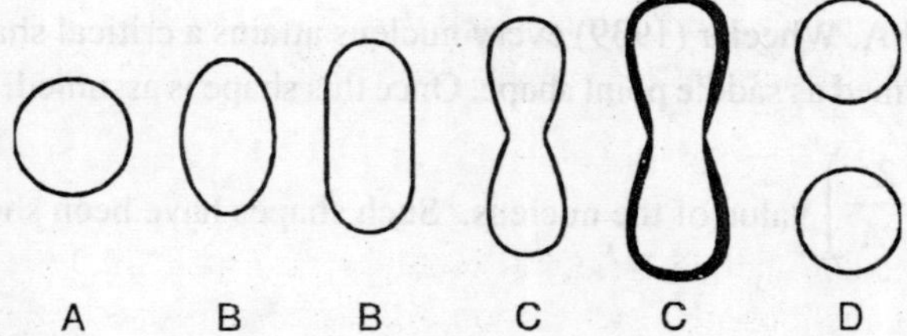

FIGURE 9.4 Possible steps in the process of nuclear fission according to the liquid-drop model.

Condition of Spontaneous Fission

A crude estimate can be obtained by assuming that the nucleus divides into two equal halves after absorbing energy. Then the change in the energy is given by

$$\Delta E = a_s A^{2/3}\left(1 - 2^{1/3}\right) + a_c\left[\frac{Z^2}{A^{1/3}}\right]\left(1 - \frac{1}{2^{2/3}}\right) \quad ...(9.12)$$

for the reaction

$$_0n^1 + X \longrightarrow Y + Z + x_0n^1$$

x being the integer which is generally two or three. For obtaining eqn. (9.12) the energy of the nucleus that undergoes a change has been assumed as sum of surface energy, $a_sA^{2/3}$, and coulomb energy, $a_cZ^2/A^{1/3}$ where a_s and a_c are surface energy and coulomb energy constants.

On simplification, eq. (9.12) reduces to

$$\Delta E = -0.26a_s A^{2/3} + 0.37a_c \frac{Z^2}{A^{1/3}} \quad ...(9.13)$$

Naturally, when $\Delta E > 0$, nucleus is unstable to fission and when $\Delta E < 0$, it is stable as some energy will be required to cause fission. Hence the condition

$$\frac{Z^2}{A} > \left(\frac{26}{37}\right) \times \left(\frac{a_s}{a_c}\right) \quad \text{(Spontaneous fission)}$$

$$< \left(\frac{26}{37}\right) \times \left(\frac{a_s}{a_c}\right) \quad \text{(Stable)}$$

and the critical value of this ratio is

$$\left(\frac{Z^2}{A}\right)_{cri} \approx \left(\frac{26}{37}\right) \times \left(\frac{14}{0.6}\right) = 16$$

$$\therefore \quad a_s \approx 14 \text{ MeV and } a_c \approx 0.60 \text{ MeV}$$

Thus it is expected that when this condition for a given nucleus is satisfied it will show spontaneous fission. But a more accurate estimate has shown that $\left(\frac{Z^2}{A}\right)_{cri} \approx 48$ for spontaneous fission. This value agrees with the experimental data.

Saddle Point Shape

According to N. Bohr and J.A. Wheeler (1939) every nucleus attains a critical shape before it breaks up into two halves. This shape is termed as saddle point shape. Once this shape is assumed, the nucleus starts breaking. This shape depends upon $\left(\frac{Z^2}{A}\right)$ value of the nucleus. Such shapes have been shown for different values of $\left(\frac{Z^2}{A}\right)$ in Fig. 9.5. In fact, saddle point shapes has been shown in terms of χ-values $\frac{Z^2}{A} = \chi \left[\frac{40\pi r_0{}^3 T}{(3e^2)}\right]$; where nuclear radius $R = r_0(A)^{1/3}$ and T is the surface tension. It is to be noted that for $\chi = 1$, the original shape of the nucleus is already a saddle point shape and its slight deformation will cause fission.

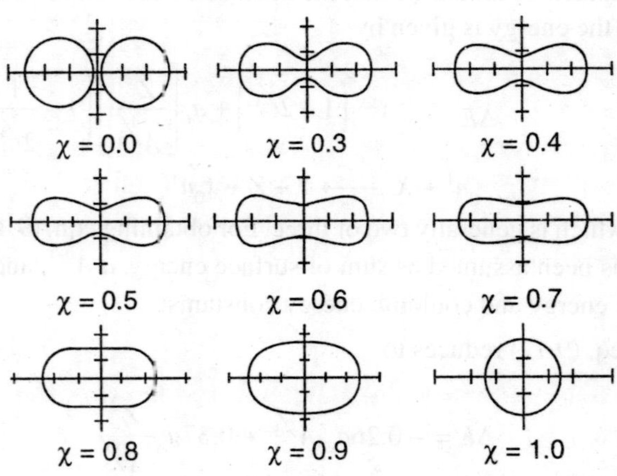

FIGURE 9.5 **Saddle point shapes for various values of χ.**

Series Chain Reaction

In a fission the product nuclei are generally neutron rich and are unstable to β-decay. So, these nuclei decay through a series of steps before these become stable. These decay processes are termed as series chain reactions. We mention here some of them.

Long Chain :

$$_{54}Xe^{140} \xrightarrow[16s]{\beta^-} {}_{55}Cs^{140} \xrightarrow[66s]{\beta^-} {}_{56}Ba^{140} \xrightarrow[12.8d]{\beta^-} {}_{57}La^{140} \xrightarrow[40h]{\beta^-} {}_{58}Ce^{140}\,(\text{Stable})$$

Short Chain :

$$_{60}Nd^{147} \xrightarrow[16s]{\beta^-} {}_{61}Pm^{147} \xrightarrow[4y]{\beta^-} {}_{62}Sm^{147}\,(10^{11}y)$$

Here Pm denotes Promethium which escaped detection at the time of discovery of fission. Another important chain is

$$_{42}Mo^{99} \xrightarrow[66h]{\beta^-} {}_{43}Tc^{99} \xrightarrow[2.2\times 10^{3}\,y]{\beta^-} {}_{44}Ru^{99}\,(\text{Stable})$$

The element technetium (Tc) does not occur in nature.

Lastly, we may mention that the number of neutrons, which are available per fission in case of U^{235}, is about 2.5 on the average. Therefore if this is not controlled properly one of these neutrons would further cause fission. In this way the whole material will explode. Such an explosive situation in which fission is self-sustaining is very dangerous. Therefore, U^{235} is to be mixed with some other but less fissionable material in such a way that the multiplying factor of neutrons remains under control. If this is less than unit, the bulk is called sub-critical when it is greater than unity it is termed as super-critical. We shall talk about it in the next chapter.

9.4 NUCLEAR FUSION

In fusion process two lighter nuclei fuse together to form a bigger nucleus. Thus fusion is inverse of fission. In fusion also lot of energy is released because of the difference in the binding energies of the lighter nuclei and the heavier product nucleus. It is the fusion process, which is the source of energy of sun and the stars. In fact, four protons fuse together to form helium nucleus releasing about 26.7 MeV of energy.

There are two ways by which four protons are assumed to be fused to form helium. One is known as proton cycle while the other Carbon or CNO cycle.

Proton Cycle

In the proton cycle (*p*-cycle) the helium is produced through the following steps :

$$_{1}H^{1} + {}_{1}H^{1} \longrightarrow {}_{1}H^{2} + \beta^{+} + \nu + \gamma$$

$$_{1}H^{2} + {}_{1}H^{1} \longrightarrow {}_{2}He^{3} + \gamma$$

$$_{2}He^{3} + {}_{2}He^{3} \longrightarrow {}_{2}He^{4} + {}_{1}H^{1} + {}_{1}H^{1}$$

Here β^+ denotes positron, ν denotes neutrino and γ denotes gamma rays.

From these reactions one find that in all four protons have been used to form one helium nucleus and in this process as per calculations about 26.7 MeV of energy is released.

Carbon Cycle

There is second method by which four protons could fuse to give one helium nucleus. In this method carbon acts as a catalyst and hence called the carbon cycle. This cycle consists of the following steps :

$$
\begin{aligned}
{}_1H^1 + {}_6C^{12} &\longrightarrow {}_7N^{13} + \gamma \quad (T_{1/2} = 9.96 \text{ m})\\
{}_7N^{13} &\longrightarrow {}_6C^{13} + \beta^+ + \nu\\
{}_6C^{13} + {}_1H^1 &\longrightarrow {}_7N^{14} + \gamma\\
{}_7N^{14} + {}_1H^1 &\longrightarrow {}_8O^{15} + \gamma \quad (T_{1/2} = 124 \text{ s})\\
{}_8O^{15} &\longrightarrow {}_7N^{15} + \beta^+ + \nu\\
{}_7N^{15} + {}_1H^1 &\longrightarrow {}_6C^{12} + {}_2He^4
\end{aligned}
$$

Clearly, carbon acts as a catalyst for converting four protons into a helium nucleus. Again about 26.7 MeV of energy is released after each cycle. It is believed that in the sun it is the proton cycle, which is the cause of solar energy while in the hotter stars it is the carbon cycle. However, in both cycles hydrogen is continuously burning and disappearing from the sun and the stars. But, amount of hydrogen in the sun is still sufficient to sustain solar energy for the next few billion years.

Since plants derive life from sun's light, coal and oil from plants, water power from sun's heat and all animals live on plants we owe our existence on earth to the solar hydrogen. Further, as there is high temperature and pressure in the stars and the binding energy increases as we go upwards in mass number (upto ${}_{26}Fe^{56}$) from helium, the stars are the best places to get heavy nuclei upto ${}_{26}Fe^{56}$ through fusion. In fact, a number of ways have been suggested in which heavy nuclei could be formed in stars starting from helium nucleus. So sun and stars are nuclear crucible to synthesis heavy nuclei.

SOLVED NUMERICAL PROBLEMS

Problem 1. *Obtain the Q-value for the reaction $Li^7 (p, \gamma) Be^8$, given $M_{Li} = 7.01822$ amu and $M_{Be} = 8.00531$ amu.*

Solution.

$$
\begin{aligned}
Q &= 931 \times [M_{\text{Li}} + M_p - M_{\text{Be}}]\\
&= 931 \times [7.01822 + 1.00814 - 8.00531] = 19.6 \text{ MeV.}
\end{aligned}
$$

Problem 2. *Find the Q-value of the reaction $F^{19}(p, \gamma)Ne^{20}$. Given, $M_F = 19.00445$ amu, $M_p = 1.00814$ amu and $M_{Ne} = 19.99876$ amu.*

Solution.

$$
\begin{aligned}
Q &= 931 \times [M_{\text{F}} + M_{\text{p}} - M_{\text{Ne}}]\\
&= 931 \times [19.00445 + 1.00814 - 19.99876] = 12.9 \text{ MeV.}
\end{aligned}
$$

Problem 3. *Find Q-value of the reaction $Be^9(\alpha, n)C^{12}$. Given $M_{Be} = 9.015$ amu, $M_\alpha = 4.00387$ amu, $M_C = 12.0038$ amu and $M_n = 1.00898$ amu.*

Solution. The given reaction is $Be^9 + He^4 \rightarrow C^{12} + {}_0n^1 + Q$.

$$
\begin{aligned}
Q &= (M_{\text{Be}} + M_\alpha - M_{\text{C}} - M_n) \times 931\\
&= (9.015 + 4.00387 - 12.0038 - 1.00898) \times 931 = 5.67 \text{ MeV.}
\end{aligned}
$$

Problem 4. *Show that for spontaneous symmetric fission $Z^2/A \geq 15$.*

Solution. Let us assume a nucleus of atomic No. Z and mass No. A undergoes a symmetric fission into two equal nuclei with atomic No. $Z/2$ and mass No. $A/2$. Now, since only coulomb energy

$= a_c\left(\frac{Z^2}{A}\right)^{1/3}$ and surface energy, given by $a_s A^{2/3}$ compete, the change in energy in this reaction will be,

$$\frac{a_c Z^2}{A^{1/3}} + a_s A^{2/3} - 2\left[a_c\frac{\left(\frac{Z}{2}\right)^2}{\left(\frac{A}{2}\right)^{1/3}} + a_s\left(\frac{A}{2}\right)^{2/3}\right] = a_s\left(1-2^{1/3}\right)A^{1/3} + a_c\left(1-\frac{1}{2^{2/3}}\right)\frac{Z^2}{A^{1/3}}$$

Taking $a_s = 13$ MeV and $a_c = 0.6$ MeV, we get the release of energy $= 0.22\,\frac{Z^2}{A^{1/3}} - 3.38A^{2/3}$. Hence condition for spontaneous fission will be $\left(\frac{Z^2}{A}\right) \geq 15$.

Note. More accurate result first reported by Bohr and Wheeler (1939) gives the condition for spontaneous fission as $(1 - 0.022\, Z^2/A) \leq 0$ *i.e.* $Z^2/A \geq 1/0.022 \approx 45$.

Problem 5. *Find the energy of slow neutron of v = 2200 m/sec.*

Solution.

$$E = \frac{Mc^2}{\left[1-\left(\frac{v}{c}\right)^2\right]^{\frac{1}{2}}} - Mc^2 = Mc^2\left[1+\frac{v^2}{2c^2}-1\right] = M\frac{v^2}{2}$$

$$= \frac{Mc^2}{2}\left(\frac{v}{c}\right)^2 \approx \frac{931}{2}\times\left(\frac{22}{3}\right)^2 \times 10^{-12}\ \text{MeV} = 0.025\ \text{eV}$$

Problem 6. *Calculate the velocity of neutrons of 14 MeV energy.*

Solution.

$$14 = \left[\frac{1}{\left(1-\frac{v^2}{c^2}\right)^{1/2}} - 1\right] \times 931 \times M_n$$

$$\therefore \quad \left(1-\frac{v^2}{c^2}\right)^{1/2} = \frac{14}{(M_n \times 931)} + 1$$

and $v = 0.17c$ [$M_n \rightarrow$ neutron mass in amu = 1.00898]

Problem 7. *Calculate the energy release in the complete fissioning of 0.5 kg of U^{235}. Given that energy release/fission = 200 MeV.*

Solution.

$$\text{No. of atoms of } U^{235} \text{ in } 0.5 \text{ kg} = \left(\frac{6\times 10^{23}}{235}\right)\times 500$$

Thus energy released after complete fissioning

$$= \left(\frac{6\times10^{23}}{235}\right)\times 500\times 200\ \text{MeV}$$

$$= \left(\frac{6\times10^{28}\times1.6\times10^{-13}}{235}\right)\text{J} = 4.1\times10^{13}\ \text{J}.$$

Problem 8. *Calculate the energy released when 1gm of U^{235} is completely fissioned. Given energy/fission = 200 MeV*

Solution. No. of atoms $= 1\times 6\times\dfrac{10^{23}}{235}$ per gm

Energy released $= \dfrac{200\times6\times10^{23}}{235}\ \text{MeV} = \left(\dfrac{200}{235}\right)\times 6\times10^{23}\times 1.6\times10^{-13}\ \text{J}$

$$= \frac{(12\times1.6\times10^{12})}{(235\times3600\times10^{3})}\ \text{kWh} = 2.27\times10^{4}\ \text{kWh}\quad\left[\because\ 1\ \text{J} = \frac{10^{-3}}{3600}\ \text{kWh}\right]$$

Problem 9. *Calculate the energy release in the fission reaction ${}_0n^1 + U^{235} \rightarrow Mo^{95} + La^{139} + 2\,{}_0n^1 + Q$. Given $M_U = 235.124$ amu, $M_{Mo} = 94.946$ amu, $M_{La} = 138.955$ amu and $M_n = 1.00898$ amu.*

Solution.

Energy release per fission $= 931\times[M_U + M_n - M_{Mo} - M_{La} - 2M_n]$ MeV

$= 931\times 0.214 = 199.25$ MeV.

Problem 10. *Find the energy released in the following fission ${}_0n^1 + U^{235} \rightarrow Xe^{136} + Mo^{98} + 2\,{}_0n^1$. Given $M_U = 235.124$ amu, $M_{Mo} = 97.905$ amu, $M_{Xe} = 135.950$ amu and $M_n = 1.00898$ amu.*

Solution. Since masses are given in amu, the energy released per fission will be given by

$$E_f = 931\times[M_U - M_{Mo} - M_{Xe} - M_n]\ \text{MeV}$$

$$= 931\times 0.26 = 242\ \text{MeV}.$$

Problem 11. *Give the nuclear reaction to produce a beam of monochromatic neutrons. Also calculate the energy of the product neutrons.*

Solution. The following α-induced reaction is very common for the production of neutrons

$$\alpha + {}_4Be^9 = c^{12} + n,$$

symbolically written as $Be^9(\alpha, n)C^{12}$; Q-value of this reaction is given by

$$Q = 931\times[4.00387 + 9.01504 - 12.0038 - 1.00898]\ \text{MeV} = 5.71\ \text{MeV}.$$

Problem 12. *Calculate the energy released in the fission reaction*

$${}_0n^1 + {}_ZX^A = {}_{Z1}X_1^{A1} + {}_{Z2}X_2^{A2} + {}_x({}_0n^1).$$

Solution. The Q-value of this reaction is

$$Q = (M_n + M_X - M_{X_1} - M_{X_2} - xM_n)\ \text{MeV}$$

where M's denote the masses in MeV. Now expressing masses in terms of the binding energies we get

$$Q = ZM_p + (A-Z)M_n - B_X - Z_1M_p - (A_1 - Z_1)M_n + B_{X_1} - Z_2M_p - (A_2 - Z_2)M_n + B_{X_2} - (x-1)M_n$$

$$= B_{X_1} + B_{X_2} - B_X$$

obtained by using the condition,

$$Z = Z_1 + Z_2 \text{ and } A + 1 = A_1 + A_2 + x$$

$$B_{X_1} = A_1 B_{X_1} \qquad B_{X_2} = A_2 \mathbf{B}_{X_1} \quad \text{and} \quad B_X = AB_{X_2}$$

where ($\mathbf{B}_{X_1}$, $\mathbf{B}_{X_2}$) and $\mathbf{B}_X$ denote the binding energies per nucleon of intermediate and heavy nuclides,

$$Q = (A_1 + A_2)\,\mathbf{B}_{X_1} - A\mathbf{B}_X = [(A_1 + A_2) \times 8 - 7.1A] \text{ MeV}$$

where we have used $\mathbf{B}_{X_1} = \mathbf{B}_{X_2} = 8$ MeV and $\mathbf{B}_X = 7.1$ MeV on the basis of well known binding energy per nucleon variation with mass no. A.

So,
$$Q = (A + 1 - x)\,8 - 7.1\,A = [0.9A - (x - 1)\,8] \text{ MeV}$$
$$= \text{energy released per fission.}$$

EXERCISE

1. Draw the graphs of $f(r)$, $g(r)$ and $h(r)$ taking $\alpha = 1$fm and increasing r in the steps of 0.1 fm.
2. (*i*) Show that deutron is loosely bound nucleus

   Given Mass of deutron = 2.0147 amu ($A = 2, Z = 1$)

   $M_p = 1.00814$ amu

   $M_n = 1.00898$ amu

   (*ii*) Calculate B/A of ${}_2He^4$, ${}_6C^{12}$, ${}_8O^{16}$. Given Mass of He = 4.00387 amu

   C = 12.00381 amu

   O = 16.00000 amu **[Ans.** 7.07 MeV, 7.67 MeV, 7.97 MeV]
3. Calculate Q-values of the reactions

   $$H^3 + H^2 \rightarrow He^4 + {}_0n^1 + Q$$
   $$Li^7 + H^1 \rightarrow Be^7 + {}_0n^1 + Q$$

   Given $M_n = 1.00898$ amu, $M_p = 1.00814$ amu, $M_{Li} = 7.0182$ amu, $M_{Be} = 7.019$ amu, and $M_t = 3.016$ amu

   [Ans. 16.62 MeV, – 1.45 MeV]
4. Calculate the $B.E.$ per nucleon of the following nuclei. Given $M_n = 1.00898$ amu, $M_p = 1.00814$ amu and 1 amu = 931 MeV

Nucleus	*Mass (amu)*
${}_1H^2$	2.0147
${}_2He^4$	4.00387
${}_6C^{12}$	12.0038
${}_8O^{16}$	16.0000
${}_{28}Fe^{56}$	55.9527
${}_{79}Au^{197}$	197.028
${}_{82}Pb^{206}$	206.037
${}_{92}U^{238}$	238.1243

[Ans. 1.13 MeV, 7.04 MeV, 7.67 MeV, 7.97 MeV, 8.76 MeV, 7.91 MeV, 7.88 MeV, 7.57 MeV]

5. Find the ratio $(Z^2/A)_{cri}$ for U^{235} and Pu^{239}. **[Ans.** 36.02, 36.97]
6. Calculate the energy released in the following fission reaction

   $$ {}_0n^1 + {}_{92}U^{235} = {}_{36}Kr^{90} + {}_{56}Ba^{144} + 2\,{}_0n^1$$
   $$= {}_{38}Sr^{94} + {}_{54}Xe^{140} + 2\,{}_0n^1$$

   (Given B/A = 8 MeV for U and 7.1 MeV for product nuclides) **[Ans.** 203.5 MeV]

7. Calculate the energy of neutrons having $v = 2200$ m/sec. **[Ans.** 0.025 eV**]**
8. Calculate the temperature of 1 MeV Neutrons. **[Ans.** 1.16×10^{10} K**]**
9. Give a note on fission reaction.
10. Calculate the energy released per fission of $_{Z}X^{A}$ taking binding energy per nucleon of heavy and intermediate nuclides as 7.1 MeV and 8 MeV respectively and 2 as the number of neutrons released per fission.
[Ans. $[8(A-1) - 7.1A]$ MeV**]**
11. Which type of nuclear reaction is responsible for solar energy ? Discuss this reaction.
12. Distinguish between Fission and Fusion.
13. Calculate the energy released in the following reaction :
$4(_1H^1) \rightarrow {}_2He^4 + 2\beta^+ + 2\nu$. Given that $M_p = 1.0081$ amu, $M_{He} = 4.0039$ amu and 1 amu = 931 MeV
[Ans. 28.53 MeV**]**
14. Explain *p*- and CNO-cycles.
15. Calculate the percentage loss of mass of a U^{235} nucleus, given that energy release per fission is 200 MeV.
(*June 2002, K.U.K.*) **[Ans.** 0.09**]**
16. A nucleus of mass 235.124 amu breaks into two nuclei of masses 135.951 amu and 97.936 amu and also a neutron of mass 1.009 amu. Calculate the energy released in MeV. **[Ans.** $0.228 \times 931 = 212.27$ MeV**]**

Chapter **10** # REACTOR PHYSICS

10.1 CROSS-SECTION, MULTIPLICATION FACTOR AND CRITICALITY

Cross-section

In any reaction, the cross-section is defined as the ratio of No. of events of particular type occurring per second to the intensity of incident particles. As intensity has a unit of cm^{-2}, the unit of cross-section will be cm^2. Since the cross-sections are small quantities these are expressed in the smaller unit called barns and 'b' in short and $1b = 10^{-24}\ cm^2$. Obviously, the total cross-section will be the sum of all types of cross-sections, σ_i. Thus

$$\sigma_t = \Sigma_i \sigma_i \qquad ...(10.1)$$
$$= \sigma_s + \sigma_f + \sigma_r$$

where σ_s = elastic scattering cross-section,

σ_f = fission cross-section,

and σ_r = radiative capture cross-section.

Sometimes sum of σ_f and σ_r is called the absorption cross-section, σ_a.

That is, $\sigma_a = \sigma_f + \sigma_r$ and $\sigma_t = \sigma_s + \sigma_a$

Further, it has been observed that for slow neutrons fission cross-sections follow $1/\nu$ law. That is $\log \sigma_f = -m \log \nu + c$, where m and c are constants.

Multiplication Factor

The multiplication factor of neutrons is defined as the number of neutrons available for fission per neutron absorbed. If ν is the number of neutrons produced per fission and $\alpha = \dfrac{\sigma_r}{\sigma_f}$ then the multiplication factor η will be given by

$$\eta = \nu \frac{\sigma_f}{\sigma_f + \sigma_r} \qquad ...(10.2)$$

Here, σ_s has been assumed as of no consequences as the fissionable material is present in bulk. Eq. (10.2) may be further written as

$$\eta = \frac{\nu}{(1+\alpha)} \qquad ...(10.3)$$

Naturally, when $\sigma_r = 0$, $\eta \approx \nu$ (as in the case of pure U^{235}) and for $\alpha = 1$, $\eta = \nu/2$.

The data of these quantities is available and has been listed in Table 10.1.

TABLE 10.1 Calculated values of multiplying factor η for different fissionable nuclides on the basis of experimental data

Quantity	U^{233}	U^{235}	*Natural U*	Pu^{239}
ν	2.51	2.44		2.89
$\alpha = \frac{\sigma_r}{\sigma_f}$	0.10	0.18		0.39
$\eta = \frac{\nu}{(1+\alpha)}$	2.28	2.07	0.38	2.08

Criticality

The criticality states whether the fissionable material is self sustaining or not. This state depends on the multiplying factor of neutrons. If this factor, η, is less than unity the state of the reactor is said to be sub-critical and the fission process is not self sustaining. If the multiplying factor, η, is greater than unity, the state of fissionable material in a reactor is termed as supercritical and this will explode in due course of time. Hence, it is very dangerous. Thus, for most of the time the material in the reactor should be kept in the sub-critical state. Such a state can be attained by mixing the supercritical fissionable material with the sub-critical one in an appropriate manner. This mixing is kept under control so that the state of the reactor may be adjusted at will from one criticality to other.

As an illustration, one can mix U^{235} ($\eta > 1$) with U^{238} ($\eta < 1$) in a suitable proportion so that η becomes equal to unity. This proportion can be calculated from the relation

$$\eta = \frac{[(N_5\sigma_{f5} + N_5\sigma_{f8}) \times \nu]}{[N_5(\sigma_{f5} + \sigma_{r5}) + N_8(\sigma_{f8} + \sigma_{r8})]} \qquad ...(10.4)$$

where N_5 and N_8 denote the numbers of atoms of U^{235} and U^{238} respectively. In the same way σ_f and σ_r denote the fission and radiative capture cross-sections of the two isotopes. As $\sigma_{r5} = 0$ while $\sigma_{f5} = 1.3b$; $\sigma_{f8} = 0.5b$, $\sigma_{r8} = 2.0\ b$ and $\nu = 2.5$ ($1b = 10^{-24}\ \text{cm}^2$)

$$\eta = \frac{[(1.3N_5 + 0.5N_8) \times 2.5]}{[1.3N_5 + 2.5N_8]} \qquad ...(10.5)$$

To have $\eta = 1$, we must solve eq. (10.5) for the ratio say $\frac{N_8}{N_5}$. This comes out to be

$$\frac{N_8}{N_5} = 39/25 \approx 3/2 \quad \text{...(10.6)}$$

That mixture should be of 60% U^{238} and 40% U^{235}.

10.2 CHAIN REACTION

The chain reaction in a fissionable material means that once a fission reaction occurs it will not stop till the entire material gets exhausted. The chain reaction in a reactor is, therefore, determined by the competition between the following processes.

(*a*) Fission capture of neutrons by the fuel atoms (like U^{235}, U^{238}).

(*b*) Non-fission capture of neutrons by the fuel atoms and other material of the reactor.

If the number of neutrons absorbed by the process (*b*) is less than or equal to the surplus neutrons produced by (*a*) the chain reaction will occur otherwise this reaction will stop.

Thus, the possibility of chain reaction will depend upon fission cross-section and non-fission capture cross-section. The non-fission capture is dominantly radiative capture. Let us denote these cross- sections as σ_f and σ_r respectively and corresponding reactions as (*n*, *f*) and (*n*, γ). As defined earlier, we now write the multiplication factor, of neutrons, η, by the equation

$$\eta = \nu \frac{\sigma_f}{(\sigma_f + \sigma_r)} = \nu \frac{\sigma_f}{\sigma_a}. \quad \text{That is, } \eta = \nu \frac{1}{(1+\alpha)}$$

where ν and α denote the no. of neutrons available after the preceding fission and the ratio $\frac{\sigma_r}{\sigma_f}$ respectively.

Obviously, when $\eta > 1$, the fission will always occur after the preceding fission and the chain reaction will occur and when $\eta < 1$, the chain reaction will stop because no neutron will be available to initiate the next fission reaction. In the former situation the reactor is said to be super critical while in the latter it is said to be sub-critical. We now give some examples :

1. Case of Pure U^{235}

In case of U^{235} we have

$$\sigma_f = \sigma_a \qquad \sigma_r = \text{negligible}$$

and $\qquad \nu = 2.5$

Hence $\qquad \eta = 2.5 > 1.$

Situation is super-critical. The chain reaction will continue and the entire fuel of U^{235} will explode in no time.

2. Case of U^{238} (Pure)

In this case, above threshold energy (= 1.4 MeV)

$$\sigma_f = 0.5b, \sigma_r = 2.0b \quad \text{and} \quad \nu = 2.5$$

So, $$\eta = \frac{2.5 \times 0.5}{2.5} = 0.5 < 1$$

Hence chain reaction will not occur and fission in pure U^{238} will stop after couple of steps.

3. Natural Uranium as Fuel

The following data is available for natural uranium

$$\sigma_{5f} = 1b, \sigma_{8f} = 0.5b, \sigma_{5r} = 0, \sigma_{8r} = 2.0b$$

$$\%U^{235} = 0.72, \%U^{238} = 99.28\%$$

and $$\nu = 2.5$$

When the fuel consists of a mixture of two isotopes the multiplication factor would be given by

$$\eta = 2.5 \frac{(N_1\sigma_{1f} + N_2\sigma_{2f})}{(N_1\sigma_{1a} + N_2\sigma_{2a})}$$

where N_1 and N_2 denote the no. of atoms of the two types of isotopes in a given fuel. Thus, for the case of natural uranium we have

$$\eta = 2.5 \times \frac{\left(\sigma_{5f} + \frac{N_8}{N_8} \times \sigma_{8f}\right)}{\left(\sigma_{5a} + \left(\frac{N_8}{N_5}\right) \times \sigma_{8a}\right)}$$

$$= 2.5 \times \frac{\left(1 + \left(\frac{99.28}{0.72}\right) \times 0.5\right)}{\left(1 + \left(\frac{99.28}{0.72}\right) \times 2.5\right)} = 2.5 \times \frac{69.94}{345.72} = 0.5 < 1$$

Therefore, chain reaction will not occur if natural Uranium is used as the fuel. So, reactor will never work with natural Uranium as a fuel unless it is enriched in U^{235} or the fast fission neutrons are appropriately thermalized.

10.3 MODULATION OF NEUTRON INTENSITY

When a beam of neutrons is incident on a bulk of fissionable material, its intensity decreases with the depth. If σ_t is the total cross-section of interaction with the material and $dN = Ndx$ is the number of atoms of the material which a neutron encounters while passing thickness dx, the loss of intensity, dI, is given by

$$\frac{dI}{I} = -\sigma_t\, dN \qquad \text{...(10.7)}$$

$$= -\sigma_t Ndx,$$

where N denotes the number of atoms of the material per unit volume. On integration one gets

$$I_x = I_0 \exp(-\sigma_t Nx) \qquad \text{...(10.8)}$$

where I_x is the intensity at x while I_0 is its value at $x = 0$ or is the incident intensity of neutrons. In case $\sigma_t = \sigma_f$

$$I_x = I_0 \exp(-\sigma_f Nx) \qquad ...(10.9)$$

Now, eq. (10.9) states that I_x neutrons have travelled through dx. Hence, mean length that each neutron travels (called mean free path) would be

$$L = \text{Total length traversed / Total number of neutrons}$$

$$= \frac{\int_0^\infty I_x dx}{I_0} = \int_0^\infty \exp(-N\sigma_t x)dx = \frac{1}{N\sigma_t}$$

For fission, the mean free path would be

$$L_f = \frac{1}{N\sigma_f} \qquad ...(10.10)$$

10.4 MODERATOR

It is known that the slow neutrons are more efficient for self sustaining fission reactions than fast neutrons. The multiplying factor for slow neutrons is found to be unity even for a natural uranium (0.7%U^{235} + 99.3%U^{238}). Thus, critical conditions can be better achieved in natural uranium by slow neutrons. However, fission neutrons (~ 2.5 per fission) are fast (2 MeV) and be slowed down all along to obtain critical conditions more effectively. This process is known as moderation and the material used for this purpose a moderator.

The process of slowing down of fission neutrons is most efficient if it is based on elastic collisions with atoms of comparable masses. So, these are hydrogen and deutron and are used in the form of water and heavy water. Through elastic collision with these atoms neutrons get thermalised (from 2 MeV to 0.025 eV) after about 20 collisions. Moreover, the capture cross-sections of neutrons (being thermalised) by hydrogen (H) and deutron (D) is quite small. Therefore, the chance that the fission neutron survives before thermalisation is very high and the fission neutrons achieve the thermal equilibrium at room temperature (~ 20°C) with moderator atoms very quickly. However, carbon being more abundant is also used as a moderator. This is taken in the form of graphite rods.

10.5 REACTOR (FIRST)

The first reactor which was commissioned in Dec. 2, 1942 used enriched natural uranium lump size piles separated by graphite spaces. This reactor was conceived and made operational by Fermi and his associates. In fact, a self sustaining fission reaction is always possible in pure U^{235} pile. However, to obtain pure sample of U^{235} from the natural uranium (containing 0.7% U^{235}) is a tough job. The other alternative is to enrich the natural sample in appropriate manner.

Also, the moderator should be free from any impurity of Boron and Cadmium as these absorbs the neutrons significantly. However, these materials being good neutron absorbers could be used to control the intensity of neutrons. In fact, these are alloyed with steel in the form of rods and arranged to be pushed inside or taken out of the pile of the uranium as per requirement. In short, the graphite spaces thermalise the fission neutrons to 0.025 eV and Boron and Cadmium rods are arranged to control the number of neutrons so as to maintain the reactor at a desired level of criticality (sub-critical/super-critical).

Further, the walls and surroundings be so designed that there is no leakage of neutron flux and thereby becomes health hazard.

10.6 TYPES OF REACTOR

A number of reactors are available depending upon their utility. For example, these are

(*i*) Research Reactors
(*ii*) Power Reactors
(*iii*) Boiling water reactors
(*iv*) Breeder Reactors

Now we describe these one by one but very briefly.

(*i*) **Research Reactor.** Every reactor is considered as an intense source of neutrons. The neutrons being uncharged interact with the matter very quickly. Because of this some reactors are designed to make the neutrons available upto any amount for research purpose etc. window is available for placing the sample under study.

For example, the given element whose investigations are required is exposed to the neutrons. The sample absorbs neutrons and become a radioactive isotope. By studying the resulting radiation one can obtain information regarding the nuclear structure of the parent sample.

Also these artificially produced radioactive isotopes are used for medical treatment and identification of trace elements present in a given sample. Some radioactive isotopes such as Co^{60} are intense source of γ-rays and are used for food preservation while their β (or electron) activity is used for research in solid state physics and industrial research.

In summary, the research reactors are used to prepare radioactive isotopes of intense source of γ- and β-rays. These radiations are then put to variety of uses.

(*ii*) **Power Reactor.** Power reactors are used to generate electricity by converting the heat energy released through chain reactions of fission. It is estimated that about 200 MeV of energy is released per fission of U^{235} or Pu^{239}. This energy appears mainly as the kinetic energy of the product nuclides and neutrons. Thus the pile of the reactor will soon become hot because of the above mentioned kinetic energy of the products (equivalent to 12000 million degree Kelvin). If this heat energy is transferred outside to run the turbine and hence the generator we will get electricity. It is obvious that this heat transfer can be achieved by circulating some coolant liquid. The heat transfer would be most efficient if this coolant has (*a*) large specific heat capacity (= product of specific heat and the density) and (*b*) large thermal conductivity. Thus, in the past water, mercury and liquid sodium had been used with increasing merit. Among these liquid coolants, the liquid sodium is most useful. Through the circulation of the liquid coolant, the heat generated in the reactor is transferred to the externally located turbine, which runs the generator.

(*iii*) **Boiling Water Reactor.** The boiling water reactors have some advantages over the other reactors. Firstly, since water is used as a liquid moderator in these reactors it will become steam after absorbing heat and is used to run the steam turbine. Thus water serves as moderator as well as heat exchanger. Secondly, the reactor become more stable in the sense that with increasing temperature water becomes less effective as moderator and hence stabilizes the performance of the reactor. Besides, the water can be taken from the river or the nearby lake or may be used again and again.

The only disadvantage is that such reactors have limited power output because of above mentioned reasons. So, the boiling water reactors can not meet the sudden demand for increased power.

(*iv*) **Breeder Reactor.** Breeder reactors are designed to manufacture fissionable nuclides like Pu^{239} and U^{233}. The natural uranium used in the reactors is expected to exhaust in due course of time. Therefore, there is a need to produce other fissionable materials side by side.

Pu^{239} can be obtained from U^{238} while Th^{232} can be converted into U^{233} through the absorption of neutrons, the breeder reactors generally use U^{238} and Th^{232} as the parent (fertile) material to manufacture Pu^{239} and U^{233} for their future use in other reactors like power reactors. Following are the reactions, which convert fertile nuclides into fissionable ones.

$$_{92}U^{238} + {_0n^1} \longrightarrow {_{92}U^{239}} \xrightarrow[23m]{\beta^-} {_{93}Np^{239}} \xrightarrow[2.4d]{\beta^-} {_{94}Pu^{239}}$$

$$_{90}Th^{232} + {_0n^1} \longrightarrow {_{90}Th^{233}} \xrightarrow[23m]{\beta^-} {_{91}Pa^{233}} \xrightarrow[27d]{\beta^-} {_{92}U^{233}}$$

These new fissionable nuclides are formed as layer or blanket of the fertile material placed in the breeder reactor. After some time the reactor get poisoned because of tendency of the product nuclei to absorb neutrons thereby obstructing the operation of the reactor. As a result the fertile material is removed and reprocessed to extract the new fuel material Pu^{239} or U^{233} as the case maybe.

It is required that the no. of atoms of the new fuel produced in the conversion should be at least equal to the no. of atoms destroyed in the original fission reaction. Otherwise the purpose of the breeder reactor will be defeated. So, the breeding ratio (= ratio of fissionable atoms produced to the fissionable atoms destroyed) ultimately decides the performance of the breeder reactor. This should be more than one. Otherwise nothing will be gained.

The reactors, which use thermal neutrons are called Thermal Breeders while those which use fast or high energy neutrons the Fast Breeders. It is also to be noted that at least two neutrons should be available one to convert fertile atom and the other to sustain the next fission process.

Lastly we show the main stages of a reactor in Fig. 10.1 which is common to all types of reactors for the conversion of nuclear energy into heat, then mechanical and lastly to electric energy through the running of a generator by the steam turbine.

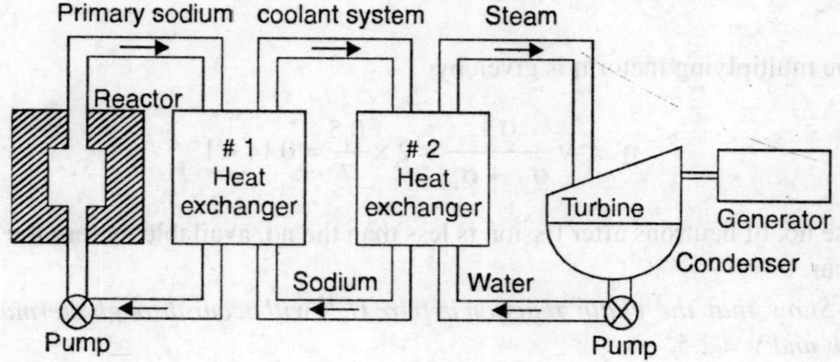

FIGURE 10.1 **Schematic representation of a liquid-metal coolant reactor showing the conversion of nuclear energy into electrical energy.**

10.7 CLASSIFICATION OF REACTORS

From the preceding discussions regarding various types of reactors one can list the following points by which reactors may be classified.

1. **Energy of the neutrons :** Three types of neutron energy are used

   (*a*) High energy (> 1 MeV) (*b*) Intermediate energy (eV-1 MeV)

   (*c*) Thermal energy (< eV)

2. **The nature of the fuel :**

   (*a*) Natural Uranium (*b*) Enriched Uranium

   (*c*) U^{233}, Pu^{239} (*d*) Th^{232}, U^{238} (in breeder reactors)

3. **Nature of coolant :**

   (*a*) Gas (*b*) Water (Hard water)

   (*c*) Liquid Metal *e.g.* Hg, Na

4. **Nature of Moderator :**

   (*a*) H_2O, D_2O (*b*) Graphite (in the form of rods)

5. **Assembly of Moderator and Fuel:**

   (*a*) Salt of UO_2 suspended in water or solution of Uranyl sulphate in water in case of Homogeneous reactor

   (*b*) Lump size of fuel and graphite rods placed alternately as in heterogeneous reactors.

6. **Purpose :**

   (*a*) Research (*b*) Power

   (*c*) Breeding of fissionable material like U^{233} and Pu^{239} from Th^{232} and U^{238} respectively.

Thus, one can easily decide the nomenclature of a reactor on the basis of above listed considerations.

SOLVED NUMERICAL PROBLEMS

Problem 1. *Show that by fast neutrons chain reaction in U^{238} will not occur. Given $\sigma_f = 0.5b$, $\sigma_t = 6.5b$ and $\nu = 2.5$.*

Solution. The multiplying factor η is given by

$$\eta = \nu \frac{\sigma_f}{\sigma_f + \sigma_r} = 2 \times \frac{0.5}{7} = 0.14 < 1$$

Thus, because no. of neutrons after fission is less than the no. available before the fission, the chain reaction will not occur.

Problem 2. *Show that the chain reaction in pure U^{235} will occur through thermal neutrons. Given $\sigma_f = 600b$, $\sigma_t = 700b$ and $\nu = 2.5$.*

Solution. The multiplying factor η is given by

$$\eta = \nu \frac{\sigma_f}{\sigma_t} = 2.5 \times \frac{600}{700} = \frac{15}{7} = 2.14 > 1$$

Hence no. of neutrons available is more than one to cause further fission. So, this process will repeat till the entire lot of U^{235} is exploded. This situation is called super-critical.

Problem 3. *Justify that a chain reaction in Pu^{239} will take place by thermal neutrons. Given σ_f = 700b, σ_t = 1000b and neutrons/fission(ν) = 2.5.*

Solution. No. of neutrons available after initial fission of Pu^{239} and which will cause next fission is given by

$$\eta = \nu \frac{\sigma_f}{\sigma_t} = 2.5 \times \frac{700}{1000} = 1.7 > 1$$

Therefore, the chain reaction will occur.

Problem 4. *In natural Uranium, the chain reaction will not sustain. Prove this statement. Given : in natural Uranium, U^{235} = 1% and U^{238} = 99%, and σ_f^{235} = 1b, σ_t^{235} = 1.2b, σ_f^{238} = 0, σ_t^{238} = 2b and neutrons/fission(ν) = 2.5.*

Solution. Multiplying factor η for a mixture is given by

$$\eta = \nu \frac{\left(N_5 \sigma_f{}^{235} + N_8 \sigma_f{}^{238}\right)}{\left(N_5 \sigma_t{}^{235} + N_8 \sigma_t{}^{238}\right)} ; \frac{N_5}{N_8} = 0.01$$

$$\eta = 2.5 \frac{(0.01 \times 1 + 0)}{(0.01 \times 1.2 + 2.0)} = 2.5 \times 10^{-2} = \frac{0.025}{2.01} = 0.01 << 1$$

So, chain reaction will not occur for which $\eta \geq 1$.

Problem 5. *Calculate the energy released when four protons fused to form one Helium nucleus. Given : M_p = 1.00814 amu* [M_{He} = 4.00387 amu and 1 amu = 931 MeV.]

Solution. Now, fusion reaction involved may be assumed as

$$4({}_1H^1) \rightarrow {}_2He^4 + 2\beta^+ + 2\nu + Q$$

Hence
$$Q = 931 \times [4 \times 1.00814 - 4.00387] = 26.71 \text{ MeV}$$

Problem 6. *How much natural Uranium is to be enriched so that the reactor becomes just critical.*

Solution. Given data is

$$\eta = 1, \sigma_{5f} = \sigma_{5a} = 1b, \ \nu = 2.5,$$
$$\sigma_{8f} = 0.5b, \quad \text{and} \quad \sigma_{8a} = 2.5b.$$

Then we have
$$1 = 2.5 \times \frac{\left(1 + \left(\frac{N_8}{N_5}\right) \times 0.5\right)}{\left(1 + \left(\frac{N_8}{N_5}\right) \times 2.5\right)}, \quad i.e. \quad \frac{2}{5} = \frac{\left(2 + \left(\frac{N_8}{N_5}\right)\right)}{\left(2 + 5\left(\frac{N_8}{N_5}\right)\right)}$$

or
$$\frac{N_8}{N_5} = \frac{6}{5} \quad i.e. \quad \frac{N}{N_5} = \frac{11}{5} \approx 2 \text{ where } N = N_5 + N_8$$

Hence, U^{235} and U^{238} should be 50% each.

EXERCISE

1. **On the basis of the given data prove that in natural uranium the chain reaction will not occur. Given data :**
$$\sigma_f = 4\ b,\ \sigma_t = 22\ b \text{ and } \nu = 2.0.$$
2. **Discuss various parts of a nuclear reactor.**
3. **Give points to classify different types of nuclear reactors.**
4. **Explain the concept of a chain reaction. Also discuss the neutron multiplication factor and the criticality of a reactor.**
5. **Explain the conversion of nuclear energy into electric energy with the help of a block diagram.**
6. **Write notes on :**
   (*a*) Nuclear reactors
   (*b*) Neutron cross-section
   (*c*) Criticality of reactors. *(Jan. 2002, K.U.K.)*
7. **Calculate the number of fission occurring per second in a Nuclear Reactor of 2 W power.**
   [Ans. 6.25×10^{10}/sec]
8. **Calculate the mass of U^{235} fuel being lost per day in 10^3 megawatt power nuclear reactor.** **[Ans. 1 kg/day]**

Chapter 11

NUCLEAR RADIATION DETECTORS

11.1 GAS-FILLED NUCLEAR DETECTORS

The gas-filled nuclear detectors are the oldest and are meant to detect nuclear radiations like α, β, and γ. These being handy and cheap are still in use. The working of these detectors is based on the fact that the nuclear radiations when pass through gases cause ionisation and the resulting ion-pairs (electron plus + ve ions) can be collected in the form of current or detectable voltage pulses.

Although the nomenclature of the detectors like ionisation chamber, proportional counter, and Geiger-Muller (GM) counters is based on the region of the operating voltage to collect the ion-pair, the geometry of the enclosing chambers containing the gas and the electronic set-up are, more or less, common. Let us discuss the theory of detectors.

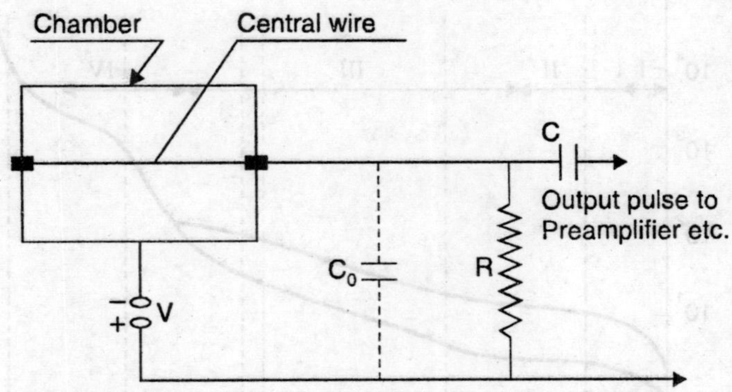

FIGURE 11.1 **Schematic representation of electronic set-up of gas-filled nuclear detector.**

First consider Fig. 11.1 which schematically represents the electronic set-up and the gas chamber.

A = Anode, a thin wire of tungsten,

K = Cylindrical metallic vessel, the cathode,

R = Resistance,

V = Applied voltage and C = capacitor.

The vessel K contains gas usually a noble gas like He, Ar, Kr, etc. at a pressure of few cm of mercury. As the nuclear radiation passes the gas through the walls of K, it gets ionized into + ve ions and electrons. As a result of the applied potential, + ve to A and – ve to K, the electrons move towards A while + ve gas ions towards the walls of K. As these approach the respective electrodes their potentials change and a voltage pulse appear across R. The shape of such a pulse varies with time in a way as shown in Fig. 11.2.

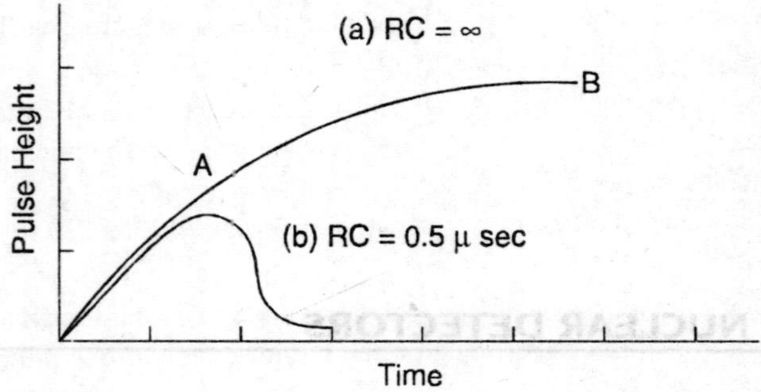

FIGURE 11.2 **The shape of electric pulse as a function of time.**

Sharp rising part (OA) is due to the fast moving electrons towards A while slow rising part (AB) is because of slow moving +ve ions towards cathode. This shape can be modified by adjusting the RC-value of the set-up. Here, C_0 denotes the inter-electrodes capacitance. If we increase the operating voltage V and measure the pulse size, P, as a function of V one obtains the following graph (Fig. 11.3).

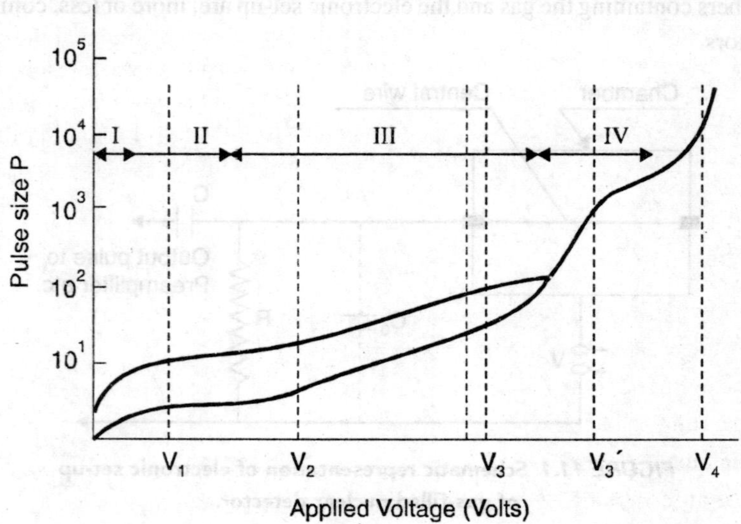

FIGURE 11.3 **Plot of pulse height, *P*, (or No. of ion pairs being collected by electrodes) vs. the applied potential, *V*.**

This graph gets divided into four distinct regions called as

I ⟶ Recombination region
II ⟶ Ionisation region
III ⟶ Proportional region
and IV ⟶ Geiger-Muller (GM) region.

Therefore, the detectors working in the region II, III, and IV are named respectively as

— Ionisation Chamber
— Proportional Counter
— G.M. Counter.

In the region I the resulting ions (electrons and + ve ions) of the gas have the finite probability to recombine to form neutral gas atoms. As the voltage V is increased the movements of the ions become faster and the probability of recombination decreases. Hence the pulse size increases. This explains the behaviour of the region I. The range of this region is about 0–20 V.

Next, as the potential V is raised further beyond a specific value, say V_1, there is no recombination and hence the pulse size depending upon the extent of ionisation caused by the nuclear radiation (called primary ionisation) becomes constant. This constancy stays within defined range *e.g.* V_1 to V_2. So, this explains the behaviour of the gas filled detectors for the region II. This range is about 20 to 100 V depending upon the nature of the gas and its pressure.

Beyond region II, the primary electrons becomes so energetic near the axial wire (where the field $E(r) = V/[r \log (b/a)]$ is very large, b being the radius of the cylindrical chamber, a the radius of the anode, V the applied potential between A and K and r the distance from the wire) that they cause avalanche of secondary ionisation through collisions with gas atoms. This phenomenon, more popular as Townsend avalanche, is known as electron multiplication. Thus in region III the pulse size will not only depend on the multiplication but will also be proportional to the extent of primary ionisation. The latter dependence is because the multiplication is independent of extent of primary ionisation. Thus, on the basis of combined effect the pulse size increases proportionally with the increasing operating voltage. This explains the behaviour of region III say from V_2 to V_3. This range is about 100 V to 500 V.

As the voltage is raised further the above mentioned proportionality gets destroyed although the pulse size still increases till the region IV where the multiplication becomes so large that the secondary ionisation spreads all along the central wire independent of the primary ionisation. As a result, the pulse size again becomes constant but for some definite range of potential. This region is now known as G.M. region its range is about 800–2000 V. The curve marked as 'b' corresponds to the case of radiation that has caused more primary ionisation than the lower case marked as 'a'. Because of higher primary ionisation curve, b, stays above the curve, a, all along.

Ionisation Chamber

For this detector following specification are commonly used.

Operating voltage – lies in region II (~ 20 V to 100 V)
Radius of the cylindrical tube (of copper) = 1cm
Radius of the Tungsten axial wire = 0.01cm
Length of the tube = 10 cm
Gas–He or Ar at few cm of Hg or air at normal atmospheric pressure.
The output pulse is extracted through additional capacitor C (Shown in Fig. 11.1).

Some times the ionisation chamber is used in D.C. current mode in which following circuit is employed (Fig. 11.4).

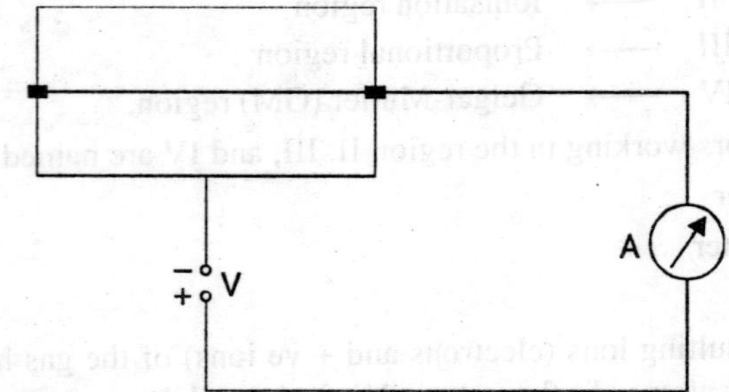

***FIGURE 11.4* Electronic circuit of ionization chamber in D.C. mode.**

The variation of D.C. current as measured by the ammeter *A* is as shown in Fig. 11.5.

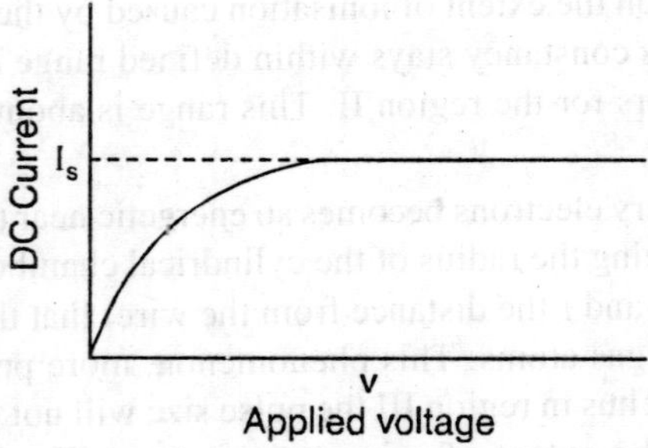

***FIGURE 11.5* Variation of D.C. current as a function of applied potential, V.**

The value of saturation current I_s, obviously determines the intensity of the incident radiation.

The ionisation chamber cannot detect weak radiations like *X*-rays, low energy β-radiation etc. It also has a large dead time.

G.M. Counter

The geometry of the tube is the same as that of ionisation chamber. However, in the case of G.M. counter, the tube may be provided with an end window of thin sheet of mica or aluminium. G.M. counter's operating voltage falls in the G.M. region about from 800 Volts to 2000 Volts. Usually one measures the counts of radiation entering the tube in a given duration. When counting rate plotted against voltage one gets the following type of graph (Fig. 11.6).

From the plateau of the graph the operating voltage is determined which should be around its middle as shown in (Fig. 11.6.) In G.M. tubes the gas used is a noble gas Ar (90%) mixed with ethyl alcohol (10%). The latter acts as the quenching gas, which helps in making the performance of the G.M. counter stable by suppressing the possibility of secondary avalanche. In fact, the +ve argon ions while moving towards the walls (cathode) exchange their charge with ethyl alcohol molecules, which become + ve ions. The ethyl alcohol ions then dissociate after interacting with the cathode. This way the possibility of secondary avalanche is completely checked. Hence there will be no false counts if some organic polyatomic gas is mixed with argon in the G.M. counter. The pressure of the gas in G.M. counter is maintained at few cm of Hg.

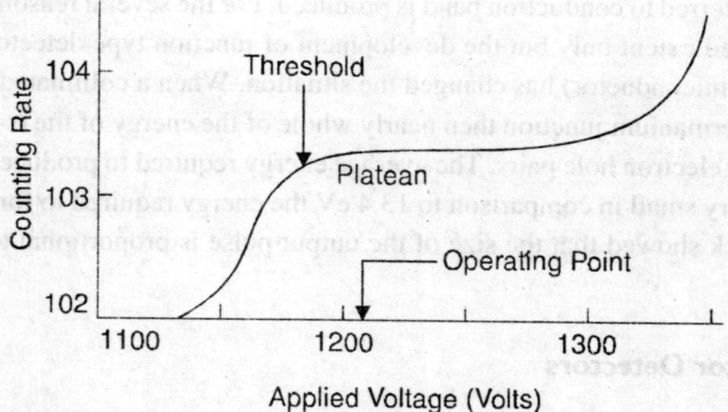

FIGURE 11.6 **Variation of counting rate as a function of applied potential, *V*.**

Unfortunately the G.M. counter has a finite time (called dead time) to recover it's original state of operation. So, it can not be used for measuring fast counting rates (radiations). The G.M. counter is mainly used to measure and detect γ-radiation, X-rays, β-rays etc.

We can now distinguish between ionization chamber and G.M. counter in the following way.

	Ionization Chamber		*G.M Counter*
1.	Operational voltage is low *e.g.* 20–100 V [Ionization Region].	1.	Operational voltage is high *e.g.* 800–2000 V [G.M Region].
2.	Gas could be air at atmospheric pressure.	2.	Gas is Argon 90% + ethyl alcohol 10% at few cm of Hg.
3.	Dead time is large.	3.	Dead time is smaller.
4.	(*a*) Output pulse depends upon Primary ionization. (*b*) No quenching gas is required	4.	(*a*) Output pulse is independent of primary ionization. (*b*) Quenching gas used in G.M. counter is 10% ethyl alcohol.
5.	Used to measure intensity and energy of the incident radiation.	5.	It is used to count the number of pulses in a given interval of time.
6.	It cannot measure weak radiation like X-rays etc.	6.	It can detect all types of weak as well as strong radiations.

Note : Electrical circuits for both are identical.

11.2 SOLID STATE NUCLEAR DETECTOR

Ionization chamber operation is not limited to gas filled ion chambers. Crystal counters are essentially the ionization chambers using solid dielectrics between parallel plate electrodes.

Principle

A crystal of AgCl (intrinsic semiconductor) is placed between two electrodes, to which a voltage is applied. When the crystal is exposed to the radiations some of the valance band electrons get energy and enter into the conduction band leaving the positive holes. Under the action of an electric field the electrons move towards the positive electrode and holes move towards negative electrode and a pulse of current equivalent to the number of electrons transferred to conduction band is produced. For the several reason, the simple solid state counter is used to a limited extent only but the development of junction type detectors (based on impurity rather than on intrinsic semiconductor) has changed the situation. When a collimated beam of α-paricles is allowed to fall on a *n-p* germanium junction then nearly whole of the energy of the α-particles, absorbed in it, is utilized in producing electron hole pairs. The average energy required to produce one electron hole pair is about 3 eV, which is very small in comparison to 13.4 eV the energy required to form an ion pair in a gas. J.D.Mayer and B.Geessick showed that the size of the output pulse is proportional to the energy of the α-paricle.

Types of Semiconductor Detectors

Solid state detectors are of following two types :

1. Crystal Detector
2. Junction type detector

1. *Crystal Detector.* It consists of thin wafer of an intrinsic semiconductor, say pure Si or Ge, whose opposite faces are covered with very thin layer of gold metal [see Fig. 11.7]. The two conducting surfaces are connected to a d.c. source of 1–20 volts and an amplifier.

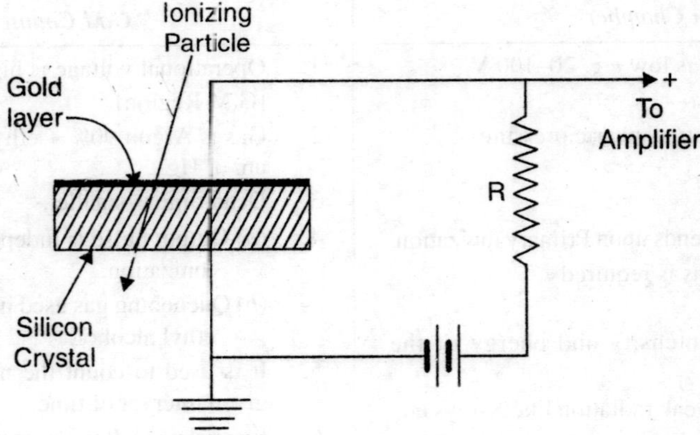

FIGURE 11.7 **The electronic set-up of the crystal detector.**

Under ordinary conditions, the electrical conductance of an intrinsic semiconductor crystal is quite small. But when an ionizing particle enters the crystal, free electrons and holes are produced which increase its conductivity. Due to mobility of electrons and holes in opposite directions, under the influence of an electric field a sudden current is established. The current pulse is amplified and then either measured or counted.

Drawbacks of Crystal Detector : The major drawback of this detector is that at ordinary room temperature it gives considerable background noise *i.e.* there are pulses of random size and frequency even in the absence of an ionizing particle. So the crystal has to be kept at low temperature.

2. *Junction type detector.* In this type of detector, a thin layer of the *n*-type is formed on the surface of a *p*-type (Si, Ge crystal) by allowing a donor impurity (Phosphorous) to diffuse into one face of the crystal at a fairly high temperature. There is a tendency for electrons to diffuse from *n*-region (where electrons are in excess) into *p*-region (where the electrons are in deficit). An equilibrium is reached when fermi level is constant throughout the material. As a result an intermediate volume around the interface will be cleared from carriers of both signs and this region is called depletion layer. This layer has low conductivity and high field hence offers ideal conditions for the detection of ionizing radiation.

When the negatively charged electrons diffuse away from the n-type region, they leave the material positively charged. Similarly, the filling of holes in the p-region generates a corresponding negative charge. The small potential difference of about a volt is established in this manner is enhanced by an additional potential of several hundred volts referred to as a reverse bias voltage. In the presence of this electric field the positive holes in the p-type side are pulled towards the negative electrodes and the electrons towards the p-n junction and thus a depletion layer is created in the p-type material. The thickness of the depletion layer increase with the reverse bias voltage and the limit is set by electrical break down in the semiconductor.

When ionizing particles enter the depletion layer region through the p-n junction side (see Fig. 11.8), as a result of energy absorption electron hole pairs are created, as in an intrinsic semiconductor. Under the influence of the reverse bias voltage the electron hole pair are swept rapidly to the electrodes and thus producing a pulse of current which can be amplified and recorded. If the ionizing particle is completely stopped in the depletion layer, so that its whole energy is utilized in producing electron hole pairs, the pulse height is proportional to the initial energy of the particle.

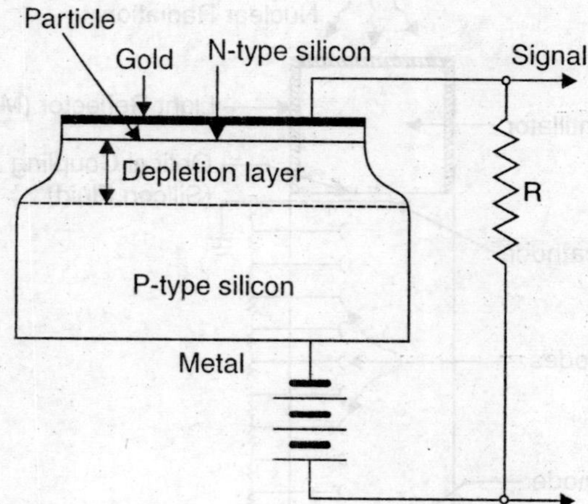

FIGURE 11.8 **The electronic set-up of a junction detector.**

Scintillation Detector

There are several substances, certain organic and inorganic, which emit light flashes or scintillation when charged particles, X-rays or gamma rays pass through them. These substances are called scintillators. In a scintillation detector these light flashes are allowed to fall on the photo cathode of photomultiplier tube and a pulse is extracted out to signal the passage of nuclear radiation. The height of this output pulse can be made proportional to the energy dissipated by the ionizing radiation in the scintillation. Thus a scintillation detector can be used not only for counting but for energy analysis also. We shall describe the operation of a scintillation detector as a step by step process.

(*a*) ***Absorption Process.*** A charged particle falling on a scintillator can dissipate all its energy in it, if the dimensions of the scintillator are large as compared to its range. The gamma-rays (X-rays) it may interact with matter in three ways : (1) Photoelectric effect, (2) Compton effect and (3) Pair production. In each of the effect electrons are produced and by successive interactions the gamma rays transform all or part of its energy into K.E. of such secondary electrons. These secondary electrons will give up their K.E. in ionization or excitation in the scintillator material.

(*b*) ***The Scintillation Process.*** The scintillator absorbs energy when it undergoes ionization and excitation by the electron. The absorbed energy appears either as heat energy or as luminescence photons. In the later process, the excited states of the scintillator material de-excite to lower states by light emission with in 10^{-8} sec. The emitted light is known as scintillation. The time dependence of the photon emission follows an exponential law $N_p = \text{const}\,(1 - e^{t/\tau})$ where N_p is the number of light photon emitted in the time t after the ionizing radiation has arrived and τ is the decay time or the time required for the emission of 63% of the photons. Each scintillator material has a characteristics mean lifetime t.

(*c*) ***Conversion of Light into Electrical Pulse.*** The light emitted by a scintillator is allowed to fall on the photo cathode of a photomultiplier tube which is optically coupled to the scintillator. As a result, depending on the geometry and optical properties of the system, certain fraction of light produced in the scintillator falls on the photo cathode of the photo multiplier tube producing photoelectrons. The structure of photo multiplier tube is such that these are several dynodes as shown in Fig. 11.9 and these are maintained at successively higher electrical potentials (about 100 volts per dynode stage).

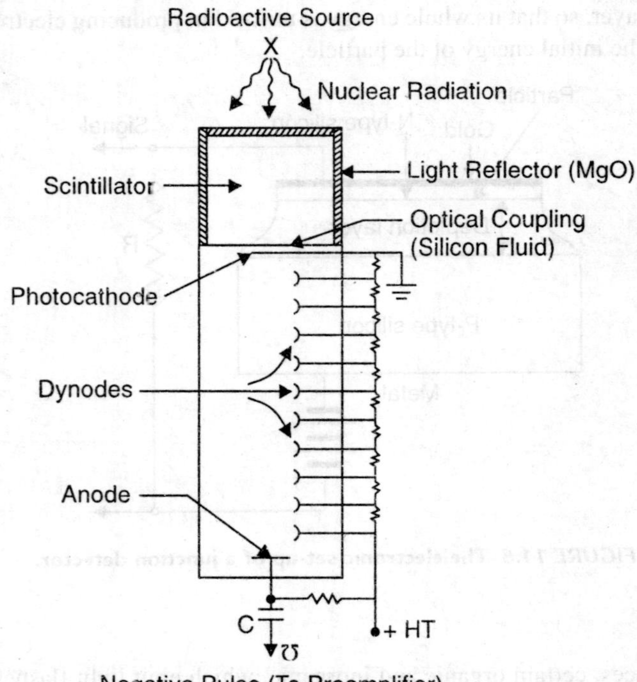

FIGURE 11.9 **Block diagram of a scintillation detector.**

Further the dynodes surfaces have a property of emitting more secondary electrons when one incident electron strike them. Thus a photoelectron emitted by a photo cathode is accelerated by the electric field to the first dynode where it produce a bunch of secondary electrons. These electrons are now accelerated towards the next dynode where they, in turn produce more electrons and this process is repeated at each dynode. In a photomultiplier tube there are usually 10 or more dynode stage and one can achieve a gain upto10^7 to 10^8 by the time the electron reach the last stage which is called the anode. In this way a burst of electrons at the anode where an electrical pulse is taken out for further analysis.

The quality of the photomultiplier tube is mostly determined by two properties :

1. The transit time or the time taken by the electrons to reach anode through dynode structure. This property is important for the timing characteristics of the photomultiplier tube.

2. The stability of the electron mutiplication factor. This property determines the energy resolution characteristics of the photomultiplier tube.

Now we prove that the amplitude of the pulse appearing at the output of the mutiplier tube is proportional to the energy of the incident nuclear radiation. The pulse height V at the output will be related to the amount of charge (Q) collected at the output through the relation

$$V = \frac{Q}{C} \quad ...(11.1)$$

where C is the capacity of the output point.

If the incident nuclear radiation has energy E_n, a fraction f_n of this energy will be dissipated in the scintillator. If this energy $f_n E_n$ is converted into light energy by the scintillator with an efficiency ε_e then an amount $\varepsilon_e f_n E_n$ is the amount of light energy produced in the scintillator. If f_p is the fraction of the photon which reach the photo cathode then an amount $f_p \varepsilon_e f_n E_n$ of light energy reaches the photo cathode. Let f_m denote the no. of photoelectrons emitted by the photo cathode when an 1.0 eV of light energy strikes it and let f_d be the fraction of photoelectrons collected by dynode. If the electron multiplication factor is M then the total no. of electrons reaching the final stage will be

$$M_e = M f_d f_m f_p \varepsilon_e f_n E_n$$

and

$$Q = M_e e = eM \ f_d f_m f_p \varepsilon_e f_n E_n.$$

Thus usually $V \propto E_n$. We thus see that pulse height in a scintillation detector is proportional to the initial energy of the incident particle.

Advantages of Scintillation Detector

Following are some advantages of Scintillation Detector.

1. Pulse height is proportional to the energy of the incident rays.
2. Dead time of the detector is very small and so high counting rate is possible.
3. Pulse duration being very short, so the resolving power is high.
4. In case of solid scintillator no. of atoms per unit volume being quite large, so a strong pulse is observed than in case of gas scintillator.

Disadvantages of Scintillation Detector

The output varies with the type of radiation and also the overall efficiency is less as the phenomenon of conversion of radiation into light and then light into electron to count the high-energy radiation is a complex phenomenon.

11.3 SOME OTHER DETECTORS

Nuclear Track Detectors

The counters we have been discussing use a sensitive volume of gas, liquid or solid with well-defined dimensions to detect the presence of a particle. Now we shall discuss four detectors that not only record the presence of the particles but indicates the path that the particle travelled as well. We call these detectors as track detectors. Four track detectors, which we shall discuss here, are the Wilson Cloud Chamber, The Bubble Chamber, Photographic Emulsion and The Spark Counter.

Wilson's Cloud Chamber

Principle. When air contained in an enclosure, saturated with water vapour, is subjected to a sudden adiabatic expansion the temperature falls and super saturation takes place leading to the condensation of the water vapour and the formation of a cloud of water droplets. The production of such a cloud is, however, impossible unless nuclei on which the water vapour may condense are provided. In ordinary air dust particle plays the role of nuclei. But in dust free air, the phenomenon of condensation and cloud formation is very difficult. Wilson, however, explained that the condensation cloud could be produced even in dust free air provided the air is capable of ionization by an ionizing agent such as α-rays, cathode rays, X-rays etc. Evidently, the ions produced act as centers of condensation.

A super cooled vapour may remain in the vapour phase and not condense. But in dusty and smoky region there are discontinuities on which condensation can begin and therefore fog or smog form rapidly.

Construction and Working. The cloud chamber illustrates the fact that charged particles are discontinuous on which condensation can be initiated. The schematic diagram of Wilson cloud chamber is shown in Fig. 11.10.

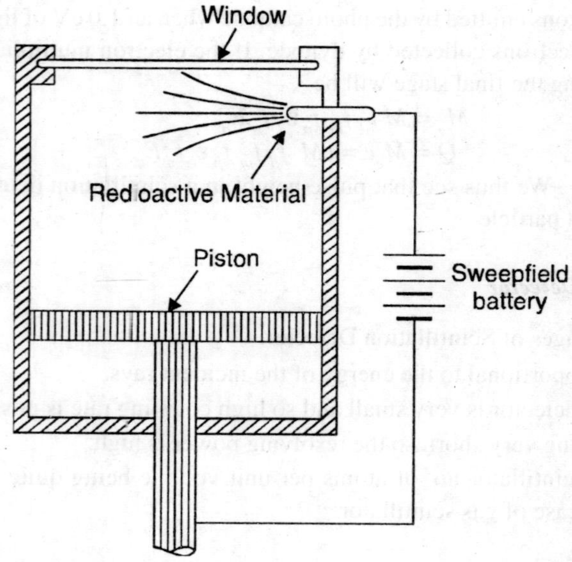

FIGURE 11.10 **Schematic arrangement of cloud chamber.**

It consists of a gas tight chamber having a large glass window, so that its volume can be illuminated and the events within the chamber can be seen or photographed. One wall of the chamber is a movable piston. The chamber contains a saturated vapour from an excess amount of some liquid, usually a mixture of alcohol and H_2O. If the volume of the chamber is suddenly increased by moving the piston, the adiabatic expansion causes cooling that renders the vapour supersaturated, unstable and likely to condense. If there are ions within the chamber, condensation occurs preferentially on them. Since we have no interest in stray ions therefore the cloud chamber is provided with electrodes, which maintain a weak electric sweep field.

Bubble Chamber

To detect very energetic particles the cloud chamber is not very suitable since the interaction of energetic particles cannot be observed in chamber. To overcome this disadvantage it is necessary to use a chamber filled with a substance having a much greater stopping power. This has lead to the development of Bubble chamber.

Principle. The basic principle of bubble chamber is the superheating of a liquid and the bubbles that form in the process of boiling, *e.g.* water boils at 100°C at standard atmospheric pressure. If the pressure is increased as in pressure cooker boiling will not begin until a higher temperature is reached. If the pressure is then suddenly reduced, boiling begins with the sudden formation of tiny bubbles that grow quickly in size and the liquid become in superheated state. If an ionizing particles traverse the liquid in a few milliseconds after the pressure is released, the ions left in the track of a particle act as condensation centers for the formation of vapour bubbles. The vapour bubbles grow at rapid rate and attain a visible size in a time of the order of 10 to 100 μ sec. Upon this principle the bubble chamber operates. The bubbles are like the droplets in a cloud chamber, visible under strong illumination.

The time sequence of events for a typical chamber is shown in Fig. 11.11.

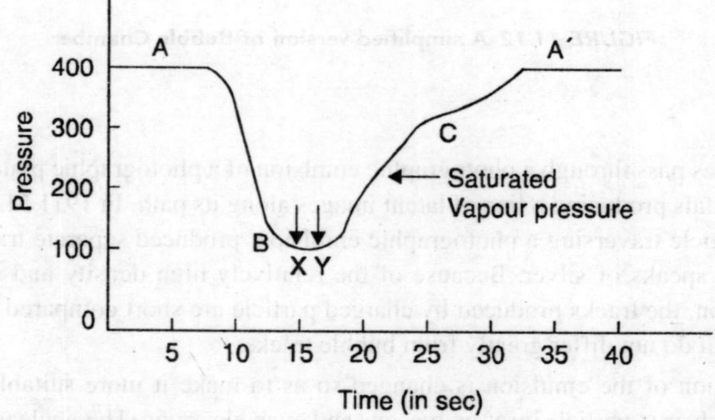

FIGURE 11.11 **Time sequence of events for a typical bubble chamber.**

At the instant *A*, the high pressure P_1 applied to the bubble chamber liquid is suddenly reduced to the lower pressure P_2. The pressure starts to rise due to the bubble formation. The liquid is recompressed at *C* and the chamber returns to its initial position at *A′*. The time interval between *A* and *A′* gives the cycling rate of the chamber. The expansion is timed so that the particles enters the chamber at *X* and bubbles are illuminated and photographed at the moment *Y*. The liquid suitable as the working substance of a chamber are transparent and having a low surface tension and a high vapour pressure. The medium in first bubble chamber was ordinary ether, diethyl ether but liquid hydrogen, helium, deuterium, propane, xenon and other liquids have been utilized successfully for different purpose.

Circuit Diagram. A simplified diagram of a bubble chamber is shown in Fig. 11.12.

A box with thick glass wall, filled with a liquid is connected to a pressure system and then heated to some predetermined temperature. High-energy particles enter the liquid through a thin window. A sudden release of valve in the pressure system is quickly followed by the flash of floodlights and snap of a camera shutter. If the chamber is operating properly and events are correctly timed, then sharply defined trails of bubbles that have formed on the paths of ions made by the traversing particles will be photographed.

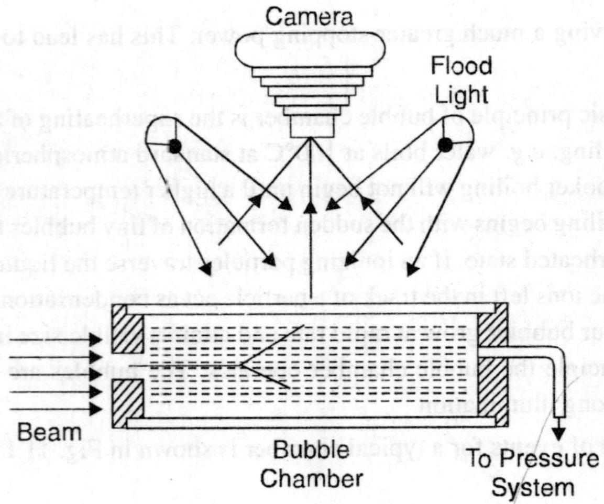

FIGURE 11.12 **A simplified version of Bubble Chamber.**

Nuclear Emulsions

When charged particles pass through a photographic emulsion of a photographic plate they interact with the tiny silver halide crystals producing a line of latent images along its path. In 1911 M. Reiganum recognized that individual α-particle traversing a photographic emulsion, produced separate tracks in it as a series of closely spaced black speaks of silver. Because of the relatively high density and stopping power of the photographic emulsion, the tracks produced by charged particle are short compared with those obtained in the cloud chamber, but do not differ greatly from bubble tracks.

The composition of the emulsion is changed so as to make it more suitable for study of various ionizing particles, such as α-particle, proton, mesons and even electrons. The nuclear emulsions differ from optical emulsions in a manner that they have considerably higher silver halide content and smaller grain size. In nuclear emulsions the thickness is greater than that of optical emulsion. The smaller the grain size, the more sensitive the emulsions to ionizing particles.

SOLVED NUMERICAL PROBLEMS

Problem 1. *A slow neutron entering a U^{235} fission chamber releases fission fragments of a total energy of 200 MeV. If the capacity of the collector system is 25 μμf, calculate the resultant pulse height.*

Solution. Since the energy required for the production of one ion pair is 35 eV.

Hence total number of ion pairs produced $N = \dfrac{200 \times 10^6}{35} = 57.14 \times 10^5$

∴ Total charge produced $= Ne = 57.14 \times 10^5 \times 1.6 \times 10^{-19}\ \text{C} = 91.42 \times 10^{-14}\text{C}$

Pulse Height $\Delta V = \dfrac{\Delta q}{C} = \dfrac{91.42 \times 10^{-14}}{25 \times 10^{-12}} = 3.65 \times 10^{-2}$

Problem 2. *A GM tube with a cathode 5.0 cm in diameter and a wire having a diameter of 0.012 cm is filled with argon to a pressure such that the mean free path is 7.8×10^{-4} cm. Calculate the value of the voltage that must be applied to just produce an avalanche.*

Solution. As we know the electric field strength is given by $E = \frac{V}{r}\left(\frac{1}{\log\left(\frac{b}{a}\right)}\right)$

Avalanche will just start when the integration of this eqn taken over one mean free path is equal to the ionization potential of the filling gas, *i.e.*

$$E_a\lambda = \frac{V\lambda}{a\log\left(\frac{b}{a}\right)} = I \quad \text{or} \quad V = Ia\frac{\log\left(\frac{b}{a}\right)}{\lambda}$$

$$V = 15.7 \times 0.006 \times 10^{-2} \times \frac{\log\left(\frac{0.025}{0.00006}\right)}{7.8} \times 10^{-6} = 730 \text{ Volts}$$

EXERCISE

1. Discuss the principle, construction and working of scintillation counter.
2. Explain in detail the solid state detectors.
3. Explain the nuclear emulsion technique in the study of ionizing radiations.
4. What is the physical phenomenon that allows Wilson cloud chamber to detect particles ? What is it for bubble chamber ?
5. Describe the basic principle and operation of scintillation spectrometer. What are its merits?
6. Describe the principle, construction and working of cloud chamber and bubble chamber.
7. An organic quenched GM tube operates at 1000 Volts and has a wire diameter of 0.2 mm. The radius of the cathode is 2 cm and tube has life-time of 10^9 counts. What is the maximum radial field ?
   [Ans. 1.89×10^6 V/m]
8. An anthracene crystal and a 10 stage photomultiplier tube are to be used as a scintillation spectrometer. It is desired that a 10 keV β-particles incident on the scintillator produce a 2 mV pulse on the photomultiplier output circuit. Which has a capacity of 120μμ*f*. What average electron multiplication per stage is required in the photomultiplier? Assuming a light collection efficiency unity and a photocathode efficiency 0.1.
   [Ans. $n = 3.16$]
9. Distinguinsh between Ionization Chamber and G.M. Counter.
10. Give the principle and the experimental setup of a gas filled nuclear detector.
11. Explain the principle, construction and working of ionization chamber.
12. Discuss salient features of Geiger-Muller detector.
13. Distinguish between Ionization Chamber and G.M. Counter.
14. Explain the phenomenon of self-quenching in G.M. Detector.
15. Show how ionisation chamber can be used as G.M. counter ? *(Jan. 2002, K.U.K.)*

Chapter 12 THERMAL PHYSICS

12.1 TEMPERATURE AND ITS MEASUREMENTS

We are familiar with the sensation which we call warmth. Our sense enables us to distinguish between different degree of warmth, to some extent and we express this by saying that a body is either hot or cold. The degree of hotness or coldness of a body is known as its temperature. Though the change in temperature of a body results from the change in the heat energy but the temperature is not the measure of the heat. A body possessing a large quantity of heat may be cold while a body possessing a small quantity of heat may be quite hot.

Temperature Measurement

When a substance receives heat, the following effects are noticed :

(*i*) Change in volume
(*ii*) Change in pressure
(*iii*) Change in electric resistance
(*iv*) Change in radiation
(*v*) Change in thermo electric e.m.f.
(*vi*) Change in colour.

Any of these changes can be used for measurement of temperature. The most accurate method of measuring temperature is by making use of either first or the last of the above effects of heat. An instrument making use of expansion of a substance is called thermometer and used for measuring the temperature of the body. Since solids, liquids and gases expand on heating, any of the three can be used to construct a thermometer. In case of solids it should be heated to a high temperature. Thus expansion of solids is used to measure high temperature. While liquids expand moderately on heating, and the thermometer making use of liquids as thermometric substance are used to measure moderate temperatures. In case of gases the expansion is very large even for a small change in temperature. As a result thermometers using gas as a thermometric substance are used to measure small change in temperature.

12.2 PLATINUM-RESISTANCE THERMOMETER

Principle

It is found that the electrical resistance of metals increases more or less uniformly with temperature. This effect is used for measuring temperature and the metal normally used is platinum.

If α = thermal co-efficient of temperature *i.e.* α is change in resistance of a wire from its original resistance at 0°C, for every degree centigrade change in temperature.

Let R_t = resistance of a given wire at t°C

R_0 = resistance of a given wire at 0°C

Then $R_t = R_0 + R_0\alpha t$ ...(12.1)

Similarly, if the resistance of the same wire at 100°C is R_{100} then

$$R_{100} = R_0 + R_0\alpha \times 100 \quad \text{... (12.2)}$$

From eqs. (12.1) and (12.2) we have

$$R_0\alpha t = R_t - R_0 \quad \text{...(12.3)}$$

and

$$R_0\alpha \times 100 = R_{100} - R_0 \quad \text{...(12.4)}$$

Dividing eq. (12.3) by eq. (12.4)

$$\frac{t}{100} = \frac{(R_t - R_0)}{(R_{100} - R_0)} \quad \text{or} \quad t = \left[\left(\frac{R_t - R_0}{R_{100} - R_0}\right)\right] \times 100 \quad \text{...(12.5)}$$

Knowing the value of R_{100}, R_0 and R_t the value of t can be calculated.

Construction

Construction of platinum resistance thermometer is shown in Fig. 12.1. It consists of a pure platinum wire wound in a double spiral on a mica frame and placed at the bottom of a glazed porcelain tube. The free ends

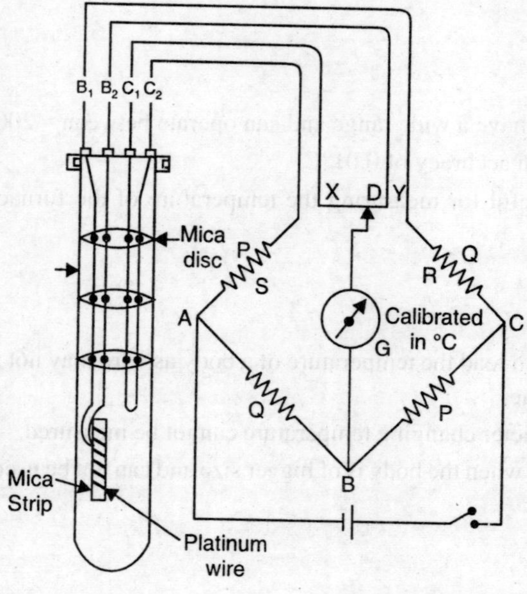

FIGURE 12.1 **Platinum resistance thermometer.**

of the wire are connected to the binding terminals B_1 and B_2. Two identical leads of the same resistance as that used for platinum wire are also provided to compensate for the resistance of the leads connecting platinum wire. The whole tube is sealed and binding terminals are provided at the top. To ensure best insulation and prevent convection losses all the leads pass through mica discs.

Working

A modified Wheat-Stone bridge is used to measure the change in resistance, of the platinum wire, used in the platinum thermometer, with temperature.

Two resistance of equal value P and Q and a standard fractional resistance box S form the three arms of the Wheat-Stone bridge. The platinum wire; whose resistance is to be determined, is connected in the fourth arm of the bridge as shown in Fig. 12.1. The compensating lead resistance is connected with the resistance S in series. A sensitive galvanometer is connected between the point B and D. A cell is connected between A and C with a key in series. A wire of ρ resistance per unit length and of total length $2l$ is connected between the two terminals X and Y of the bridge.

The thermometer is dipped in an ice bath and the jockey is adjusted so that the galvanometer gives no deflection when at D. Let O be the middle point of the bridge wire and the balanced point D be x cm to the left of O for some suitable values of S.

Then, $XD = l - x$ and $DY = l + x$

Since $P = Q$, the resistance of the arm AD should be equal to the resistance of the arm CD

$$\therefore \quad R + r + (l - x)\rho = S + r + (l + x)\rho \qquad ...(12.6)$$

where r is the resistance of compensating lead.

From eq. (12.6), we have

$$R = S + 2x\rho \qquad ...(12.7)$$

Thus by knowing the values of S, ρ and x the resistance of the platinum wire can be determined at particular temperature and by substituting the value of resistance at 0°C, 100°C and t°C the value of t can be calculated through eq. (12.5).

Merits

(*i*) This thermometer have a wide range and can operate between – 200°C and 1200°C.

(*ii*) It can work with an accuracy of 0.01°C.

(*iii*) It is especially useful for measuring the temperature of the furnace sitting quite far from the furnace.

Demerits

(*i*) It takes long time to read the temperature of a body as wire may not attain the temperature of the body in a short time.

(*ii*) With this thermometer changing temperature cannot be measured.

(*iii*) It can be used only when the body is of bigger size and can not be used to measure the temperature at a point.

12.3 THERMOELECTRIC THERMOMETER

Principle

When two wires of different metals are joined together in a loop an emf is produced if the junction are kept at different temperature. The emf so produced is known as thermoelectric emf and its magnitude depends upon the difference of the temperature between the hot and cold junctions. A simple arrangement as shown in Fig. 12.2, can demonstrate this effect. The hot junction can be heated by lamp flame while other junction can be kept cold by giving an ice bath. Because of a temperature difference between the two junctions a deflection is seen in the galvanometer.

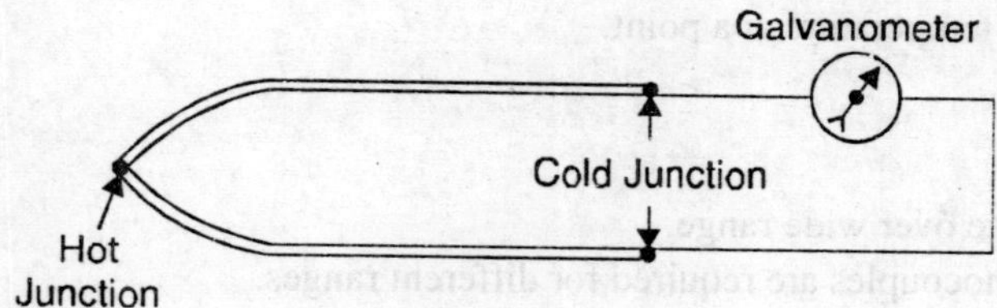

FIGURE 12.2 **Thermocouple.**

Construction

Thermoelectric thermometer construction is shown in Fig. 12.3. It consists of two different types of wires joined at one end by welding. To provide insulation the two wires are enclosed in a tube of glass or clay. These wires are passed through mica disc to keep them in position. The whole arrangement is enclosed in porcelain or quartz tube. The free ends of the wires are connected to two terminals ; in between them is connected a calibrated galvanometer.

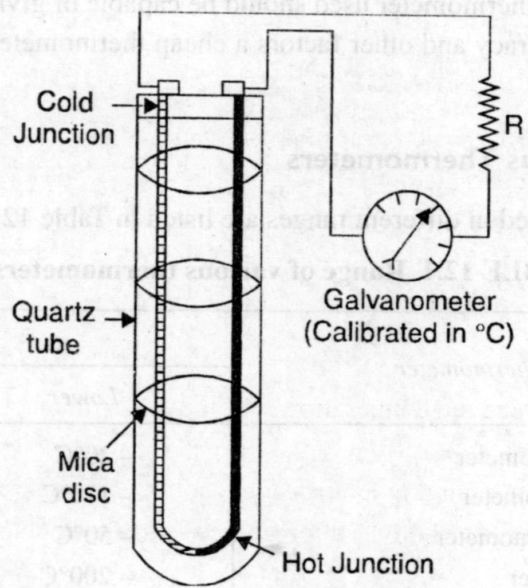

FIGURE 12.3 **Thermoelectric thermometer.**

Working

The body whose temperature is to be found is brought in contact with the welded junction (hot junction) of the thermometer. After some time when the junction attains the temperature of the body, the deflection in galvanometer is noted. This reading directly gives the temperature of the body. By choosing suitable metallic wires, with the help of this thermometer, temperature up to 300°C can be measured.

Merits

(*i*) It has a very wide operation range from 200°C to 300°C.
(*ii*) Rapidly varying temperature can be measured with this thermometer.
(*iii*) It is cheap and easy to construct.
(*iv*) It can measure temperature at a point.

Demerits

(*i*) It is not accurate over wide range.
(*ii*) Different thermocouples are required for different ranges.
(*iii*) For each thermocouple galvanometer requires fresh calibration.

12.4 FACTORS FOR THE SELECTION OF A THERMOMETER FOR PARTICULAR USE

1. **Range.** The thermometer should be such that it can measure the required temperature of the body.
2. **Availability.** If a particular required thermometer is not available a linear thermometer may be used.
3. **Accuracy.** The selecting of the thermometer should be in view of accuracy required.
4. **Time of response.** The thermometer used should be capable of giving reading quickly.
5. **Cost.** If in view of accuracy and other factors a cheap thermometer can work we should not use costly thermometer.

Temperature Range of Various Thermometers

Which thermometers should be used in different ranges are listed in Table 12.1.

TABLE 12.1 Range of various thermometers

S.No.	*Type of Thermometer*	*Range*	
		Lower	*Higher*
1	Mercury Thermometer	– 20°C	333°C
2	Alcohol Thermometer	– 110°C	55°C
3	Bimetallic Thermometer	– 50°C	550°C
4	Gas Thermometer	– 200°C	1600°C
5	Platinum Resistance Thermometer	– 200°C	1500°C
6	Thermoelectric Thermometer	– 200°C	3000°C
7	Pyrometers	– 200°C	3200°C

SOLVED NUMERICAL PROBLEMS

Problem 1. *The resistance of a platinum wire of a platinum resistance thermometer at the ice point is 5 ohm and at the steam point 5.23 ohm. When this thermometer is inserted in a hot bath, the resistance of the platinum wire is found 5.795 ohm. Calculate the temperature of the bath.*

Solution. Given $R_0 = 5 \text{ ohm}, R_{100} = 5.213 \text{ ohm}, R_t = 5.795 \text{ ohm}$

We know that $$t = \left[\frac{R_t - R_0}{R_{100} - R_0}\right] \times 100 = \left[\frac{(5.795 - 5)}{(5.23 - 5)}\right] \times 100$$

$$= 0.795 \times \frac{100}{0.23} = 345.65°\text{C}$$

Problem 2. *The resistance of platinum wire thermometer is found to be 5.5 ohm and 7.624 ohm at 0°C and 100°C, respectively. The resistance of the wire is found to be 14.45 ohm when brought in contact with a hot body. Calculate the temperature of the body.*

Solution. Given $R_0 = 5.5 \text{ ohm}, R_{100} = 7.624 \text{ ohm}, R_t = 14.45 \text{ ohm}$

We know that $$t = \left(\frac{R_t - R_0}{R_{100} - R_0}\right) \times 100 = \frac{8.95}{2.124} \times 100 = 421.37°\text{C}$$

Problem 3. *The resistance of wire in a platinum thermometer is 2.4 ohm and 3 ohm at 0°C and 100°C respectively. What will be the resistance of the wire when placed in cold bath of 190 K.*

Solution. Given $R_0 = 2.4 \text{ ohm}, R_{100} = 3 \text{ ohm}, R_t = ?, t = 190 \text{ K} = -83°\text{C}$

We know that $$t = \left(\frac{R_t - R_0}{R_{100} - R_0}\right) \times 100$$

or $$R_t - R_0 = \frac{t}{100} \times (R_{100} - R_0) = -\frac{83}{100} \times (3 - 2.4) = \frac{-83}{100} \times 0.6 = -0.498$$

or $$R_t = R_0 - 0.498 = 2.4 - 0.498 = 1.902 \text{ ohm.}$$

EXERCISE

1. Discuss the different factors you will take into consideration before choosing a particular thermometer for a particular use.
2. Explain the construction and function of platinum resistance thermometer.
3. Explain the construction and function of thermoelectric thermometer.
4. What precautions should be taken in measuring temperature with a thermoelectric thermometer.
5. The resistance of the wire in a platinum thermometer is 2.2 ohm and 2.8 ohm at 0°C and 100°C respectively. What will be the resistance of the wire when placed in cold bath of 200 K. **[Ans.** 1.762 ohm]
6. The resistance of platinum wire thermometer is found to be 5.0 ohm and 7.0 ohm at 0°C and 100°C, respectively. The resistance of the wire is found to be 15.0 ohm when brought in contact with a hot body. Calculate the temperature of the body. **[Ans.** 500°C]
7. Human body temperature is 98.4°F and boiling point of water is 212°F. Express these temperatures in centigrade. **[Ans.** 36.89°C and 100°C]

Chapter **13**

HEAT RADIATIONS

13.1 BLACK BODY AND BLACK BODY RADIATIONS

A black body is one which absorbs all the heat radiation falling on it.

When radiation are allowed to fall on such a body they are neither reflected nor transmitted. Such a body after absorbing the incident radiation *i.e.* on heating starts emitting radiation which are independent of the nature of the body and only depend on the temperature of the black body. These radiation are known as black body radiation. It is called a black body due to the fact that whatever may be the colour of the incident radiation, the body appears black. In practice there exist no perfectly black body. However, a body showing close approximation to a perfectly black body can be constructed as follows.

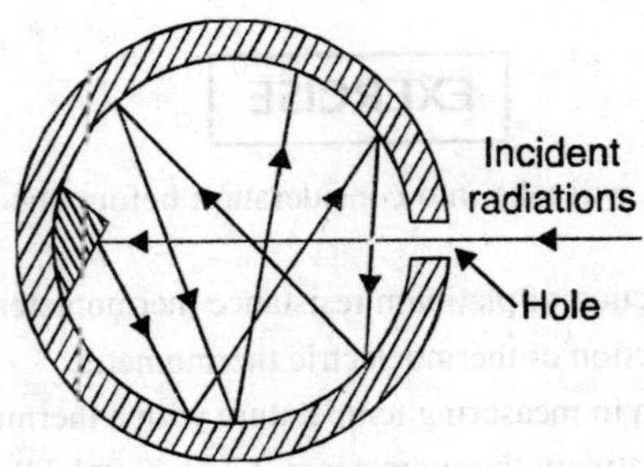

FIGURE 13.1 **Black body.**

Take a hollow sphere with a fine hole and a pointed projection in front of the hole and coat it with lamp black on its inner surface as shown in Fig. 13.1. When the radiation enter in such a body through the hole they suffer multiple reflection and are completely absorbed. Such a body acts as a black body absorber.

When such a body is heated through a high temperature it starts emitting radiation called black body radiation through the whole as shown in Fig. 13.2.

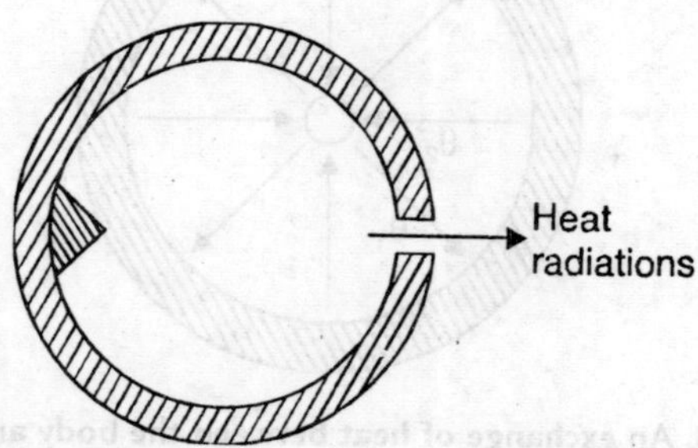

FIGURE 13.2 **Black body radiation.**

13.2 LAWS OF BLACK BODY RADIATION

Some of the characteristics of black body radiation have been very useful in understanding various aspects of heat and thermodynamics and have been widely used in solving related problems. These can be summarized in the form of the following laws ;

(*i*) Kirchhoff's law

(*ii*) Stefan's law

(*ii*) Wien's law

Kirchhoff's Law

According to this law at a given temperature the ratio of the emissive power to the absorptive power for all bodies is constant and is equal to the emissive power of a perfectly black body. This shows that bad absorber are bad emitters and good absorbers are good emitters.

Let e = emissive power of the body

a = absorptive power of the body

E = emissive power of the black body.

Then according to Kirchhoff's law

$$e/a = E \qquad ...(13.1)$$

or emissive power of a body = absorptive power of body × emissive power of a black body.

Proof. Consider an enclosure whose walls are maintained at a uniform temperature. Place a body inside the enclosure.

Let emissive power of the body $= e_\lambda$

Absorptive power of the body $= a_\lambda$

Now an exchange of heat will take place between the body and the surrounding enclosure Fig. 13.3.

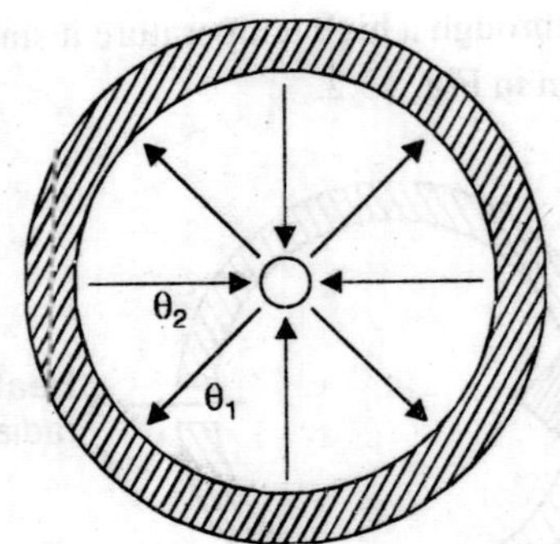

FIGURE 13.3 **An exchange of heat between the body and surrounding.**

Let Q = quantity of heat radiation incident on the surface of body in one second.

Q_1 = quantity of heat absorbed by the surface.

Q_2 = quantity of the heat reflected by surface.

Then heat emitted by body in one second $= e_\lambda d\lambda$

Total heat falling on the surface $= Q$

Total heat given out by the surface $= (Q - Q_1) + e_\lambda d\lambda$

Under equilibrium conditions the total heat falling must be equal to the total heat given out *i.e.*

$$(Q - Q_1) + e_\lambda d\lambda = Q \qquad ...(13.2)$$

but

$$a_\lambda = \frac{Q_1}{Q} \quad \text{or} \quad Q_1 = a_\lambda Q \qquad ...(13.3)$$

$$(Q - a_\lambda Q) + e_\lambda d\lambda = Q$$

or

$$a_\lambda Q = e_\lambda d\lambda$$

or

$$\left(\frac{e_\lambda}{a_\lambda}\right) = \frac{Q}{d\lambda} \qquad ...(13.4)$$

For a perfectly black body absorptive power $a_\lambda = 1$ and emissive power $= E_\lambda$ and hence from eq. (13.4) we have

$$\frac{E_\lambda}{1} = \frac{Q}{d\lambda} \qquad ...(13.5)$$

From eqs.(13.4) and (13.5) $\left(\frac{e_\lambda}{a_\lambda}\right) = E_\lambda$ = constant for all substances, which is Kirchhoff's Law.

Stefan-Boltzmann Law

According to this law total rate at which a black body emits heat radiation is proportional to the 4th power of its absolute temperature.

Let E = heat energy radiated per unit area of the black body.

T = temperature of black body.

According to Stefan's law

$$E \propto T^4 \qquad ...(13.6)$$

or

$$E = \sigma T^4 \qquad ...(13.7)$$

Here σ is the proportionality constant and known as Stefan's constant and its value is 5.672×10^{-2} $W/m^2/K^4$. If the body is not perfectly black body and has emissivity, e, the above law becomes

$$E = \sigma\, e\, T^4 \qquad \text{...(13.8)}$$

Suppose a black body having temperature T_1K is surrounded by another black body at T_2K, then

Rate of loss (or gain) by 1st body $= \sigma T_1^4$

Rate of loss (or gain) by 2nd body $= \sigma T_2^4$

∴ Net rate of loss (or gain) of heat energy per unit area of the surface is given by

$$E_{net} = \sigma(T_1^4 - T_2^4) \qquad \text{...(13.9)}$$

Wein's Law

This law states that the wavelength corresponding to maximum energy is inversely proportional to the absolute temperature.

The amount of energy radiated by a black body is not uniformly distributed over all the wavelengths emitted by the body but is maximum for a particular wavelength. The value of this wavelength is different for different temperature and varies inversely with temperature.

If λ_m = wavelength of most intense radiation

T = temperature of black body

$$\lambda_m \propto \frac{1}{T} \quad \text{or} \quad \lambda_m T = \text{constant (0.2892 cm K)} \qquad \text{...(13.10)}$$

Wein has also shown that the maximum energy is directly proportional to the fifth power of the temperature *i.e.*

$$E_m \propto T^5 \quad \text{or} \quad E_m = \text{constant} \times T^5$$

The constant value was determined by Wein and gave the following relation

$$E_\lambda = c_1 \lambda^{-5} \exp\left(\frac{-c_2}{\lambda T}\right) \qquad \text{...(13.11)}$$

where c_1 and c_2 are constants. This equation gives Wien's law of distribution of energy.

Determination of Stefan's Constant

The experimental arrangement is shown in Fig. 13.4. It consists of a hollow copper hemisphere B having a diameter of about 25 cm with its inner surface uniformly blackened and fitted in a metal box M, which serves as a steam chamber. The arrangement is fitted on a wooden plateform AB, with a small hole of 2.5 cm at the center. B is heated to a steady temperature which can be measured accurately by two thermometers T_1 and T_2 inserted in M. The blackened face of B, thus constitute a black body at constant temperature say θ_1K (temperature of steam). A silver disc D is filled in the hole and is supported by an ebonite handle. The heat radiated by the black body is received by silver disc D blackened on the top surface and placed in the hole. The ebonite handle can be used to put the silver disc in the hole or taking it out when required. One junction of a thermocouple is fixed at the bottom of the disc while the other is placed in a test tube containing oil and placed in a beaker W filled with water. A galvanometer is introduced in the circuit and in order that different parts of the lead wires may not be at different temperature, these are packed in cotton wool in vessel C. A resistance box R is introduced in the circuit to check the deflection of the galvanometer.

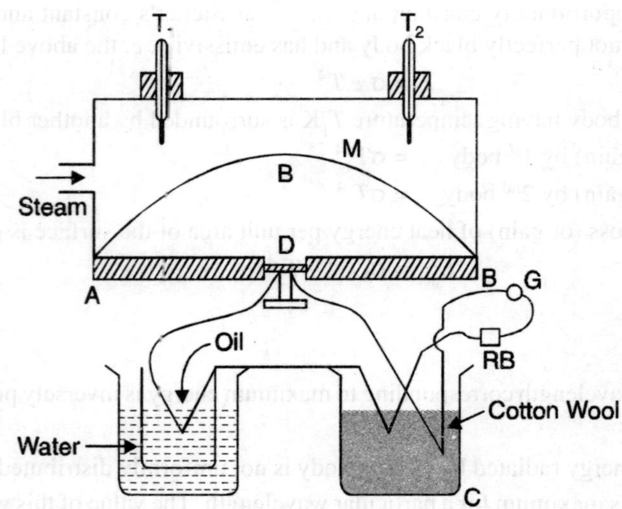

FIGURE 13.4 **Experimental determination of Stefan's constant.**

When the black body is at constant temperature θ, and silver disc is introduced in the hole, it will absorb radiation emitted by black body.

Let θ_2 = **temperature of silver disc**

R_1 = **amount of radiation absorbed by silver disc/area/sec at temperature** θ_1.

R_2 = **amount of radiation emitted by silver disc/area/sec at temperature** θ_2.

Then from Stefan's law

$$R_1 = \sigma\theta_1^{\,4} \qquad R_2 = \sigma\theta_2^{\,4}$$

$$R_1 - R_2 = \sigma(\theta_1^{\,4} - \theta_2^{\,4}) \qquad \text{...(13.12)}$$

Since some of the heat will be absorbed by silver disc, it will gain energy equal to $(R_1 - R_2)\,A$ **where** A = **area of disc.**

Again if m **is its mass and** S **specific heat**

$d\theta/dt$ = **rate of change of temperature.**

Then
$$mS\,\frac{d\theta}{dt} = (R_1 - R_2)\frac{A}{J}$$

Substituting for $R_1 - R_2$ **from eq. (13.12)**

$$m.S.\,\frac{d\theta}{dt} = \frac{\sigma A}{J}\left(\theta_1^4 - \theta_2^4\right).$$

or
$$\sigma = \frac{mSJ}{A\,(\theta_1^{\,4} - \theta_2^{\,4})}\cdot\frac{d\theta}{dt} \qquad \text{...(13.13)}$$

In eq. (13.13) m, S, A **are constants of silver disc** θ_1 **corresponds to the temperature of steam in absolute unit and** J **is mechanical equivalent of heat. So if** θ_2 **and** $d\theta/dt$ **are known,** σ **can be found.**

Planck's Law

In the classical theory we assume that the radiation are continuous ***i.e.*** **the energy changes of radiator takes place continuously has led to inconsistent results. So the classical idea of continuous energy changes must be replaced by discrete or discontinuous energy changes. We may take up the energy changes as integral multiple of a small unit of energy known as quantum. The electrical linear oscillator which constitute the radiation of**

a black body may vibrate only with integral energy values nE where $n = 0, 1, 2, 3.....$ where E represents the elementary quantum.

Applying the laws of probability to such oscillators, an expression for the black body radiation can be obtained in the form

$$dE = \frac{8\pi hc\lambda^{-5}d\lambda}{\left(e^{\frac{hc}{k\lambda T}} - 1\right)}$$

This is Planck's formula for thermal radiation.

Photon Gas (Derivation of Planck's Law)

Interaction of e.m. radiation with matter led to the conclusion that electromagnetic radiation are composed of discrete energy particles called photons. Each photon carries an energy quantum $h\nu$ and momentum h/λ where ν is the frequency and λ is the wavelength of the radiation.

It is assumed that photons are indistinguishable and number of photons may have same energy.

These photons are considered as Boson since they obey Bose-Einstein Statistics.

Since in this case, photons can either be emitted or absorbed so total number of photons is not constant i.e. $\sum dn_i \neq 0$

In Bose-Einstein statistics the number of particles distribution n_i is given by

$$n_i = \frac{g_i}{\left(e^{\alpha+\beta Ei} - 1\right)} \qquad ...(13.14)$$

$\alpha = 0$ because particles are indistinguishable and $\beta = \frac{1}{kT}$ where k is Boltzmann constant.

$$n_i = \frac{g_i}{\left(e^{\alpha+\beta Ei} - 1\right)} \qquad ...(13.15)$$

where g_i = degeneracy of each energy level.

In terms of energy eq. (13.15) can be written as

$$dn = \frac{g(E)dE}{\left(e^{\beta E} - 1\right)} \qquad ...(13.16)$$

Since energy of photon $E = h\nu$ or we can replace $g(E)dE$ by $g(\nu)d\nu$ where $g(\nu)d\nu$ = number of modes of oscillations in the frequency range $d\nu$.

Now we can calculate the number of states in a black body radiation in the frequency range ν and $\nu + d\nu$ by finding the spherical volume bound by spheres of radii $\frac{h(\nu + d\nu)}{c}$ and $\left(\frac{h\nu}{c}\right)$

The volume of spherical shell is given by $= \left(\frac{4\pi h^3}{c^3}\right)(\nu^2 d\nu)$ ...(13.17)

where $h^3 = V$ is the volume of phase space.

$\therefore$ No. of states in the frequency range ν and $\nu + d\nu$ are given by

$$g(\nu)\, d\nu = \left(\frac{4\pi V}{c^3}\right)\nu^2\, d\nu$$

There are two states of polarization of radiation

$$\therefore g(\nu)\,d\nu = \left(\frac{8\pi V}{c^3}\right)\nu^2\,d\nu$$

$$dn = \left(\frac{8\pi V}{c^3}\right)\frac{\nu^2 d\nu}{(e^{\beta/h\nu}-1)} \qquad ...(13.18)$$

for dn photons in frequency range ν and $\nu + d\nu$, the energy is equal to $h\nu\, dn$ and the energy per unit volume

$$= \frac{h\nu dn}{V} \qquad ...(13.19)$$

So the energy density distribution for black body radiation is given by

$$E(\nu) = \left(\frac{h\nu}{V}\right)\left(\frac{dn}{d\nu}\right) \qquad ...(13.20)$$

Putting the value of dn from eq. (13.18) in eq. (13.20) we get

$$E(\nu) = \left(\frac{8\pi V}{c^3}\right)\left(\frac{1}{V}\right)\frac{h\nu^3 d\nu}{\left(e^{\beta/h\nu}-1\right)d\nu}$$

$$E(\nu) = \left(\frac{8\pi h\nu^3}{c^3}\times\frac{1}{\left(e^{\beta/h\nu}-1\right)}\right) \qquad ...(13.21)$$

Eq. (13.21) represents the Planck's law for black body radiation.
From eq. (13.21), we have

$$E(\nu)d\nu = \left(\frac{8\pi h\nu^3}{c^3}\times\frac{1}{\left(e^{\beta/h\nu}-1\right)}d\nu\right) \qquad ...(13.22)$$

Also $\quad c = \nu\lambda \quad$ or $\quad \nu = \frac{c}{\lambda} \quad$ or $\quad d\nu = -\frac{c}{\lambda^2}\,d\lambda$

Now eq. (13.22) can be written as

$$E(\lambda)\,d\lambda = \left(\frac{8\pi h\nu^3}{c^3}\right)\left(\frac{c}{\lambda^2}\right)\left[\frac{1}{\left(e^{\beta/h\nu}-1\right)}\right]d\lambda$$

$$E(\lambda)\,d\lambda = \left(\frac{8\pi h\nu^3}{\nu^3\lambda^3}\right)\left(\frac{c}{\lambda^2}\right)\left[\frac{1}{\left(e^{\frac{h\nu}{kT}}-1\right)}\right]d\lambda \qquad \left(\because\ \beta = \frac{1}{kT}\right)$$

$$E(\lambda) = \left(\frac{8\pi hc}{\lambda^5}\right)\left[\frac{1}{\left(e^{\frac{h\nu}{kT}}-1\right)}\right] \qquad ...(13.23)$$

Eq. (13.23) represents the energy density for wavelength λ of the black body radiation.

Limitations :

(*i*) For small temperature, λT is small

$\therefore$ $e^{hc/\lambda kT} >> 1$ and so 1 can be neglected in the denominator of eq. (13.23)

$$\therefore \qquad E(\lambda) = \left(\frac{8\pi hc}{\lambda^5}\right)e^{-hc/\lambda kT} \qquad \text{...(13.24)}$$

(*ii*) For high temperature, λT is large

$$\therefore \qquad e^{\frac{hc}{\lambda kT}} = 1 + \frac{hc}{\lambda kT} + \left(\frac{hc}{\lambda kT}\right)^2\left(\frac{1}{2!}\right) + \ldots\ldots\ldots\ldots = 1 + \frac{hc}{\lambda kT}$$

$$\text{or} \qquad E_\lambda \, d\lambda = \frac{8\pi kT}{\lambda^4} d\lambda \qquad \text{...(13.25)}$$

which is Rayleigh-Jeans's law. From above it is clear that Planck's law is the perfect law while all other laws follow from it.

13.3 SOLAR CONSTANT

Solar constant is defined as the amount of heat energy received per second per unit area of a perfectly black surface placed at a mean distance of the earth from the sun, in the absence of earth's atmosphere, the surface being held perpendicular to the direction of sun's rays. Its value is found to be 0.32 kcal/S-m^2 or 1.34 kW/m.

The total energy emitted by the sun does not reach earth a part of it is absorbed by the atmosphere and apart of sun's radiation are scattered by the air, clouds, dust particles etc.

Determination of Solar Constant by Angstrom's Pyroheliometer

We can determine the solar constant by angstrom's pyroheliometer. Such a pyroheliometer is shown in Fig. 13.5 and consists of two exactly identical blackened strips S_1 and S_2 made of magnin or constantan mounted side

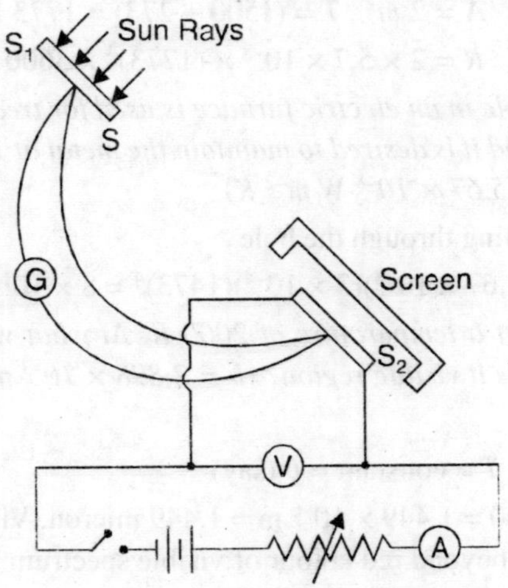

FIGURE 13.5 **Pyroheliometer.**

by side inside a tube in such a way that any of the two can be exposed to the solar radiation while other completely shielded by a screen. A thermocouple of copper and constantan is kept very close to the backs of strips S_1 and S_2.

When strip S_1 is exposed to sun rays its temperature rises, as a result the temperature of the one of the junction of thermocouple placed behind S_1 also rises and the galvanometer G shows a deflection. Now current is passed through strip S_2 which is shielded by a screen and as a result the temperature of strip S_2 increases. The current through strip S_2 is so adjusted as to make its temperature equal to that of S_1. When the temperature of both strips equals the galvanometer shows no deflection. Under these conditions the energy supplied electrically to S_2 is equal to the heat energy received by the strip S_1 from the sun.

If V is the voltage applied to strip S_2 and I is the current, then heat supplied to it per second will be

$$= VI/J \qquad ...(13.26)$$

Now if area of exposed surface is A and absorptive power is a and solar constant is S, then heat energy absorbed by S_1 will be $= aSA$ ...(13.27)

From eqs. (13.26) and (13.27), we have

or
$$S = VI/aJA \qquad ...(13.28)$$

With the help of eq. (13.28) the value of solar constant can be calculated.

SOLVED NUMERICAL PROBLEMS

Problem 1. *A sphere of radius 5.0 cm at 1200°C is suspended in an enclosure at 500°C. Find the rate of loss of heat of the sphere assuming it to behave as 50% of black body ($\sigma = 5.7 \times 10^{-5}$ erg/cm²/sec/K⁴)*

Solution. From Stefan's law $R = eA\sigma(T_1^4 - T_2^4)$

Now
$$A = 4\pi r^2 = 4\pi(5)^2 = 4 \times 3.142 \times 25 = 314.2 \text{ cm}^2 \quad \text{or} \quad e = 0.50$$

$$R = 0.50 \times 314.2 \times 5.7 \times 10^{-12} \times (1200^4 - 500^4)\ K^4 = 3.6 \times 10^3 \text{ watt}$$

Problem 2. *The surface of a furnace is at 15000°C. How much heat is radiated by 2.0 m² of this surface in one hour. Assuming it to be a black body. ($\sigma = 5.7 \times 10^{-8}$ W/m²/K⁴)*

Solution. From Stefan's law $R = A\sigma T^4 \times 60 \times 60$

$$A = 2\ m^2,\ T = (1500 + 273) = 1773 \text{ K}$$

$$R = 2 \times 5.7 \times 10^{-8} \times (1773)^4 \times 3600 = 4.05 \times 10^9 \text{ Joules}$$

Problem 3. *A very small hole in an electric furnace is used for treating metals acts as a black body. If the hole has an area 300 mm²; and it is desired to maintain the metal at 1200°C, how much energy travels per second through thin hole. ($\sigma = 5.67 \times 10^{-8}$ W m⁻² K)*

Solution. The energy travelling through the hole

$$E = \sigma AT^4 = (5.67 \times 10^{-8})(3 \times 10^{-4})(1473)^4 = 8 \times 10^{13} \times 10^{-12} = 80\text{J/s}.$$

Problem 4. *A furnace is at a temperature of 2000 K. At what wavelength would you expect the greater intensity of the radiation? Is it visible region? ($b = 2.898 \times 10^{-3}$ mK)*

Solution. From Wien's law

$$\lambda_m T = \text{constant} = b \text{ (say)}$$

$\lambda_m = T/b = 2.898 \times 10^{-3}/2000 = 1.449 \times 10^{-6}$ m = 1.449 micron. Visible region is between 4000Å to 8000 Å so the above wavelength is beyond red colour of visible spectrum.

Problem 5. *If the sun's surface radiates heat 6.3×10^7 Wm⁻². Calculate the temperature of the sun's surface assuming it to be a black body. (Take $\sigma = 5.7 \times 10^{-8}$ Wm⁻²K⁻⁴.)*

Solution. We know that the heat energy radiated by $1m^2$ surface of the sun in one second is

$$H = \sigma T^4$$

Here $H = 6.30 \times 10^7\ Wm^{-2}$ and $\sigma = 5.7 \times 10^{-8}\ Wm^{-2}K^{-4}$.

∴ $6.3 \times 10^7 = 5.7 \times 10^8 \times T^4$

or $T^4 = 6.3 \times 10^7 \times 10^8/5.7 = 6.3 \times 10^{15}/5.7$

or $T = 5766$ K.

Problem 6. *Calculate the surface temperature of the sun and moon given that λ_m = 4573 A.U. and 14μ respectively, λ_m being the wavelength of the maximum intensity emission.*

Solution. We know $\lambda_m T = 0.2898$

For sun $\lambda_m = 4573 \times 10^{-10}$ or 4573×10^{-3} cm

∴ $4573 \times 10^8 \times T = 0.2892$

or $T = 0.2892 \times 10^8/4573 = 6324$ K or 6051°C

For moon $\lambda_m = 14 \times 10^{-6}$ m or 1400×10^{-6} cm

∴ $1400 \times 10^{-5} \times T = 0.2892$

or $$T = \frac{0.2892 \times 10^6}{1400} = 206.57 \text{ K.}$$

EXERCISE

1. What is black body ? What are the characteristics of a black body ?
2. State and explain Kirchhoff's and Stefan's law of heat radiation.
3. Explain the solar constant and method for determining the temperature of the sun.
4. Write short notes on
   (*i*) Wien's displacement law,
   (*ii*) Energy distribution in black body spectrum
   (*iii*) Kirchhoff's law.
5. Explain in detail the Planck's energy distribution law of black body radiation.
6. The black body at 0°C radiates heat at the rate of 315 W/m². How much heat is radiated in 1 minute by a black sphere of diameter 10 cm at 1000°C ? **[Ans.** 2.8×10^5 J]
7. A spherical black body 5 cm radius is maintained at a temperature of 327°C. What is the power radiated ? **[Ans.** 231 W]
8. A sphere of radius 60 cm at 1200°C is suspended in an enclosure at 500°C. Find the rate of loss of heat of the sphere assuming it to behave as 60% of a black body. **[Ans.** 3.1×10^9J]

Chapter 14

THERMOELECTRICITY

14.1 SEEBECK EFFECT

In 1826, Seebeck discovered the thermoelectric effect, which after his name is also known as Seebeck effect.

The phenomenon of production of an e.m.f. causing an electric current to flow in a thermocouple, when its two junctions are kept at two different temperatures, is known as Seebeck effect. To study this effect, two wires of different materials, say copper and iron are joined at their ends so as to form two junctions. A sensitive galvanometer is included in the circuit as shown in Fig. 14.1. This arrangement is called a copper-iron thermocouple. When one junction of the thermocouple is kept hot and the other cold, the galvanometer gives deflection indicating the production of current in the arrangement. The current so produced is called thermoelectric current. Further the continuous flow of current in the thermocouple indicates that there must be a source of e.m.f. in the circuit, which is causing flow of current. This e.m.f. is called thermoelectric e.m.f.

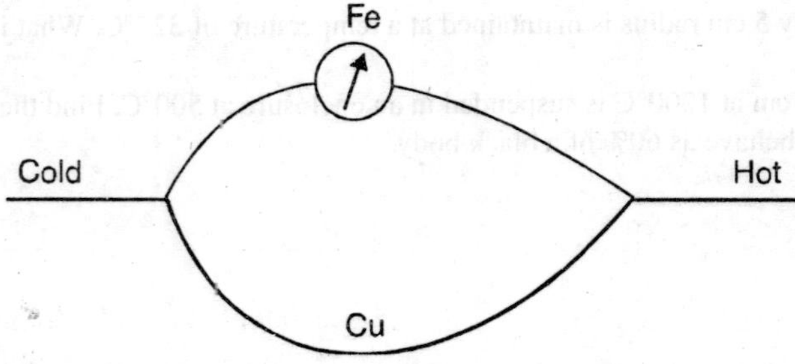

FIGURE 14.1 **Fe-Cu Thermocouple.**

Origin of Thermo-e.m.f.

When two metallic wires of different materials are joined at their ends to form a thermocouple, electrons from metal having greater electron density diffuse into the other with lower value of electron density. Due to this a small potential difference is established across the junction of two metals. The potential difference so established is called contact potential and its value depends on the temperatures of the junction. If the two junctions are at the same temperature the contact potential at the two junctions will be equal. But when one of the two junctions is heated, then diffusion of electrons at hot junction becomes more and like wise the contact potential at the hot junction also becomes more than that at the cold junction. Hence, when the two junctions of a thermocouple are at different temperature, a net e.m.f., called thermo-e.m.f. is produced.

Variation of Thermoelectric e.m.f. with Temperature

The graph showing the variation of thermo-e.m.f. with the temperature of hot junction, while the temperature of cold junction is kept at 0°C, is parabola and is shown in Fig. 14.2. Keeping the cold junction at 0°C if the temperature of hot junction is gradually increased, for small temperature difference between the two junctions, the thermo-e.m.f. is directly proportional to the difference of temperature. As the temperature difference increases, the thermo-e.m.f. also increases and becomes maximum at a temperature T_n called neutral temperature. This neutral temperature is a constant for a given thermocouple. When the temperature of the hot junction is increased beyond the neutral temperature the thermo-e.m.f. starts decreasing and at a particular temperature T_i, called temperature of inversion it becomes zero. Beyond the temperature of inversion the thermo-e.m.f. again starts increasing but in the opposite direction. The temperature of inversion depends on the temperature of cold junction and is as much above the neutral temperature as the cold junction is below it.

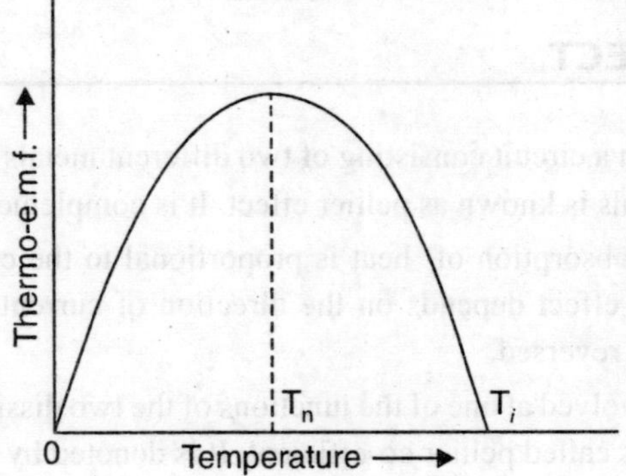

FIGURE 14.2 **A graph showing the variation of thermo-e.m.f. with temperature of hot junction.**

14.2 MEASUREMENT OF THERMOELECTRIC E.M.F.

The thermoelectric e.m.f. of a thermocouple can be determined with the help of potentiometer using a standard cell. As the thermo-e.m.f. is of the order of microvolt, the potentiometer is modified to give a potential gradient of this order along the wire of the potentiometer. The circuit diagram is shown in Fig. 14.3.

The potentiometer wire *AB* is of one meter length and of about 4 ohm resistance and is connected in series with a constant high resistance, a rheostat and a battery *E*. The potentiometer gradient is first determined. With the help of rheostat the current in the circuit is adjusted to such a value that the potential difference across the fixed resistance becomes equal to the e.m.f. of the standard cell as indicated by no deflection in the galvanometer. The current in the potentiometer wire will be equal to the ratio of the e.m.f. of the standard cell and fixed resistance. This current multiplied by the resistance per unit length of wire *AB* gives the potential gradient.

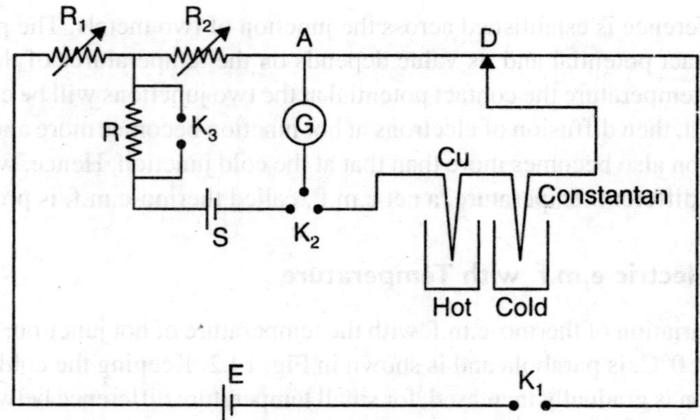

FIGURE 14.3 **Potentiometer with standard cell used for measuring thermo-e.m.f.**

The thermocouple whose thermoelectric e.m.f. is to be determined is connected in the circuit as shown in Fig. 14.3. The cold junction is kept immersed in a hot bath. For any temperature, thermo-e.m.f. can be found by noting the length at which null point is observed and multiplying by the potential gradient of potentiometer wire.

14.3 THE PELTIER EFFECT

When electric current is passed in a circuit consisting of two different metals heat is evolved at one junction and absorbed at the other end. This is known as peltier effect. It is complementary to seebeck effect.

The rate of evolution or absorption of heat is proportional to the current and the peltier effect is therefore reversible. The peltier effect depends on the direction of current. If the direction of current is changed, the peltier effect is also reversed.

The energy absorbed or evolved at one of the junctions of the two dissimilar metals when one ampere of current flows for one second is called peltier co-efficient. It is denoted by π and expressed in volts.

Determination of Peltier Co-efficient (π)

The peltier co-efficient can be measured by the following simple experiment.

Immerse one of the junctions of the two metals of which the peltier co-efficient is to be determined in a calorimeter containing water (Fig. 14.4). Pass a current through the thermocouple which can be measured with the help of an ammeter. Let it be I, for t secs. The resultant change in temperature is recorded by the thermometer.

If H_1 = amount of heat produced then $JH_1 = I^2RT + \pi It$ ...(14.1)

where J = Joule's constant

R = Resistance of the thermocouple

π = Peltier co-efficient of the junction.

Then according to Joule's effect energy evolved = I^2RT Joules and due to Peltier effect energy evolved at the junction = π/t Joule

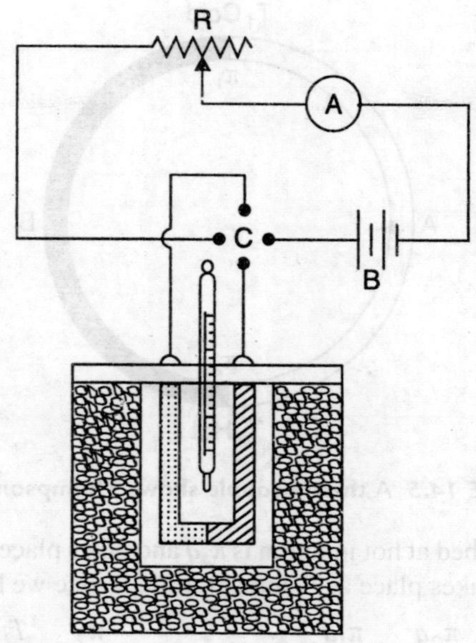

FIGURE 14.4 **Experimental determination of π (the Peltier Co-efficient).**

Now reverse the direction of the current and pass the same current I for the same time, then the junction immersed will be cooled and the temperature of the water will decrease. If H_2 is the net heat developed in this case, then

$$JH_2 = I^2RT - \pi It \qquad ...(14.2)$$

Subtracting Eq. (14.2) from Eq. (14.1) we get

$$J(H_1 - H_2) = 2\pi It$$

or

$$\pi = J\frac{(H_1 - H_2)}{2It} \qquad ...(14.3)$$

As the value of H_1 and H_2 can be calculated by reading the rise in temperature of the water in both the cases and measuring the mass of the water and water equivalent of the calorimeter.

14.4 THOMPSON EFFECT

According to Thompson effect when a current flows in a conductor whose different parts are at different temperature there is an evolution or absorption of heat throughout the conductor.

Consider a thermocouple of metals A and B (Fig. 14.5) with its hot and cold junctions at temperatures T_2 and T_1 respectively. When a current flows in the circuit, across the hot junction the heat energy is absorbed, at cold junction heat is evolved. Let π_2 and π_1 be the peltier co-efficient in volts of hot and cold junctions respectively. If q is the charge which flows through the thermocouple, then an amount of heat energy $\pi_2 q$ (Joule) will be absorbed at hot junction and an amount of heat energy $\pi_1 q$ will be evolved at cold junction. The difference of energy evolved and absorbed $(\pi_2 q - \pi_1 q)$ is used for driving the current through the circuit. As π_2 and π_1 are equal to the potential difference at hot and cold junction, respectively, the thermo-e.m.f. 'e' is developed in the thermocouple will be given by $e = \pi_2 - \pi_1$.

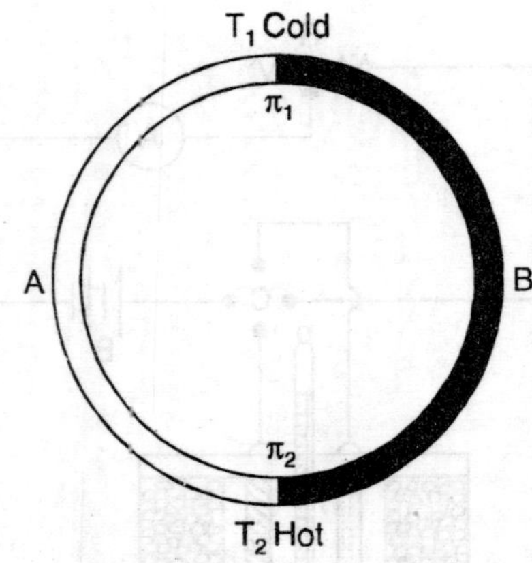

FIGURE 14.5 **A thermocouple shows Thompson's effect.**

The amount of heat absorbed at hot junction is $\pi_2 q$ and takes place at temperature T_2 while the energy given out to the sink is $\pi_1 q$ and takes place at T_1 temperature. Hence we have from thermodynamics

$$\frac{\pi_2 q}{T_2} = \frac{\pi_1 q}{T_1} \quad \text{or} \quad \frac{\pi_2}{\pi_1} = \frac{T_2}{T_1}$$

or
$$\frac{(\pi_2 - \pi_1)}{\pi_1} = \frac{(T_2 - T_1)}{T_1}$$

Now we know $e = \pi_2 - \pi_1$, the total e.m.f. in the circuit

$$e = \left(\frac{\pi_1}{T_1}\right)(T_2 - T_1) \qquad ...(14.4)$$

If the cold junction temperature is kept constant *i.e.* (T_1 constant) the peltier co-efficient π_1 will also be constant and the e.m.f.

$$e \propto (T_2 - T_1) \qquad ...(14.5)$$

The above relation shows that the thermocouple e.m.f. is directly proportional to the difference of temperature of the two junction.

Thompson Co-efficient

The Thompson co-efficient of a conductor is defined as the amount of energy absorbed or evolved when one ampere of current flows for one second throughout the conductor maintained at a difference of temperature of 1°C and is expressed in volts. If σ is Thompson co-efficient *i.e.* it is the e.m.f. due to unit difference of temperature, then the total e.m.f. between points at T_1 and T_2 will be $\int_{T_1}^{T_2} \sigma dT$.

If the current passed is I and is for time t then total e.m.f. will be. $\int_{T_1}^{T_2} \sigma\, ItdT.$ Now if two metals A and B form a thermocouple (Fig. 14.6) and have σ_a and σ_b as Thompson co-efficients and I current passes for time t seconds the equation for the energy absorbed will be

or
$$eIt = \pi_2 It - \int_{T_1}^{T_2} \sigma_a\, Itdt - \pi_1\, It - \int_{T_1}^{T_2} \sigma_b\, ItdT \text{ or } e = \pi_2 - \pi_1 - \int_{T_1}^{T_2} \sigma_a\, dT + \int_{T_1}^{T_2} \sigma_b\, dT \qquad ...(14.6)$$

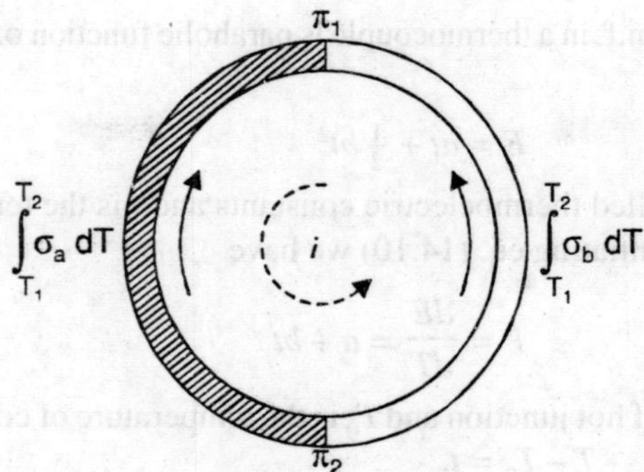

FIGURE 14.6 **A thermocouple used for determination of Thompson's Co-efficient.**

14.5 THERMOELECTRIC POWER

The rate of change of e.m.f. with temperature is called as thermoelectric power and denoted by P. Consider a thermocouple of iron and copper as shown in Fig. 14.7. Let the temperature of cold and hot junctions be T and $T + dT$ respectively and the peltier co-efficient for cold and hot junctions be π and $\pi + d\pi$, respectively.

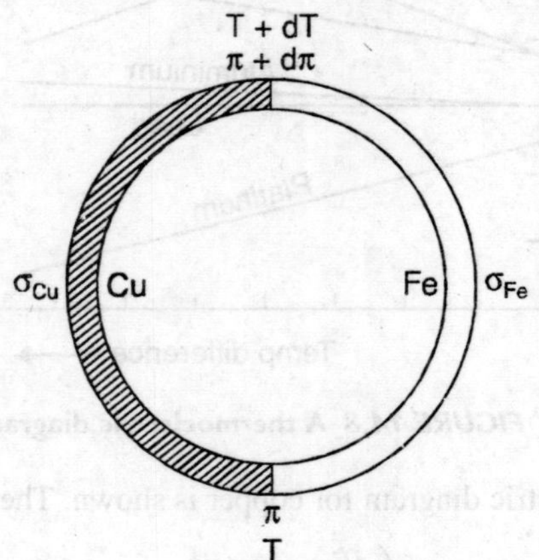

FIGURE 14.7 **Total e.m.f. in a thermocouple.**

For small temperature difference dT of the junction, from relation

$$E = \pi_2 - \pi_1 + \int_{T_1}^{T_2} (\sigma_{Cu} - \sigma_{Fe})\, dT$$

We have $$dE = d\pi + (\sigma_{Cu} - \sigma_{Fe})dT \qquad ...(14.7)$$

$\therefore$ $$\frac{dE}{dT} = \frac{d\pi}{dT} + (\sigma_{Cu} - \sigma_{Fe}) \qquad ...(14.8)$$

Hence $$P = \frac{dE}{dT} = \frac{d\pi}{dT} + (\sigma_{cu} - \sigma_{Fe}) \qquad ...(14.9)$$

We have seen that the e.m.f. in a thermocouple is parabolic function of temperature and can be expressed by the following equation

$$E = at + \frac{1}{2}bt^2 \qquad ...(14.10)$$

where a and b are constants called thermoelectric constants and t is the temperature difference between the hot and cold junctions. Differentiating eq. (14.10) we have

$$P = \frac{dE}{dT} = a + bt \qquad ...(14.11)$$

If T is the temperature of hot junction and T_0 is the temperature of cold junction which is kept constant then

$$T - T_0 = t.$$

From eq. (14.11) we conclude that the variation of thermoelectric power with temperature is straight line curve and the curve is called thermoelectric diagram. Such curves for different metals are shown in Fig. 14.8. In a thermoelectric diagram, we can represent the total e.m.f. and also e.m.f. due to peltier and Thompson effect.

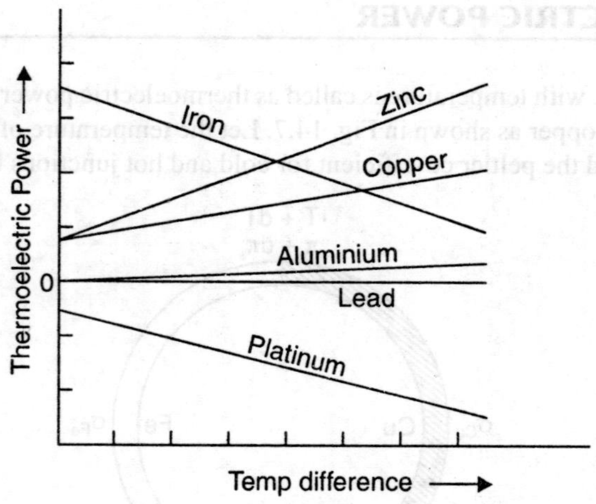

FIGURE 14.8 **A thermoelectric diagram.**

In Fig. 14.9, a thermoelectric diagram for copper is shown. The area under a small strip LM

$$= \left(\frac{dE}{dT}\right) . dT = P . dt = dE$$

where dE represents the e.m.f. acting round the thermocouple while the difference of temperature of the junctions is dT. In order to get the total e.m.f. acting in the circuit integrate dE over the limits T_1 to T_2. Hence

$$E = \int_{T_1}^{T_2} \left(\frac{dE}{dT}\right) dT = \text{area } ABNP$$

At the neutral temperature t_n°C or T_nK we have for thermoelectric power

$$\frac{dE}{dT} = a + bt_n = 0 \qquad \text{[at neutral temperature e.m.f. is maximum]}$$

$$\therefore \qquad t_n = T_n - 273 = \frac{a}{b}$$

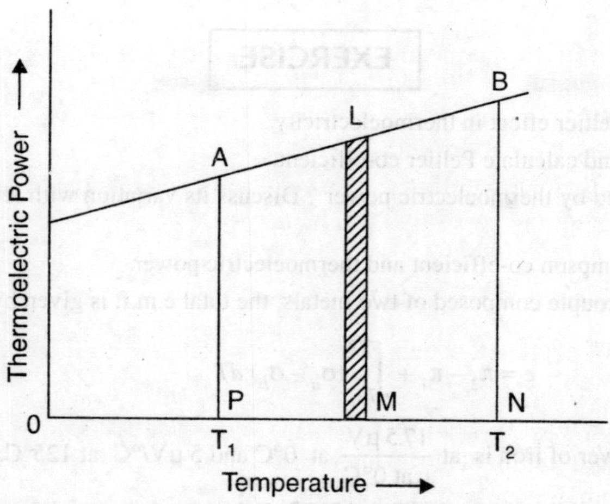

FIGURE 14.9 **Thermoelectric power for copper.**

Similarly, at temperature of inversion (t_1) the seebeck e.m.f. is zero, hence

$$E = at_1 + \tfrac{1}{2}bt_1^2 = 0 \quad \text{or} \quad t_1(a + \tfrac{1}{2}bt_1) = 0$$

$$\therefore \quad a + \tfrac{1}{2}bt_1 = 0 \quad \text{or} \quad t_1 = -2\frac{a}{b} = 2t_n$$

SOLVED NUMERICAL PROBLEMS

Problem 1. *The e.m.f. in lead iron thermocouple, one junction of which is at 0°C is given by E = 1784 t – 2.4 t^2 where t is the temperature in °C. Find the neutral temperature and π and σ.*

Solution. As $E = a + \frac{1}{2}bt^2$

We have $a = 1784$ and $b = -4.8$

Now $t_n = -\dfrac{a}{b} = -\dfrac{1784}{-4.8} = 371.67°\text{C}$

Peltier co-efficient $\pi = T\left(\dfrac{dE}{dT}\right) = T[a + b(T - 273)] = 3094.4T - 4.8T^2$

Thompson co-efficient $\sigma = T\dfrac{d}{dT}(a + bt)$

or $\sigma = T \,.\, b = -4.8T.$

Problem 2. *Calculate the neutral temperature of an iron silver thermocouple. The values of a and b are 1775 and –0.092 for iron and for silver these values are 2.72 and 0.016 respectively.*

Solution. We know, $a_{\text{Fe-Ag}} = a_{\text{Fe-Pb}} - a_{\text{Ag-Pb}}$

$\therefore$ $a_{\text{Fe-Ag}} = 17.75 - 2.72 = 15.03$

Similarly, $b_{\text{Fe-Ag}} = b_{\text{Fe-Pb}} - b_{\text{Ag-Pb}}$

$\therefore$ $b_{\text{Fe-Ag}} = -0.092 - 0.016 = -0.108$

$\therefore$ $t_n = -\dfrac{a}{b} = \dfrac{15.03}{0.108} = 139.16°\text{C}$

EXERCISE

1. Explain Seebeck and Peltier effect in thermoelectricity.
2. Explain Peltier effect and calculate Peltier co-efficient.
3. What do you understand by thermoelectric power ? Discuss its variation with temperature and thermoelectric diagram.
4. Define Peltier and Thompson co-efficient and thermoelectric power.
5. Show that in a thermocouple composed of two metals, the total e.m.f. is given by

$$e = \pi_2 - \pi_1 + \int_{T1}^{T2} (\sigma_a - \sigma_b)\, dT$$

6. The thermoelectric power of iron is at $\frac{17.5\ \mu\text{V}}{\text{c at } 0°\text{C}}$ at 0°C and 5 μV/°C at 125°C. The thermoelectric power of cadmium is 3 μV/°C at 0°C and 15 μV/°C at 150°C. Calculate the neutral temperature of the iron cadmium junction. **[Ans. 80.55°C]**

Chapter **15** *ULTRASONICS*

15.1 ULTRASONIC WAVES

The sound waves having frequencies above the 20 kHz *i.e.* audible range are called ultrasonic waves or supersonic waves. The wavelengths of ultrasonic waves are very small as compared to audible sound. Most of the applications of the ultrasonic waves have been possible on account of their small wavelength. The sound waves which have frequencies less than 20 Hz are called infrasonic waves.

There are two important methods used for the generation of ultrasonic waves :

(*i*) Magnetostriction method
(*ii*) Piezo-electric method

(*i*) Magnetostriction Method

Principle

Magnetostriction method is used when frequencies up to 100 kHz are needed. It is based upon the phenomenon of magnetostriction *i.e.* when a rod of ferromagnetic material, such as iron or nickel, is placed in a magnetic field parallel to its length, a small extension or contraction occur. This change of length is independent of the sign of the field but depends on magnitude of the field and nature of material.

Experimental Arrangement

An experimental arrangement for producing ultrasonic waves is shown in Fig. 15.1. There is a short nickel rod which is clamped at the center. This rod is permanently magnetized in the beginning by passing a direct current in the coil which is wrapped round the rod. There are two other coils L_1 and L_2 which are wrapped round the rod as shown in Fig. 15.1. The coil L_2 is connected in the plate of a valve V while L_1 is connected in the grid circuit. The frequency of the oscillating plate circuit adjusted with the help of a variable condenser

C connected across the coil L_2. The DC milli ammeter(mA) reads the plate current. When the frequency of the plate circuit is the same as the natural frequency of the rod, resonant vibrations are produced in the surrounding medium. The vibrations are maintained due to coupling provided by coil L_1. When the resonant oscillations start, milli ammeter records the maximum current. By adjusting the length of the rod and the capacity of the condenser, high frequency oscillations of different frequencies are obtained.

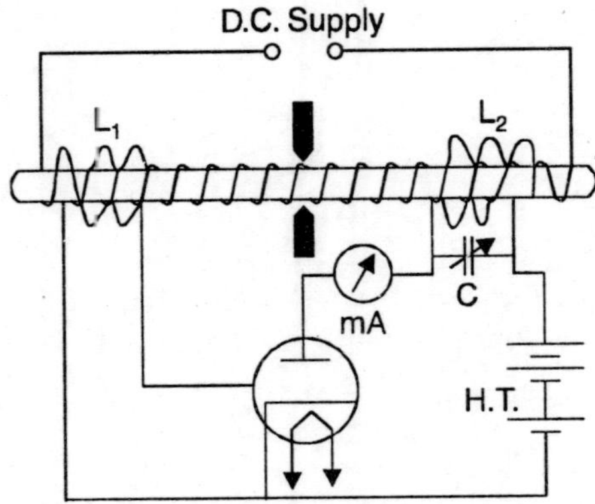

FIGURE 15.1 **An experimental arrangement of magnetostriction method for producing ultrasonic waves.**

(ii) Piezo-electric method

This method is used for the production of waves of frequencies greater than 100 kHz.

Principle

This method is based on the piezo-electric effect. According to this effect when certain crystals like quartz, rochelle salt, tourmaline etc. are stretched or compressed along certain axis (known as mechanical axis) an electric potential difference is produced along a perpendicular axis (known as electrical axis). The converse of this effect is also true. If the frequency of electric oscillations coincide with natural frequency of the crystal, the vibrations will be of large amplitudes. This phenomenon is utilized for the production of ultrasonic waves.

Experimental arrangement

The experimental arrangement is shown in Fig. 15.2. The high frequency alternating voltage which is applied to the crystal is obtained by Hartley oscillatory circuit. The Hartley circuit consists of a tuned oscillator circuit (inductance L_1 and a variable condenser C_1 in parallel). One end of the tuned circuit is connected to the plate of valve V while the other is connected to the grid. The coil L_1 is tapped near the center and joined to the cathode. The quartz crystal Q is connected in parallel of variable condenser C_1.

The proper grid bias is obtained by means of grid leak resistor R_g and grid condenser C_g. The DC voltage is applied to the plate through radio frequency choke. The radio frequency choke prevents the radio frequency current to pass through high tension battery. C_b is a blocking capacitor which prevents the direct current to pass through the track circuit while by passes the radio frequency current. The capacity of the variable condenser C_1 is adjusted so that the frequency of the oscillating circuit is tuned to the natural frequency of the crystal. Now the quartz crystal is set into mechanical vibrations and ultrasonic waves are produced.

The ultrasonic waves upto a frequency of 500 kHz with a crystal of moderate size can be produced by this method. However, frequencies up to 15×10^7 Hz can be produced by tourmaline crystal.

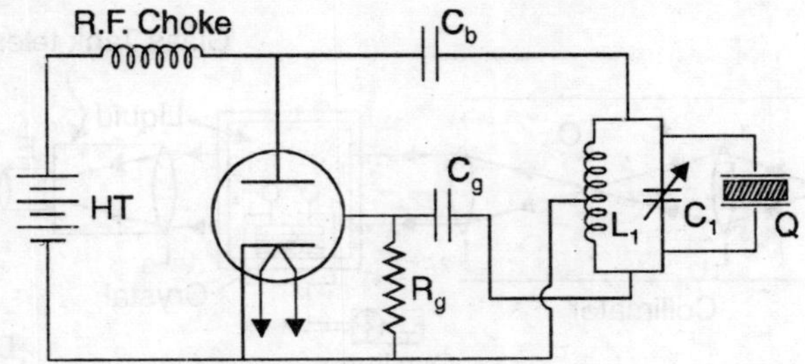

FIGURE 15.2 **An experimental arrangement of Piezo-electric method for producing ultrasonic waves.**

The velocity of quartz along X-direction is given by

$$v = \sqrt{\frac{Y}{\rho}}$$

for quartz $\quad Y = 7.9 \times 10^{10}\ \text{Nm}^{-2}$ and $\rho = 2650\ \text{kg m}^{-3}$

$$\therefore \quad v = \sqrt{\frac{7.9 \times 10^{10}}{2650}} = 5450\ \text{m/s} \qquad ...(15.1)$$

if t – thickness of quartz slab in meters then

$$v = n\lambda = n(2t)$$

where n is the frequency

$$\therefore \quad n = \frac{v}{2t} = \frac{2725}{t}\ \text{Hz}$$

if t is expressed in mm, then

$$n = \left(\frac{2725}{t}\right) 1000 = \frac{2725000}{t}\ \text{Hz}$$

or

$$n = \frac{2725}{t}\ \text{kHz}$$

So for a thickness of 1 mm of quartz crystal

$$n = 2725\ \text{kHz} = 2.725\ \text{MHz}.$$

When, by adjusting variable capacitor C_1 of tank circuit the crystal is made to vibrate as its natural frequency, then the frequency of oscillatory circuit gives the frequency of vibration of quartz crystal. Thus

$$N = \frac{1}{2\pi\sqrt{(L_1 C_1)}}$$

15.2 WAVELENGTH OF ULTRASONIC WAVES

We can find the wavelength and velocity of ultrasonic waves in liquids by diffraction method.

When ultrasonic waves are propagated in a liquid, the density varies from layer to layer due to periodic variation of pressure. If under this condition monochromatic light is passed through the liquid at right angles to the waves, the liquid behaves as a diffraction grating. Such a grating is known as acoustical grating. The grating behaves in the same way as a ruled grating. Hence the method can be used for finding the wavelength and velocity of ultrasonic waves in the liquid.

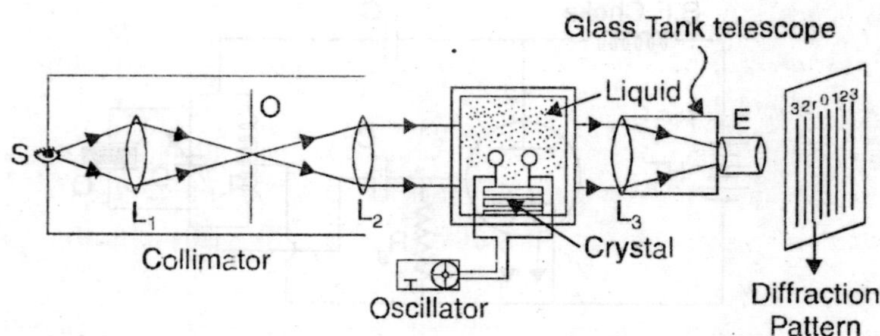

FIGURE 15.3 **An experimental arrangement for determining wavelength of ultrasonic waves.**

The experimental arrangement is shown in Fig. 15.3. The light from a monochromatic source of light *S*, is focussed on the narrow slit *O* by lens L_1. The light from the slit now passes through lens L_2 which turns it into a parallel beam. This parallel beam then passes through the ultrasonic cell. This cell consists of a rectangular glass tank containing the liquid. The crystal is dipped into the liquid and is connected to the oscillating waves which travel through the liquid. The waves are reflected from the walls of the cell and form a stationary wave pattern consisting of condensation and rarefactions which acts as acoustical grating. The crystal is placed in the cell in such a way that the acoustical grating is formed in a perpendicular direction of the propagation of light. Now light emerging from the cell is focussed by lens L_3 and can be seen through a telescope.

When the crystal is at rest a single image of the slit is observed. But when ultrasonic waves are produced in the liquid by excitation of crystal the intensity of the center image decreases and number of diffracted images appear on either side of the center. The angular separation θ between the direct image of the slit and the diffracted images of any order say *n* is measured. Applying the theory of diffraction grating the wavelength of the ultrasonic waves can be calculated. Here the grating element is the same as the wavelength of ultrasonic waves. Let it be λ_c. If λ is the wavelength of the monochromatic light used, then

$$\lambda_c \sin\theta_n = n\lambda$$

or

$$\lambda_c = \frac{n\lambda}{\sin\theta_n} \quad ...(15.2)$$

In this way, λ_c can be calculated. If *N* be the frequency of ultrasonic waves the velocity, *v*, in liquid can be calculated by the formula

$$v = N\lambda_c = N\left(\frac{n\lambda}{\sin\theta_n}\right) \quad ...(15.3)$$

15.3 APPLICATION OF ULTRASONIC WAVES

Following are some important applications of Ultrasonic waves.

1. ***Detection of iceberg and submarines etc.*** A sharp ultrasonic beam is directed in various directions into the sea. The reflection of waves from any direction shows the presence of some reflecting body in the sea.

2. *Depth of sea.* We know that ultrasonic waves are highly energetic and show a little diffraction effect. Thus they can be used for finding the depth of the sea. The time interval between sending the waves and receiving the reflected waves from the bottom of the sea is recorded. As the velocity of the waves is known, hence the depth of the sea can be estimated.

3. *Cleaning and clearing.* These waves can be used for cleaning utensils, washing clothes, removing dust and soot from the chimney.

4. *Direction signalling.* The ultrasonic waves can be concentrated into a sharp beam due to smaller wavelength and hence can be used for signalling in a particular direction.

5. *Soldering and metal cutting.* Ultrasonic waves can be used for drilling and cutting processes in metals. These waves can also be used for soldering.

6. *Formation of Alloys.* The constituents of alloys having widely different densities can be kept mixed uniformly by a beam of ultrasonics. Thus it is easy to get alloy of uniform composition.

7. *Ultrasonics in metallurgy.* To irradiate molten metals which are in process of cooling so as to refine the grain size and to prevent the formation of cores and to release trapped gases, the ultrasonic waves are used.

8. *Destruction of lower life.* The animals like rats, frogs, fishes etc. can be killed or injured by high intensity ultrasonics.

9. *Treatment of neuralgic pain.* The body parts effected due to neuralgic or rheumatic pains on being exposed to ultrasonics get great relief from pain.

10. *Detection of abnormal growth.* Abnormal growth in the brain, certain tumors which cannot be detected by X-rays can be detected by ultrasonic waves.

SOLVED NUMERICAL PROBLEM

Problem 1. *A quartz crystal of thickness of 0.001 m is vibrating at resonance. Calculate the fundamental frequency. Given Y for quartz = 7.9×10^{10} N/m² and ρ = 2650 kg/m³.*

Solution. We know that

$$v = \sqrt{\frac{Y}{\rho}} \quad \text{putting the numerical values we get}$$

$$= \sqrt{\frac{7.9 \times 10^{10}}{2650}} = 5461 \text{ m/s}$$

For the fundamental mode of vibration the thickness should be equal to λ/2. Hence

$$\lambda = 2t = 2 \times 0.001 = 0.002 \text{ m}$$

now

$$v = n\lambda \quad \text{or} \quad n = \frac{v}{\lambda}$$

$$n = \frac{5461}{0.002} = 2.730 \times 10^6 \text{ Hz} = 2730 \text{ kHz}$$

EXERCISE

1. **Define ultrasonics. Describe Piezo-electric method for their production. How are these used for sonar (depth sounding) exploration ?**
2. **Explain the phenomenon of magnetostriction. How will you produce high frequency sound waves with its help ?**
3. **What are ultrasonic waves ? Describe some of their applications.**
4. **What are ultrasonic waves ? Give an account of their properties. How are they produced are detected ?**
5. **Write an essay on ultrasonic and their uses.**
6. **A quartz crystal of thickness of 0.002 m is vibrating at resonance. Calculate the fundamental frequency. Given Y for quartz = 7.9×10^{10} N/m^2 and ρ for quartz = 2650 kg/m^3.** **[Ans. 1365 kHz]**

Chapter 16 LASER

16.1 INTRODUCTION

LASER is an acronym that stands for "Light Amplification by Stimulated Emission of Radiation". It is essentially an optical amplifier producing highly directional, monochromatic and coherent light.

First laser was built by Maiman, in 1960. Its output can either be a continuous beam of low to medium power or a short twist of intense radiation delivering million of watts of power.

16.2 MAIN FEATURES OF LASER LIGHT

Directionality

Laser emit radiation only in one direction *i.e.* they have high degree of beam directionality. This directionality of laser beam is usually expressed in terms of full angle beam divergence which is twice the angle that the outer edge of the beam makes with axis of the beam.

Monochromaticity

Perhaps the most important property of the laser light is its extraordinary monochromaticity as compares to any conventional monochromatic source. No laser can produce absolutely monochromatic light. One can make better and better approximation to the ideal monochromatic light *e.g.* He-Ne laser is 10^7 times more monochromatic than ordinary discharge lamp and about 10^5 times more than a cadmium source. No conventional source of light can approach this high degree of monochromaticity present in the output beam of a typical laser.

Coherence

Laser is regarded as the first truly coherent light source. Coherence is a measure of the degree of phase correlation that exits in the radiation field cf a light source at different locations and different times.

Independent concepts of coherence

Following are two independent concepts of coherence.

(*a*) *Temporal Coherence (Longitudinal Coherence):* It is that type of coherence in which the field at a point and the field at the same point at a later time are correlated.

(*b*) *Spatial (Transverse) Coherence :* Two fields at two different points on a wave front of a given electromagnetic wave are said to possess spatial coherence if a constant phase difference is maintained over any time.

Measurement of coherence

The ability to measure coherence is connected to the relation between coherence and interference.

A measure of fringe contrast, called fringe visibility, *V*, is defined as follows

$$V = \frac{(E_{max} - E_{min})}{(E_{max} + E_{min})}$$

where E_{max} = energy of a bright fringe and E_{min} = energy of a neighbouring dark fringe. If interference is produced by coherent beams of equal amplitude, then $E_{min} = 0$ and $V = 1$. Further, if interference is produced by non-coherent beams, then $E_{max} = E_{min}$ and $V = 0$. Hence, fringe visibility is intimately related to coherence.

Coherence time and coherence length

Coherence time (t_c) is that time during which ideal sinusoidal emission continues from a light source. During this time, the correct phase of the laser beam at a given point in space can be predicted.

Coherent length, l_c, is the distance travelled by light waves in time t_c, *i.e.*

$$l_c = ct_c, \quad c \text{ being the velocity of light.}$$

Intensity

Laser gives out light into a narrow beam and its energy is concentrated in a small region and, therefore, its intensity is very high. Even a one-watt laser appears many thousands time more intense than a 100 watt filament lamp.

Laser photons emitted per sec = $P/h\omega$

$P = 10^{-3}$ watt to 10^9 watts and $h\omega = 10^{-19}$ Joule

Range of laser photons emitted per sec = 10^{16} to 10^{28}.

16.3 EINSTEIN'S CO-EFFICIENTS

Einstein showed that interaction of radiation with matter could be explained in terms of three basic processes *viz.*, stimulated absorption, stimulated emission and spontaneous emission [see Fig. 16.1].

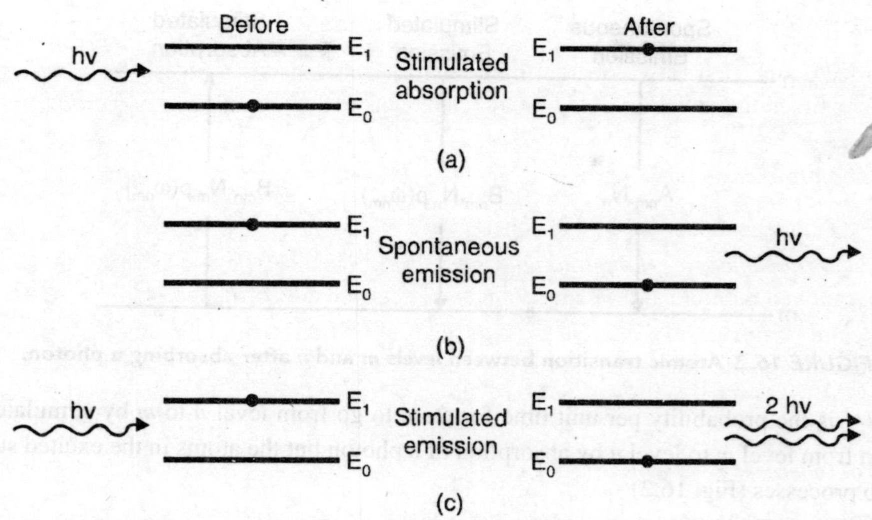

***FIGURE 16.1* Different modes of interaction of radiation with matter.**

(*i*) ***Stimulated Absorption.*** It occurs when radiations containing photon of energy $\hbar\,\omega = E_1 - E_0$, is incident on matter having ground state energy E_0 and arbitrary excited state energy E_1. The resonant photon of energy $h\omega$ raises the atom from energy E_0 to E_1. In the process, the photon itself is absorbed.

(*ii*) ***Spontaneous emission.*** It takes place when atoms are already in an excited state. No external radiation is required to start the emission process. During this emission, an atom in an excited state E_1 spontaneously gives up its energy and falls to E_0 releasing a photon of energy $\hbar\,\omega = E_1 - E_0$. The photon is emitted in a random direction. Even if external radiation is present, the photon is emitted in a direction, which is absolutely not related to the direction of external radiation.

(*iii*) ***Stimulated emission.*** It requires the presence of external radiation. When an incident photon of resonant energy $\hbar\,\omega = E_1 - E_0$ passes by an atom in an excited state E_1, it stimulates the atom to drop to the ground state E_C. During this process, the atom releases a photon of the same energy, direction, phase and polarization as that of the passing by photon. This produces two identical photons in place of one. So there is an increase in the intensity of the incident beam. This process of stimulated emission makes the amplification of light in lasers.

The transition rate, R_{mn}, is assumed to be proportional to

(*i*) B_{mn}, the probability of absorption per unit time

(*ii*) $\rho(\omega_{nm})$, the density of radiation

(*iii*) N_m, the number of atoms per unit volume in lower level, *m*.

$$R_{mn} = \rho(\omega_{nm})\, B_{mn} N_m \qquad \text{...(16.1)}$$

After excitation there is a finite probability A_{nm} per unit time that an atom returns to level, *m*, by spontaneously emitting a photon of energy $h\omega_{nm}$. The number of atoms spontaneously emitting radiation per unit time is

$$R_{nm}\,(\text{sp.}) = A_{nm} N_n \qquad \text{...(16.2)}$$

N_n is the number of atom per unit volume in level, *n*. To account for the discrepancy Einstein incorporated stimulated emission with transition rate R_{nm} being given by

$$R_{nm}(\text{st.}) = \rho(\omega_{nm})\, B_{nm} N_n \qquad \text{...(16.3)}$$

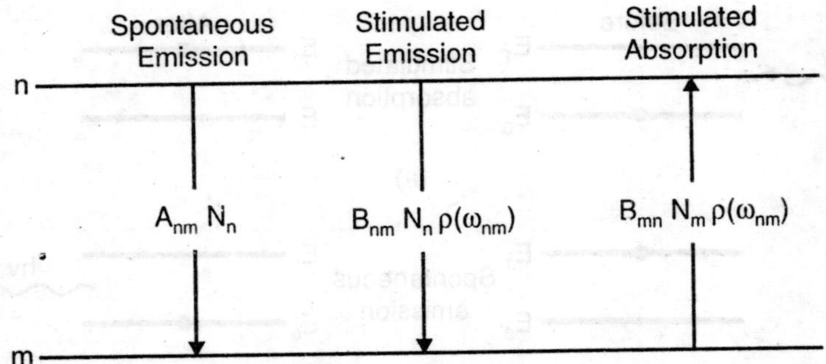

FIGURE 16.2 **Atomic transition between levels *m* and *n* after absorbing a photon.**

Here B_{nm} is the probability per unit time for atom to go from level n to m by stimulated emission. Hence atom go from level m to level n by absorption of a photon but the atoms in the excited state return to level m by two processes (Fig. 16.2)

(*i*) Stimulated Emission

(*ii*) Spontaneous Emission

So, the equation for the rate of change of N_n is

$$\frac{dN_n}{dt} = R_{mn} - R_{mn}(\text{sp.}) - R_{nm}(\text{st.})$$

Substituting from eqs. (16.1), (16.2) and (16.3) we get

$$\frac{dN_n}{dt} = \rho(\omega_{nm})\, B_{mn} N_m - A_{nm} N_n - \rho(\omega_{nm})\, B_{nm} N_n$$

in equilibrium condition $$\frac{dN_n}{dt} = 0$$

∴ $$\rho(\omega_{nm}) B_{mn} N_m - A_{nm} N_n - \rho(\omega_{nm}) B_{nm} N_n = 0$$

or $$A_{nm} N_n = \rho(\omega_{nm}) [B_{mn} N_m - B_{mn} N_n]$$

or $$\rho(\omega_{nm}) = \frac{A_{nm} N_n}{[B_{mn} N_m - B_{nm} N_n]}$$

or $$\rho(\omega_{nm}) = \frac{A_{nm}}{\left[B_{mn}\dfrac{N_m}{N_n} - B_{nm}\right]} \qquad \text{...(16.4)}$$

Now using Boltzmann relation and assuming $g_m B_{mn} = g_n B_{nm}$ and then comparing the expression with Planck's law of Black Body radiation, we obtain

$$\rho(\omega_{nm}) = \frac{A_{nm}}{B_{nm}} \left[\exp\left(-\frac{\hbar\omega_{nm}}{kT}\right)\right]$$

and $$\frac{A_{nm}}{B_{nm}} = \frac{\hbar\omega^3{}_{nm}}{\pi^2 c^3}$$

The quantities A_{nm}, B_{nm} and B_{mn} are called Einstein's co-efficients. If there are no degenerate levels, then $g_m = g_n = 1$

Then, $$B_{mn} = B_{nm}$$

16.4 POPULATION INVERSION

For a laser to operate, it is essential that $N_n > N_m$ *i.e.* [Number of atoms per unit volume in upper level > that of in lower level]. This condition is called population inversion.

The stimulated emission co-efficient B_{nm} and the stimulated absorption coefficient B_{mn} are equal at least for non degenerate energy level. This equality shows that stimulated emission and stimulated absorption processes are inverse in so far as their occurrence is concerned. But the rates differ because

$$\frac{dN_n}{dt} = \rho(\omega_{nm})\, B_{nm} N_n \quad \text{and} \quad \frac{dN_m}{dt} \quad \rho(\omega_{nm}) B_{mn} N_m$$

of different population densities N_n and N_m. If $N_n > N_m$ and a radiation field interacts with atoms, then stimulated emission exceeds stimulated absorption and photons will be added to the fields and these identical photons may cause stimulated emission from the other (excited) atoms. The process is repeated when these photons pass over excited atoms. The net result thus is that stimulated emission becomes stronger and stronger. Some of energy (or photons) may be absorbed by unexcited atoms, which are, therefore, raised to the excited state. Thus this act of stimulated absorption reduces the number of photons. Now if $N_m > N_n$ stimulated absorption exceeds stimulated emission and photons will be removed from the field. So $N_n > N_m$ leads to increase in $\rho(\omega_{nm})$ *i.e.* amplification and $N_m > N_n$ leads to decrease in $\rho(\omega_{nm})$ *i.e.* attenuation. Thus net effect depend upon which of these two processes is dominated. To make stimulated emission dominant population inversion is created. The excited incident photons raise the atom from the lower energy level to higher one. Most of the electrons reaching higher (E_2) level will make non-radiative transition to metastable (E_3) level. The probability of transition ($E_2 \rightarrow E_3$) is much higher than for ($E_2 \rightarrow E_1$) and excess energy ($E_2 - E_3$) is used up in heating the substance (Fig. 16.3). In this way, number of electrons in E_3 are made to exceed the number of electrons in E_1. This is called population inversion.

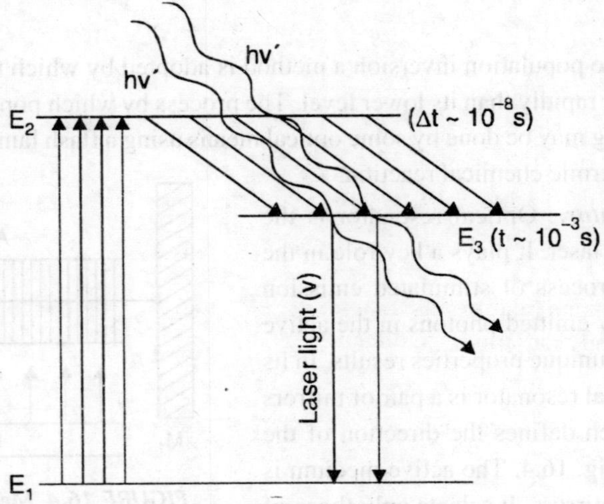

FIGURE 16.3 **The mechanism of population inversion.**

Laser or Optical Pumping

It is the process by which necessary population inversion is obtained *i.e.* atoms are raised from lower level to upper level. It involves a large number of energy level with complex excitation process. The optical pumping is carried out by flash lamp or arc lamp. Other methods of pumping are electrical, chemical and gas dynamic pumping. Population difference between two state or levels depends upon

(*a*) τ, decay time of upper level.

(*b*) $\rho(\omega_{21})$ the density of radiation

$$\Delta N = \frac{N_t}{(1 + 2B_{12}\tau\rho(\omega_{21}))} \qquad [N_t = N_1 + N_2]$$

To maintain this given population difference ΔN, the pump power absorbed by the material per unit volume is

$$\frac{dP}{dV} = \hbar\,\omega_{21} B_{12}\rho(\omega_{21}) N_t \Delta N$$

or

$$\frac{dP}{dV} = \hbar\,\omega_{21} B_{12}\rho(\omega_{21}) \frac{N_t}{[1 + 2B_{12}\tau\rho(\omega_{21})]}$$

16.5 MAIN COMPONENT OF A LASER

The laser system has three main components—(1) Active medium (2) Pump (3) Resonator as shown in Fig. 16.4.

1. *Active Medium.* The active medium may be an atomic or a molecular gas, for example, He-Ne, CO_2, a liquid dye solution or a solid such as ruby rod, semiconductor etc. The important characteristics of a laser medium are that it is capable of decaying radiatively from its higher energy level E_2 to a lower energy level E_1 such that $E_2 - E_1$ lies in the optical region and population inversion between these two levels is achieved by some means. The atom (ions or molecules) of the medium involved in this process are called active species.

2. *Pump.* To create population inversion a method is adopted by which the upper level of the active species is populated more rapidly than its lower level. The process by which population inversion attained is called pump. The pumping may be done by some optical means using a flash lamp, by electrical discharge or by using a suitable exothermic chemical reaction.

3. *Optical resonator.* Optical resonator is the principal component of a laser. It plays a key role in the sense that it guids the process of stimulated emission induced by spontaneously emitted photons in the active medium so that a laser of unique properties results. In its simplest version the optical resonator is a pair of mirrors set on an optic axis which defines the direction of the laser beam as shown in Fig. 16.4. The active medium is placed in between these mirrors. It selects only those of the emitted photons which move along its axis rejecting all others.

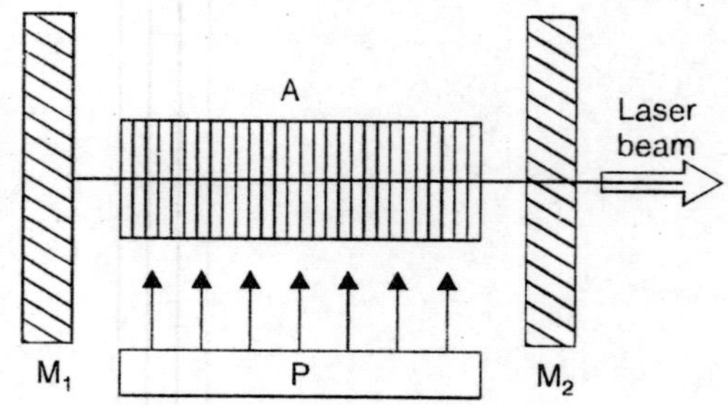

FIGURE 16.4 **Main component of laser.**

16.6 RUBY AND HE-NE LASERS

Ruby Laser

Its main part is a ruby rod due to this it is known as a ruby laser. Ruby is a crystal of aluminium oxide (Al_2O_3) with an addition of 0.03 to 0.05 % of chromium oxide (Cr_2O_3) in which Cr ions replace some of Aluminium ions in crystal lattice. These crystals are grown in special furnaces and are given the shape of a rod (Fig. 16.5).

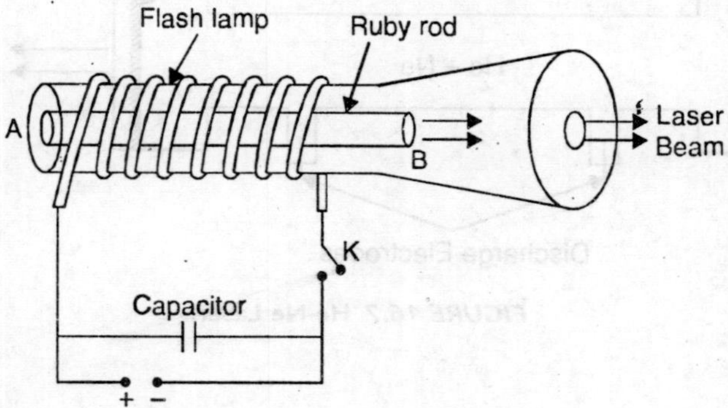

FIGURE 16.5 **Ruby Laser.**

Flat end faces of rod are made strictly parallel and are silvered so that the surface of one face *A* becomes fully reflecting and that of the other face *B* partially reflecting. A narrow intense flash emerges from partially silvered end which lasts for about a microsecond. Several flashes are emitted in a millisecond. Parts of this light gets reflected back into the crystal and helps to stimulate the remaining Cr^{+3} ions to emission. A chain reaction is, therefore, set up resulting in high intensity output. The energy levels that are shown in Fig. 16.6 correspond to the levels of Chromium ions which consists of two bands E_2 and E_1 respectively centered at $\lambda = 4200$ Å and $\lambda = 5500$ Å and are called Blue B and Green Bands. These bands are themselves nearly 1000 Å wide.

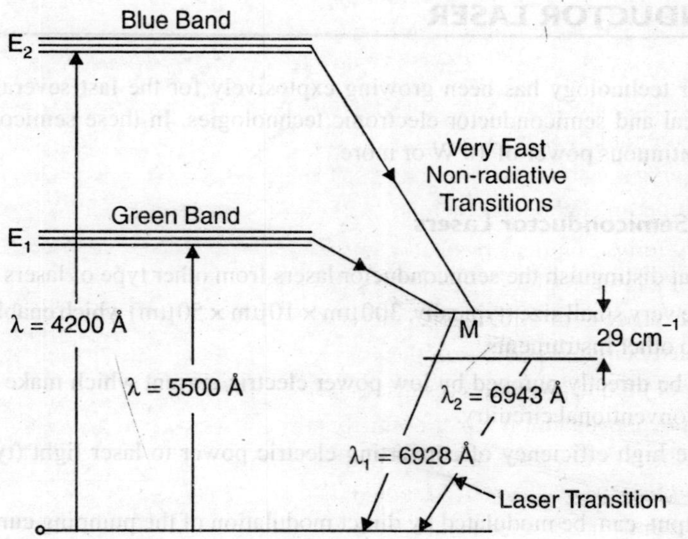

FIGURE 16.6 **Energy levels of Chromium ions.**

Helium-Neon Laser

The He-Ne laser consists of a mixture of He and Ne in the ratio ranging from 10:1 to 5:1, the optimum ratio is found to be 7:1, kept inside a long narrow discharge tube 0.25 m in length and about 1 mm in diameter. The pressure inside the tube is 1 mm of Hg. The gas system is enclosed between a set of plane mirrors (Fig. 16.7).

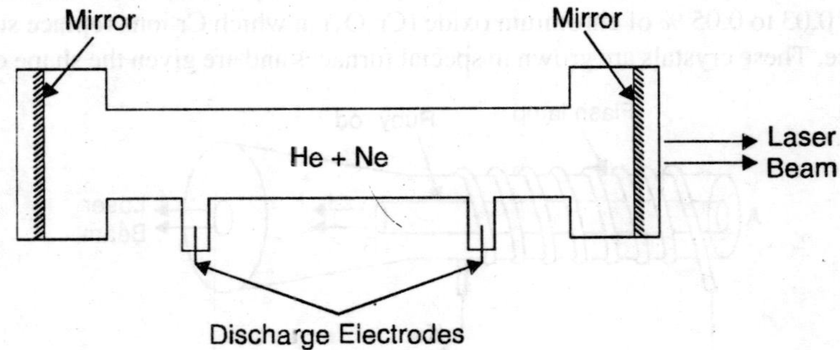

FIGURE 16.7 **He-Ne Laser.**

Principle and Working

When an electric discharge is produced in gas, the atom in it are excited. These collide with atoms of helium and excite them to metastable state for sufficiently long time (this state lie 20.16 eV above the ground state). They lose their energy to neon atoms through collisions. Consequently, neon atoms get excited to a metastable state which is 20.66 eV above the ground state. The difference in these energies of metastable states *i.e.* 0.05 eV is taken from the kinetic energy of helium atom. Thus helium atoms helps to build up a large population in metastable state of neon with energy 20.66 eV with the result that any spontaneously emitted photon can triggers laser action. Thus He-Ne laser gives a continuous laser beam unlike ruby laser which gives pulsed laser beam.

16.7 SEMICONDUCTOR LASER

Semiconductor Laser technology has been growing explosively for the last several years. It is based on combination of optical and semiconductor electronic technologies. In these semiconductors lasers power have increased to continuous power of 10 W or more.

Main Features of Semiconductor Lasers

The main features that distinguish the semiconductor lasers from other type of lasers are

1. They have very small size (typically, 300 μm × 10 μm × 50 μm) which enables it to be incorporated easily into other instruments.
2. They can be directly pumped by low power electric current which make it possible to derive it with the conventional circuitry.
3. They have high efficiency of converting electric power to laser light (typically, more than 20 per cent).
4. Their output can be modulated by direct modulation of the pumping current at rates exceeding 10 GHz. This is of major importance in high data rate optical communication systems.
5. They have semiconductor based manufacturing technology which leads it self to mass production.

With these features accounting for the high degree of applications and by economic standards, the semiconductor lasers are the most important of all lasers.

Theory of Semiconductor Laser

The most common way to the population inversion in a semiconductor material is by joining a *p*-type and an *n*-type materials together. The contact that occur is known as *p-n* junction (Fig. 16.8).

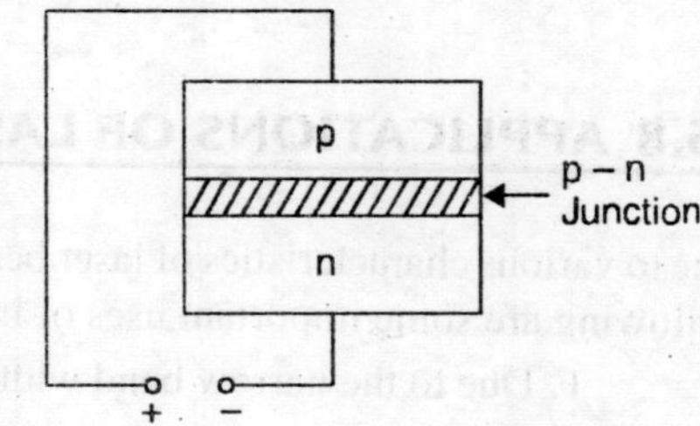

FIGURE 16.8 Semiconductor *p-n* junction in forward bias.

If a d.c. voltage is applied to a semiconductor chip having *p-n* junction in the forward bias, then electrons will be injected from the *n*-side into the junction area, while holes will enter the junction from the *p*-side.

The electrons and holes will recombine in the junction area and may emit the surplus energy as radiation (as in the case of GaAs). Hence a laser action may be realised if the material is in a suitable optical resonator. In many other materials, however, the energy is emitted in the form of heat and they are of no use for laser action (as in case of Si and Ge). The wavelength of light energy emitted depends on the activation energy required to free the electrons. In GaAs it is ~ 1.4 eV and photons emitted have wavelength 0.9 μm.

Condition of Laser Action

Let us consider semiconductor at $T = 0\,K$. (zero *K*), as shown in Fig. 16.9 (*a*). The valence band is shown to be completely filled with electrons. If some valence electrons are raised to conduction band (Fig. 16.9 (*b*)), then the semiconductor is not in thermal equilibrium. However, the equilibrium will be attained in each band in a very short time (10^{-3} s) during which the electrons in the conduction band will have dropped to the lowest level in this band and any electron near the top of the valence band will also have dropped to the lowest unoccupied level, leaving the top of valence band filled with holes.

This means that there has been achieved a population inversion between the conduction and valence band. The electrons in the conduction band will fall back into the valence band, recombine with holes and emit photon, in this process. If this happens in the presence of another photon of the same energy, then it will cause stimulated emission of recombination radiation which will give rise to a laser oscillation if placed in a suitable resonator. It is also evident from Fig. 16.9 (*b*) that the frequency of the emitted radiation must satisfy the condition

$$(E_g < h\nu < (F_c - F_v)$$

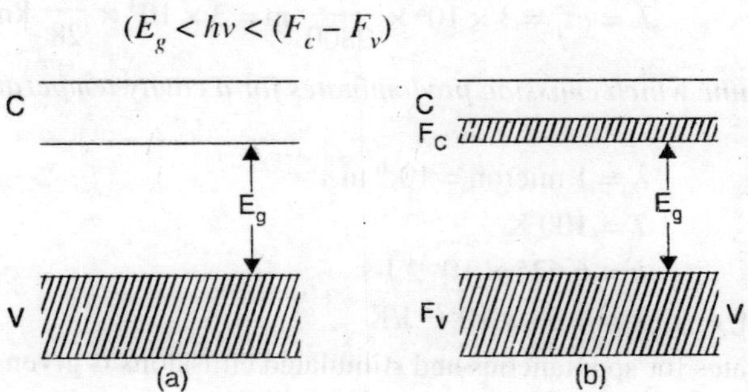

FIGURE 16.9 (*a*) Energy levels in semiconductor at 0 K (*b*) Energy levels when electrons are excited to conduction band.

which determine the gain band width of the semiconductor, where F_c and F_v are called Quasi-Fermi levels of the conduction and valence bands respectively. These levels separate the zones of fully occupied and completely empty levels of the two bands.

If E_c and E_v are the corresponding energies of the conduction and valence band then laser condition becomes

$$F_c - F_v > E_c - E_v = h\nu$$

16.8 APPLICATIONS OF LASER

Due to various characteristics of laser beam, the lasers have wide applications in different branches of science. Following are some important uses of lasers.

1. Due to the narrow band width, lasers are used in microwave communication.
2. The laser beam have become a means of communication between earth and moon or other satellites because of narrow angular spread. The earth moon distance has been measured by lasers.
3. The storage capacity for information in computers is greatly improved by using lasers.
4. The lasers have wide industrial applications *e.g.* blasting holes in diamond and hard steels.
5. Lasers have wide medical applications for example in the treatment of detached retinas and the human and animal cancers.
6. Lasers may serve as a war-weapon for military.
7. As chemical applications lasers are used to initiate or hasten certain chemical reactions.
8. High power lasers can be used in nuclear fusion reactions.
9. Holography is another important application of laser.

SOLVED NUMERICAL PROBLEMS

Problem 1. *A laser beam has a bandwidth of 2800 Hz. Obtain the coherence length.*

Solution. Given $\Delta\nu = 2800$ Hz

Coherence time $\tau_c = \dfrac{1}{2800}$ sec

Coherence length $\lambda = c\tau_c = 3 \times 10^8 \times \dfrac{1}{2800}$ m $= 3 \times 10^3 \times \dfrac{1}{28}$ km $= 107$ km.

Problem 2. *Determine which emission predominates for a cavity temperature of 100 K and for one micron wavelength.*

Solution. Given $\lambda = 1$ micron $= 10^{-6}$ m

$T = 100$ K

Planck's Constant $h = 6.625 \times 10^{-34}$ J-s

Boltzmann Constant $k = 1.38 \times 10^{-23}$ J/K

Ratio of transition rates for spontaneous and stimulated emissions is given by

$$\frac{A_{21}}{B_{21}}\rho(\omega) = \exp\left(\frac{hc}{\lambda kT}\right) - 1 = \exp(143)$$

Obtained by putting the values of h, c, k, λ and T.

Since, $\exp(143) >> 1, A_{21} >> B_{21}\rho(\omega)$, hence spontaneous emission predominates over the stimulated emission.

EXERCISE

1. Explain the concept of directionality, monochromaticity and coherence as applied to a laser.
2. Distinguish between spontaneous and stimulated emission.
3. What are Einstein's co-efficients ? Show that they are interrelated.
4. What is population inversion ? Why is it necessary for laser action ? How is it achieved ?
5. Give in detail the construction, principle and working of He-Ne laser.
6. Is ruby laser a three-level laser ? Discuss with diagrams the principle and working of a ruby laser.
7. Calculate coherence length for a laser beam for which bandwidth $\Delta\nu = 10^{11}$ Hz.

Chapter 17 SIMPLE HARMONIC MOTION

17.1 INTRODUCTION

Oscillatory or vibratory motion is one of the important type of motion occurring in nature. In this type of motion, a body has an equilibrium position, and when displaced from equilibrium it experiences a restoring force, which causes it to undergo a back and forth motion about the equilibrium position. The motion of a simple pendulum and that of a mass suspended from a spring, are the most common examples of oscillatory motion. The oscillatory motion of an object can be placed in the class of motions where object moves under the influence of variable forces. Since, a variable force may vary by infinite number of ways, therefore, it is impossible to derive a general expression for the motion of an object under the influence of a variable force. However, there exists a particular class of problems, which occurs frequently in practical situations, where the force and displacements are related by Hooke's law ($F = -kx$). Such a force is known as elastic restoring force. The type of motion, under the influence of an elastic restoring force and in the absence of all friction, is called simple harmonic motion abbreviated as S.H.M.

Of all the oscillatory motions, the most important is simple harmonic motion, because, besides being the simplest motion to describe mathematically, it constitutes a rather accurate description of many oscillations found in nature. The particles of medium vibrate simple harmonically in many kinds of wave motions. The electric and magnetic fields in any electromagnetic wave also oscillate simple harmonically. So, we concentrate on the simple harmonic motion in this chapter.

17.2 BASIC CONCEPTS

Basic concepts of simple harmonic motion can be introduced quite easily through an idealized model for this motion. The idealized model for simple harmonic motion is shown in Fig. 17.1.

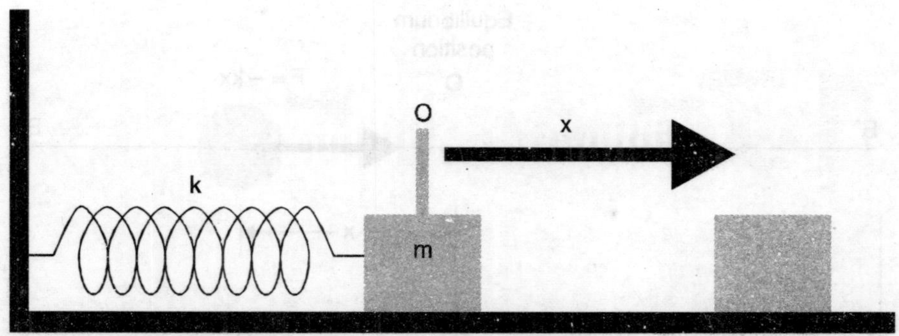

FIGURE 17.1 **Arrangement of periodic motion of mass, *m*, along X-axis.**

In Fig. 17.1, a body of mass, *m*, is attached to one end of a perfectly elastic spring, whose other end is held stationary. It is assumed here that the body of mass, *m*, can move without any friction along a straight line on the horizontal platform. The position of the body at any instant is described by *x*-coordinate, taking the origin ($x = 0$) as the equilibrium position, where the spring is neither stretched nor compressed. The position *x* is taken positive to the right of the equilibrium position and negative to the left of the equilibrium position. When the body is displaced towards right then the spring is stretched and it exerts a force on the body towards left *i.e.* towards the equilibrium position. When the body is displaced towards left then the spring is compressed and it again exerts a force on the body but now towards right *i.e.* towards the equilibrium position. Thus the sign of the *x*-component of force on the body is always opposite to that of *x*-itself. Now, since the spring is assumed to be perfectly elastic, therefore,

$$F = -kx \qquad \text{...(17.1)}$$

where *k* is the force constant for the spring. This is Hooke's law.

Now suppose the body is displaced through some small distance from the equilibrium position towards right and released. The body moves under the action of restoring force of the spring given by eq. (17.1). The acceleration during this motion is not constant because the accelerating force becomes smaller as the body approaches the equilibrium position. When the body reaches the equilibrium position, the restoring force has decreased to zero as $x = 0$; but because of the velocity that has been acquired, the body overshoots the equilibrium position and continues to move towards the left. As soon as the equilibrium position is passed, the restoring force again comes into play, directed now towards the right. The body's speed thus decreases at a rate which increases with increasing distance from *O*. It therefore comes to rest at some point to the left of *O*, and repeats its motion in the opposite direction. If there were no loss of energy by friction, this to-and-fro motion would continue indefinitely, once it had been started. This type of motion under the influence of restoring force given by eq. (17.1) in the absence of all frictional forces is called simple harmonic motion.

The positions of the body at different instants are shown in Fig. 17.2, when it is moving simple harmonically.

The one round trip motion, say from *O* to *B* to *O* to *B′* and back to *O* is known as a complete vibration or complete cycle. The time(T) required to complete one vibration is called the time period of SHM. The number of complete vibrations per unit time is known as the frequency (ν) of SHM. Evidently the frequency is the reciprocal of the time period or $\nu = 1/T$. The SI unit of frequency is cycle per second (cps) or Hertz (Hz). The maximum displacement from the equilibrium position on either side is known as the amplitude of the motion.

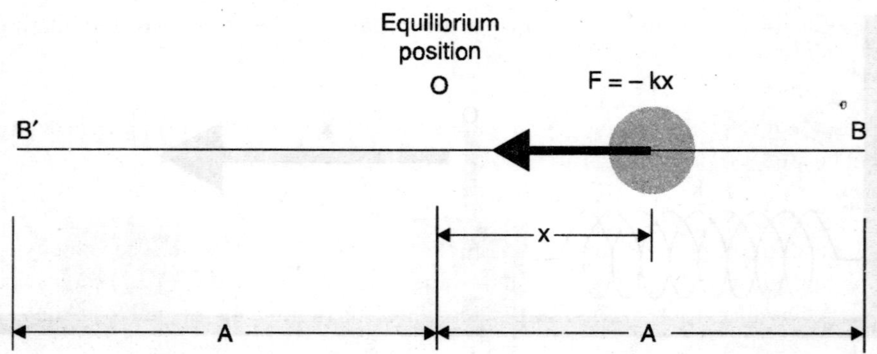

FIGURE 17.2 **Representation of positions of the body w.r.t. time, *t*.**

17.3 EQUATIONS OF SHM

From the definition of SHM, we know that the force, *F*, acting on the particle of mass, *m*, moving simple harmonically at any instant, *t*, is directly proportional to its displacement from its equilibrium position and is always directed towards equilibrium position *i.e.* opposite to the displacement. So,

$$F = -kx \qquad \text{...(17.1)}$$

where *x* is the displacement of the particle from its equilibrium position and *k* the constant of proportionality also called force constant as its value distinguishes one SHM from the other.

Also, from Newton's second law of motion, we have

$$F = ma \qquad \text{...(17.2)}$$

***a* being the acceleration.**

Comparing eqs. (17.1) and (17.2), we get

$$ma = -kx$$

or $$a = -(k/m)x \qquad \text{...(17.3)}$$

It is obvious from eq. (17.3) that the acceleration at each instant *t* is proportional to the negative of the displacement, *x*, at that instant.

Since, the acceleration in rectilinear motion is $a = d^2x/dt^2$, therefore, eq. (17.3) can be written in the following form

$$d^2x/dt^2 = -(k/m)x$$

or $$d^2x/dt^2 + \omega^2 x = 0 \qquad \text{...(17.4)}$$

with $\omega^2 = k/m$ **= force constant/mass**

or $\omega = \sqrt{(k/m)}$ **which is related to time period and hence frequency.**

This is a second order differential equation, whose general solution may be written as

$$x = a \sin(\omega t) + b \cos(\omega t)$$

where *a* and *b* are arbitrary constants.

This solution is equivalent to $$x = A \sin(\omega t + \phi) \qquad \text{...(17.5)}$$

obtained by putting $a = A \cos \phi$ and $b = A \sin \phi$ in the preceding equation. It is quite easy to verify that this expression for *x* [Eq. (17.5)] satisfies eq. (17.4). From eq. (17.5) it is obvious that the position *x* would be repeated after every interval *T* such that $\omega T = 2\pi$. Then, $\omega = 2\pi/T = 2\pi\nu$ so that *T* is the time period and ν the frequency. Also, all these ω, T, ν depend on *k* through $\omega = \sqrt{(k/m)}$.

The velocity of the particle executing SHM, at any instant, t, is obtained by differentiating eq. (17.5) w.r.t. t

$$v = dx/dt = A\omega \cos(\omega t + \phi) \quad ...(17.6)$$

This expression for velocity can be rewritten in terms of the position of the particle at any instant by using the fact that $\cos(\omega t + \phi) = [1 - \sin^2(\omega t + \phi)]^{1/2}$.

That is,
$$v = \omega\sqrt{(A^2 - x^2)} \quad ...(17.7)$$

when the particle is at extreme position, $x = A$ from eq. (17.5) and hence $v = 0$ there. The velocity is maximum at equilibrium position when $x = 0$ and the maximum value of velocity is $A\omega$. Eq. (17.7) shows kinematically that the kinetic energy of the particle at equilibrium position is maximum and is equal to $mA^2\omega^2/2$. Assuming potential energy to be zero at equilibrium position this will also be the total energy, say E.

Again, since $v = \omega\sqrt{(A^2 - x^2)}$, the kinetic energy E_k at any position x would be given by $E_k = (m/2)\,\omega^2(A^2 - x^2)$. Also, since total energy in SHM is constant of time and

Total energy = kinetic energy + potential energy

we have,
$$E = E_k + E_p$$
$$(m/2)A^2\omega^2 = (m/2)\,\omega^2(A^2 - x^2) + E_p, \quad \because\ E = \frac{m}{2}A^2\omega^2.$$

Hence,
$$E_p = (m/2)\,\omega^2 x^2 = kx^2/2 \quad ...(17.8)$$

Obviously, potential energy E_p is minimum at the equilibrium position ($x = 0$), where kinetic energy, E_k, is maximum and maximum at the extreme position $[=(m/2)\omega^2A^2]$ where E_k is minimum *i.e.* zero.

The frequency, ν, number of complete vibrations per unit time, is the reciprocal of the period T and $\omega T = 2\pi$ mentioned earlier below Eq. (17.5).

It is therefore given by $\nu = 1/T = \omega/2\pi$

ω is therefore given by $\omega = 2\pi\nu$, called angular frequency. It is to be noted that ω, T and ν all depend on k for a particle executing particular SHM.

In eq. (17.5), the sine function can never be greater than + 1 or less than – 1, so x can never be greater than A or less than $-A$; this confirms that A is the amplitude of the motion. The constant ϕ determines the position of the body at time $t = 0$. If $\phi = 0$, then at $t = 0$, $x = 0$ and the body starts at the origin. But if $\phi = \pi/2$ then at $t = 0$, $x = A$ and the body starts at its maximum positive displacement. In other words while $(\omega t + \phi)$ determine state of motion of the particle ϕ determines its initial state. $(\omega t + \phi)$ is also called the phase angle or simply phase.

Energy in SHM

The total energy of a particle executing SHM at any point consists of kinetic energy as well as potential energy.

$$E = E_k + E_p$$

As mentioned earlier, the expressions of these energies are

$$E_k = (m/2)\,\omega^2(A^2 - x^2) = k(A^2 - x^2)/2$$
$$E_p = (m/2)\,\omega^2 x^2 = kx^2/2$$

and
$$E = (m/2)\,\omega^2 A^2 = kA^2/2,$$

all at the displacement, x, from the equilibrium position.

Therefore, we may say that during an oscillation, there is a continuous exchange of kinetic and potential energies mutually. In moving away from the equilibrium position, potential energy increases at the expense of kinetic energy ; the reverse happens when the particle moves towards the equilibrium position.

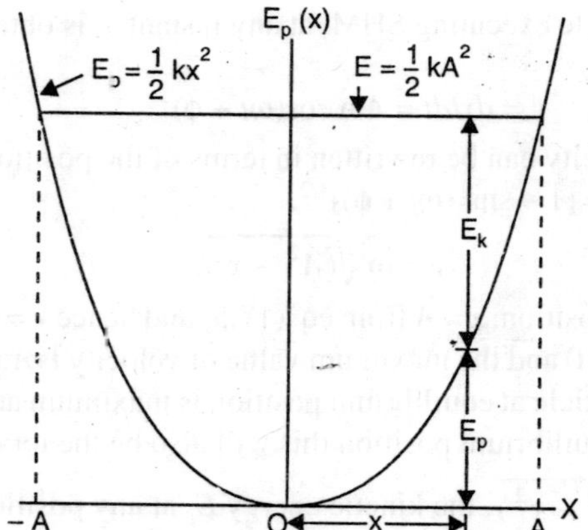

FIGURE 17.3 **Energy relations in SHM.**

Fig. 17.3 shows the potential energy $E_p = kx^2/2$ represented by a parabola. For a given total energy E, represented by the horizontal line, the limits of oscillation are determined by its intersections with the potential energy curve. Since the parabola E_p is symmetric the limits of oscillation are at equal distances $\pm A$ from O. At any point x the kinetic energy E_k is given by the distance between the curve $E_p(x)$ and the line E.

17.4 THE SIMPLE PENDULUM

A simple pendulum consists of a point particle of mass m suspended from a point O by an inextensible string of length l and of negligible mass in a uniform gravitational field (Fig. 17.4).

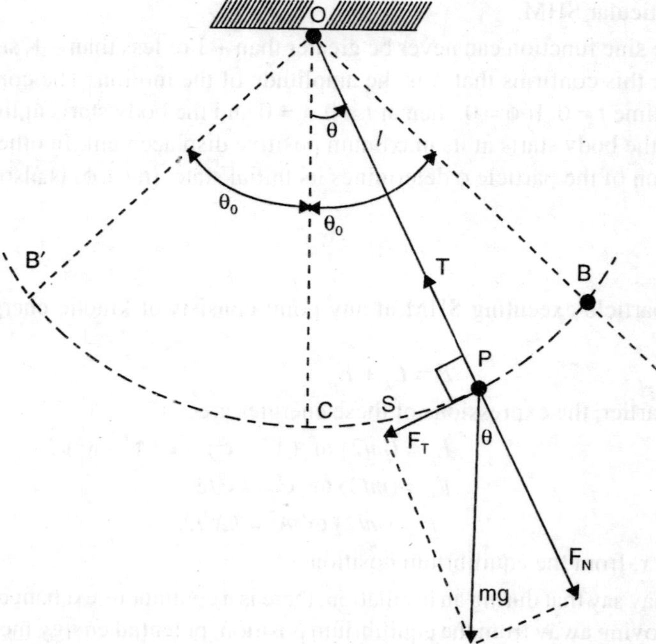

FIGURE 17.4 **Oscillatory motion of a pendulum ; *OP* = *OC* = *l*.**

When the particle is pulled aside to a position B such that the string makes an angle θ_0 with the vertical OC and then released, it oscillates about the equilibrium position C between B and the symmetric position B' as shown in Fig. 17.4.

To determine the nature of oscillations, we must write the equation of motion. The particle moves in an arc of a circle of radius l. When the particle is at position P then various forces acting on the particle are as shown in Fig. 17.4. These are weight (mg) of the particle in vertically downward direction and the tension T along the string. The component of weight along the string, F_N, is exactly balanced by the tension T as there is no motion along the radius of the arc. So the force causing motion is the tangential component of the weight, F_T, which is, from Fig. 17.4, given as

$$F_T = -\,mg \sin \theta$$

where the minus sign appears because it is opposite to the displacement $x = CP$.

Hence the equation of motion is

$$F_T = m\, a_T$$

Since the particle moves along a circle of radius l, therefore,

$$a_T = l\, d^2\theta/dt^2 \qquad [\because \text{ arc} = x = l\,\theta \text{ and } d^2x/dt^2 = l\, d^2\theta/dt^2]$$

The equation for the motion thus becomes

$$ml\, d^2\theta/dt^2 = -\,mg \sin \theta$$

or

$$d^2\theta/dt^2 + (g/l) \sin \theta = 0$$

This equation is not of the same type as eq. (17.4) because of the presence of sin θ. However, if the angle θ is small, which is true if the amplitude of the oscillation is small, we may use the approximation that $\sin \theta \sim \theta$ (in radians)

Hence the preceding equation may be written as

$$d^2\theta/dt^2 + (g/l)\,\theta = 0 \quad \text{or} \quad d^2\theta/dt^2 = -\,(g/l)\theta$$

This is a differential equation identical to eq. (17.4) with x replaced by θ. Thus we may conclude that, within our approximation the angular motion of the pendulum is simple harmonic with $\omega^2 = g/l$. The angle θ can thus be expressed in the form $\theta = \theta_0 \sin(\omega t + \alpha)$.

Then using $T = 2\pi/\omega$, we can express the time period of the oscillation as

$$T = 2\pi \sqrt{(l/g)}$$

we note that the time period is independent of the mass of the particle.

For larger amplitude, the approximation $\sin\theta \sim \theta$ is not valid, hence the motion of the simple pendulum is no longer simple harmonic. In that case the formula for the time period depends on the amplitude θ_0. In this case, the general formula for the time period can be expressed in terms of maximum angular displacement θ_0, by the following infinite series

$$T = 2\pi \sqrt{(l/g)}\left[1 + \frac{1^2}{2^2}\frac{\sin^2 \theta_0}{2} + \left(\frac{1^2}{2^2}\right)\left(\frac{3^2}{4^2}\right)\frac{\sin^4 \theta_0}{2} + \ldots\ldots\right]$$

{the mathematical derivation of this expression is beyond the scope of present book}

The variation of T with the amplitude θ_0, expressed in terms of the period $T_0 = 2\pi \sqrt{l/g}$ corresponding to very small amplitudes, is illustrated in Fig. 17.5.

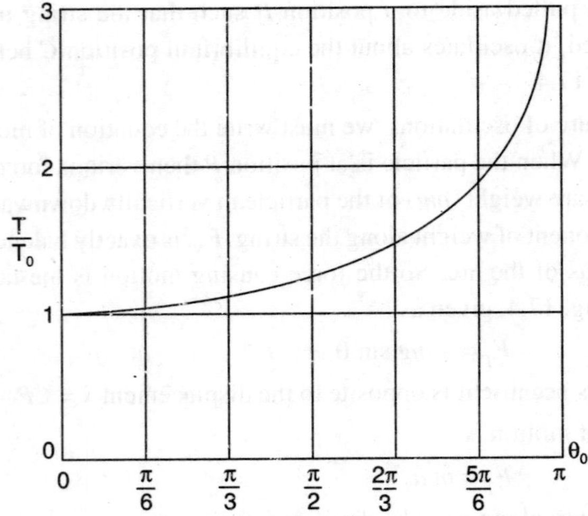

FIGURE 17.5 **Variation of the period of a pendulum, *T*, with the amplitude, θ_0.**

The time may be computed to any desired degree of precision by taking enough terms in the infinite series. For small amplitudes it is enough to take only the first corrective term, and even then substitute $\theta_0/2$ for sin $\theta_0/2$, resulting in

$$T = 2\pi\sqrt{(l/g)}\ [1 + 1/16\ \theta_0^2]$$

where θ_0 must be expressed in radians. This is a sufficient approximation for most practical situations. In fact, the corrective term $\theta_0^2/16$ amounts to less than 1% for amplitudes less than 23°.

17.5 MOTION OF A BODY SUSPENDED FROM A COIL SPRING

Consider a perfectly elastic spring of force constant *k* and no-load length *l* is suspended freely from a support as shown in Fig. 17.6.

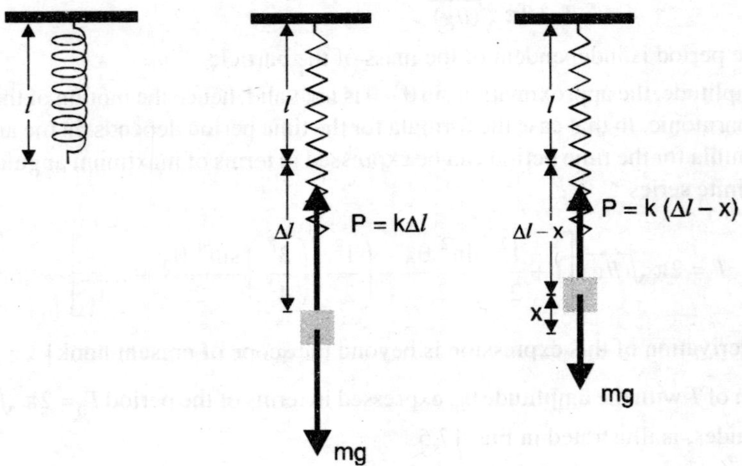

FIGURE 17.6 **SHM of mass, *m*, in the vertical direction.**

When a body of mass m is suspended from the lower end of the spring, it extends and comes to an equilibrium position. Let Δl be the extension in the spring in equilibrium position as shown in Fig. 17.6 (*b*). In equilibrium position the upward force P exerted by the spring is equal to the weight of the body, mg.

But $$P = k\Delta l,$$

i.e. $$k\Delta l = mg$$

when the body is at a distance x above its equilibrium position, then the extension in the spring is $\Delta l - x$ [see Fig. 17.6 (*c*)]. Hence, the upward force, which the spring exerts on the body, is $k(\Delta l - x)$ and the resultant force F on the body is

$$F = k(\Delta l - x) - mg = -kx$$

The resultant force is therefore proportional and opposite to the displacement of the body from its equilibrium position and if set in vertical motion the body oscillates with an angular frequency $\omega = \sqrt{k/m}$ as explained earlier.

17.6 DAMPED OSCILLATIONS

So far we have discussed the simple harmonic motion with constant amplitude. However, in various practical situations a vibrating body such as a spring or a pendulum oscillates with an amplitude that gradually decreases and eventually it stops. That is, the oscillatory motion is said to be damped.

To explain the damping dynamically, we may assume that, in addition to the elastic restoring force $F = -kx$, another force, opposite to the velocity, acts. This force can be written as $F' = -\lambda v$, where λ is a constant and v is the velocity. The negative sign is due to the fact that F' is opposite to v. However, it is important to mention here that other types of damping forces – proportional to higher power of the velocity or having other different physical relationship may also be present in actual physical situations.

The resultant force on the body is $F + F'$, and its equation of motion is

$$ma = -kx - \lambda v$$

or $$md^2x/dt^2 + \lambda\, dx/dt + kx = 0$$

This equation is customarily written as

$$d^2x/dt^2 + 2\gamma\, dx/dt + \omega_0^2 x = 0$$

with $2\gamma = \lambda/m$ and $\omega_0^2 = k/m$, the natural angular frequency without damping.

When damping is small or $\gamma < \omega_0$, then the solution of this equation is

$$x = Ae^{-\gamma t} \sin(\omega t + \alpha)$$

where A and α are arbitrary constants to be determined by the initial conditions and that

$$\omega = \sqrt{(\omega_0^2 - \gamma^2)} = \sqrt{[k/m - \lambda^2/(2m^2)]} \quad \text{[note that } \omega_0 > \gamma \text{ for small damping]}$$

It indicates that the effect of damping is to decrease the frequency of the oscillations.

The amplitude of the oscillations is no longer constant of time as it is given by $Ae^{-\gamma t}$. Because of the negative exponent, the amplitude decreases as t increases, resulting in a damped motion. Fig. 17.7 shows how this changes with time t.

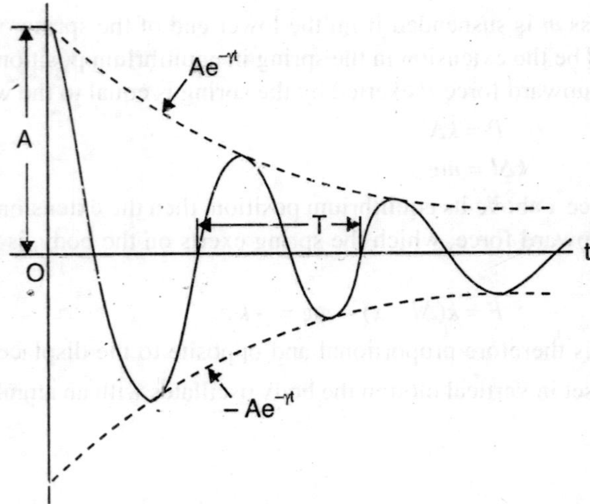

FIGURE 17.7 **Damped oscilliations, $x\ Ae^{-\gamma t} \sin(\omega t + \alpha)$; $\alpha = \frac{\pi}{2}$.**

If the damping is very large, γ may become larger than ω_0 and ω becomes imaginary. In this case there are no oscillations and the particle, if displaced and released, gradually approaches the equilibrium position without crossing it or at most crossing it once. The energy lost by the particle in damped oscillations is absorbed by the surrounding medium.

17.7 FORCED OSCILLATIONS

If a particle is subjected to an external oscillatory force along with the elastic restoring force, then the vibrations of the particle are said to be forced vibrations. We may observe forced oscillations in various practical situations. A tuning fork placed on a resonating box, absorption of electromagnetic waves by an antenna are the common examples of such situations.

In order to study the problem of forced oscillations, dynamically, let $F = F_0 \cos \omega_f t$ be the oscillating applied force, its angular frequency being given by ω_f. Assume that $F = -kx$ and $F = -\lambda v$ be the elastic restoring force and the damping force, respectively, acting on the particle.

Now, the equation of the motion can be written as

$$ma = -kx - \lambda v + F_0 \cos \omega_f t$$

or making the substitutions $v = dx/dt$, $a = d^2x/dt^2$, we have

$$m\, d^2x/dt^2 + \lambda\, dx/dt + kx = F_0 \cos \omega_f t$$

which, if we again make $2\gamma = \lambda/m$ and $\omega_0^2 = k/m$, may be written in the form

$$d^2x/dt^2 + 2\gamma\, dx/dt + \omega_0^2 x = F_0/m \cos \omega_f t \qquad ...(17.9)$$

Instead of solving this differential equation by standard techniques, we make use of our physical intuitions. It seems logical that the particle will be forced to oscillate with the angular frequency ω_f of the applied force.

Thus, we shall try as a possible solution of eq. (17.9) an expression of the form

$$x = A \sin(\omega_f t - \phi) \qquad ...(17.10)$$

where, for convenience, a negative sign has been given to the initial phase ϕ.

Now, eq. (17.10) can be written as

$$x = A \cos \phi \sin \omega_f t - A \sin \phi \cos \omega_f t \qquad ...(17.11)$$

Differentiating it w.r.t. t, we have

$$dx/dt = A\, \omega_f \cos \phi \cos \omega_f t + A\, \omega_f \sin \phi \sin \omega_f t \qquad ...(17.12)$$

again differentiating eq. (17.12) w.r.t. t, we obtain

$$d^2x/dt^2 = A\, \omega_f^2 \sin \phi \cos \omega_f t - A\, \omega_f^2 \cos \phi \sin \omega_f t \qquad ...(17.13)$$

Now, substituting eqs. (17.11), (17.12) and (17.13) in eq. (17.9) and then comparing the coefficients of $\sin\omega_f t$ and $\cos\omega_f t$, respectively on both sides, we obtain

$$2\gamma A\, \omega_f \sin \phi + A[\omega_0^2 - \omega_f^2] \cos \phi = 0 \qquad ...(17.14)$$

and

$$A[\omega_f^2 - \omega_0^2] \sin \phi + 2\gamma A\, \omega_f \cos \phi = F_0/m \qquad ...(17.15)$$

Eq. (17.14) can now be written as

$$2\gamma A\, \omega_f \tan \phi + A(\omega_0^2 - \omega_f^2) = 0,$$

which gives us

$$\tan \phi = (\omega_f^2 - \omega_0^2)/2\gamma\omega_f \qquad ...(17.16)$$

From eq. (17.16), it is quite easy to show that

$$\sin \phi = (\omega_f^2 - \omega_0^2)/[(\omega_f^2 - \omega_0^2)^2 + 4\gamma^2\omega_f^2]^{1/2}$$

and

$$\cos \phi = 2\gamma\omega_f/[(\omega_f^2 - \omega_0^2)^2 + 4\gamma^2\omega_f^2]^{1/2}$$

Inserting the values of $\cos \phi$ and $\sin \phi$, in eq. (17.15), we obtain

$$A = (F_0/m)/\,[(\omega_f^2 - \omega_0^2)^2 + 4\gamma^2\omega_f^2]^{1/2} \qquad ...(17.17)$$

Note that both the amplitude A and the initial phase ϕ are no longer arbitrary quantities, but fixed quantities that depend on the frequency ω_f of the applied force. Mathematically, this means that we have obtained a particular solution of the differential equation, but not the general one.

The variation of amplitude, A, with the frequency, ω_f, of the applied force for a given value of λ is shown in Fig. 17.8.

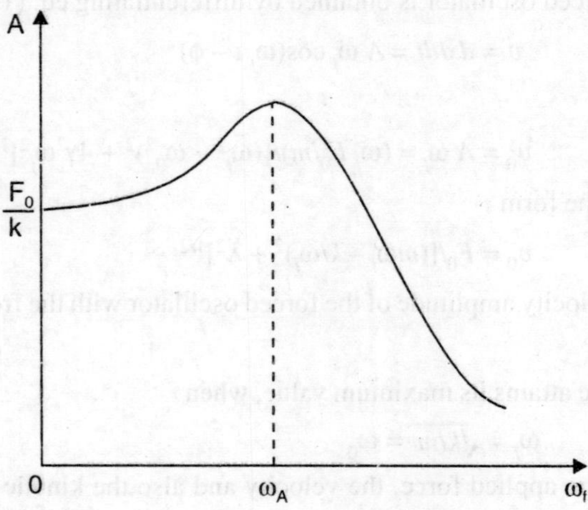

FIGURE 17.8 **Variation of amplitude with the frequency of the applied force.**

The amplitude has a pronounced maximum when the denominator in eq. (17.17) has its minimum value. This occurs for the frequency ω_A, given by

$$\omega_A = \sqrt{(\omega_0^2 - 2\gamma^2)} = \sqrt{(k/m - \lambda^2/2m^2)} \qquad ...(17.18)$$

when the frequency ω_f of the applied force is equal to ω_A, it is said that there is amplitude resonance. The smaller the damping the more pronounced the resonance, and when λ is zero or $\gamma = 0$, the resonance amplitude is infinite and occurs at $\omega_A = \omega_0 = \sqrt{k/m}$.

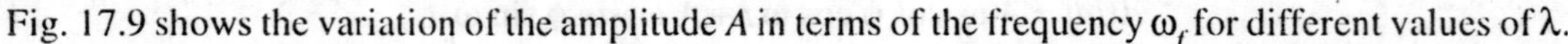

Fig. 17.9 shows the variation of the amplitude A in terms of the frequency ω_f for different values of λ.

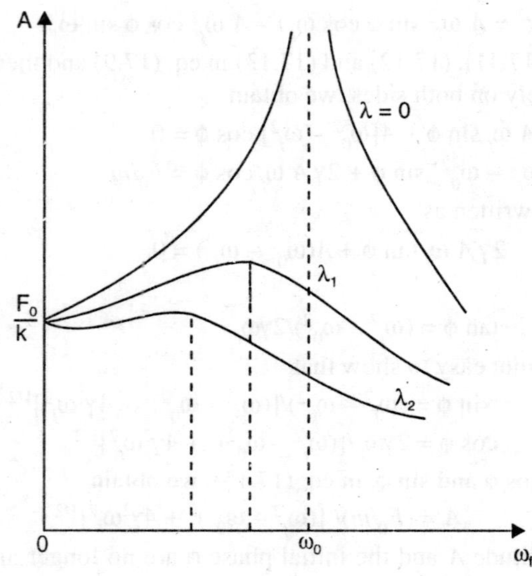

FIGURE 17.9 **Variation of the amplitude of the forced oscillations with the damping (in the figure, λ_2 is larger than λ_1).**

The velocity of the forced oscillator is obtained by differentiating eq. (17.10) w.r.t. t,

$$v = dx/dt = A\,\omega_f \cos(\omega_f t - \phi) \qquad \text{...(17.19)}$$

the velocity amplitude v_0 is

$$v_0 = A\,\omega_f = (\omega_f F_0/m)/[(\omega_f^2 - \omega_0^2)^2 + 4\gamma^2\omega_f^2]^{1/2}$$

which can also be written in the form

$$v_0 = F_0/[(m\omega_f - k/\omega_f)^2 + \lambda^2]^{1/2} \qquad \text{...(17.20)}$$

The variation of the velocity amplitude of the forced oscillator with the frequency of the applied force is shown in Fig. 17.10.

The velocity amplitude attains its maximum value, when

$$\omega_f = \sqrt{k/m} = \omega_0 \qquad \text{...(17.21)}$$

At this frequency of the applied force, the velocity and also the kinetic energy of the oscillator are maximum, and it is said that there is energy resonance.

It is quite easy to show that when $\omega_f = \omega_0$ then $\phi = 0$ *i.e.* force and velocity are in same phase. Since the rate of work done on the oscillator by the applied force is $F.\,v$, and this quantity is always positive when F and v are in same phase.

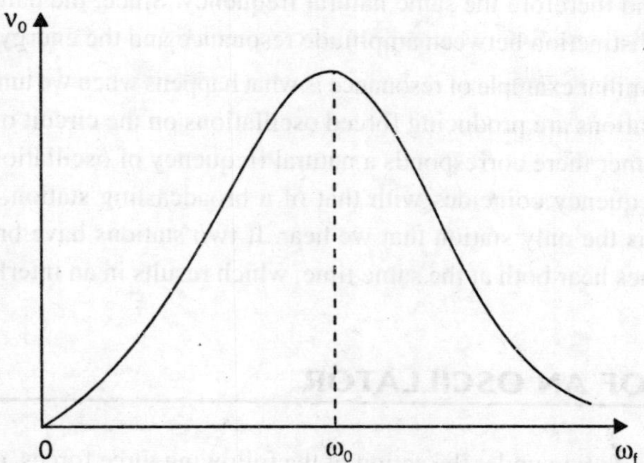

FIGURE 17.10 **Variation of velocity amplitude of forced oscillation with the frequency of the applied force.**

Therefore, at energy resonance the energy transfer from the applied force to the forced oscillator is maximum.

When the damping is very small there is no great difference between the frequencies corresponding to amplitude resonance and to energy resonance.

Amplitude resonance can be illustrated nicely through a simple experiment. Consider the situation shown in Fig. 17.11.

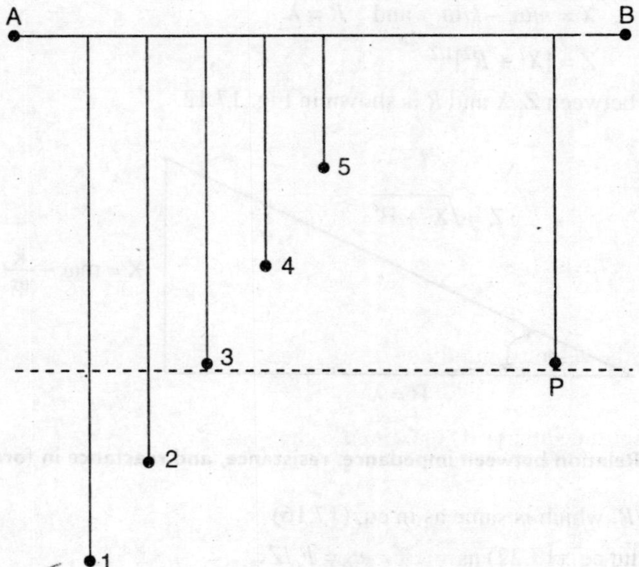

FIGURE 17.11 **Amplitude resonance in pendulum motion.**

If we set pendulum *P* in motion, the others will also start oscillating because of their coupling. However, of the five pendulums forced into oscillations, the one oscillating with greatest amplitude is number 3, which has the same length as P and therefore the same natural frequency. Since, the damping is negligible in this case, therefore there is no distinction between amplitude resonance and the energy resonance.

Perhaps, the most familiar example of resonance is what happens when we tune a radio to a broadcasting station. All broadcasting stations are producing forced oscillations on the circuit of the receiver at all times. But to each setting of the tuner there corresponds a natural frequency of oscillation of the electric circuit of the receiver. When this frequency coincides with that of a broadcasting station, the energy absorption is maximum, and hence this is the only station that we hear. If two stations have broadcast frequencies very close together, we sometimes hear both at the same time, which results in an interference effect.

17.8 IMPEDANCE OF AN OSCILLATOR

Consider an oscillator is oscillating under the action of the following three forces, oscillating applied force $F = F_0 \cos \omega_f t$, elastic force $-kx$ and damping force $-\lambda v$. We know that the velocity amplitude v_0 of this oscillator is given as

$$v_0 = A\,\omega_f = (\omega_f F_0/m)/[(\omega_f^2 - \omega_0^2)^2 + 4\gamma^2\omega_f^2]^{1/2}$$

$$v_0 = F_0/[(m\omega_f - k/\omega_f)^2 + \lambda^2]^{1/2} \qquad \text{...(17.22)}$$

The denominator of above equation is called the impedance of the oscillator, and is designated by Z, That is

$$Z = [(m\omega_f - k/\omega_f)^2 + \lambda^2]^{1/2}$$

Similarly, the reactance X and the resistance R are defined by

$$X = m\omega_f - k/\omega_f \quad \text{and} \quad R = \lambda$$

$$\therefore \qquad Z = [X^2 + R^2]^{1/2}$$

The relationship between Z, X and R is shown in Fig. 17.12.

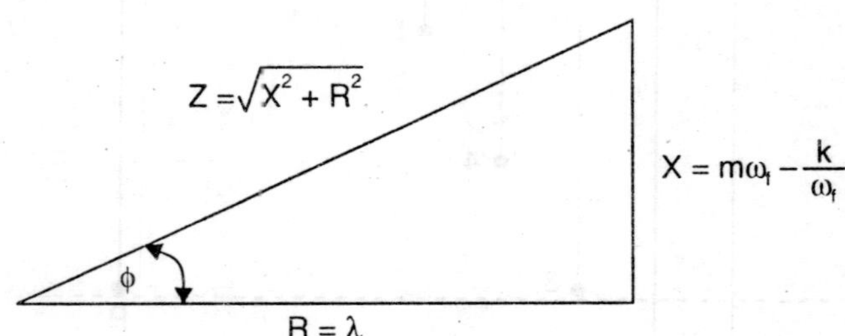

FIGURE 17.12 **Relation between impedance, resistance, and reactance in forced oscillations.**

Hence $\tan \phi = X/R$, which is same as in eq. (17.16)

Now we may write eq. (17.22) as $\qquad v_0 = F_0/Z$

and the velocity at any instant is given as $\qquad v = v_0 \cos(\omega_f t - \phi)$

The force $F = F_0 \cos \omega_f t$ and the velocity v may now be represented by rotating vectors as shown in Fig. 17.13.

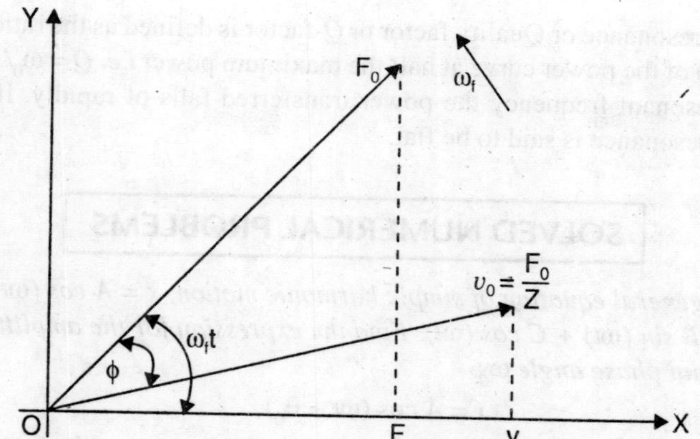

FIGURE 17.13 **Relation between the force and velocity rotating vectors in forced oscillators.**

Note that if ϕ is positive, the rotating vector v_0 lags the rotating vector $\mathbf{F}_0$ and if ϕ is negative, rotating vector v_0 leads $\mathbf{F}_0$. When there is energy resonance, $\phi = 0$, and v_0 and $\mathbf{F}_0$ have the same direction. The power transferred to the oscillator is

$$P = Fv = (F_0^2/Z) \cos \omega_f t \cos (\omega_f t - \phi)$$

This expression can also be rewritten as

$$P = (F_0^2/Z) [\cos^2 \omega_f t \cos \phi - \cos \omega_f t \sin \omega_f t \sin \phi]$$

We are more interested in the average power, P_{av}, since this is what counts when we are computing the energy absorbed by the oscillator in a certain time.

Since $(\cos^2 \omega_f t)_{av} = 1/2$ and $(\cos \omega_f t \sin \omega_f t)_{av} = 0$

Therefore $P_{av} = (F_0^2/2Z) \cos \phi = 1/2\ F_0 v_0 \cos \phi = F_0^2 R/2Z^2 = 1/2\ R v_0^2$ [from eq. (17.12)]

This verifies that the maximum transfer of energy occurs when v_0 is a maximum, since R is fixed.

At energy resonance, $\phi = 0$ and $Z = R$, resulting in $(P_{av})_{res} = F_0^2/2R$

the relation between the ratio $P_{av}/(P_{av})_{res}$ and ω_f is shown in Fig. 17.14.

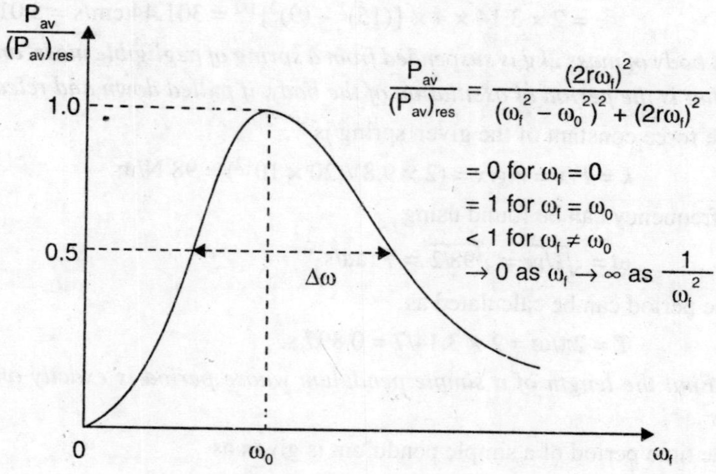

FIGURE 17.14 **Plot of $P_{av}/(P_{av})_{res}$ vs ω_f.**

The sharpness of resonance or Quality factor or Q-factor is defined as the ratio of resonant frequency ω_0 to the bandwidth $(\Delta\omega)$ of the power curve at half the maximum power *i.e.* $Q = \omega_0/\Delta\omega$. A resonance is said to be sharp, if around resonant frequency the power transferred falls of rapidly. If the power transferred decreases gradually the resonance is said to be flat.

SOLVED NUMERICAL PROBLEMS

Problem 1. *The general equation of simple harmonic motion, $x = A \cos(\omega t + \theta_0)$ can be written in the equivalent form $x = B \sin(\omega t) + C \cos(\omega t)$. Find the expression for the amplitude B and C in terms of amplitude A and the initial phase angle ω_0.*

Solution. Given $\dot{x} = A \cos(\omega t + \theta_0)$

It can be written as $x = A[\cos(\omega t)\cos(\theta_0) - \sin(\omega t)\sin(\omega_0)]$

or $x = -A \sin(\theta_0)\sin(\omega t) + A\cos(\theta_0)\cos(\omega t)$

Now, comparing it with $x = B\sin(\omega t) + C\cos(\omega t)$, we have

$$B = -A\sin(\theta_0) \text{ and } C = A\cos(\theta_0).$$

Problem 2. *A body is vibrating with simple harmonic motion of amplitude 15 cm and frequency 4Hz. Compute (i) the maximum values of the acceleration and velocity. (ii) the acceleration and velocity when the coordinate is 9 cm.*

Solution. Given, Amplitude A = 15 cm and Frequency ν = 4Hz.

(*i*) The maximum acceleration is given by the relation

$$| a_{max} | = \omega^2 A = 4\pi^2\nu^2 A = 4 \times (3.14)^2 \times (4)^2 \times 15 = 9465.2 \text{ cm/s}^2 = 94.6 \text{ m/s}^2.$$

and maximum velocity is given by the relation

$$| v_{max} | = \omega A = 2\pi\nu A = 2 \times 3.14 \times 4 \times 15 = 376.8 \text{ cm/s} = 3.77 \text{ m/s}.$$

(*ii*) The acceleration at any position x in simple harmonic motion is given as

$$a = \omega^2 x = 4\pi^2\nu^2 x = 4 \times (3.14)^2 \times (4)^2 \times 9 = 5679.1 \text{ cm/s}^2 = 56.8 \text{ m/s}^2.$$

and the velocity at any position x is given as

$$v = \omega (A^2 - x^2)^{1/2} = 2\pi\nu(A^2 - x^2)^{1/2}$$

$$= 2 \times 3.14 \times 4 \times \{(15)^2 - (9)^2\}^{1/2} = 301.44 \text{ cm/s} = 3.01 \text{ m/s}.$$

Problem 3. *A body of mass 2kg is suspended from a spring of negligible mass, and is found to stretch the spring 20 cm. What is the period of oscillation of the body, if pulled down and released.*

Solution. The force constant of the given spring is

$$k = F/x = mg/x = (2 \times 9.8)/(20 \times 10^{-2}) = 98 \text{ N/m}.$$

The angular frequency can be found using

$$\omega = \sqrt{k/m} = \sqrt{98/2} = 7 \text{ rad/s}$$

Now, the time period can be calculated as

$$T = 2\pi/\omega = 2 \times 3.14/7 = 0.897 \text{ s}.$$

Problem 4. *Find the length of a simple pendulum whose period is exactly one second at a point where $g = 9.80$ m/s^2.*

Solution. The time period of a simple pendulum is given as

$$T = 2\pi\sqrt{l/g}$$

This equation gives us that $l = (T^2 g)/(4\pi^2) = (1 \times 9.8)/(4 \times 3.14 \times 3.14) = 0.248$ m.

EXERCISE

1. A small block is executing simple harmonic motion in a horizontal plane with an amplitude of 10 cm. At a point 6cm away from equilibrium, the velocity is 24 cm/s. What is the period? What is the displacement when velocity is ± 12 cm/s ? **[Ans.** 2.09 s, 9 17 cm]
2. A body of mass 5 kg hangs from a spring and oscillates with a period of 0.5 s. How much will the spring shorten when the body is removed ? **[Ans.** 6.21 cm]
3. Discuss the motion of mass m on which an oscillating force $F = F_0 \sin \omega t$ is acting.
4. Derive an expression for the time period of a simple pendulum.
5. Show that the motion an object suspended from a vertical spring is simple harmonic motion.
6. Discuss in detail the phenomenon of resonance. What do you mean by Q-factor ? *(K.U. Jan., 2003)*
7. What do you mean by damped oscillations ?
8. What are forced oscillations ? Show that the velocity amplitude in forced oscillations is maximum when the frequency of the external force matches with the natural frequency of the oscillator.
9. Show that in forced oscillations amplitude resonance occurs at a frequency $\omega_A = \sqrt{(k/m - \lambda^2/(2m^2))}$.
10. The maximum velocity and acceleration for a S.H.M. are 1 ms^{-1} and 1.57 ms^{-2}. Find the time period. **[Ans.** 4 s]
11. Show that the damping forces make the pendulum slower.

Chapter 18 ELECTROMAGNETIC WAVES

18.1 INTRODUCTION

Mutually coupled, sinusoidal electric and magnetic fields, constitutes what is known as electromagnetic wave. These waves are different from the mechanical waves in the sense that they requre no material medium for their propagation. Now-a-days, the electromagnetic waves play an essential role in modern technology, particularly. in telecommunication. This chapter is, therefore, devoted to explore various important properties of these waves.

18.2 BASIC IDEAS OF ELECTROMAGNETIC WAVE THEORY

First of all, J.C.Maxwell predicted the existence of electromagnetic waves in 1865. He wrote all the basic laws of electricity and magnetism in a consistent set of four equations, now known as Maxwell's equations. He introduced the concept of displacement current to remove the then existing inconsistency in the laws of electricity and magnetism. Besides introducing symmetry in physical laws, the concept of displacement current was found to be crucial to the existence of electromagnetic waves. Maxwell's equations yielded wave equations for E and B, when the displacement current was taken into considerations. Though Maxwell successfully explained the existence of electromagnetic waves on theoretical ground but gave no clues as to how such waves might be generated or detected. However, he postulated that if such waves could ever be produced, they would travel through free space with the speed of light. The real existence of electromagnetic waves was discovered thirty two years later by Heinrich Rudolf Hertz, a German Physicist.

18.3 PRODUCTION OF ELECTROMAGNETIC WAVES

The pioneer experiment to produce electromagnetic waves was performed by a German scientist, Heinrich Hertz in 1888. The experimental arrangement used by him is shown in Fig. 18.1. It consists of two spheres Q and Q' located near the ends of two straight rods A and B separated by a spark gap S.

The two rods A and B are connected to an induction coil I. Due to the application of high voltage across the rods A and B, the air inside the spark gap get ionized and as a résult sparks jump across the gap giving rise to oscillating currents in AB. This is an oscillating circuit, it can be seen from the fact that the spheres Q and Q' form the plates of a capacitor and the rods form the inductance. This circuit is known as transmitter.

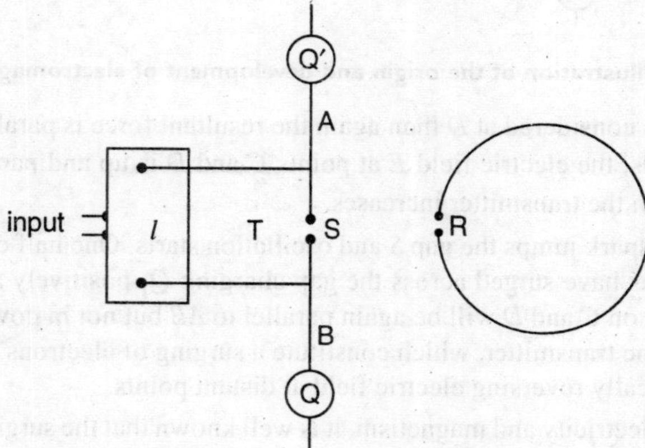

FIGURE 18.1 Experimental arrangement to demonstrate the production and detection of electromagnetic waves.

In Fig. 18.1, the loop of a single wire with a tiny spark gap R is known as receiver or detector. This circuit too is an oscillating circuit with the spark gap as a capacitance C and the loop as an inductance L. The frequency of transmitter can be changed by sliding the spheres Q and Q' along the rods A and B. When the frequency of transmitter is adjusted such that it becomes equal to the frequency of detector (*i.e.* resonance occurs) then sparks appear in the gap R. The appearance of the sparks in the gap R reveals the presence of electromagnetic waves.

With this apparatus, Hertz was able to produce and transmit electromagnetic waves through a distance of several hundred feet. Besides this, he also demonstrated the reflection, refraction and interference of electromagnetic waves.

18.4 EXPLANATION OF THE PRODUCTION OF ELECTROMAGNETIC WAVES

To visualize the production of electromagnetic waves in the experiment discussed in previous section, consider the schematic diagram in Fig. 18.2. Let us suppose that initially the rod A and sphere Q_1 is charged negatively and the rod B and sphere Q_2 is charged positively. Now, a small charge placed at C will experience the forces **a** and **b** as shown in Fig. 18.2. By symmetry these two forces are equal in magnitude and their resultant CE is parallel to AB.

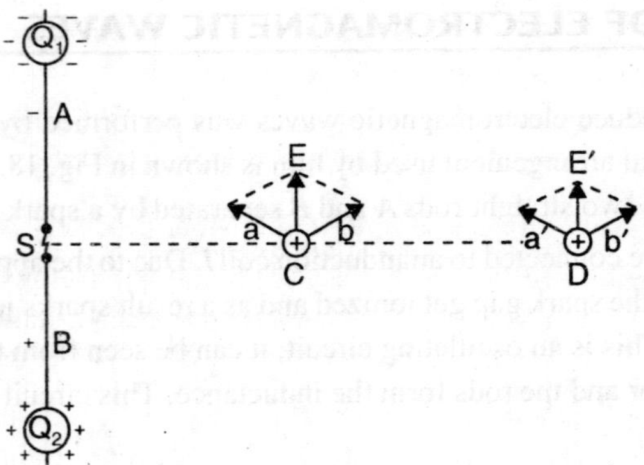

FIGURE 18.2 **Illustration of the origin and development of electromagnetic waves.**

If the small charge is considered at D then again the resultant force is parallel to AB but now weaker in magnitude. In other words, the electric field E at points C and D is up and parallel to AB, decreasing in intensity as the distance from the transmitter increases.

Now suppose that a spark jumps the gap S and oscillation starts. One half cycle later the condition is shown in Fig. 18.2, electrons have surged across the gap charging Q_1 positively and Q_2 negatively. In this situation, the resultant force on C and D will be again parallel to AB but not in downward direction. Thus, it is seen how oscillations in the transmitter, which constitute a surging of electrons back and forth between A and B, give rise to a periodically reversing electric field at distant points.

From basic laws of electricity and magnetism, it is well known that the surging of electrons in AB give rise to magnetic field, along with electric field, at C and D. When the electron current is down then using Fleming's left hand rule, it is quite easy to show that the magnetic intensity H at C or D is inward perpendicular to the plane of the page. The direction of magnetic intensity reverses *i.e.* outward perpendicular to the plane of page, when the electron current is up. The surging of the charges therefore gives rise to a periodically reversing magnetic intensity, the direction of which is at right angle to the electric intensity at the same point. Thus, these periodically electric and magnetic field coupled together in perpendicular directions give rise to the production of electromagnetic waves.

If an electrical conductor is located at some distant point the free electrons within it will, as such waves go by, experience up and down forces tending to set them in oscillation. If the conductor is an oscillatory circuit whose natural frequency is that of the passing waves, resonance will occur and large currents will be set up.

18.5 MAXWELL'S EQUATIONS

Nearly one and half century ago, a Scottish scientist James Clerk Maxwell brought together and extended four basic laws dealing with electromagnetism. These four laws deal with electrostatics, magnetostatics, varying electric field and varying magnetic field. Here, we present, along with the derivations, all the four Maxwell's equation in their simplest form.

Electrostatics

We start with the experimental result, which is usually called Coulomb's law. This law states that the force of interaction between any two point-charges q_1 and q_2, separated by a distance r from each other, is directly proportional to the product of charges and inversely proportional to the square of distance between them.

Mathematically, in S.I. Units, this law is usually written as

$$F = (1/4\pi\varepsilon_0)\, q_1 q_2/r^2 \quad ...(18.1)$$

The factor $1/4\pi\varepsilon_0$ is a constant introduced to get the units right.

In order to visualize the interaction between two static charges, the concept of electric field was introduced. The electric field intensity at a point in space around a static charge is defined as the force experienced by a unit positive charge placed at that point. More precisely, $\mathbf{E} = \underset{q \to 0}{\text{Lt}}\ \mathbf{F}/q$. Combining the definition of electric field intensity and Coulomb's law we obtain the following expression for the electric field intensity at any arbitrary point P, which is at distance r apart from a static charge q.

$$E = (1/4\pi\varepsilon_0)\, q/r^2 \quad ...(18.2)$$

The intensity of electric field may also be defined in terms of the number of hypothetical lines of electric force. The number of lines of force passing through a unit area at right angle to the direction of lines is equal to electric field intensity. The more intense the field the higher concentration of lines of force used to represent it.

Comparing both the definitions of electric field intensity, it can be concluded that the number of lines of force, which a charge q produces per unit area is given by $E = (1/4\pi\varepsilon_0)\, q/r^2$, so the intensity varies inversely as the square of the distance. In Fig. 18.3, a sphere l of radius r is drawn by the position of charge q as center. One may now ask, how many lines of force penetrate it.

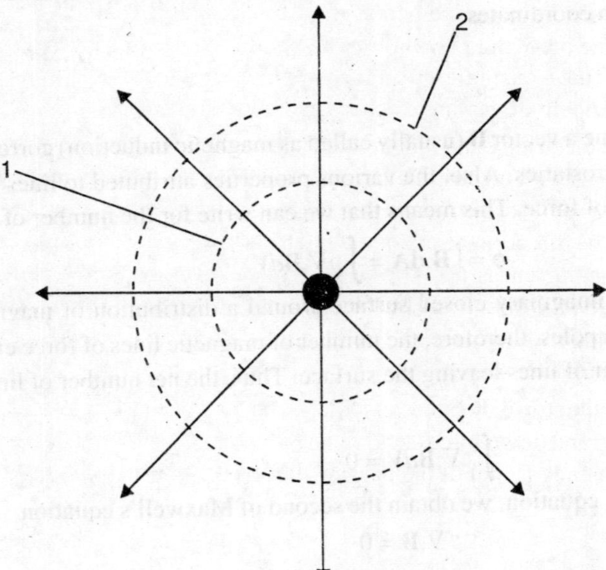

FIGURE 18.3 **Diagram to show the independence of the number of lines of force from distance *r*.**

The answer is obviously $N = [(1/4\pi\varepsilon_0)\, q/r^2][4\pi r^2] = (q/\varepsilon_0)$. This is independent of distance, so the total number of lines of force must be constant all the way from the charge to infinite. Similar result can be obtained through considering another imaginary sphere 2, shown in Fig. 18.3.

Hence, the number of lines of force leaving an electrical charge q is given by $N = (q/\varepsilon_0)$. Now, this idea can be extended to an array of charges $q_1, q_2, q_3, \ldots, q_n$. The number of lines penetrating an imaginary closed surface surrounding this array of charges is obviously given by

$$N = (1/\varepsilon_0) \sum_{i=1}^{n} q_i \quad ...(18.3)$$

In doing the addition lines going outwards through the surface are counted as positive while those going inwards as negative.

Since, the surface integral $\int \mathbf{E.dA}$ represents the flux of **E** across the surface **A**, therefore, we can write

$$N = \int \mathbf{E.dA} = (1/\varepsilon_0)\,\Sigma_{i=1}{}^{n} q_i \qquad ...(18.4)$$

This result is usually referred to as Gauss's law. If the charge distribution is continuous throughout the volume contained by the surface, instead of being discrete, then we must replace $\Sigma_{i=1}{}^{n} q_i$ by an integral $\int \rho dV$. Here ρ is the electrical charge density per unit volume and dV is an element of volume.

Gauss's law therefore becomes

$$\int \mathbf{E.dA} = 1/\varepsilon_0 \int \rho dV \qquad ...(18.5)$$

Comparing this form of Gauss's law with a well known mathematical result known as Gauss's theorem [*i.e.* $\int_V \nabla.\mathbf{E}dV = \int \mathbf{E.dA}$] we obtain

$$\int_V \nabla.\mathbf{E}dV = 1/\varepsilon_0 \int_V \rho dV \qquad ...(18.6)$$

Differentiating both sides leads to the following equation

$$\nabla.\mathbf{E} = \rho/\varepsilon_0 \qquad ...(18.7)$$

This is the statement of the first of Maxwell's equation.

In this equation, the vector ∇ (sometimes called 'nabla' and sometimes 'del') is defined as $i\,\partial/\partial x + j\,\partial/\partial y + k\,\partial/\partial z$ in cartesian coordinates.

Magnetostatics

In magnetostatics we define a vector **B** (usually called as magnetic induction) corresponding to electric field intensity vector **E** in electrostatics. Also, the various properties attributed to lines of electrical force can be applied to magnetic lines of force. This means that we can write for the number of magnetic lines of force

$$\phi = \int \mathbf{B.dA} = \int_V \nabla.\mathbf{B}dV$$

Now, consider an imaginary closed surface around a distribution of magnets. Since, each magnet consists of north and south poles, therefore, the number of magnetic lines of force entering into the surface is always equal to the number of lines leaving the surface. Thus, the net number of lines through the surface is zero *i.e.* $N = 0$

or

$$\int_V \nabla.\mathbf{B}dV = 0$$

Differentiating this equation, we obtain the second of Maxwell's equation

$$\nabla.\mathbf{B} = 0 \qquad ...(18.8)$$

At the earlier stages of scientific development, the magnetism and electricity were considered completely independent from each other. However, in 1819 Oersted observed that a wire carrying current produced deflection in magnetic compass needle, placed in the neighbourhood of the wire. Thus, the current was source of magnetic field. Two French scientists named Biot and Savart gave the quantitative relation between the current flowing along a wire and the magnetic field it produces, first of all.

Mathematically, Biot and Savart relation is written as

$$dB = (\mu_0 i/4\pi r^3)\,(dl \times \mathbf{r}) \qquad ...(18.9)$$

Here, $d\mathbf{B}$ is the magnetic flux density at a point P, which is at a distance r from a small current element of length dl of the wire carrying current i.

The application of Biot and Savart law to a very long wire gives us the following expression for the total magnetic field

$$B = \mu_0 i/2\pi r \qquad ...(18.10)$$

When we draw an imaginary closed surface around the wire and find out the flux of lines of force through this surface. It comes out to be zero because of the fact that every line coming out of the surface also have an entry point. Since the net flux through a surface is defined as $\nabla.\mathbf{B}$, therefore, the nature of magnetic field produced by the current is same as that of a permanent magnetic dipole *i.e.* $\nabla.\mathbf{B} = 0$.

Now, let us evaluate the line integral of magnetic field **B** over a closed curve *C*, shown in Fig. 18.4. It can be seen from Fig. 18.4 (*a*) and (*b*) that

$$\mathbf{B}.d\mathbf{S} = BdS \cos\phi = Bdl = Brd\theta$$

Using eq. (18.10), we can write

$$\int \mathbf{B}.d\mathbf{S} = \int_0^{2\pi} \mu_0 \frac{i}{2\pi r} rd\theta = \mu_0 i \qquad ...(18.11)$$

We can easily extend this equation to a continuous distribution of currents. In this case, we define a current density **j** per unit area such that $i = \int \mathbf{j}.d\mathbf{A}$, with $d\mathbf{A}$ as a small area element.

Inserting the current $i = \int \mathbf{j}.d\mathbf{A}$ in eq. (18.11), we have

$$\int \mathbf{B}.d\mathbf{S} = \mu_0 \int \mathbf{j}.d\mathbf{A} \qquad ...(18.12)$$

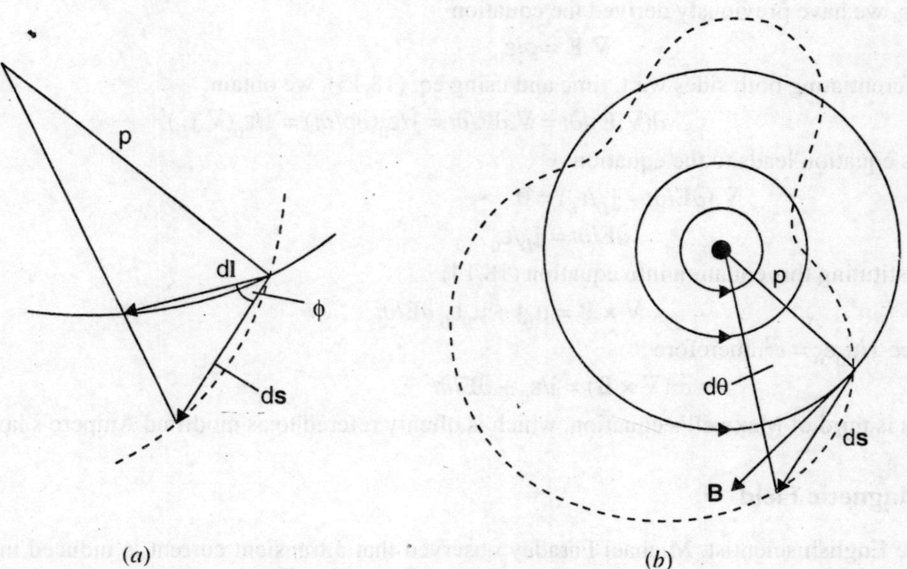

FIGURE 18.4 **(*a*) Relation between *dS* and *dl*. (*b*) Relation between *r*, *dl* and *d*θ.**

Comparing this equation with a well known mathematical result known as Stoke's theorem [$\int \mathbf{B}.d\mathbf{S} = \int \nabla \times \mathbf{B}.d\mathbf{A}$], we obtain

$$\int \nabla \times \mathbf{B}.d\mathbf{A} = \mu_0 \int \mathbf{j}.d\mathbf{A}$$

Differentiating this equation, we get

$$\nabla \times \mathbf{B} = \mu_0 \mathbf{j} \qquad ...(18.13)$$

this is known as Ampere's law, after the nineteenth century French scientist.

Varying Electric Field

The equation of continuity for electric charge is written as

$$\nabla.\mathbf{j} = -\partial\rho/\partial t$$

but by taking divergence on both sides of Ampere's law, we obtain

$$\nabla.\mathbf{j} = 0 \qquad \text{[because the divergence of the curl of a vector is always zero]}$$

Comparing this result with the equation of continuity we see that Ampere's law implies that $\partial\rho/\partial t = 0$. In other words, it only holds for steady currents, where the rate of flow of charge is constant.

So, Ampere's law require an extension for the cases where $\partial\rho/\partial t \neq 0$. Thus, Maxwell extended this law through the introduction of an extra term having the dimensions of current density in the right hand side of this equation *i.e.*

$$\nabla \times \mathbf{B} = \mu_0 (\mathbf{j} + \mathbf{j}_D) \qquad ...(18.14)$$

Maxwell called the new term $\mathbf{j}_D$ as displacement current.

Now, taking divergence on both sides of eq. (18.14), we have

$$\nabla.(\nabla \times \mathbf{B}) = 0 = \mu_0 (\nabla.\mathbf{j} + \nabla.\mathbf{j}_D)$$

This means that $\nabla.\mathbf{j}_D = -\nabla.\mathbf{j}$

However, we have seen that $\nabla.\mathbf{j} = -\partial\rho/\partial t$, so it follows that

$$\nabla.\mathbf{j}_D = \partial\rho/\partial t \qquad ...(18.15)$$

Also, we have previously derived the equation

$$\nabla.\mathbf{E} = \rho/\varepsilon_0$$

Differentiating both sides w.r.t. time and using eq. (18.15), we obtain

$$\partial(\nabla.\mathbf{E})/\partial t = \nabla.\partial\mathbf{E}/\partial t = 1/\varepsilon_0(\partial\rho/\partial t) = 1/\varepsilon_0(\nabla.\mathbf{j}_D)$$

This equation leads to the equation

$$\nabla.(\partial\mathbf{E}/\partial t - \mathbf{j}_D/\varepsilon_0) = 0$$

or

$$\partial\mathbf{E}/\partial t = \mathbf{j}_D/\varepsilon_0$$

Substituting this equation into equation (18.14)

$$\nabla \times \mathbf{B} = \mu_0\mathbf{j} + \mu_0\varepsilon_0\, \partial\mathbf{E}/\partial t \qquad ...(18.16)$$

Since $1/\mu_0\varepsilon_0 = c^2$, therefore

$$c^2(\nabla \times \mathbf{B}) = \mathbf{j}/\varepsilon_0 + \partial\mathbf{E}/\partial t \qquad ...(18.17)$$

This is third of Maxwell's equation, which is oftenly refered to as modified Ampere's law.

Varying Magnetic Field

In 1831, the English scientist, Michael Faraday observed that a transient current is induced in a circuit if (*i*) the steady current flowing in a neighbouring is turned on or off, (*ii*) the neighbouring circuit with a steady current flowing is moved relative to the first circuit, (*iii*) a permanent magnet is brought into or out of the circuit. The direction of induced current was found so as to oppose the cause of its production (This is known as Lenz's law, after the Russian scientist who suggested it). Faraday interpreted the production of induced current in terms of change in magnetic flux linked to the circuit. Also, the induced current produced its own magnetic field which acts in a direction opposite to the original field.

Actually, the changing magnetic flux induces an electric field around the circuit, the line integral of which is called electromotive force(e.m.f.), ε, which in turn induces the observed current. Hence the induced current is a secondary effect : the value depends on the resistance of the circuit as per Ohm's law.

Mathematically, the induced e.m.f. can be written as

$$\varepsilon = -\,d\phi/dt \qquad ...(18.18)$$

ϕ being the number of magnetic lines of force or magnetic flux, and the minus sign reflects Lenz's law. Now, we can write eq. (18.18) in the same form in which the other Maxwell equations are written. From the definition of e.m.f., we have

$$\varepsilon = \int_c \mathbf{E}.d\mathbf{l} \qquad ...(18.19)$$

Using Stoke's theorem

$$\int_c \mathbf{E}.d\mathbf{l} = \int_A \nabla \times \mathbf{E}.d\mathbf{A} \qquad ...(18.20)$$

Now, the number of lines of force ϕ can be related to magnetic field **B** through the following relation.

$$\phi = \int_A \mathbf{B}.d\mathbf{A}$$

Differentiating it both sides w.r.t. time, we get

$$d\phi/dt = \int_A \partial\mathbf{B}/\partial t.d\mathbf{A} \qquad ...(18.21)$$

the partial derivation of magnetic field on right hand side is taken because of the fact that the magnetic field depends on time as well as on position.

Combining eqs. (18.18), (18.19), (18.20) and (18.21), we obtain

$$\int_A \nabla \times \mathbf{E}.d\mathbf{A} = -\int_A \partial\mathbf{B}/\partial t.d\mathbf{A}$$

or

$$\nabla \times \mathbf{E} = -\,\partial\mathbf{B}/\partial t \qquad ...(18.22)$$

This is Maxwell's fourth equation.

Maxwell's Equations

The following differential equations connecting $\vec{E}, \vec{D}, \vec{B}$ and $\vec{H}$ with their sources ρ and $\vec{J}$ are known as Maxwell's equations (in SI system of units).

$$\vec{\nabla} \times \vec{E} + \frac{\partial}{\partial t}\vec{B} = 0\,; \quad \vec{B} = \mu_0\,\vec{H}$$

$$\vec{\nabla}.\vec{D} = \rho\,; \ \vec{D} = \epsilon_0\,\vec{E}$$

$$\vec{\nabla} \times \vec{H} - \frac{\partial}{\partial t}\vec{D} = \vec{J} \quad \text{and} \quad \vec{\nabla}.\vec{B} = 0$$

ρ and $\vec{J}$ denote charge density (C/m^3) and current density (Amp/m^2) respectively. The units of the field vectors as well as of sources are given in Table 18.1.

TABLE 18.1 Units of field vectors and the sources

Quantity	*Unit*
$\vec{E}$	$\frac{\text{Volt}}{\text{meter}} = \frac{\text{V}}{\text{m}}$
$\vec{D}$	$\frac{\text{Amp} - \text{sec}}{(\text{meter})^2} = \frac{\text{Coulomb}}{(\text{meter})^2} = \frac{\text{C}}{\text{m}^2}$
$\vec{B}$	$\frac{\text{Weber}}{(\text{meter})^2} = \frac{\text{w}}{\text{m}^2}$ or Tesla
$\vec{H}$	$\frac{\text{Amp.}}{\text{meter}} = \frac{\text{Amp.}}{\text{m}}$
ρ	$\frac{\text{Coulomb}}{(\text{meter})^3} = \frac{\text{C}}{\text{m}^3}$
$\vec{J}$	$\frac{\text{Amp.}}{(\text{meter})^2} = \frac{\text{Amp.}}{\text{m}^2}$

In source free space (air or vacuum) the Maxwell's equations become

$$\vec{\nabla} \times \vec{E} + \frac{\partial}{\partial t}\vec{B} = 0$$

$$\vec{\nabla} . \vec{D} = 0 ; \rho = 0$$

$$\vec{\nabla} \times \vec{H} - \frac{\partial}{\partial t}\vec{D} = 0 ; \vec{J} = 0$$

and

$$\vec{\nabla} . \vec{B} = 0$$

Lastly, $\vec{E}, \vec{D}, \rho, \vec{J}$ are termed as electric quantities while $\vec{B}$ and $\vec{H}$ the magnetic quantities.

Integral Form of Maxwell's Equations

1.
$$\vec{\nabla} \times \vec{E} + \frac{\partial}{\partial t}\vec{B} = 0$$

$$\therefore \quad \vec{\nabla} \times \vec{E} = -\frac{\partial}{\partial t}\vec{B}$$

Integrating both sides over a given surface, S, we get

$$\int_S \vec{\nabla} \times \vec{E} . d\vec{S} = -\frac{d}{dt}\int_S \vec{B} . d\vec{S}$$

Using Stokes Theorem, we get

$$\oint_{\text{loop}} \vec{E} . \vec{dl} = -\frac{d}{dt}\int_S \vec{B} . d\vec{S}$$

or $$\oint_l \vec{E}\cdot d\vec{l} = -\frac{d}{dt}\phi \quad \because \quad d\phi = \vec{B}\cdot d\vec{S}$$

where l denotes the length of the loop enclosing S and ϕ denotes the magnetic flux.

2. $$\vec{\nabla}\cdot\vec{D} = \rho$$

$\therefore$ $$\int_V \vec{\nabla}\cdot\vec{D}\, dV = \int_V \rho\, dV$$

Using Divergence Theorem on the left we get

$$\oint_S \vec{D}\cdot d\vec{S} = \int_V dq = q$$

As $$\vec{D} = \varepsilon_0 \vec{E}$$

$$\oint_S \vec{E}\cdot d\vec{S} = q, \text{ total charge.}$$

3. $$\vec{\nabla}\times\vec{H} - \frac{\partial}{\partial t}\vec{D} = \vec{J}$$

Integrating both sides over a given surface S, we get

$$\int_S \nabla\times H\cdot dS = \frac{d}{dt}\int \vec{D}\cdot d\vec{S} + \int_S \vec{J}\cdot d\vec{S}$$

Using Stokes' Theorem on the left, we obtain

$$\oint_l \vec{H}\cdot d\vec{l} = \varepsilon_0 \frac{d}{dt}\int_S \vec{E}\cdot d\vec{S} + I$$

where I denotes the total current through S.

4. $$\vec{\nabla}\cdot\vec{B} = 0$$

So, $$\int_V \vec{\nabla}\cdot\vec{B}\, dV = 0$$

Using Divergence Theorem, we get

$$\oint_S \vec{B}\cdot d\vec{S} = 0$$

or $\phi = 0$ denoting that the total flux across a closed surface always vanishes.

18.6 PLANE ELECTROMAGNETIC WAVES

In previous section, we have derived all the four Maxwell's equations starting from their respective experimental basis. Here, we present the solution of Maxwell's equation in a source free space, namely $\rho = 0$ and $\mathbf{j} = 0$. Maxwell's equation in empty space acquire the following form

$$\nabla.\mathbf{E} = 0 \qquad ...(18.23)$$

$$\nabla.\mathbf{B} = 0 \qquad ...(18.24)$$

$$\nabla\times\mathbf{B} = 1/c^2\,\partial\mathbf{E}/\partial t \quad \text{or} \quad \vec{\nabla}\times\vec{\mathbf{H}} = \frac{\partial\vec{\mathbf{D}}}{\partial t} \quad \because \quad \frac{1}{c^2} = \mu_0\varepsilon_0 \qquad ...(18.25)$$

$$\nabla \times \mathbf{E} = -\partial \mathbf{B}/\partial t \qquad ...(18.26)$$

Here, the first two equations imply that the flux of **E** and that of **B** out of any volume is zero and last two equations imply that **E** and **B** fields are coupled together. Indeed it is this ability of each field to generate other field which form the basis for the existence of electromagnetic waves.

To prove the existence of electromagnetic waves, consider eq. (18.26) *i.e.*

$$\nabla \times \mathbf{E} = -\partial \mathbf{B}/\partial t$$

Taking curl on both sides, we have

$$\nabla \times (\nabla \times \mathbf{E}) = -\partial (\nabla \times \mathbf{B})/\partial t \qquad ...(18.27)$$

Since the two operators, $\partial/\partial t$ and ∇ are independent of each other, therefore, they can be interchanged with each other.

Substituting eq. (18.25) into eq. (18.27), we get

$$\nabla \times (\nabla \times \mathbf{E}) = -1/c^2\, \partial^2 \mathbf{E}/\partial t^2 \qquad ...(18.28)$$

Using vector identity

$$\mathbf{A} \times (\mathbf{B} \times \mathbf{C}) = \mathbf{B}\,(\mathbf{A}.\mathbf{C}) - \mathbf{C}\,(\mathbf{A}.\mathbf{B})$$

the double curl on the right hand side of eq.(18.28) becomes

$$\nabla \times (\nabla \times \mathbf{E}) = \nabla\,(\nabla.\mathbf{E}) - (\nabla.\nabla)\,\mathbf{E}$$

Finally, using $\nabla.\mathbf{E} = 0$, we obtain

$$\nabla^2 \mathbf{E} = 1/c^2\, \partial^2 \mathbf{E}/\partial t^2 \qquad ...(18.29)$$

Similarly by eliminating **E** in eqs. (18.25) and (18.26) we get

$$\nabla^2 \mathbf{B} = 1/c^2\, \partial^2 \mathbf{B}/\partial t^2 \qquad ...(18.30)$$

Comparing eqs. (18.29) and (18.30) with the classical wave equation

$$\nabla^2 u - 1/v^2\, \partial^2 u/\partial t^2 \qquad ...(18.31)$$

where, v is the velocity of the wave. We conclude that these equations are vector wave equations with c as the velocity of propagation of wave. One may easily observe that the displacement current term in Maxwell's equation is essential in the derivation of these vector wave equations. If we consider waves propagating in one direction, say x, then the two vector wave equations can be written as a set of six scalar wave equations in the cartesian components of the two fields.

For the x-component of the **E** field, we have

$$\nabla^2 E_x = 1/c^2\, \partial^2 E_x/\partial t^2 \qquad ...(18.32)$$

with similar equations for the y and z components of the **E** field and for the x, y and z components of **B** field.

Since the magnitude and direction of the field do not vary over an unbound plane wave front, therefore

$$\partial[\text{field component}]/\partial y = 0 \text{ and } \partial[\text{field component}]/\partial z = 0 \qquad ...(18.33)$$

obviously, $\partial^2[\text{field component}]/\partial y^2 = 0$ and $\partial^2[\text{field component}]/\partial z^2 = 0$ ...(18.34)

hence 3-D wave equation (18.32) reduce to the set of 1-D wave equations

$$\partial^2 E_x/\partial x^2 = 1/c^2\, \partial^2 E_x/\partial t^2,\ \partial^2 E_y/\partial x^2 = 1/c^2\, \partial^2 E_y/\partial t^2 \quad \text{and} \quad \partial^2 E_z/\partial x^2 = 1/c^2\, \partial^2 E_z/\partial t^2$$

Further $\nabla.\mathbf{E} = 0$ gives us $\partial E_x/\partial x = 0$.

Now $\partial^2 E_x/\partial x^2 = 1/c^2\, \partial^2 E_x/\partial t^2$ provides that $\partial E_x/\partial t = 0$

Hence E_x does not vary with position as well as time. Therefore, E_x is a static uniform field. Earlier, we set the magnitude of static field to zero, so that $E_x = 0$.

Similarly it is quite easy to show that $B_x = 0$. Thus, plane electromagnetic waves are transverse in nature with electric and magnetic field vectors at right angles to the direction of propagation.

In order to decide the orientation of **E** and **B** vectors, assume that **E** vector always points in y-direction *i.e.* $E_z = 0$. As **E** and **B** fields are linked together through equations (18.25) and (18.26), so we are now not free to assign any arbitrary direction to **B** field.

Substituting the conditions (18.33) into eq. (18.26) gives us

$$\partial E_z/\partial x = \partial B_y/\partial t \quad \text{and} \quad \partial E_y/\partial x = -\partial B_z/\partial t \qquad ...(18.35)$$

while eqs. (18.33) and (18.25) gives

$$-\partial B_z/\partial x = -1/c^2\, \partial E_y/\partial t \text{ and } \partial B_y/\partial x = 1/c^2\, \partial E_z/\partial t \qquad ...(18.36)$$

Inserting $E_z = 0$ in eqs. (18.35) and (18.36), we get

$$\partial B_y/\partial t = 0 \quad \text{and} \quad \partial B_y/\partial x = 0.$$

Thus B_y is also uniform static field giving rise to $B_y = 0$

It tells us that if $E_z = 0$ then $B_y = 0$.

Taking $\partial/\partial t$ of eq. (18.35) and $\partial/\partial x$ of eq. (18.36) and using the fact that

$$\partial^2/\partial t\partial x = \partial^2/\partial x\partial t, \text{ we have}$$

$$\partial^2 B_z/\partial x^2 = 1/c^2\, \partial^2 B_z/\partial t^2 = \mu_0\varepsilon_0\, \partial^2 B_z/\partial t^2 \qquad ...(18.37)$$

Thus the E and B vectors in the wavefront of a plane electromagnetic wave are oriented at right angle to each other.

If we take $\partial/\partial x$ of eq. (18.35) and $\partial/\partial t$ of eq. (18.36), we have

$$\partial^2 E_y/\partial x^2 = 1/c^2\, \partial^2 E_y/\partial t^2 = \mu_0\varepsilon_0\, \partial^2 E_y/\partial t^2 \qquad ...(18.38)$$

The fundamental solutions of these equations are

$$B_z = B_0 \sin[2\pi/\lambda\,(x - ct)] = B_0 \sin(kx - \omega t) \qquad ...(18.39)$$

$$E_y = E_0 \sin[2\pi/\lambda\,(x - ct)] = E_0 \sin(kx - \omega t) \qquad ...(18.40)$$

Here λ is the wavelength and c the velocity of plane wave represented by these equations. Wave propagates in +ve x-direction if x is taken as –ve and propagates in –ve x-direction if +ve sign is taken.

Equations (18.39) and (18.40) are the solutions of equations (18.37) and (18.38) respectively, it can be easily verified by substituting (18.39) and (18.40) in these equations respectively. Also, this substitution gives us that

$$E_0 = cB_0.$$

Finally, it may be mentioned here that the Maxwell's equations are linear, *i.e.* to say that they contain **E** and **B** and their derivatives to the first power only. Also Maxwell's equations are homogeneous, *i.e.* to say that every term contain either an **E** or a **B**. Equations of this type have an important property that the sum of two different solutions is also a solution.

18.7 CONSERVATION OF ELECTROMAGNETIC ENERGY AND POYNTING VECTOR

Conservation of Electromagnetic Energy

Let us consider the Maxwell's equations in the SI system of units.

$$\vec{\nabla} \times \vec{E} + \frac{\partial}{\partial t}\vec{B} = 0\,; \; \vec{\nabla} \times \vec{H} - \frac{\partial}{\partial t}\vec{D} = \vec{J}$$

Multiplying the Ist equation by $\vec{H}$ and 2nd by $\vec{E}$ and substracting the two we get

$$\vec{H}\,.\,\nabla \times \vec{E} + \vec{H}\,.\,\frac{\partial}{\partial t}\vec{B} - E\,.\,\vec{\nabla} \times \vec{H} + E\,.\,\frac{\partial}{\partial t}\vec{D} = -\vec{E}\,.\,\vec{J}$$

As $\vec{B} = \mu_0 \vec{H}$ and $\vec{D} = \varepsilon_0 \vec{E}$ (in air/vacuum) and $\vec{\nabla}.(\vec{E} \times \vec{H}) = \vec{H}.\vec{\nabla} \times \vec{E} - \vec{E}.\vec{\nabla} \times \vec{H}$, we get

$$\vec{\nabla}.(\vec{E} \times \vec{H}) + \left[\mu_0 \vec{H}.\frac{\partial}{\partial t}\vec{H} + \varepsilon_0 \vec{E}.\frac{\partial}{\partial t}\vec{E}\right] = -\vec{E}.\vec{J}$$

or
$$\vec{\nabla}.(\vec{E} \times \vec{H}) + \frac{1}{2}\frac{\partial}{\partial t}(\vec{H}.\vec{B} + \vec{E}.\vec{D}) = -\vec{E}.\vec{J}$$

Taking volume integral of both sides

$$\int_V \vec{\nabla}.(\vec{E} \times \vec{H})\,dV + \frac{1}{2}\frac{d}{dt}\int_V (\vec{H}.\vec{B} + \vec{E}.\vec{D})\,dV = -\int_V \vec{E}.\vec{J}\,dV$$

Using divergence theorem for the Ist term we get

$$\oint_A (\vec{E} \times \vec{H}).d\vec{A} + \frac{d}{dt}W_{em} = -\int_V \vec{E}.\vec{J}\,dV$$

where A is the surface that encloses the entire volume V.

where W_{em} = Electromagnetic energy

$$= \frac{1}{2}\int_V (\vec{H}.\vec{B} + \vec{E}.\vec{D})\,dV$$

The expression $\vec{E} \times \vec{H}$ is called Poynting vector and is denoted by $\vec{S}$. So, we get

$$\oint_A \vec{S}.d\vec{A} + \frac{d}{dt}W_{em} = -\int_V \vec{E}.\vec{J}\,dV$$

If in the region V there is no current, the R.H.S. will vanish and

$$\oint_A \vec{S}.d\vec{A} + \frac{d}{dt}W_{em} = 0$$

or
$$\frac{d}{dt}W_{em} = -\oint_A \vec{S}.d\vec{A}$$

That is, rate of change of electromagnetic energy stored in the volume V is equal to rate of energy flow into the volume across the surface A. This relation is also called the conservation of electromagnetic energy and Poynting vector, $\vec{S}$, may be interpreted as the rate of flow of energy through unit area (having units = watt/m^2).

Poynting Vector

The Poynting vector, $\vec{S}$, is defined as $\vec{S} = \vec{E} \times \vec{H}$ (SI system of units)

This vector represents the rate of flow of electromagnetic energy through unit area. So, it may be used to determine the intensity of sun's radiation on earth. Radiation from a distance source is always a plane wave for which electric field and magnetic induction (as mentioned above) are given as

$$\vec{E} = \hat{j}\,E_0 \sin(kx - \omega t)$$

and
$$\vec{B} = \hat{k}\ B_0 \sin(kx - \omega t)\ ;\ B_0 = \sqrt{\mu_0 \varepsilon_0}\ E_0$$

where $\omega = 2\pi\nu$, ν being the frequency and $k = \dfrac{2\pi}{\lambda}$, the wave number.

These expressions hold for monochromatic plane wave radiations travelling along X-axis. Further, $\vec{E}$ and $\vec{B}$ are both perpendicular to the direction of radiation and are also mutually perpendicular. Now

$$\vec{S} = \vec{E} \times \vec{H} = \frac{1}{\mu_0} \vec{E} \times \vec{B}$$

$$= \frac{1}{\mu_0} (\hat{j} \times \hat{k}) \sqrt{\mu_0 \varepsilon_0}\ E_0^2 \sin^2 (kx\text{-}\omega t)$$

$$= \hat{i} \sqrt{\frac{\varepsilon_0}{\mu_0}}\ E_0^2 \sin^2 (kx - \omega t)$$

Here, $\hat{i}$, $\hat{j}$, $\hat{k}$ are unit vectors along *X*-, *Y*-and *Z*-axis respectively. Thus, we have

$$|\vec{S}| = \sqrt{\frac{\varepsilon_0}{\mu_0}}\ E_0^2 \sin^2 (kx - \omega t)$$

Now, one cannot find the instantaneous value of the magnitude of $\vec{S}$ as *t* is never known. Also, its measurement will always take a finite interval of time. Hence one can either calculate the maximum values, $|\vec{S}|_{max}$ or average values $|\vec{S}|_{av}$ as

$$|\vec{S}|_{max} = \sqrt{\frac{\varepsilon_0}{\mu_0}}\ E_0^2$$

and
$$|\vec{S}|_{av} = \sqrt{\frac{\varepsilon_0}{\mu_0}} \frac{E_0^2}{2} \quad \text{because } \langle \sin^2 (kx - \omega t) \rangle_{av} = \frac{1}{2}$$

The latter relation may be called as the intensity of radiation (plane wave). Further, for any type of wave motion the root mean square values of *E* and *B* would be

$$E_{rms} = \frac{E_0}{\sqrt{2}},$$

$$B_{rms} = \frac{B_0}{\sqrt{2}} = \sqrt{\mu_0 \varepsilon_0} \frac{E_0}{\sqrt{2}} = \sqrt{\mu_0 \varepsilon_0}\ E_{rms}$$

and
$$H_{rms} = \frac{B_{rms}}{\mu_0} = \sqrt{\frac{\varepsilon_0}{\mu_0}}\ E_{rms}$$

18.8 WAVE GUIDES

A hollow metallic cylinder, without end surfaces, is called a wave guide. A wave guide can have any arbitrary cross-section (Fig. 18.5). But the wave guides with rectangular and circular cross-sections are in common use.

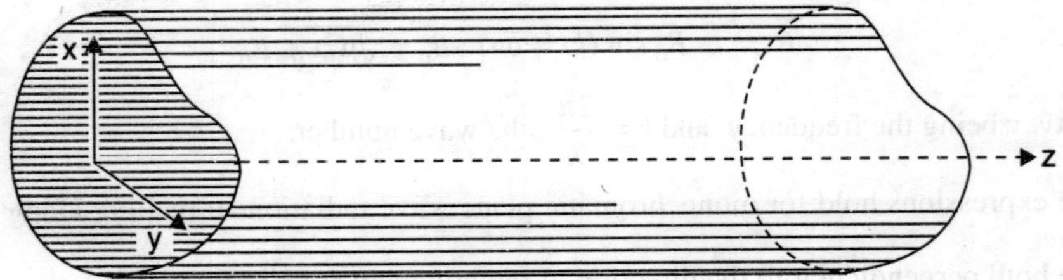

FIGURE 18.5 **A wave guide with arbitrary cross-section.**

Through a wave guide microwave electromagnetic radiations can be transmitted with relatively little attenuation. The dominating transmission modes in a wave guide are transverse electric (*TE*) and transverse magnetic (*TM*) modes. In *TE* modes the electric vector of the field has no component in the direction of propagation *i.e.* when the wave is propagating along *z*-axis, then $E_z = 0$ every where ; boundary condition $\partial B_z/\partial n|_s = 0$ here $\partial/\partial n$ is the normal derivative at a point on the surface *s*.

While in *TM* modes the magnetic vector of the field has no component in the direction of propagation *i.e.* when the wave is travelling along *z*-axis then (see Appendix-V for more details)

$B_z = 0$ every where ; boundary condition $E_z|_s = 0$

It is quite worthy to mention here that the *TEM* (transverse electromagnetic) mode cannot exist inside a single, hollow, cylindrical conductor of infinite conductivity. The surface is an equipotential ; the electric field therefore vanishes inside. It is necessary to have two or more cylindrical surfaces in order to support the *TEM* mode. The familiar co-axial cable and the parallel wire transmission line are structures for which this is the dominant mode. The co-axial cables are discussed in detail in next section.

18.9 CO-AXIAL CABLE

A cable consisting of a central conductor surrounded by an insulator, which in turn is surrounded by an earthed sheath of another conductor. The outer conductor is in mesh form and is further surrounded by an insulator usually referred as outer insulator Fig. 18.6. The inner and outer conductor are separated by low dielectric insulator such as polythene and teflon. The central conductor and the outer conductor are co-axial *i.e.* have a common axis, hence the name co-axial cable.

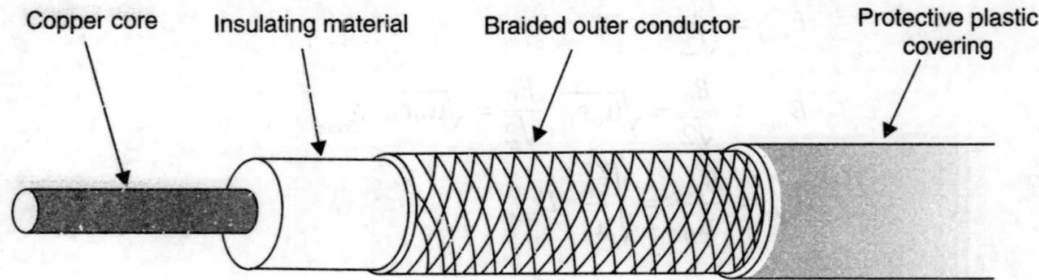

FIGURE 18.6 **A co-axial cable.**

Since, they produce no external fields and are not influenced by them *i.e.* they do not suffer from radiation problems, therefore, they are used to transmit high frequency signals (1 to 20 GHz). TV cable operators to interconnect a transmitter and an earthed antenna commonly use them. Due to their well protection their transmission efficiency is very good.

SOLVED NUMERICAL PROBLEMS

Problem 1. *A parallel plate capacitor of area 30 cm² and plate separation of 2.0 mm is charged initially to 50 μC. After sometime, the medium between the plates gets slightly conducting and the plate lose charge at the rate of 1.0×10^{-7} C/s. What is the displacement current ?*

Solution. The displacement current is given as

$$I_d = \varepsilon_0\, d\phi_E/dt$$

Using Gauu's law $\phi_E = q/\varepsilon_0$

Therefore, $I_d = dq/dt$

Hence, $I_d = 1.0 \times 10^{-7}$ A.

Problem 2. *A capacitor of two circular plates each of radius 15 cm and separated by 0.1 mm is given. Calculate the displacement current if $dV/dt = 2 \times 10^{10}$ V/s.*

Solution. The displacement current is given as

$$I_d = \varepsilon_0\, d\phi_E/dt$$

or $I_d = \varepsilon_0\, d(EA)/dt = \varepsilon_0\, Ad\,(E)/dt = \varepsilon_0\, Ad(V/d)/dt = (\varepsilon_0\, A)/d\; d(V)/dt = (\varepsilon_0\, \pi r^2)/d\; d(V)/dt$

Therefore,

$$I_d = (8.85 \times 10^{-12} \times 3.14 \times (15 \times 10^{-2})^2 \times 2 \times 10^{10})/(0.1 \times 10^{-3}) = 125.05\ A.$$

Problem 3. *A plane electromagnetic wave has a maximum electric field of 3.0×10^{-4} V/m. Find the maximum magnetic field.*

Solution. We know that electric and magnetic fields in a plane electromagnetic wave are related as

$$E = cB$$

Therefore, $B = E/c = (3.0 \times 10^{-4})/(3.0 \times 10^{8}) = 1.0 \times 10^{-12}$ T.

Problem 4. *In a plane electromagnetic wave, the electric field oscillates sinusoidal at a frequency of 3.0×10^{11} Hz and amplitude 50V/m. What is the wavelength of the wave? What is the amplitude of the oscillating magnetic field ?*

Solution. We know that $c = \nu\lambda$

or $\lambda = c/\nu = (3 \times 10^{8})/(3.0 \times 10^{11}) = 10^{-3}$ m = 1 mm.

Also, $c = E_0/B_0$

or $B_0 = E_0/c = 50/(3 \times 10^{8}) = 16.66 \times 10^{-8}\ T = 1.67 \times 10^{-7}$ T.

Note : Amplitude of magnetic intensity $= H_0 = \dfrac{B_0}{\mu_0} = 1.33 \times 10^{-1}$ amp/m.

Problem 5. *For a plane wave propogation of light prove that $B_0 = E_0/c$.*

Solution. For a plane wave propogation of light along Z-axis, electric and magnetic fields are given by

$$\vec{E} = \hat{i}\, E_0 e^{i(kZ - \omega t)}$$

and
$$\vec{B} = \hat{j}\left(\frac{ik}{i\omega}\right) E_0 e^{i(kZ - \omega t)} = \hat{j}\,\frac{k}{\omega}\, E_0 e^{i(kZ - \omega t)}$$

Here $i = \sqrt{-1}$ and $\hat{i}, \hat{j}$ are unit vectors along X- and Y-axis.

Thus, $B_0 = \frac{kE_0}{\omega} = \sqrt{\mu_0\varepsilon_0}\,\frac{\omega}{\omega}\,E_0$ because $k = \sqrt{\varepsilon_0\mu_0} \times \omega = \sqrt{\mu_0\varepsilon_0}\,.\,E_0$

But $\sqrt{\mu_0\varepsilon_0} = \frac{1}{c}$, c being the velocity of light.

So, $B_0 = \frac{E_0}{c}$.

Problem 6. *Prove that* $\sqrt{\varepsilon_0\mu_0} = \frac{1}{c}$.

Solution. Since $\mu_0 = 4\pi \times 10^{-7}\,\frac{n}{(\text{Amp})^2}$, and $\frac{1}{4\pi\varepsilon_0} = 9 \times 10^9\,\frac{n\,m^2}{\text{coul}^2}$,

$$\sqrt{\varepsilon_0\mu_0} = \left(\frac{4\pi \times 10^{-7}}{4\pi \times 9 \times 10^9}\right)^{1/2} \text{sec/m} = \frac{1}{3 \times 10^8} = \frac{1}{c}.$$

Problem 7. *If the intensity of sun light on earth is 1500 watt/m². Calculate E_0 and B_0. Also find their root mean square values.*

Solution. For a plane wave radiation from the sun the intensity is given by the average value of Poynting vector as

$$S_{av} = \frac{1}{2}\frac{E_0 \times B_0}{\mu_0} = \frac{E_0^2}{2\mu_0 c} \qquad \because\ B_0 = \frac{E_0}{c}$$

So, $$1500 = \frac{E_0^2}{2\mu_0 c}$$

or $$E_0 = (2 \times 1500 \times \mu_0 \times c)^{1/2}$$

By putting $\mu_0 = 4\pi \times 10^{-7}$ and $c = 2.998 \times 10^8$, we get

$$E_0 = 10.6 \times 10^2 \text{ V/m}$$

Then, $B_0 = \frac{E_0}{c} = \frac{10.6 \times 10^2}{2.998 \times 10^8} = 3.54 \times 10^{-6}$ Tesla

Their rms values are $E = \frac{E_0}{\sqrt{2}} = 7.5 \times 10^2$ V/m

and $B = \frac{B_0}{\sqrt{2}} = 2.51 \times 10^{-6}$ Tesla.

EXERCISE

1. A capacitor is made of two circular plates, each of diameter 24 cm and separated by 5×10^{-3} m. The capacitor is charged by an external source supplying the charging current 0.2 *A*. What is the capacitance and the rate of change of potential between the plates? Calculate the displacement current across the plates.
   [Ans. 80 *pf*, 2.5×10^9 *V/s*, 0.2 *A*]
2. In a plane electromagnetic wave, the electric field oscillates sinusoidal with a frequency of 2×10^{10} Hz and amplitude 48 *V/m*. What is the wavelength of the electromagnetic wave? Calculate the amplitude of the oscillating magnetic field.
   [Ans. 1.5×10^{-2} m, 10×10^{-8} *T*]
3. Discuss Hertz experiment to produce and detect electromagnetic waves.

4. Explain the phenomenon of the production of electromagnetic waves.
5. Explain the inconsistency in Ampere's law.
6. State and prove modified Ampere's law.
7. Prove that for a plane electromagnetic wave $E/B = c$ and $c = 1/\sqrt{\mu_0 \varepsilon_0}$.
8. Explain the term displacement current.
9. What is a wave guide ? Define Poynting vector. *(K.U. Jan., 2003)*
10. Discuss in detail co-axial cable.
11. Show that electromagnetic waves are transverse in nature.
12. Write note on Wave guides and co-axial cables. *(K.U. Jan., 2003)*
13. In a plane wave radiation the amplitude of electric field is 10^{-4} V/m. Calculate the amplitude of magnetic intensity and the average value of poynting vector. **[Ans.** 2.65×10^{-7} **Amp/m ;** 1.33×10^{-11} **Watt/m²]**

$$\left[\textbf{Hint. } S_{av} = \frac{1}{2}(E_0 H_0) = \frac{E_0 B_0}{2\mu_0} \text{ watt/m}^2\right]$$

14. Calculate average value of poynting vector in the solved problem 4. **[Ans.** 3.32 **Watt/m²]**
15. If the intensity of sun's radiation on earth is 2 cal/cm²-min, find electric field and the magnetic induction. **[Ans.** $E_{rms} = 7.22 \times 10^2\ v/m$ **;** $B_{rms} = 2.41 \times 10^{-6}\ w/m^2$**]**
16. If the amplitude of electric field E_0 is 10^3 *V/m*, calculate the amplitude of magnetic induction and magnetic intensity. *(K.U. Dec., 2006)* **[Ans.** $B_0 = 3.33 \times 10^{-6}$ **T,** $H_0 = 2.65$ **Amp/m]**

Chapter **19** QUANTUM MECHANICS

19.1 DIFFICULTIES WITH CLASSICAL PHYSICS

The classical concepts had been obtained through the study of the dynamics of bigger objects (or macroscopic systems). Therefore, these may not be found valid when we study the behaviour of smaller objects like atoms and molecules or material particles like electron, neutron and protons. In the early part of the twentieth century a lot of experimental data involving atoms, molecules etc. became available which could not be analyzed on the basis of classical concepts. In addition to these more areas were discovered where the classical laws did not work. As time passed the feeling to develop a new mechanics was more and more strengthened and hence came the pioneer work of Schrodinger, Heisenberg, Dirac etc under the new name, the quantum mechanics. Before we give the postulates and concepts of quantum mechanics let us list some of the points where classical physics was not found working.

The first point is regarding the stability of atoms and the diatomic molecules and also their properties. The atoms and molecules contain both positive and negative charges obviously must be in motion all the time. How then atoms and molecules are stable? Besides, these emit radiation of selective frequencies on excitation. All these aspects of atoms and molecules could not be explained by classical physics. Even the subsequent modifications of classical approach did not save the situation.

The second area where classical mechanics stood in conflict was the specific heat of solids particularly at low temperature. The classical conclusions could not explain the experimentally measured values of specific heat. The calculated values were higher than measured ones by a factor of 100.

As another illustration of the failure of classical mechanics one may consider the behaviour of light. It is well known that light behaves like waves while showing phenomenon like interference, and diffraction and it behaves as consisting of particles called photons while exhibiting phenomenon of photoelectric effect and Compton scattering. The photons contain energy, $h\nu$, where ν is the frequency of light.

The studies by C. Davisson and L.H. Germer (1927) and of G.P. Thomson (1927–30) on the diffraction of electron beam from thin films and the subsequent work using beams of neutrons, protons etc. established beyond doubt that material particles such as electrons, neutrons, protons also behave like waves. That is, dual character of wave motion and particle became fundamental. Naturally, classical concepts can not explain these simultaneously ; it was also established that for such material particles the wave motion require that the wavelength, $\lambda = \frac{h}{p}$, p and h being the linear momentum and Planck's constant and the energy $E = h\nu$ where ν is the frequency of the associated wave. Further, a fraction of photon is never observed.

As another instance of inadequacy of classical physics one may consider the spectrum of black body radiation. The observed energy spectrum is continuous and has a peak at wavelength depending upon the temperature of the hot body. It is independent of the shape and size of the black body. These properties could not be understood on the basis of classical laws.

Finally, classical physics failed to explain that how a photon interferes with itself (as shown by interference phenomenon) and how a photon polarized obliquely to the optic axis of Tourmaline crystal could pass through it (as observed experimentally). Besides, classical physics also failed to explain spin ½ of electrons, protons and neutrons. It is usually said that spin ½ is quantum mechanical concept and that it has no classical analogue.

19.2 WAVE PACKET AND GROUP VELOCITY

To explain the dual behaviour of the material particles like electron, proton, neutrons etc. an attempt was made earlier on. It consisted in associating a wave packet with the moving particles so that the velocity of the wave packet as a whole makes the velocity of the particle. Such a wave packet is constructed by the superposition of harmonic waves whose wavelength λ or the wave number k (= $1/\lambda = 2\pi/\lambda$) lies within a small interval say Δk for '.. That is the function $\psi(x, t) = \int_{\Delta k} a(k)\, e^{i(kx - \omega t)}\, dk$ constructed by adding progressive harmonic waves $e^{i(kx - \omega t)}$ with appropriate amplitudes $a(k)$ would represent one dimensional wave packet. It is assumed that $a(k)$ vanishes outside the interval Δk and is peaked in the middle. Here, $k = p/\hbar$ and $\hbar\omega = E$ where p denotes the linear momentum and E the energy of the material particle. For a force free motion $E = p^2/2m$ (non relativistic).

Now we give some examples of such a wave packet. Suppose $a(k)$ varies with k as in Fig. 19.1 (*a*) which has a peak at $k = k_0$. Then the shape of $\psi(x, t)$ at a particular instant say $t = 0$ will be found as of the type

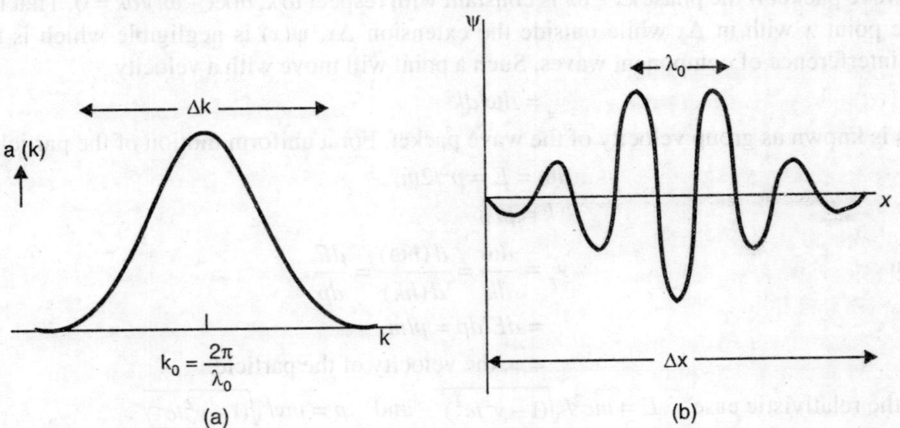

FIGURE 19.1 **(*a*) Plot of $a(k)$ vs k, k_0 denotes maximum of $a(k)$; (*b*) Shape of $\psi(x)$ vs x, Δx represents extension of wave packet.**

as shown in Fig. 19.1 (*b*). In this figure Δx indicates the extension of $\psi(x)$ so that the position of the particle is somewhere within Δx.

As a second example we choose $a(k)$ of the form as in Fig. 19.2 (*a*). $a(k)$ has been taken as constant within Δk. Then the calculations show that $\psi(x)$ would be of the type as in Fig. 19.2 (*b*).

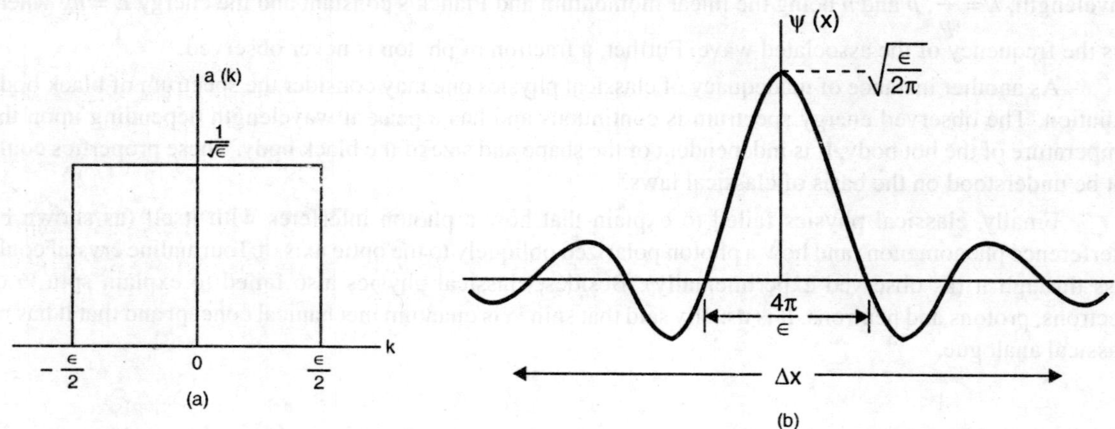

FIGURE 19.2 **(*a*) Plot of *a*(*k*) *vs k* ; (*b*) Shape of ψ(*x*) *vs x* ; Δ*x* represents the extension of the wave packet.**

Phase Velocity

For each component harmonic wave $e^{i(kx - \omega t)}$, $(kx - \omega t)$ is the phase. Its specific value would move so that

$$kdx - \omega dt = 0$$

i.e.
$$dx/dt = \omega/k$$

or
$$v_{ph} = \omega/k$$

This is called the phase velocity of the individual components.

Group Velocity

Within the wave packet if the phase $kx - \omega t$ is constant with respect to k, $\partial(kx - \omega t)/\partial k = 0$. That is $x - d\omega/dk\, t = 0$ at some point x with in Δx while outside the extension Δx, $\psi(x)$ is negligible which is the result of destructive interference of component waves. Such a point will move with a velocity

$$v_g = d\omega/dk$$

This is known as group velocity of the wave packet. For a uniform motion of the particle of mass m

$$\hbar\omega = E = p^2/2m$$

and
$$k = p/\hbar$$

Then
$$v_g = \frac{d\omega}{dk} = \frac{d(\hbar\omega)}{d(\hbar k)} = \frac{dE}{dp}$$

$$= dE/dp = p/m$$

$$= v, \text{ the velocity of the particle.}$$

For the relativistic case $E = mc^2/\sqrt{(1 - v^2/c^2)}$ and $p = mv/\sqrt{(1 - v^2/c^2)}$

or
$$E^2 = m^2c^4/(1 - v^2/c^2) = m^2c^4 + m^2v^2c^2/(1 - v^2/c^2)$$

or
$$E^2 = m^2c^4 + p^2c^2$$

Therefore, $v_g = dE/dp = pc^2/E$

$= [mvc^2/\sqrt{(1-v^2/c^2)}]/[mc^2/\sqrt{(1-v^2/c^2)}]$

[obtained using values of p & E]

$= v$, again same as the velocity of the particle. [m = rest mass]

Note. Calculations also show that $\Delta x \,.\, \Delta k = 1$.

or $\Delta x \,.\, \Delta p = \hbar$ which is the Heisenberg Uncertainity Principle. Here x and p are canonical conjugate variables. We have similar relations for other pairs of conjugate variables such as $\Delta E \,.\, \Delta t = \Delta J \,.\, \Delta\theta = \hbar$; J being the angular momentum.

19.3 QUANTUM MECHANICS

On the basis of several experiments such as interference of monochromatic light through double slits, diffraction through single slit, diffraction of electron beam, Compton scattering, photoelectric effect etc., the following facts have been established.

1. Propagation of light is through light particles called photons each carrying an energy = $h\nu$ where ν is the frequency of the given light and h is Planck's constant having a value 2π (1.0545) × 10^{-34} J sec.
2. The motion of particles like electrons, protons, neutrons etc can be described by waves with average wavelength, λ, given by $\lambda = h/p$, p being the linear momentum of the particle. In fact, superposition of harmonic waves having slightly different wavelengths around above mentioned λ, could describe the classical motion of the particle.
3. The simultaneous measurements of the pair of dynamical variables, which are conjugate to each other, are not possible. In other words, measurements of one would change the value of the other. However, in classical mechanics one assumes that measurements of such a pair of quantities is possible at the same time. This experimental fact is now known as ***Heisenberg Uncertainty*** principle which is expressed as

$$\Delta x \,.\, \Delta p_x \geq \hbar \;;\; \hbar = h/2\pi$$

$$\Delta y \,.\, \Delta p_y \geq \hbar$$

$$\Delta z \,.\, \Delta p_z \geq \hbar$$

$$\Delta t \,.\, \Delta E \geq \hbar \qquad \Delta \text{ denotes uncertainty.}$$

Thus, the product of uncertainty in the measurements of such a pair of variables (conjugate to each other) is at least of the order of $\hbar$.

For example, if we have complete knowledge of the momentum of the particle, we will have complete ignorance of its position. Also, for exact measurement of the energy one requires infinite interval of time for its measurement.

The association of photons with light radiation and of waves with motion of particles like electrons, protons, neutrons etc. (called dual nature) ultimately led to the emergence of new mechanics now known as quantum mechanics.

Schrodinger Equation

After the above discussion we may now derive the differential equation for the motion of a particle of mass m, linear momentum p and energy E. The wave function $\psi(x, y, z, t)$ satisfying such a differential equation would contain all the plausible information regarding the motion. While deriving this equation we assume complete knowledge of linear momentum of particle and hence complete ignorance of its position at any instant t and that the equation is linear and the coefficients do not depend on dynamic quantities like, p, E, ν etc. This differential equation is called Schrodinger Equation. For simplicity, we shall involve motion only along x-direction such that $\psi(x, t)$ is independent of y, z, coordinates. We shall also remember some known equations such as for transverse waves along a stretched string or for sound waves in a gaseous medium. We also assume that ψ may be a complex function so that $\psi^*.\psi$ is a real quantity.

So, let us start from an equation of the type known for waves along stretched string

$$\partial^2\psi/\partial t^2 = \gamma\, \partial^2\psi/\partial x^2$$

where γ is a constant. This constant should not depend on E, p etc. as we want the differential equation of general nature. Some function like sin $(kx - \omega t)$, cos $(kx - \omega t)$, Exp$[i(kx - \omega t)]$, Exp $[-i(kx - \omega t)]$ are well known in the literature. These describe harmonic waves with

$$k = 2\pi/\lambda = \frac{1}{\lambda} \text{ ; wave number,}$$

$$\lambda = h/p,$$

and $\qquad \omega = 2\pi\nu \quad \text{or} \quad E = h\nu = \hbar\omega$

We particularly substitute $\psi(x, t)$ = Exp $[i(kx - \omega t)]$ in the above differential equation. After simplification we obtain

$$\omega^2 = \gamma k^2$$

i.e.

$$\begin{aligned} \gamma &= \omega^2/k^2 \\ &= E^2/p^2 \\ &= E/2m \end{aligned}$$

where we have used $E = p^2/2m$ in case of force free motion of the particle of mass m.

Now, since γ depends on E, the above differential equation is not the one we were looking for.

Next, as $\partial/\partial t$ results in the multiplication of ω or E and $\partial/\partial x$ in the multiplication of k or p and the fact that $E = p^2/2m$ we now try the following differential equation

$$\partial\psi/\partial t = \gamma\, \partial^2\psi/\partial x^2$$

If we substitute

$$\psi = \text{Exp}\,[i(kx - \omega t)]$$

we obtain

$$i\omega = \gamma k^2$$

i.e.

$$\begin{aligned} \gamma &= i\omega/k^2 \\ &= iE\hbar/p^2 \\ &= i\hbar/2m \qquad (\because\ E = p^2/2m) \end{aligned}$$

Now, γ is independent of E or p and so the preceding equation seems to be quite appropriate. It is rewritten as

$$\partial\psi/\partial t = i\hbar/2m\, \partial^2\psi/\partial x^2$$

or

$$i\hbar\, \partial\psi/\partial t = -\,\hbar^2/2m\, \partial^2\psi/\partial x^2$$

we can easily extend this equation for three dimensional motion by changing

$$\psi\,(x, t) \to \psi\,(x, y, z, t) \quad \text{or} \quad \psi\,(\mathbf{r}, t),$$

$$\mathbf{p} = \hbar\,\mathbf{k} \text{ such that}$$

$$|\,\mathbf{k}\,| = p/\hbar = k,$$

and

$$\partial^2/\partial x^2 \to \partial^2/\partial x^2 + \partial^2/\partial y^2 + \partial^2/\partial z^2 = \nabla^2$$

As a result we obtain $\quad i\hbar\, \partial\psi(\mathbf{r}, t)/\partial t = -\,\hbar^2/2m\, \nabla^2\psi(\mathbf{r}, t)$

which is known as Schrodinger equation for force free motion of the particle of mass m and energy $E = p^2/2m$. This can be seen by substituting $\psi(\mathbf{r}, t) = \text{Exp}\,[i(\mathbf{k}.\mathbf{r} - \omega t)]$ in the differential equation. In fact, we get

$$E\,\psi(\mathbf{r}, t) = p^2/2m\;\psi(\mathbf{r}, t)$$

or

$$E = p^2/2m.$$

The preceding differential equation

$$i\hbar\,\partial\psi(\mathbf{r}, t)/\partial t = -\,\hbar^2/2m\,\nabla^2\psi(\mathbf{r}, t)$$

is known as the Non-relativistic Time Dependent Schrodinger equation for the force free motion of a particle of mass m as we have seen above that energy $E = p^2/2m$.

Note : The wave motion in a string satisfies $\dfrac{T}{m}\dfrac{\partial^2\psi}{\partial x^2} = \dfrac{\partial^2\psi}{\partial t^2}$.

When a force also acts on the particle the total energy E of the particle is given by

$$E = p^2/2m + V(\mathbf{r}, t),$$

where $V(\mathbf{r}, t)$ denotes the potential energy of the particle. So the Schrodinger equation for the motion in an external field may be written as

$$i\hbar\,\partial\psi(\mathbf{r}, t)/\partial t = -\,\hbar^2/2m\,\nabla^2\psi(\mathbf{r}, t) + V(\mathbf{r}, t)\,\psi(\mathbf{r}, t).$$

Special Case of Time Independent Schrodinger Equation

When potential energy does not depend on the time variable, t, we have $V(\mathbf{r}, t) \rightarrow V(\mathbf{r})$

In such a situation we may write $\psi(r, t) = A(t)\,\varphi(r)$,

where $A(t)$ is a function of t while $\varphi(\mathbf{r})$ is a function of $\mathbf{r}$. As a result

$$i\hbar\,\varphi(\mathbf{r})\,\partial A(t)/\partial t = [-\,\hbar^2/2m\,\nabla^2\varphi(\mathbf{r}) + V(\mathbf{r})\,\varphi(\mathbf{r})]\,A(t)$$

For the case of simple harmonic wave function

$$\psi(\mathbf{r}, t) = \text{Exp}\,[i(\mathbf{k}.\mathbf{r} - \omega t)]$$

$$\varphi(\mathbf{r}) = \text{Exp}\,[i(\mathbf{k}.\mathbf{r})]$$

and

$$A(t) = \text{Exp}\,[-\,i\,\omega t]$$

the preceding equation yields

$$\hbar\omega\,\varphi(\mathbf{r}) = -\,\hbar^2/2m\,\nabla^2\varphi(\mathbf{r}) + V(\mathbf{r})\,\varphi(\mathbf{r})$$

(obtained by canceling the common factor $e^{-i\omega t} = A(t)$ from both sides). As $\hbar\omega = E$,

$$E\varphi(\mathbf{r}) = [-\,\hbar^2/2m\,\nabla^2 + V(\mathbf{r})]\,\varphi(\mathbf{r})$$

This equation is known as Time Independent Schrodinger Equation. Also,

$$\psi(\mathbf{r}, t) = \varphi(\mathbf{r})\,\text{Exp}\,[-\,i\,\omega t]$$

19.4 CONCEPT OF OPERATORS

It is well known that the state of the particle undergoes change when measurements of a dynamical variable like position, linear momentum etc. are performed. This act of measurement is therefore denoted by an operator in quantum mechanics. Also, such an operator will be the operator for that particular variable which is being measured. Mathematically this is expressed as

$$P_x\psi(\mathbf{r}, t) = p_x\varphi(\mathbf{r}, t)$$

where $\varphi(\mathbf{r}, t)$ represents the new state of the motion and p_x the result of the measurement of x-component of linear momentum,

P_x being the corresponding operator.

In the above discussion we have noticed that when

$$\psi(\mathbf{r}, t) = \text{Exp}\,[i(\mathbf{k.r} - \omega t)]$$

$$i\hbar\, \partial\psi/\partial t = E\psi$$

and $\quad -i\hbar\, \partial\psi(\mathbf{r}, t)/\partial x = \hbar k_x \psi(\mathbf{r}, t) = p_x \psi(\mathbf{r}, t) \qquad (\because \mathbf{p} = \hbar\mathbf{k}, \mathbf{k.r} = \sum k_x x)$

along with

$$-i\hbar\, \partial\psi(\mathbf{r}, t)/\partial y = p_y \psi(\mathbf{r}, t),$$

$$-i\hbar\, \partial\psi(\mathbf{r}, t)/\partial z = p_z \psi(\mathbf{r}, t)$$

Thus, $\quad i\hbar \dfrac{\partial}{\partial t} \rightarrow$ energy operator

and $\quad -i\hbar\, \nabla \rightarrow$ linear momentum operator

Similarly, it can be proved that **r** denotes the position operator and $V(\mathbf{r}, t)$ the potential energy operator in the present approach called the 'Schrodinger Picture'.

19.5 NORMALIZATION

When $\int \psi^* \psi\, d\mathbf{r} = 1$, (unity) $\psi(\mathbf{r}, t)$ is said to be normalized *i.e.*, probability of the presence of a particle is unity. For instance, if $\psi(\mathbf{r}, t) = 1/\sqrt{V}\; e^{i(\mathbf{k.r} - \omega t)}$ where V denotes the entire volume.

$$\int_V \psi^*\psi\, d\mathbf{r} = 1/V \int_V d\mathbf{r} = 1,$$

In the general case, when $\int \psi^*(\mathbf{r}, t)\, \psi(\mathbf{r}, t)\, d\mathbf{r} = N$, $1/\sqrt{N}\; \psi(\mathbf{r}, t)$ will be said to be normalized wave function for the motion of the particle. Here (*) denotes the complex conjugation.

Average Value of a Variable

Since the probability of the particle to be at some position **r** varies as a function of this position at any instant of time, we define the average value, p, of a dynamical variable by

$$\langle p \rangle = \int \psi^*\, p_{op}\, \psi\, d\mathbf{r} \quad \text{or} \quad \int \psi^*\, p_{op}\, \psi\, dxdydz$$

where p_{op} is the operator of the given dynamical variable. For instance,

$$\langle x \rangle = \int \psi^*\, x\, \psi\, dxdydz;$$

$$\langle y \rangle = \int \psi^*\, y\, \psi\, dxdydz,$$

and $\quad \langle z \rangle = \int \psi^*\, z\, \psi\, dxdydz$

are the average values of x-, y- and z- components of the particle position at time t.

Similarly,

$$\langle \mathbf{p} \rangle = \int \psi^*\, (-i\hbar\, \nabla\, \psi)\, dxdydz,$$

and $\quad \langle E \rangle = \int \psi^*\, (i\hbar\, \partial\psi/\partial t)\, dxdydz = \int \psi^*\, [-\hbar^2/2m\, \nabla^2 + V(r, t)]\, \psi dxdydz$

will be the average values of linear momentum and total energy respectively.

From the above discussion it is obvious that in general the variable (or physical quantity) may not have well defined value as is assumed in classical mechanics but it has all possible values with statistical distribution. That is, the values of a dynamical variable like position, energy etc of a particle occur with varying probabilities in the general case. So, in quantum mechanics one calculates only the average values also called expectation values. Only these values are to be compared with the measured values.

19.6 POSTULATES OF QUANTUM MECHANICS

The development of quantum mechanics was based on experimental evidences on :

(*i*) Dual character of light as well as material particles like *e*, *n*, *p*, etc.

(*ii*) Heisenberg's uncertainty principal between pair of conjugate variables such as position coordinates and momentum components; angles and angular momenta; energy and time etc. Infact, complete knowledge of one leads to complete ignorance of the other.

(*iii*) The superposition of well defined states called superposition principle.

(*iv*) The act of measurement may change the state of the system.

Only when these facts were established, a need was felt to depart from Newtonian and Maxwellian mechanics (Classical mechanics, Classical electrodynamics) and to develop a new mechanics for the study of system of atomic and molecular sizes. This new mechanics has been since known as quantum mechanics.

We shall now give fundamental postulates of quantum mechanics without bothering about its mathematical scheme.

1. The arbitrary state of a dynamical system is to be represented by a vector say $|\psi>$, in the abstract vector space [and its components in terms of position coordinates of the system say *a* set *q*, are $<q|\psi>$ or $\psi(q)$. It is called wave function in Schrodinger picture. It is complex function.]

2. In order to obtain the scalar product of these abstract vectors, an identical vector space is assumed. In this space, called space of conjugate imaginary vector, an arbitrary vector is to be written as $<\phi|$ and the scalar product with $|\psi>$ is written as $<\phi|\psi>$ such that $<\phi|\psi> = \overline{<\psi|\phi>}$, bar denoting the complex conjugation. Usually, this scalar product is a complex quantity. Further, $<\phi|$ is the conjugate imaginary vector to $|\phi>$. Naturally, $<\phi|\phi>$ or $<\psi|\psi>$ would be a real quantity.

3. These vectors can be added linearly like ordinary vectors as $|\psi> = C_1|\psi_1> + C_2|\psi_2>$ where C_1 and C_2 are numbers, real or complex. That is the state of the system represented by vector $|\psi>$ is resulted from the superposition of states represented by $|\psi_1>$ and $|\psi_2>$. A parallel equation for conjugate imaginary vector is

$$<\phi| = C_1<\phi_1| + C_2<\phi_2|$$

[Note conjugate imaginary to $C|\psi>$ to $C^*<\psi|$ where C^* is complex conjugate of C.]

If there exits $C_1|\psi_1> + C_2|\psi_2> = 0$.

for some C_1 and C_2, $|\psi_1>$ and $|\psi_2>$ are said to be dependent, otherwise independent.

Length (or magnitude) of any vector $|\psi>$ is given by $\sqrt{<\psi|\psi>}$. If this is a unity, the vector will be normalised or a unit vector.

[Dirac called $|\psi>$ as a ket vector and $<\psi|$ as bra vector *i.e.* spaces of ket vectors and bra vectors].

4. It is experimental fact that when measurement is made on the system, its state under goes a change. As a result, the subsequent measurement may give a different value. Alternatively, if we have a no. of identical systems, they will give different values of the dynamical quantity when measurements are made on them. In a rare case the system may be in a particular state that on measurement we get same value again and again. In the former situation, the system is said to be in arbitrary state while in the later situation in eigen state w.r.t. that quantity. It is also for this reason that the dynamical quantity being measured be represented by operators. In the schrodinger picture, position

coordinates operators are same as x, y, z etc. while momenta operators are $-i\hbar\frac{\partial}{\partial x}, -i\hbar\frac{\partial}{\partial y}, -i\hbar\frac{\partial}{\partial z}$ etc. where $\hbar$ denotes Planck's constant divided by 2π, $i\hbar\frac{\partial}{\partial t}$ denotes the energy operator. Since the measured value are real, these operators should be Hermitian. Then, eigen value equation is given by

$$p \mid p_m > = p_m \mid p_m >$$

where p_m is the eigen value of the operator p and $\mid p_m >$ its eigen vector.

In the Schrodinger picture the unit vectors are represented by $\mid q >$ just as **i, j, k** in ordinary space and these are infinite in number and are eigen vectors of coordinate operators. It forms a complete set such that

$$\psi > = \int < q \mid \psi > \mid q > dq$$

or $\mid \psi > = \int \psi(q) \mid q > dq$ taking $< q \mid \psi >$ as simply $\psi(q)$, a function of q.

So, $\psi(q) = < q \mid \psi >$ are the component of the vector $\mid \psi >$ along $\mid q >$ direction. Conjugate imaginary vectors would be $< q \mid$. Also

$$< \phi \mid q > = \overline{< q \mid \phi >} = \phi^*(q) \text{ so that}$$

$$< \phi \mid \psi > = \int < \phi \mid q > < q \mid \psi > dq$$

$$< \phi \mid \psi > = \int \phi^*(q)\, \psi(q)\, dq$$

Which is the scalar product between ket vectors $\mid \phi >$ and $\mid \psi >$

5. Since the operators represent the dynamical variables of a system and every measurement may change state of system and so the values of variables, the product of two operators, in general, do not commute. Only when for a pair of variables the measurement of one does not disturb the value of the other, this product of corresponding operator would commute.
6. $\psi^*(q).\psi(q)$ represents the probability density of the particles presence at the position q while $\psi^*(q).\psi(q)\, dq$ represents the probability of the particle's presence in the elemental volume dq located at the position q. Also, $\psi^*(q).\psi(q) = \mid \psi(q) \mid^2$.
7. Quantum mechanics calculates only the average value of a dynamic variable. If A_{op} is the operators of variable A, the average value of A will be given by

$$< A > = \int \psi^* A_{op}\, \psi\, dq \qquad \text{(in Schrodinger picture)}$$

i.e., the average values of **r**, **p**, and E would be respectively given as

$$< \mathbf{r} > = \int \psi^*(\mathbf{r})\mathbf{r}\, \psi(\mathbf{r})\, dxdydz$$

$$< \mathbf{p} > = \int \psi^*(\mathbf{r})\, (-i\hbar\, \nabla\psi(\mathbf{r}))\, dxdydz$$

and $$< E > = \int \psi^*(\mathbf{r}, t)\left(-i\hbar\frac{\partial}{\partial t}\psi(\mathbf{r}, t)\right) dxdydz$$

8. The state is defined by the possible values of physical quantities, which are simultaneously measurable along with the total energy of the system. The corresponding operators will all commute mutually. The set of states each corresponding to these possible set of values are called basic states and the corresponding vectors, the basic vectors. Any state of the concerned system is described in terms of these basic states to calculate the average value of required quantity.
9. $\psi(\mathbf{r})$ is to be found by solving the time independent Schrodinger equation for the given system. For a conservative system of single particle, this equation is

$$-\frac{\hbar^2}{2m}\nabla^2\psi(\mathbf{r}) + V(\mathbf{r})\,\psi(\mathbf{r}) = E\,\psi(\mathbf{r})$$

where $$\nabla^2 = \frac{\partial}{\partial x^2} + \frac{\partial}{\partial y^2} + \frac{\partial}{\partial z^2}$$

$$V(\mathbf{r}) \equiv V(x, y, z) = \text{Potential energy}$$

and $$E = \text{total energy}$$

The solution of this equation, $\psi(\mathbf{r})$, contains all information regarding the system.

In the end we may mention that when energies and momenta are large the uncertainties in the conjugate variables become negligible and the quantum mechanical calculations agree with the corresponding results in classical mechanics. This principle is known as correspondence principle. And that the success of quantum mechanics is to be seen by the extent of agreement between quantum mechanical predictions and the experimental values.

19.7 PHYSICAL SIGNIFICANCE OF $\psi(\mathbf{r}, t)$

$\psi(\mathbf{r}, t)$ is the function satisfying the Schrodinger equation $-\frac{\hbar^2}{2m}\nabla^2\psi + V\psi = i\hbar\frac{\partial\psi}{\partial t}$ where $\mathbf{r}$ denotes the set of all coordinates of a given system. This function may be a complex function. It is expected that such a function contains all informations regarding the system for which the above mentioned schrodinger equation was written. Therefore, all properties of that system are to be derived with the help of ψ.

For instance $\psi^*\psi = |\psi|^2$ represents the probability density of the system to be at the position $\mathbf{r}$. It is a +ve quantity. Similarly, $|\psi|^2\,dxdydz$ ($\equiv |\psi|^2\,d\mathbf{r}$) denotes the probability of the system to be present in the volume $dxdydz$ located at $P(x, y, z)$. Also, $\int_V |\psi|^2\,dxdydx$ would give the probability to find the system (or particle) some where in the entire volume, V. Whenever, $\int_V |\psi|^2\,dxdydz = 1$, ψ will be said as Normalised. In case,

$\int_V |\psi|^2\,dxdydz = N$, where N is a positive number, ψ will not be called as normalised. To normalise this function we divide ψ by $\sqrt{N}$. Thus, $\frac{\psi}{\sqrt{N}}$ will be a normalised function in which case $\frac{1}{\sqrt{N}}$ is called normalisation constant.

Further, ψ also helps in estimating the average value of physical quantity. Infact average of A is given by

$$<A> = \int \psi^* A_{op}\,\psi\,dxdydz,$$

Here ψ is already normalised. For instance, average values of $\mathbf{r}$, $\mathbf{p}$, E are calculated respectively by

$$<\mathbf{r}> = \int \psi^*\mathbf{r}\,\psi\,dxdydz,$$

$$<\mathbf{p}> = \int \psi^*(-i\hbar\nabla)\,\psi\,dxdydz,$$

$$<E> = \int \psi^*\left(-i\hbar\frac{\partial}{\partial t}\right)\psi\,dxdydz,$$

This is how one learns the physical behaviour of the system with the help of $\psi(\mathbf{r}, t)$. All these calculations pertain to particular instant of time.

19.8 STATIONARY STATES

When potential V is conservative $i\hbar\frac{\partial\psi}{\partial t}$ can be written as equal to $E\psi$ *i.e.* $i\hbar\frac{\partial\psi}{\partial t} = E\psi$ where E is a real number, called the total energy of the system. This equation can be integrated to $\psi(\mathbf{r}, t) = \psi_0(\mathbf{r})\, e^{-\frac{iEt}{\hbar}}$ with $\psi_0(\mathbf{r})$ denoting $\psi(\mathbf{r}, t)$ at $t = 0$. So, the wave function can be split up into the product of space part $\psi_0(\mathbf{r})$ and time dependent part $e^{-\frac{iEt}{\hbar}}$. The Schrodinger Equation will get simplified to $-\frac{\hbar^2}{2m}\nabla^2\psi_0(\mathbf{r}) + V(\mathbf{r})\,\psi_0(\mathbf{r}) = E\psi_0(\mathbf{r})$, a time independent Schrodinger Equation.

Besides the above simplification, we also note that $|\psi(\mathbf{r}, t)|^2 \Rightarrow |\psi_0(\mathbf{r})|^2$, a time independent probability density for finding the system at the position $\mathbf{r}$. Similarly, $|\psi|^2\, dxdydz$ as well as $\int_V |\psi|^2\, dxdydz$ and $\int \psi^*\, A_{op}\, \psi dxdydz$ become constant of time that is all these quantities become constant or stationary. Hence, such a state of the system expressible as $\psi(\mathbf{r}, t) = \psi_0(\mathbf{r})\, e^{-\frac{iEt}{\hbar}}$ is called Stationary State. For example, states of atomic electron in an atom and nuclear states in shell model are said to be stationary states. Note that A_{op} denotes the operator of some physical quantity A *e.g.* $\mathbf{r}$, $-i\hbar\nabla$, $i\hbar\frac{\partial}{\partial t}$.

19.9 EHRENFEST'S THEOREM

Ehrenfest's theorem concerns the wave packet representation of the particle motion. It states that the relation between weighted averages of position and linear momentum vectors agree with those used in classical mechanics. For example, one can prove that

$$d<x>/\,dt = <p_x>/\,m \text{ etc.}$$

and

$$d<p_x>/\,dt = -<\partial V/\partial x> \text{etc.}$$

Let us first find the rule of change of weighted average of x-component of the position vector.

$$d<x>/dt = d\,(\int \psi^* x\psi\, dxdydz)/dt$$

$$= \int \partial\psi^*/\partial t x\, \psi\, dxdydz + \int \psi^* x\, \partial\psi/\partial t\, dxdydz$$

Now use $i\hbar\, \partial\psi/\partial t = -\hbar^2/2m\, \nabla^2\psi + V\psi$

and $\quad -i\hbar\,\partial\psi^*/\partial t = -\hbar^2/2m\, \nabla^2\psi^* + V\psi^*$

$$d<x>/dt = -1/i\hbar \int x\psi\,[-\hbar^2/2m\, \nabla^2\psi^* + V\psi^*]\, dxdydz + 1/i\hbar \int \psi^* x\,[-\hbar^2/2m\nabla^2\psi + V\psi]\, dxdydz$$

$$= (1/i\hbar)\,[\int \psi^* x\,[-\hbar^2/2m\, \nabla^2\psi]\, dxdydz - \int \psi\, x\,[-\hbar^2/2m\, \nabla^2\psi^*]\, dxdydz]$$

on integrating 2nd term by parts

$$= (-1/i\hbar)\,[\hbar^2/2m \int \nabla(x\psi).\nabla\psi^* dxdydz]$$

[obtained by neglecting the integrated term because ψ and ψ^* vanish at large distances]

Integrating it again by parts we get

$$= (1/i\hbar)\,.\,\hbar^2/2m \int \psi^* \nabla^2(x\psi)\, dxdydz$$

$$= (1/i\hbar)\,.\,\hbar^2/2m\,[\int \psi^* x \nabla^2\psi\, dxdydz + 2\int \psi^* \partial\psi/\partial x\, dxdydz]$$

Putting this value for the 2nd term in the R.H.S. of the preceding equation we get

$$d\langle x\rangle/dt = [\int \psi^*(-i\hbar\,\partial/\partial x)\,\psi\; dxdydz]/m$$

$$= [\int \psi^* p_x \psi\; dxdydz]/m = \langle p_x\rangle/m$$

This relation is similar to that in classical mechanics. But here it is for average values.

Next we prove $d\langle p_x\rangle/dt = -\langle \partial V/\partial x\rangle = F_x$, x-component of external force.

That is, $d\langle p_x\rangle/dt = d(\int\psi^*(-i\hbar\,\partial/\partial x)\,\psi\; dxdydz)/dt$

$$= \int\partial\psi^*/\partial t\,(-i\hbar\,\partial/\partial x)\,\psi\; dxdydz + \int\psi^*(-i\hbar\,\partial/\partial x)\,\partial\psi/\partial t\; dxdydz$$

$$= 1/i\hbar[-\int(-\hbar^2/2m\,\nabla^2\psi^* + V\psi^*)\,(-i\hbar\,\partial\psi/\partial x)dxdydz$$

$$+ 1/i\hbar\int\psi^*\,(-i\hbar\,\partial/\partial x)(-\hbar^2/2m\,\nabla^2\psi + V\psi)\; dxdydz$$

$$= -\hbar^2/2m\int\nabla^2\psi^*\,\partial\psi/\partial x\; dxdydz + \hbar^2/2m\int\psi^*\,\partial/\partial x\,\nabla^2\psi\; dxdydz$$

$$+ \int V\psi^*\,\partial/\partial x\,\psi\; dxdydz - \int\partial\,(V\psi)/\partial x\,\psi^*\; dxdydz$$

$$= \hbar^2/2m\int[\psi^*\,(\partial/\partial x)\nabla^2\psi - \nabla^2\psi^*\,\partial\psi/\partial x]\; dxdydz$$

$$+ \int\psi^*\,(-\partial V/\partial x)\,\psi\; dxdydz$$

Since the first two terms in the right hand side can be seen as cancelling each other (through integration by parts) we have

$$d\langle p_x\rangle/dt = \langle -\partial V/\partial x\rangle$$

Thus we prove that $d\langle p_x\rangle/dt = -\langle F_x\rangle$, similar to that in classical mechanics.

19.10 APPLICATIONS OF SCHRODINGER WAVE EQUATION

Schrodinger wave equation has been applied to determine the energy spectra of simple systems. In fact, Schrodinger's approach is useful when the potential involved has a simple form, for example, a spherically symmetric. The energy spectra of hydrogen atom, or hydrogen like atoms, simple harmonic oscillator, a particle in a box and square well potential, particle motion through a potential step etc. has been very successfully found. In complex cases one has to resort to perturbation theory where the above mentioned simple cases stand as the starting point. We now discuss the simple applications involving one dimension and total energy as constant.

(*a*) **Potential Step.** Let us take the case of a motion of a particle under the potential step

$$V(x) = 0 \text{ for } x < 0$$

$$= V_0 \text{ for } x > 0$$

The corresponding time independent Schrodinger equations in one dimensions would be

$$-(\hbar^2/2m)\,d^2\psi(x)/dx^2 + V(x)\psi(x) = E\psi(x)$$

That is for $x < 0$ it is

$$-(\hbar^2/2\text{m})\,d^2\psi(x)/dx^2 = E\psi(x)$$

or $$d^2\psi(x)/dx^2 + \alpha^2\psi(x) = 0\;;\; x < 0,\ \alpha = [(2mE)/\hbar^2]^{1/2}$$

and $$d^2\psi(x)/dx^2 - \beta^2\psi(x) = 0\;;\; x > 0,\ \beta = [(2m(V_0 - E))/\hbar^2]^{1/2}\;;\; E < V_0$$

E being the total energy of the particle of mass m.

The obvious solutions in the two region would be

$$\psi_1(x) = A\sin(\alpha x) + B\cos(\alpha x)\;;\; x < 0$$

and $$\psi_2(x) = C\,e^{-\beta x} + D\,e^{\beta x}\;\;;\; x > 0$$

To find unknown constants A, B, C, and D we use continuity conditions of $\psi(x)$ *i.e.*

$$\psi_1(x = 0) = \psi_2(x = 0)$$

which gives

$$B = C + D,$$

Now, for large x $\psi_2 \to \infty$ unless $D = 0$. So, we assume $D = 0$ and hence

$$B = C$$

The other continuity condition requires

$$(d\psi_1(x)/dx)_{x=0} = (d\psi_2(x)/dx)_{x=0}$$

This yields

$$A\alpha = -C\beta \quad i.e., \quad C = -A(\alpha/\beta) = B$$

Both B and C becomes zero as $\beta \to \infty$ or $V_0 \to \infty$

Thus, required solutions are

$$\psi_1(x) = A \sin(\alpha x) \; ; x < 0$$

and

$$\psi_2(x) = -A(\alpha/\beta)\, e^{-\beta x} \; ; x > 0$$

Here, A remains undetermined. However, it can be obtained from normalization of the wave function

$$\int_{-\infty}^{+\infty} |\psi|^2\, dx = 1$$

Also, we assumed in the above discussion that

$$E > 0$$

For the case $E < 0$, the sin and cos will become sinh and cosh without affecting the final result. The important thing to note is that for $V_0 = \infty$, $B = C = 0$ and solution will be

$$\psi(x) = A \sin(\alpha x) \; ; x < 0$$
$$= 0 \; ; x \geq 0$$

(b) Potential Step. Now we take the case of potential step of height V_0 such that $E > V_0$ as shown in Fig. 19.3.

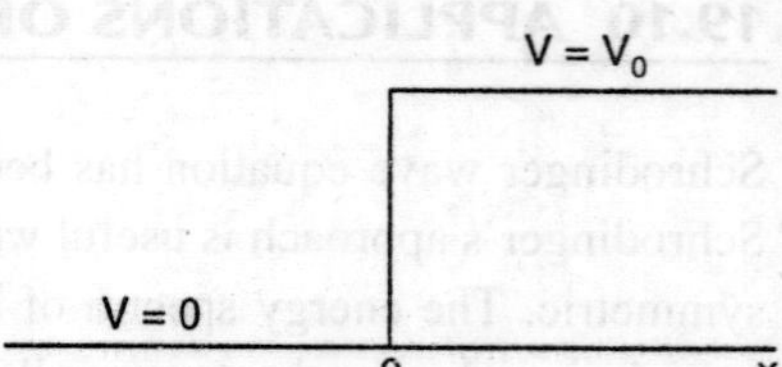

FIGURE 19.3 **Potential step of height V_0.**

Such a potential step is written as

$$V(x) = 0 \; ; x < 0$$
$$= V_0 \; ; x > 0$$

Then time independent Schrodinger equation would be

$$-(\hbar^2/2m)\, d^2\psi(x)/dx^2 = E\psi(x) \text{ for } x < 0$$

and

$$-(\hbar^2/2m)\, d^2\psi(x)/dx^2 + V_0(x)\psi(x) = E\psi(x) \text{ for } x > 0$$

In the region $x < 0$, the equation becomes

$$d^2\psi(x)/dx^2 + \alpha^2\psi(x) = 0 \; ; \alpha = [(2mE)/\hbar^2]^{1/2}$$

with a solution

$$\psi_1(x) = A\, e^{i\alpha x} + B\, e^{-i\alpha x}$$

while in the region $x > 0$, equation would be

$$d^2\psi(x)/dx^2 + \gamma^2\psi(x) = 0 \; ; \; \gamma = [(2m(E - V_0))/\hbar^2]^{1/2}$$

with a solution

$$\psi_2(x) = C\, e^{i\gamma x} + D\, e^{-i\gamma x}$$

We have chosen exponential form instead of sin and cos function for convenience.

The continuity conditions require that

$$\psi_1 = \psi_2 \text{ at } x = 0 \qquad i.e. \qquad A + B = C + D$$

and also

$$(d\psi_1(x)/dx)_{x=0} = (d\psi_2(x)/dx)_{x=0} \qquad i.e. \qquad i\alpha(A - B) = i\gamma(C - D)$$

The relations in A, B, C, and D may be rewritten as

$$C - B = A - D \quad \text{and} \quad C + (\alpha/\gamma)B = (\alpha/\gamma)A + D$$

which may be solved to get

$$C = 1/[1 + (\alpha/\gamma)].(2\alpha/\gamma)\,.\, A + D\,.\,[\{1 - (\alpha/\gamma)\}/\{1 + (\alpha/\gamma)\}]$$

$$= 2\alpha/(\alpha+\gamma).\ A + (\gamma-\alpha)/(\gamma+\alpha)\ .\ D$$

and $$B = [(\alpha/\gamma) - 1]/[1 + (\alpha/\gamma)]\ .\ A + 2D/[1 + (\alpha/\gamma)]$$

$$= (\alpha-\gamma)/(\alpha+\gamma)\ .\ A + 2\gamma/(\alpha+\gamma)\ .\ D$$

With the substitutions of these constants we get ψ_1 and ψ_2 as

$$\psi_1(x) = A\ [e^{i\alpha x} + \{(\alpha-\gamma)/(\alpha+\gamma)\}e^{-i\alpha x} + 2\gamma/(\alpha+\gamma)\ De^{-i\alpha x}]\ ;\ x < 0$$

and $$\psi_2(x) = A\ 2\alpha/(\alpha+\gamma)\ e^{i\gamma x} + D\ [\{(\gamma-\alpha)/(\gamma+\alpha)\}\ e^{i\gamma x} + e^{-i\gamma x}];\ x > 0$$

Now, we suppose that particle motion is from left to right along x-axis. As a result, ψ_1 the solution in the region $x < 0$ consists of an incident wave $Ae^{i\alpha x}$ and a reflected wave $A\ \{(\alpha-\gamma)/(\alpha+\gamma)\}e^{-i\alpha x}$. In the region $x > 0$ since we will have only transmitted particles one has to take $D = 0$ to get a solution

$\psi_2(x) = A\ 2\alpha/(\alpha+\gamma)\ e^{i\gamma x}$, representing transmitted particles. So, we have an interesting result. That is, by taking an incident beam with unit flux, $A = 1$ we have reflected particles described by $\{(\alpha-\gamma)/(\alpha+\gamma)\}$ $e^{-i\alpha x}$ with flux $\{(\alpha-\gamma)/(\alpha+\gamma)\}^2$ and the transmitted particles described by $2\alpha/(\alpha+\gamma)\ e^{i\gamma x}$ with a transmitted flux $= [2\alpha/(\alpha+\gamma)]^2$. For this reason one calls

$$\{(\alpha-\gamma)/(\alpha+\gamma)\} = R, \text{ as Reflection coefficient}$$

and $$2\alpha/(\alpha+\gamma) = T, \text{ as Transmission coefficient}$$

(*c*) **Square Well Potential.** The one dimensional potential of the form

$$V(x) = 0 \text{ for } -a < x < a$$

$$= V_0 \text{ for } |x| > a$$

is known as square well potential. It is shown in Fig. 19.4

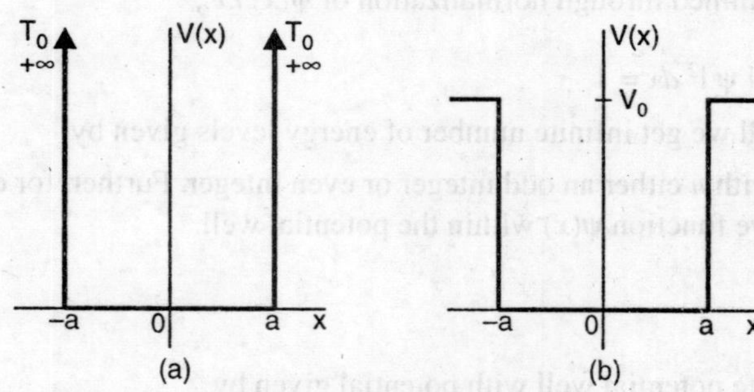

FIGURE 19.4 **(*a*) Potential well as perfectly rigid and impenetrable walls ; (*b*) Finite potential well of height V_0.**

We first consider the case of infinite potential step(well) such that

$$V(x) = 0\ ;\ -a < x < a$$

$$= \infty\ ;\ |x| > a$$

Within the well the potential is zero and the Schrodinger equation would be

$$-(\hbar^2/2m)\ d^2\psi(x)/dx^2 = E\psi(x)\ ;\ -a < x < a$$

or $$d^2\psi(x)/dx^2 + \alpha^2\psi(x) = 0$$

having a general solution

$$\psi(x) = A \sin(\alpha x) + B \cos(\alpha x)\ ;\ \alpha = [(2mE)/\hbar^2]^{1/2}$$

The wave function, on other hand, at $x = \pm a$ would be zero as $x = \pm a$ forms impenetrable walls. So, the continuity conditions on the wave function will be

$$\psi(x = a) = 0 = A \sin(\alpha a) + B \cos(\alpha a)$$

and also $$\psi(x = -a) = 0 = -A \sin(\alpha a) + B \cos(\alpha a)$$

From these two equations we obtain

$$A \sin(\alpha a) + B \cos(\alpha a) = -A \sin(\alpha a) + B \cos(\alpha a)$$

i.e. $$A \sin(\alpha a) = 0 \quad \text{as well as}$$

$$B \cos(\alpha a) = 0$$

Now for a given αa both $\sin(\alpha a)$ as well as $\cos(\alpha a)$ can not vanish and at the same time we do not want both A and B to vanish. Thus we get two types of solutions as

when $$A = 0 \text{ and } B \neq 0$$

$$\cos(\alpha a) = 0 \; ; \; \alpha a = n\pi/2, \; n = \text{odd integer}$$

The other class of solutions are

when $$B = 0 \quad \text{and} \quad A \neq 0$$

$$\sin(\alpha a) = 0 \; ; \; \alpha a = n\pi/2, \; n = \text{even integer}$$

Thus energy E is given by

$$(2mE)/\hbar^2 = \alpha^2 = n^2\pi^2/4a^2$$

i.e. $$E = \pi^2 n^2 \hbar^2/8ma^2 \text{ with } n \text{ even (odd) integer}$$

Corresponding solutions are

$$\psi(x) = B \cos(n\pi x/2a) \; ; \; n = \text{odd integer}$$

and $$\psi(x) = A \sin(n\pi x/2a) \; ; \; n = \text{even integer}$$

A or B may be determined through normalization of $\psi(x)$ *i.e.*,

$$\int_{-a}^{+a} |\psi|^2 \, dx = 1$$

Thus for infinite well we get infinite number of energy levels given by

$E_n = \pi^2 n^2 \hbar^2/8ma^2$ with n either an odd integer or even integer. Further, for each n there will be $n - 1$ number of nodes of the wave function $\psi(x)$ within the potential well.

Finite Potential Well

Fig. 19.4 (*b*) represents finite potential well with potential given by

$$V(x) = 0 \text{ for } -a < x < a$$

$$= V_0 \text{ for } |x| > a$$

So the solution for the region $|x| < a$ will remain unaltered while for the other region $|x| > a$ the Schrodinger equation would be

$$-(\hbar^2/2m)\, d^2\psi(x)/dx^2 + V_0(x)\psi(x) = E\psi(x) \; ; \; E < V_0.$$

or $$d^2\psi(x)/dx^2 - \beta^2\psi(x) = 0 \; ; \; \beta = [(2m(V_0 - E))/\hbar^2]^{1/2}$$

The general solution will be

$$\psi(x) = C e^{-\beta x} + D e^{\beta x}$$

Since $\psi(x)$ should not become ∞ as $x \to \pm\infty$, we must require

$$D = 0 \quad \text{for} \quad x > a$$

and $$C = 0 \quad \text{for} \quad x < -a$$

Thus, the appropriate solutions are

$$\psi_2(x) = C e^{-\beta x} \; ; \; x > a$$

and $$= D e^{\beta x} \; ; \; x < -a$$

Next, as usual we impose on the wave function the conditions of continuity *i.e.* ψ and $d\psi/dx$ be continuous at $x = \pm a$. As in the region $|x| < a$, the solution, as found earlier, is

$$\psi_1(x) = A \sin(\alpha x) + B \cos(\alpha x),\ \alpha = [(2mE)/\hbar^2]^{1/2}$$

So, by condition of continuity at $x = a$, we get

$$A \sin(\alpha a) + B \cos(\alpha a) = C e^{-\beta a} \quad ...(19.1)$$

and

$$\alpha A \cos(\alpha a) - \alpha B \sin(\alpha a) = -\beta C e^{-\beta a} \quad ...(19.2)$$

while at $x = -a$ we obtain an other set of equations

$$-A \sin(\alpha a) + B \cos(\alpha a) = D e^{-\beta a} \quad ...(19.3)$$

and

$$\alpha A \cos(\alpha a) + \alpha B \sin(\alpha a) = \beta D e^{-\beta a} \quad ...(19.4)$$

From equations (19.1) and (19.3) we obtain

$$2A \sin(\alpha a) = (C - D) e^{-\beta a} \quad ...(19.5)$$

as well as

$$2B \cos(\alpha a) = (C + D) e^{-\beta a} \quad ...(19.6)$$

On the other hand from eqs. (19.2) and (19.4) we obtain

$$2\alpha A \cos(\alpha a) = -\beta(C - D) e^{-\beta a} \quad ...(19.7)$$

as well as

$$2\alpha B \sin(\alpha a) = \beta(C + D) e^{-\beta a} \quad ...(19.8)$$

Simplifying eqs.(19.5) and (19.7) we get, unless $A = 0$ and $C = D$,

$$\alpha \cot(\alpha a) = -\beta \quad ...(19.9)$$

while from eqs. (19.6) and (19.8), unless $B = 0$ and $C = -D$, we get

$$\alpha \tan(\alpha a) = \beta \quad ...(19.10)$$

From the results of eqns. (19.9) and (19.10) it is clear that both cannot be valid simultaneously*. That is either eq. (19.9) with which $B = 0$ and $C = -D$ or eqn. (19.10) with $A = 0$ and $C = D$ are the appropriate results to determine energy spectra of a particle enclosed in the potential well of height V_o.

If $\alpha \cot(\alpha a) = -\beta$, the solutions will be

$$\psi_1(x) = A \sin(\alpha x);\ |x| < a$$

and

$$\psi_2(x) = C e^{-\beta x},\ x > a$$

$$= D e^{\beta x},\ x < -a\ (B = 0, C = -D)$$

A graphical method is used to determine the energies E for a given V_o and range a.

(*d*) **Potential Barrier.** Let us consider the case of square potential barrier of the form

$$V(x) = 0 \quad \text{for} \quad x < 0 \quad \text{and} \quad x > a$$

$$= V_0 \quad \text{for} \quad 0 < x < a$$

as shown in Fig. 19.5.

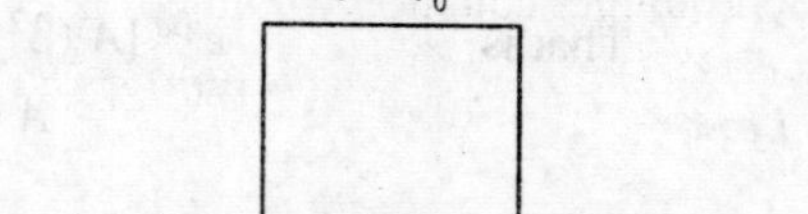

FIGURE 19.5 **A square potential barrier with height, V_0, and range *a*.**

Let E be the total energy of the particle travelling from left to right. As a result the Schrodinger equations for the regions, $x < 0$ and $x > a$ will be

$$-(\hbar^2/2m)\, d^2\psi(x)/dx^2 = E\psi(x)$$

or

$$d^2\psi(x)/dx^2 + \alpha^2\psi(x) = 0,\ \alpha = [(2mE)/\hbar^2]^{1/2}$$

---

*Because, then $\tan^2 \alpha a = -1$ leading α, to an imaginary quantity and β, *a* negative quantity contrary to our assumptions.

and for the region $0 < x < a$, the equation will be

$$-(\hbar^2/2m)\, d^2\psi(x)/dx^2 + V_o(x)\psi(x) = E\psi(x)$$

or

$$d^2\psi(x)/dx^2 + \beta^2\psi(x) = 0\ ;\ \beta = [(2m(E - V_o))/\hbar^2]^{1/2}$$

Since particles move from left to right, the solution for the region $x < 0$ will be

$$\psi_1 = A\, e^{i\alpha x} + B\, e^{-i\alpha x}$$

and for the region $x > a$

$$\psi_3 = T\, e^{i\alpha x}$$

where as the solution of the Schrodinger equation for the internal region $0 < x < a$ would be

$$\psi_2 = C\, e^{i\beta x} + D\, e^{-i\beta x}\ ;\ E > V_0$$

Now apply condition of continuity of ψ at $x = 0$ between ψ_1 and ψ_2 to get

$$\psi_1(x = 0) = \psi_2(x = 0)$$

i.e.

$$A + B = C + D \qquad \text{...(19.11)}$$

and

$$(d\psi_1(x)/dx)_{x=0} = (d\psi_2(x)/dx)_{x=0} \text{ yielding}$$

$$i\alpha(A - B) = i\beta(C - D)$$

or

$$\alpha/\beta\ (A - B) = C - D \qquad \text{...(19.12)}$$

These eqns. (19.11) and (19.12) on simplification give

$$C = A/2\ (1 + \alpha/\beta) + B/2\ (1 - \alpha/\beta) \qquad \text{...(19.13)}$$

and

$$D = A/2\ (1 - \alpha/\beta) + B/2\ (1 + \alpha/\beta) \qquad \text{...(19.14)}$$

Next, we apply condition of continuity of wave function at $x = a$ *i.e.*,

$$\psi_2(x = a) = \psi_3(x = a)$$

i.e.

$$C\, e^{i\beta a} + D\, e^{-i\beta a} = T\, e^{i\alpha a} \qquad \text{...(19.15)}$$

and

$$(d\psi_2(x)/dx)_{x=a} = (d\psi_3(x)/dx)_{x=a}$$

i.e.

$$i\beta C\, e^{i\beta a} - i\beta D\, e^{-i\beta a} = i\alpha T\, e^{i\alpha a} \qquad \text{...(19.16)}$$

Divide eqn. (19.15) by eqn. (19.16) to get

$$C\, e^{i\beta a} + D\, e^{-i\beta a} = \beta/\alpha\ C\, e^{i\beta a} - \beta/\alpha\ D\, e^{-i\beta a}$$

Which further gives

$$C/D\ e^{i\beta a}(-1 + \beta/\alpha) = (1 + \beta/\alpha)\, e^{-i\beta a}$$

Now substitute the value of C and D from eqn. (19.13) and (19.14) to get

$$e^{i\beta a}(\beta - \alpha)\ [A/2\ (1 + \alpha/\beta) + B/2\ (1 - \alpha/\beta)] = (\alpha + \beta)\, e^{-i\beta a}\ [A/2\ (1 - \alpha/\beta) + B/2\ (1 + \alpha/\beta)]$$

That is

$$e^{i\beta a}\ [A\ (\beta^2 - \alpha^2) + B\ (\beta - \alpha)^2] = e^{-i\beta a}\ [A(\beta^2 - \alpha^2) + B\ (\beta + \alpha)^2]$$

i.e.

$$A\ (\beta^2 - \alpha^2)\ 2i \sin(\beta a) = B\ [(\beta + \alpha)^2 e^{-i\beta a} - (\beta - \alpha)^2 e^{i\beta a}]$$

$$= B\ [-(\beta^2 + \alpha^2)\ 2i \sin(\beta a) + 4\alpha\beta \cos(\beta a)]$$

Thus,

$$B = [A\ (\alpha^2 - \beta^2) \sin(\beta a)]\ /\ [(\beta^2 + \alpha^2) \sin(\beta a) + 2i\alpha\beta \cos(\beta a)] \qquad \text{...(19.17)}$$

Taking $A = 1$ *i.e.*, the beam of incident particles of unit flux, we get

$|B|^2$ = Reflected flux

$$= [(\alpha^2 - \beta^2)^2 \sin^2(\beta a)]\ /\ [(\beta^2 + \alpha^2)^2 \sin^2(\beta a) + 4\alpha^2\beta^2 \cos^2(\beta a)]$$

$$= [(1 - \mu^2)^2 \sin^2(\beta a)]/[(1 + \mu^2)^2 \sin^2(\beta a) + 4\mu^2 \cos^2(\beta a)]$$

where $\mu = \beta/\alpha = [(E - V_o)/E]^{1/2} = (1 - V_o/E)^{1/2}$

Another important quantity is T, the transmission co-efficient. From eqn. (19.15) we get its value as

$$T = e^{-i\alpha a}\ [C\, e^{i\beta a} + D\, e^{-i\beta a}]$$

Putting the expression of C and D [eqns. (13) and (14)] we get with $A = 1$

$$T = e^{-i\alpha a}\,[\cos(\beta a) + \alpha/\beta\, i \sin(\beta a) + B(\cos(\beta a) - i\,\alpha/\beta \sin(\beta a))]$$

and using eqn. (19.17) for B,

$$= e^{-i\alpha a}\,\{\cos(\beta a) + \alpha/\beta\, i \sin(\beta a) + [\cos(\beta a) - i\,\alpha/\beta \sin(\beta a)]\,[(1-\mu^2)\sin(\beta a)]/[(1+\mu^2)\sin(\beta a) + 2i\mu\cos(\beta a)]\}$$

$$= e^{-i\alpha a}\,[2i\mu/\{(1+\mu^2)\sin(\beta a) + 2i\mu\cos(\beta a)\}]$$

Hence, the transmission particle flux will be given by

$$|T|^2 = [4\mu^2/\{(1+\mu^2)^2\sin^2(\beta a) + 4\mu^2\cos^2(\beta a)\}]$$

$$= 1 \text{ as } \mu \to 1 \text{ or } E \to \text{large.}$$

It also becomes unity for $\beta a = n\pi$ and the barrier is said to be transparent. Also, for $E \to V_o$, both μ and β becomes small. As a result, $\sin(\beta a) \to \beta a = \mu\alpha a$, $\cos(\beta a) \to 1$ and

$$|T|^2 \Rightarrow [4\mu^2/\{\mu^2\alpha^2a^2 + 4\mu^2\}]$$

$$= 1/[1 + \alpha^2a^2/4] = [1 + (\alpha a/2)^2]^{-1}$$

Lastly, when $E < V_o$, $\beta = i[(2m(V_o - E))/\hbar^2]^{1/2} = iK$ and $\mu = \beta/\alpha = iK/\alpha$. Putting these values in $|T|^2$ we get

$$|T|^2 = (2\alpha/K)^2/[(1 - \alpha^2/K^2)^2 \sinh(Ka) + (2\alpha/K)^2\cosh^2(Ka)]$$

Here, we have used $\sin(\beta a) = i\,\sinh(Ka)$ and $\cos(\beta a) = \cosh(Ka)$. The following graph in Fig 19.6 shows the variation of $|T|^2$ vs E/V_o.

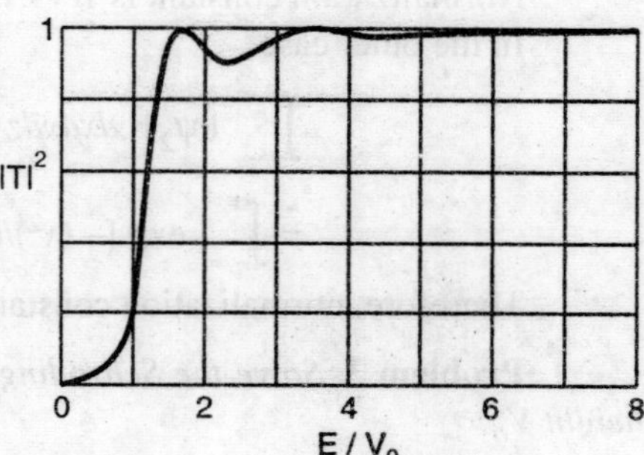

FIGURE 19.6. **Transmission of a square potential barrier as a function.**

SOLVED NUMERICAL PROBLEMS

Problem 1. *Solve* $\nabla^2\psi(x, y, z) + 2m/\hbar^2[E - V(x, y, z)]\,\psi(x, y, z) = 0$. *Also, find the region where the wave function* ψ *is oscillatory.*

Solution. Let $\quad 2m/\hbar^2\,[E - V] = k^2$ in the region $E > V$

Then $\quad \nabla^2\psi + k^2\psi = 0$

i.e. $\quad \partial^2\psi/\partial x^2 + \partial^2\psi/\partial y^2 + \partial^2\psi/\partial z^2 + (k_x^2 + k_y^2 + k_z^2)\,\psi = 0$

This can be split into three equations by using $\psi = \psi_1(x)\,\psi_2(y)\,\psi_3(z)$

so that

$$\partial^2\psi_1/\partial x^2 = -k_x^2\psi_1$$

$$\partial^2\psi_2/\partial y^2 = -k_y^2\psi_2$$

$$\partial^2\psi_3/\partial z^2 = -k_z^2\psi_3$$

These can easily be solved as

$$\psi_1 = c_1 \exp(\pm ik_x x)$$

$$\psi_2 = c_2 \exp(\pm ik_y y)$$

$$\psi_3 = c_3 \exp(\pm ik_z z)$$

Hence $\quad \psi = c \exp(\pm i\mathbf{k.r})$

where $\quad c = c_1c_2c_3$, a constant

In this region where $E > V$, ψ is oscillatory.

For $E < V$ we write $\quad 2m/\hbar^2\,[V - E] = K^2$, so that

$\nabla^2\psi = K^2\psi$ having solution of the form

$$\psi = c \exp(\pm \mathbf{k.r}),$$

decreasing or increasing in size as r increases.

Problem 2. *Find the probability density as well as the normalization constant of*

$$\psi_1 = e^{ik.r} \text{ and } \psi_2 = \exp(-r^2/2\sigma^2),\ \sigma \text{ is constant.}$$

Solution. $|\psi_1|^2 = 1$ = Probability density, which is same for all positions **r**. Hence the position of the particle is not clearly defined.

In the second case

$|\psi_2|^2 = \exp(-r^2/\sigma^2)$, depends on r. That is the probability density decreases to zero exponentially as r increases and is maximum at $r = 0$.

Normalization

Since $\int_V |\psi_1|^2\, dxdydz = V$

Normalization constant is $1/\sqrt{V}$ *i.e.* $\psi_1/\sqrt{V}$ is normalized function.

In the other case

$$\int_{-\infty}^{+\infty} |\psi_2|^2\, dxdydz = \int_{-\infty}^{+\infty} \exp[-(x^2 + y^2 + z^2)/\sigma^2]\, dxdydz$$

$$= \int_{-\infty}^{+\infty} \exp[-(x^2)/\sigma^2]dx \int_{-\infty}^{+\infty} \exp[-(y^2)/\sigma^2]dy \int_{-\infty}^{+\infty} \exp[-(z^2)/\sigma^2]dz = (\sigma\sqrt{\pi})^3$$

Therefore, normalization constant is $1/(\sigma\sqrt{\pi})^{3/2}$.

Problem 3. *Solve the Schrodinger equation for the motion of the particle in a potential barrier of height* V_0.

Solution. Let E be the energy and $V = 0$ for $x < 0$ and is equal to V_0 for $x > 0$. Thus $x = 0$ divides the two regions of potential. In the region $x < 0$ the Schrodinger equation (one dimensional) is given by

$$-\hbar^2/2m\, \partial^2\psi(x)/\partial x^2 = E\psi(x)$$

i.e.
$$d^2\psi/dx^2 + 2m/\hbar^2\, E\psi = 0$$

putting $2mE/\hbar^2 = k^2$, we obtain the general solution as

$$\psi_1 = A \sin(kx) + B\cos(kx)$$

valid for $x < 0$.

For the region $x > 0$, we will have $d^2\psi/dx^2 + 2m/\hbar^2\,(E - V_0)\,\psi = 0$

which gives a general solution of the type

$\psi_2 = Ce^{-Kx} + De^{Kx} = Ce^{-Kx}$ as D is to be zero so that ψ_2 does not become infinite as $x \to \infty$, here $K^2 = 2m/\hbar^2\,(V_0 - E)$ and $E < V_0$.

For these solutions to be continuous we must have

$$\psi_1(x = 0) = \psi_2(x = 0)$$

i.e.
$$B = C$$

and
$$(d\psi_1/dx)_{x=0} = (d\psi_2/dx)_{x=0}$$

i.e.
$$Ak = -CK \quad \text{or} \quad C = -Ak/K \to 0 \text{ as } K \to \infty$$

Thus, the solutions are
$$\psi_1 = A\sin(kx) - Ak/K\cos(kx)\ ;\ x < 0$$

and
$$\psi_2 = -Ak/K\, e^{-kx}\ ;\ x > 0$$

when the potential barrier becomes infinity *i.e.*

$$V_0 \to \infty$$
$$K \to \infty$$

So that,
$$B = C = 0$$

Then the only solution of interest would be $\psi = A\sin(kx)$, for $x < 0$ and this wave function vanishes for $x \geq 0$.

Special Case of Rigid Walls

When the particle moves with in rigid walls at $x = a$ and $-a$ the wave function would vanish for $|x| \geq a$ while for $|x| < a$ the solution would remain as

$$\psi_1 = A \sin(kx) + B \cos(kx)$$

so that
$$\psi_1(x = a) = 0 = A \sin(ka) + B \cos(ka)$$

and
$$\psi_1(x = -a) = 0 = -A \sin(ka) + B \cos(ka)$$

i.e.
$$A \sin(ka) + B \cos(ka) = -A \sin(ka) + B \cos(ka)$$

or
$$A \sin(ka) = 0$$

Also, then
$$B \cos(ka) = 0$$

Now, both sin (ka) and cos (ka) cannot vanish for fixed values of ka. So, we take sin (ka) = 0 as well as $B = 0$. Thus, $\psi_1(x) = A \sin(kx)$ is the appropriate solution.

Next, since sin (ka) = 0

$$ka = \pm n\pi,$$

or
$$k^2 = n^2\pi^2/a^2$$

$$2mE/\hbar^2 = n^2\pi^2/a^2$$

Hence, possible energy values are given by

$$E = \hbar^2 n^2\pi^2/2ma^2$$

where n is any integer positive or negative except zero.

More appropriately written as
$$E_n = \hbar^2 n^2\pi^2/2ma^2$$
$$\propto n^2$$

lowest level is at
$$E_1 = \hbar^2\pi^2/2ma^2$$

Problem 4. *The finite potential step is defined by $V(x) = 0$ for $|x| < a$ and $V(x) = V_0$ for $|x| > a$. It is one dimensional problem with a particle of mass m executing motion within this potential step. We will have two Schrodinger equations*

Solution.
$$-\hbar^2/2m\, d^2\psi(x)/dx^2 = E\psi(x)$$

i.e.
$$d^2\psi/dx^2 + 2mE/\hbar^2\, \psi = 0 \text{ for the region } |x| < a$$

and
$$d^2\psi/dx^2 - 2m(V_0 - E)/\hbar^2\, \psi = 0 \text{ for } |x| > a \text{ and } E < V_0.$$

The solution of first equation may be written as

$$\psi_1(x) = A \sin(\alpha x) + B \cos(\alpha x)$$

where
$$\alpha = \sqrt{2mE/\hbar^2}$$

and of the second equation
$$\psi_2(x) = Ce^{-\beta x} + De^{\beta x}$$

where
$$\beta = \sqrt{2m(V_0 - E)/\hbar^2} > 0$$

D can be set as equal to zero for $x > a$ and we also assume $B = 0$ because of other considerations.

These two solutions must join each other continuously at the boundary of the two regions. That is

$$\psi_1 = \psi_2$$

as well as
$$d\psi_1/dx = d\psi_2/dx \text{ at } |x| = a.$$

These conditions require that

$$A \sin(\alpha a) = Ce^{-\beta a}$$

and
$$\alpha A \cos(\alpha a) = -\beta Ce^{-\beta a}$$

on dividing these, we get
$$\alpha \cos(\alpha a) = -\beta$$

This result may be used to determine energy levels graphically knowing '*a*' and V_0 because

$$\alpha = \sqrt{2mE/\hbar^2}$$

$$\beta = \sqrt{2m(V_0 - E)/\hbar^2}$$

and with

$$\alpha a = \xi, \quad \beta a = \eta^2$$

$$\xi^2 + \eta^2 = 2mV_0 a^2/\hbar^2$$

Such a condition will always be satisfied for variety of values of 'V_0' and '*a*' provided $V_0 a^2$ is kept constant.

EXERCISE

1. Derive Schrodinger equation. How would you obtain time independent Schrodinger equation ? (*K.U. June, 2007*) (*K.U. Jan., 2003*)
2. Show that if the potential energy $V(r)$ is changed every where by a constant, the time independent wave functions are unchanged. What is the effect on energy levels ?
3. What is the meaning of wave packet? What for the idea of wave packet was proposed ? Obtain the expression of group velocity.
4. Show directly that $<p_x>$ is real for a wave packet.
5. Solve the problem of a particle motion in a finite potential step. Also find energy levels of its bound states.
6. Show that for a three dimensional wave packet $d < x^2 >/dt = 1/m[< p_x x > + < x p_x >]$.
7. What are stationary states? Explain.
8. Prove that group velocity is given by $v_g = \dfrac{dv}{d(1/\lambda)} = -\lambda^2 \dfrac{dv}{d\lambda}$ (*K.U. July, 2003*)

   [**Hint.** Since $\omega = 2\pi\nu$ and $k = 2\pi/\lambda$, one gets from $v_g = d\omega/dk$ the required result]
9. Give the physical significance of ψ (**r**, *t*).
10. What are the difficulties with classical physics ? (*K.U. June, 2007; K.U. July, 2003*)
11. For non-dispersive medium prove that group velocity is same as phase velocity. (*K.U. June, 2005*)

    [**Hint.** For non-dispersive medium $\omega \propto k$]
12. Prove that phase velocity is 50% of group velocity for a free classical motion. (*K.U. June, 2004*)

$$\left[\textbf{Hint. } v_p = \frac{E}{p} = \frac{\frac{p^2}{2m}}{p} = \frac{p}{2m}\,;\; v_g = \frac{dE}{dp} = \frac{p}{m} \right]$$

Chapter 20 CRYSTAL STRUCTURE

20.1 INTRODUCTION

The matter present in the universe is conveniently divided into three forms or states: the solid state, the liquid state and the gaseous state. The solid state of matter constitutes a very small part of the universe. However, in the physical world around us, we largely encounter the solid state matter. Hence the solid state of matter attracted the attention of early physicists. As a result of their pioneering efforts a vast field of knowledge, now known as condensed matter physics, evolved. Various modern technologies have direct bearings on this branch of physics. The miracles of modern technology like computers, electronic gadgets etc. could be possible only after the exploration of the properties of solid state semiconductor devices. The maglev trains come into existence because of the discovery of the phenomenon of superconductivity. Because of its applications in modern technology, the condensed matter physics or the solid-state physics has become the integral part of every technical course.

The solid-state physics may be defined as that branch of physics, which deals with the study of the structural and behavioral properties of the solid state of matter. This chapter is largely devoted to the study of the structural properties of solids.

20.2 THE SOLID STATE

The solid state is a condensed state of matter characterized by interatomic or intermolecular interactions of sufficient strength to give the material a definite volume and shape, which change relatively little with applied stress. The separation of adjacent atoms in a solid is of the same order of magnitude as the diameter of the electron cloud around each atom *i.e.,* $\sim 2 \times 10^{-10}$ m or 2.0 Å. In this state of matter each atom is confined to a definite small space between neighboring atoms. While each atom in a solid is confined to a definite space within the body, it is in a state of motion within that space. As the temperature decreases this motion becomes slower and slower until at absolute zero, – 273°C, all atomic motion ceases. At absolute zero, however, the atoms of certain solids are still vibrating. This residual energy of vibration is known to be an inherent property of some solids, which cannot be utilized or taken away.

Many solid materials can exist in different forms ; a common example is the element carbon. The black soot deposited on the bottom of a frying pan by a smoky campfire is nearly pure carbon. The so called "lead" in a pencil is not lead at all, but mainly a different form of carbon known as graphite. Diamond is also a form of solid carbon. Basically, the soot, the graphite and the diamond consists of pure carbon, still they show remarkably different mechanical, thermal, electrical and optical properties under the same physical conditions. This difference in their properties mainly attributed to the arrangements of carbon atoms in the solid structures. Depending on the arrangements of atoms in a solid, they can be classified into two classes namely the crystalline solids and the amorphous solids. Some of the important points of difference between these two types of solids are given in Table 20.1.

TABLE 20.1 Difference between crystalline and amorphous solids

Crystalline Solids	*Amorphous Solids*
1. In these solids the constituents atoms or molecules or ions have an infinite, regular and periodic arrangement.	1. In these solids the constituents atoms or molecules or ions have an infinite, irregular and random arrangement.
2. They have long range order in their structure.	2. They may have a short range order in their structure.
3. These are anisotropic in nature *i.e.*, their physical properties are different in different directions.	3. These are isotropic in nature *i.e.*, their physical properties are same in all the directions.
4. They have a sharp melting point because all the bonds between atoms are equally strong and break simultaneously at melting temperature.	4. They do not have a sharp melting point. The weakest bond breaks first and then others hence the solid melts gradually.

In the example of carbon mentioned above, the soot is an amorphous solid while the graphite and the diamond are crystalline solids. Graphite and diamond both are crystals but the orderly arrangements of atoms are different. Due to this different orderly arrangements of atoms their physical properties differ from each other.

20.3 THE CRYSTALLINE SOLID

A crystal is formed by infinite repetition of identical structural units or building blocks in space. The structural unit or the building block is commonly known as basis. The basis of a crystal may contain a single atom or a group of atoms. Analogically, a single brick or a group of bricks may be considered as the basis of a wall. In many simple crystals like copper, iron etc. the basis consists of a single atom while in certain inorganic and organic solids the basis is comprised by a group of atoms. The number of atoms in the basis of different crystals may vary from 100 to 10000. In order to form a crystal the basis is attached to every point of a space lattice, which is being defined in the next section. Logically,

$$\text{Crystal} = \text{Lattice} + \text{Basis}$$

Lattice is, therefore, a regular and periodic arrangement of points, each point denoting the position of the basis.

20.4 SPACE LATTICE AND TRANSLATION VECTORS

A periodic, regular and infinite arrangement of points in three-dimensional space is known as space lattice. The space lattice is characterized by three fundamental translation vectors **a**, **b** and **c** such that the arrangement of lattice points looks similar when viewed from points **r** and **r′** that are related through the relation

$$\mathbf{r}' = \mathbf{r} + u\mathbf{a} + v\mathbf{b} + w\mathbf{c} \qquad \text{...(20.1)}$$
$$= \mathbf{r} + \mathbf{T}$$

with u, v, and w as arbitrary integers. A space lattice may also be defined as the set of points **r′** for all values of integers u, v and w. It should be kept in mind that the space lattice is a mathematical abstraction only, introduced as an aid to study crystal structure conveniently. **T** is sometimes called as lattice displacement vector.

The lattice and translation vectors **a**, **b** and **c** are said to be primitive if any two points **r** and **r′** from which the arrangement of points looks the same always satisfy eqn. (20.1) with a suitable choice of integers u, v and w. Any vector obtained through a combination of primitive translation vectors **a**, **b** and **c** corresponding to a given set of integers u, v and w *i.e.*, $T = ua + vb + wc$, is referred to as lattice translation vector.

20.5 UNIT CELL AND PRIMITIVE CELL

The unit cell of a space lattice may be defined as the smallest closed portion of the lattice such that the lattice is formed by over and again three dimensional repetition of this smallest closed portion. The unit cell may be

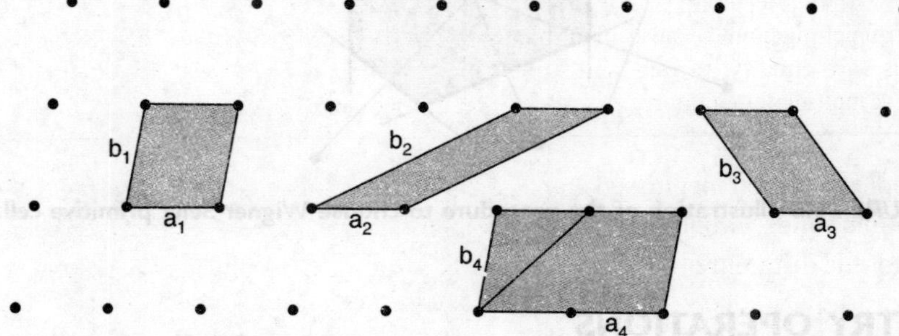

FIGURE 20.1 **Primitive and non-primitive cells in two dimensions.**

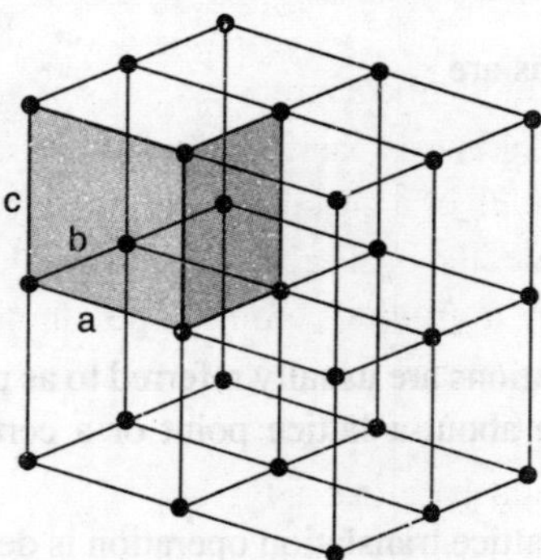

FIGURE 20.2 **Primitive cell of a space lattice in three dimensions.**

regarded as the basic building block of the lattice. The unit cells in two dimensional lattice are shown in Fig. 20.1 and those in three dimensional lattice are given in Fig. 20.2. In general a unit cell of two dimensional lattice is a parallelogram and that of a three dimensional lattice is a parallelopiped.

A unit cell containing lattice points only at the corners is known as a primitive cell. The parallelogram numbered 1, 2, and 3 in Fig. 20.1 and the parallelopiped in Fig. 20.2 defined by primitive axes ($\mathbf{a}_1$, $\mathbf{b}_1$), ($\mathbf{a}_2$, $\mathbf{b}_2$), ($\mathbf{a}_3$, $\mathbf{b}_3$) and (**a**, **b**, **c**) are primitive cells of two dimensional and three dimensional lattices, respectively. A primitive cell is always a minimum area (2-dimensional lattice) or volume (3-dimensional lattice) cell. There is a density of one lattice point per primitive cell. It can be easily verified through the following considerations. A two dimensional primitive cell has four lattice points on four corners of a parallelogram. Each corner of the parallelogram is shared by four other parallelograms. Hence there is one lattice point per parallelogram. This idea can be extended to the three dimensional case where each corner of the primitive cell (parallelopiped) is shared by eight parallelopipeds.

An alternative way to choose a primitive cell consists of the following steps.

(*i*) Draw lines to connect a given lattice point to all the nearby lattice points

(*ii*) At the midpoint and normal to these lines, draw new lines or planes.

The smallest area or volume enclosed in this way is known as Wigner-Seitz primitive cell. This procedure is illustrated in Fig. 20.3.

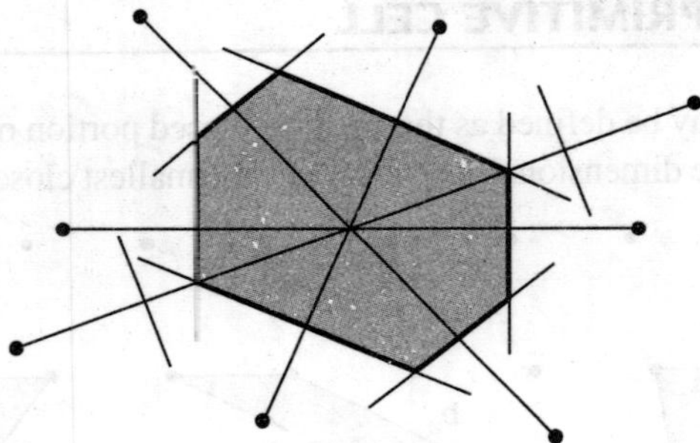

FIGURE 20.3 **Illustration of the procedure to choose Wigner-Seitz primitive cell.**

20.6 SYMMETRY OPERATIONS

Symmetry operations of a crystals are those operations, which when applied to a crystal structure carry it into itself.

The general symmetry operations are

(*i*) Translation operation

(*ii*) Rotation operation

(*iii*) Reflection operation

(*iv*) Compound operation

The rotation and reflection operations are usually referred to as point operations. It may be possible to apply rotation and reflection operation about a lattice point or a certain special point so that the crystal structure remains invariant.

(*i*) **Translation Operation.** A lattice translation operation is defined as the displacement of a crystal parallel to itself. Mathematically, the translation operation is $\mathbf{r}' = \mathbf{r} + \mathbf{T}$, where

$$\mathbf{T} = u\mathbf{a} + v\mathbf{b} + w\mathbf{c}.$$

(*ii*) **Rotation Operation.** A rotation operation is the rotation of lattice through a specified angle about an axis passing through a point in the lattice which carry the crystal into itself. The axis about which the crystal rotates is known as rotation axis. Generally, there exist lattices having one-, two-, three-, four- and six-fold rotation axes, corresponding to rotations by $2\pi/1$, $2\pi/2$, $2\pi/3$, $2\pi/4$ and $2\pi/6$ radians and by integral multiples of these angles. These rotation axes are denoted by symbols 1, 2, 3, 4, and 6 respectively. It may be mentioned here that the lattices with five- or seven-fold rotation axes, corresponding to rotation by $2\pi/5$ or $2\pi/7$ radians, do not exist. There is difference between molecular symmetry and that of lattice of the crystal. A single molecule with these symmetries can exist but the infinite arrangement of such molecules loose these symmetries. One conspicuous example of the loss of symmetry due to infinite arrangement of molecules with five-fold symmetry is presented in Fig. 20.4.

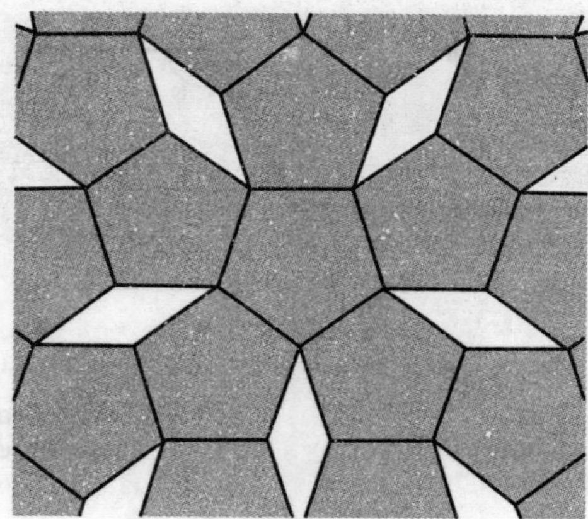

FIGURE 20.4 **Illustration showing the non-existence of five fold axis of symmetry.**

(*iii*) **Reflection Operation.** It is the mirror reflection of lattice about a plane through a lattice point. This operation is usually denoted by m.

(*iv*) **Compound Operation.** A compound operation is a suitable combination of translation and point operations such that the crystal remains invariant under the combined effect of individual operations. A typical example of the compound operation is that of inversion. The inversion operation is made up of a rotation by π radians followed by reflection in a plane normal to the rotation axis; the total effect is to replace $\mathbf{r}$ by $-\mathbf{r}$.

Some of the important definitions closely related to symmetry operations are given below.

(*a*) **Plane of Symmetry.** An imaginary plane passing through a crystal so that the points on one side of it are the mirror images of the points on its other side.

(*b*) **Axis of Symmetry.** An imaginary line passing through a crystal so that if the crystal is rotated about this line as the axis of rotation, the crystal presents the same appearance a number of times during the complete rotation.

(*c*) **Reflection Plane.** These are imaginary planes so that if a crystal is cut along such a plane through a lattice point one half of the reflected crystal reproduces the other.

20.7 TYPES OF TWO DIMENSIONAL LATTICES

There exist five distinct types of lattices (one oblique and four special) in two dimensions. A lattice with arbitrary translation vectors **a** and **b** and invariant under the rotation of π and 2π radians about any lattice point is called as oblique lattice [see Fig. 20.1].

The lattice which is made invariant under rotations of $2\pi/3$, $2\pi/4$, or $2\pi/6$ or under mirror reflection by imposing certain specific restrictions on **a**, **b** and φ (the angle between **a** and **b**) is known as special lattice. Since, there exist four distinct types of restrictions therefore four distinct types of special lattices exist. These special lattices in two dimensions are shown in Fig. 20.5. The specified restrictions on **a**, **b** and φ for the different types of special lattices are also given in Fig. 20.5.

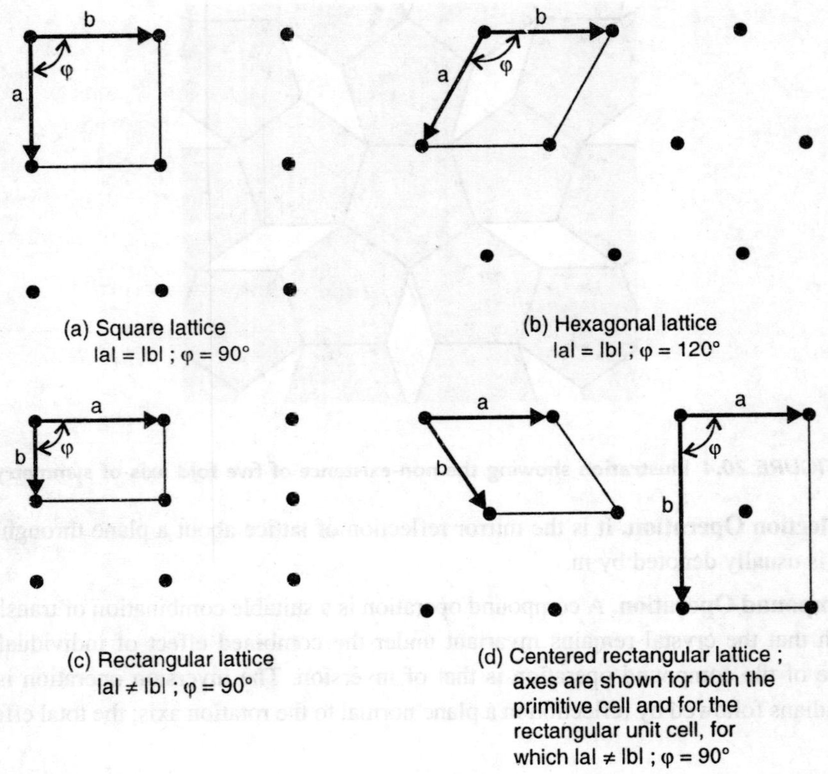

FIGURE 20.5 **Special lattices in two dimensions.**

The invariance under the rotations of $2\pi/4$ and $2\pi/6$ give the square and the hexagonal lattices respectively. The invariance under mirror reflection requires two distinct types of lattices ; rectangular lattice and centered rectangular lattice [see Fig. 20.5].

The common name to all the five distinct types of lattices in two dimensions is Bravais lattice. A brief summary of all the Bravais lattices in two dimensions is given in Table 20.2.

TABLE 20.2 A brief summary of Bravais lattices in two dimensions

Sr. No.	*Lattice Type*	*Conventional Unit Cell*	*Restrictions on* **a, b** *and* φ	*Symmetry Operations*
1.	Oblique	Parallelogram	$a \neq b$, $\varphi \neq 90°$	2
2.	Square	Square	$a = b$, $\varphi = 90°$	4
3.	Hexagonal	60° rhombus	$a = b$, $\varphi = 120°$	6
4.	Rectangular	Rectangle	$a \neq b$, $\varphi = 90°$	m
5.	Centered Rectangular	Rectangle	$a \neq b$, $\varphi = 90°$	m

20.8 THREE DIMENSIONAL LATTICE TYPES

There are 14 distinct types of lattices (one general and 13 special) in three dimensions as shown in Fig. 20.6. Conveniently, these are classified into seven systems summarized in Table 20.3. The axes **a**, **b**, **c** and angles α, β, γ used in Table 20.3 are defined in Fig. 20.7. The magnitude a, b and c are known as lattice parameters. It can be clearly observed from Fig. 20.6 that the cell chosen are not always primitive cell. Sometimes, a conventional cell is preferred over primitive cell because of its obvious connection with symmetry operations.

TABLE 20.3 Systems of three dimensional lattice types

System	*Number of Lattices*	*Restrictions on Conventional Cell Axes and Angles*
Triclinic	1	$a \neq b \neq c$ $\alpha \neq \beta \neq \gamma$
Monoclinic	2	$a \neq b \neq c$ $\alpha = \gamma = 90° \neq \beta$
Orthorhombic	4	$a \neq b \neq c$ $\alpha = \beta = \gamma = 90°$
Tetragonal	2	$a = b \neq c$ $\alpha = \beta = \gamma = 90°$
Cubic	3	$a = b = c$ $\alpha = \beta = \gamma = 90°$
Trigonal	1	$a = b = c$ $\alpha = \beta = \gamma < 120°, \neq 90°$
Hexagonal	1	$a = b \neq c$ $\alpha = \beta = 90°, \gamma = 120°$

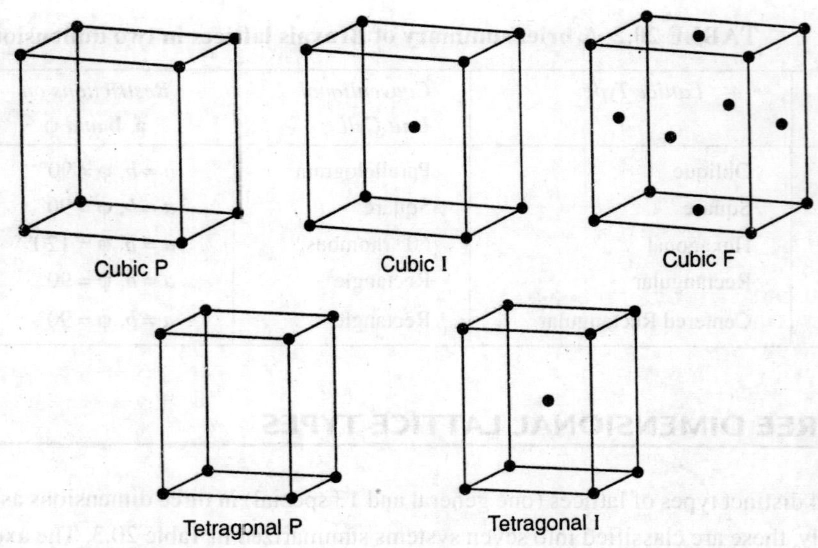

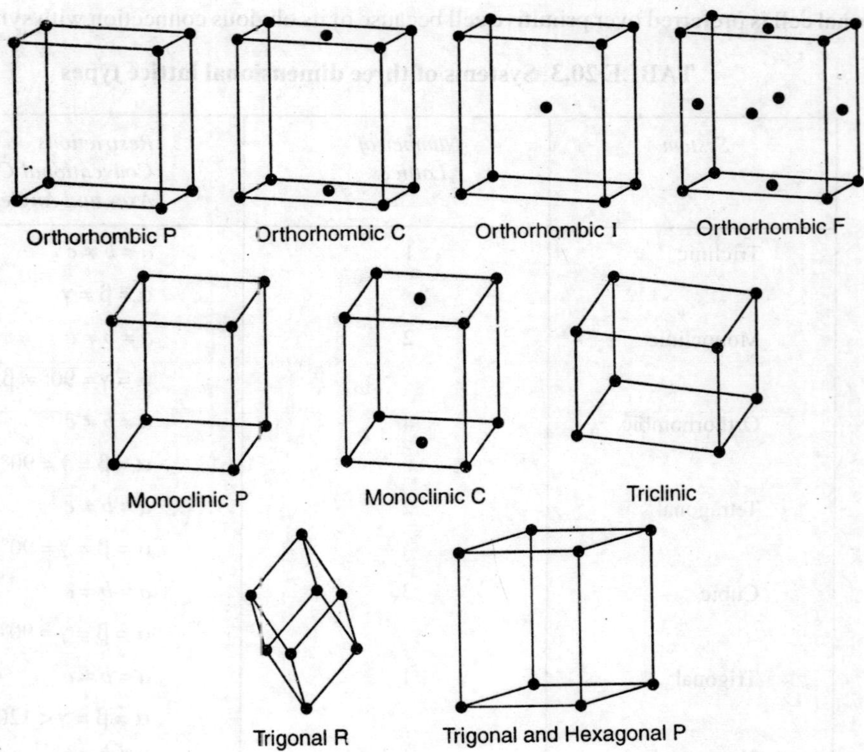

FIGURE 20.6 **Conventional cells of all the fourteen distinct types of lattices in three dimensions.**

In cubic system there are three lattices; simple cubic(*sc*), face centered cubic(*fcc*), and body centered cubic(*bcc*) lattice. These have been discussed more extensibly in the literature and so require special attention. It is obvious that eight cells share every corner of a cubic lattice while two cells share every face. Since simple cubic lattice contains lattice points on the corners only, therefore, it has a density of one lattice point per cell. The body centered cubic lattice contains lattice points on all the corners as well as a lattice point inside the body hence its density is two lattice points per cell. Similarly, the face centered cubic lattice contains lattice points on all the corners as well as in the center of each face hence its density is four lattice points per cell.

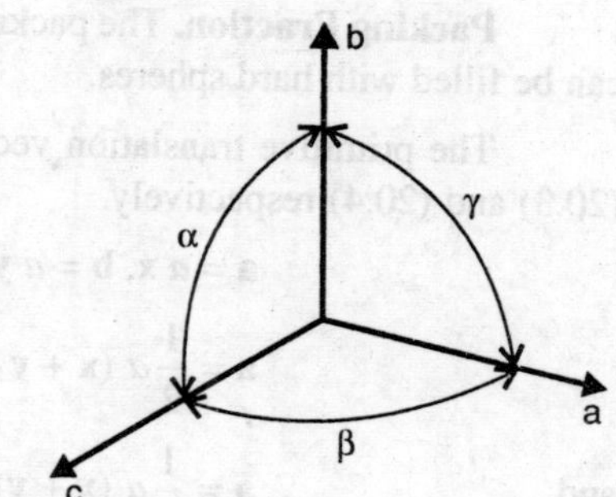

FIGURE 20.7. ***a, b, c* are crystal axes and α, β, γ are angles between them.**

Fig. 20.8 shows these cubic space lattices containing one, two and four lattice points per unit cell respectively.

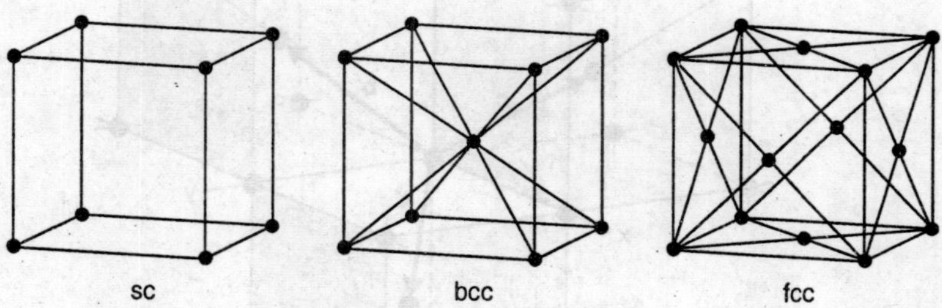

FIGURE 20.8 **Various types of cubic lattices.**

Some useful properties of these cubic lattices of side 'a' have been listed in Table 20.4.

TABLE 20.4. Properties of cubic lattices

Property	*Simple*	*Body Centered*	*Face Centered*
Volume of conventional cell	a^3	a^3	a^3
Lattice points/cell	1	2	4
Volume of primitive cell	a^3	$a^3/2$	$a^3/4$
Lattice points/volume	$1/a^3$	$2/a^3$	$4/a^3$
Number of nearest neighbors	6	8	12
Nearest neighbor's distance	a	$\frac{\sqrt{3}}{2}a$	$a/\sqrt{2}$
Number of second nearest neighbors	12	6	6
Packing fractions, f	$\pi/6$	$\frac{1}{8}.\pi.\sqrt{3}$	$\frac{1}{6}.\pi.\sqrt{2}$

Packing Fraction. The packing fraction, *f*, is the maximum proportion of the available volume which can be filled with hard spheres.

The primitive translation vectors of *sc* lattice, *bcc* lattice, and *fcc* lattice, are given by eqs. (20.2), (20.3) and (20.4) respectively.

$$\mathbf{a} = a\,\mathbf{x},\ \mathbf{b} = a\,\mathbf{y},\ \mathbf{c} = a\mathbf{z} \quad \text{...(20.2)}$$

$$\mathbf{a} = \frac{1}{2}a\,(\mathbf{x} + \mathbf{y} - \mathbf{z}),\ \mathbf{b} = \frac{1}{2}a\,(-\mathbf{x} + \mathbf{y} + \mathbf{z}),\ \mathbf{c} = \frac{1}{2}a\,(\mathbf{x} - \mathbf{y} + \mathbf{z}) \quad \text{...(20.3)}$$

and

$$\mathbf{a} = \frac{1}{2}a\,(\mathbf{x} + \mathbf{y}),\ \mathbf{b} = \frac{1}{2}a\,(\mathbf{y} + \mathbf{z}),\ \mathbf{c} = \frac{1}{2}a\,(\mathbf{z} + \mathbf{x}) \quad \text{...(20.4)}$$

In these equations *a* being the cube edge and **x**, **y** and **z** are the unit vectors along the atomic co-ordinate axes shown in Figs. 20.9 and 20.10.

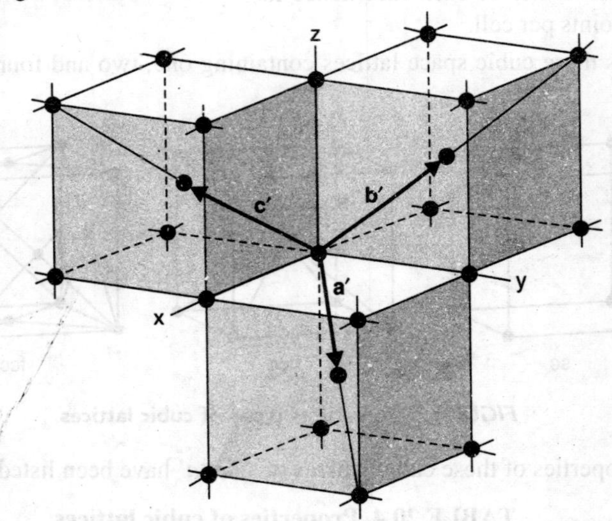

FIGURE 20.9 **Primitive translation vectors of bcc lattice.**

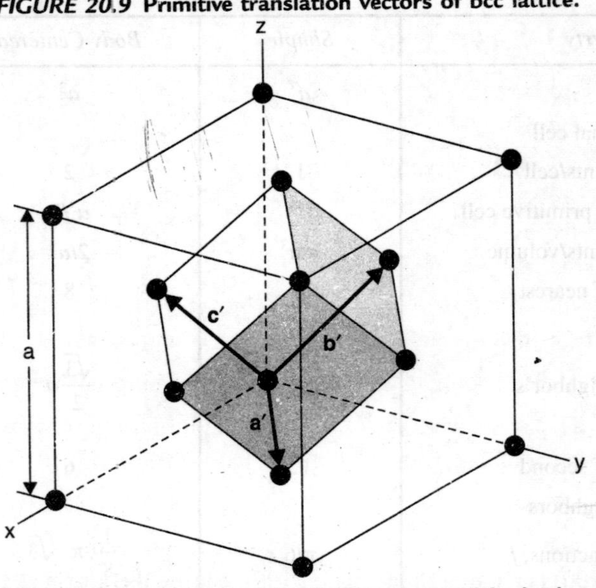

FIGURE 20.10 **Primitive translation vectors of fcc lattice.**

From the expression of **a**, **b**, and **c**, the primitive cell axes, one can calculate the volume of respective unit cells using volume = **a**.(**b** × **c**).

For instance the volume of *sc* primitive unit cell will be

$$V_{sc} = \mathbf{a}.(\mathbf{b} \times \mathbf{c}) = a^3\,[\mathbf{x}.(\mathbf{y} \times \mathbf{z})] = a^3\,[\mathbf{x}\,.\,\mathbf{x}] = a^3$$

The volume of *fcc* primitive cell will be given as

$$V_{fcc} = a^3/8\;[(\mathbf{x} + \mathbf{y})\,.\,(\mathbf{y} + \mathbf{z}) \times (\mathbf{z} + \mathbf{x})]$$

$$= a^3/8\;[(\mathbf{x} + \mathbf{y})\,.\,(\mathbf{x} - \mathbf{z} + \mathbf{y})] = a^3/8\;[1 + 1] = a^3/4$$

Similarly volume of *bcc* primitive cell

$$V_{bcc} = a^3/8\;[(x + y - z)\,.\,(-x + y + z) \times (x - y + z)]$$

$$= a^3/8\;[(x + y - z)\,.\,(x + y - z + x + y + z)]$$

$$= a^3/8\;[(x + y - z)\,.\,(2x + 2y)] = a^3/8\;[2 + 2] = a^3/2.$$

20.9 MILLER INDICES

A crystal lattice may be supposed to be an aggregate of a set of equidistant parallel planes through lattice points. These planes are known as lattice planes or cleavage planes. The position and orientation of lattice planes can be specified by the position of three non-collinear points lying on the plane. However, if the three points lies on the crystal axes then only three numbers, corresponding to positions of three points on these axes with respect to a given origin, are sufficient to specify the plane. To analyze a crystal structure the three points (or indices) are determined through the following procedure.

1. Find the intercepts on the axes **a, b, c** in terms of the lattice constants *a, b, c*. The axes may be primitive or non-primitive.

2. Take the reciprocals of these numbers and then reduce to three integers having the same ratio, usually the smallest three integers. The result is enclosed in parenthesis : (*hkl*).

The indices (*hkl*) so determined are referred to as Miller indices. The above mentioned procedure to determine Miller indices is illustrated in Fig. 20.11. The plane shown in this figure intercepts the **a, b, c** axes at 3**a**, 2**b**, 2**c**. The reciprocals of these numbers are 1/3, 1/2, 1/2. The smallest three integers having the same ratio are 2, 3, 3 thus the Miller indices of the plane are (2 3 3). If a plane intercepts an axis on negative side of origin then the corresponding index is negative and is represented by putting a minus sign over the index *e.g.*, $(\bar{h}\;k\;l)$, $(h\;k\;\bar{l})$ etc. The index corresponding to an intercept at infinity is zero (1/∞ = 0). The lattice planes parallel to *XY* plane have *X* and *Y* intercepts at infinity thus Miller indices of these planes are (001). Similarly the planes parallel to *YZ* and *ZX* planes have (001) and (010) as Miller indices respectively. Miller indices of cube faces of a cubical crystal are (100), (010), (001), $(\bar{1}\,0\,0)$, $(0\,\bar{1}\,0)$ and $(0\,0\,\bar{1})$. Planes equivalent by symmetry may be denoted by curly brackets around indices *e.g.*, {100} represent the set of cube faces. The plane (200) is parallel to the plane (100), cutting the **a** axis at 1/2 a [see Fig. 20.12].

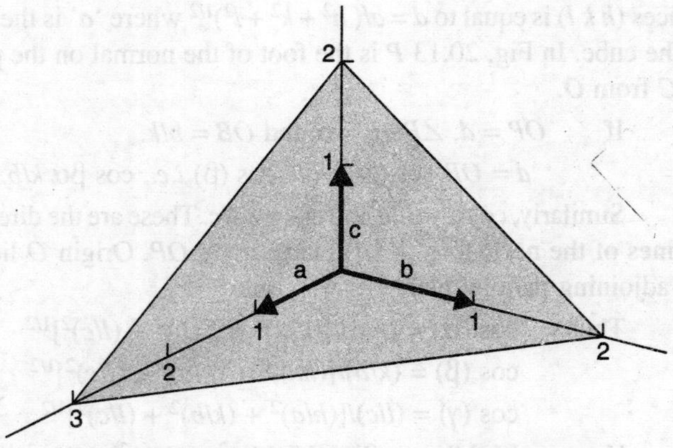

FIGURE 20.11 **Illustration of the method to determine Miller indices.**

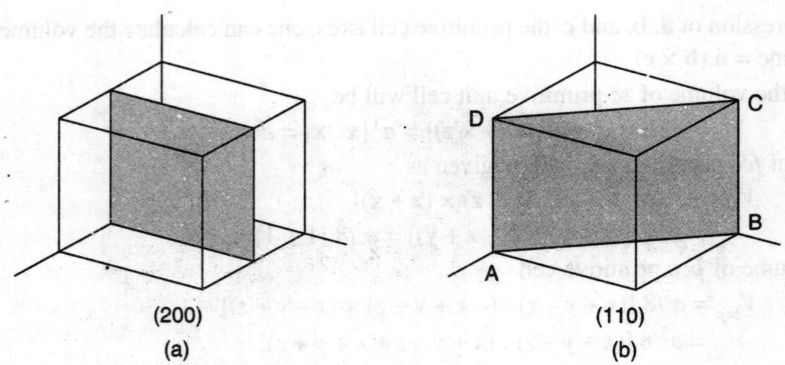

FIGURE 20.12 **(*a*) Plane shown as shaded cuts the *a*-axis at a distance *a*/2 and is parallel to *b*-*c* plane. The Miller indices of this plane are (200) ; (*b*) Plane having Miller indices (110). It cuts a and b axis at a distance *a*.**

Similarly the plane *ABCD* which cuts **a** and **b** axis at A and B such that OA = *a* and OB = *b*, and is parallel to **c**-axis, the intercepts are 1, 1, and ∞. The reciprocals are 1, 1, and 0. Hence the Miller indices of such a crystallographic plane would be (110) [see Fig. 20.12 (*b*)].

It is to be noted that Miller indices represent normals to these planes. In case of cubic crystals such identification become easier. For example, the direction [*h k l*] becomes normal to the plane (*h k l*) and the angle θ between two planes (*h k l*) and (*h′ k′ l′*) is given by

$$\cos(\theta) = (hh' + kk' + ll')/[(h^2 + k^2 + l^2)(h'^2 + k'^2 + l'^2)]^{1/2}$$

Also, distance between two parallel planes having Miller indices (*h k l*) is equal to $d = a/(h^2 + k^2 + l^2)^{1/2}$ where '*a*' is the edge of the cube. In Fig. 20.13 *P* is the foot of the normal on the plane *ABC* from *O*.

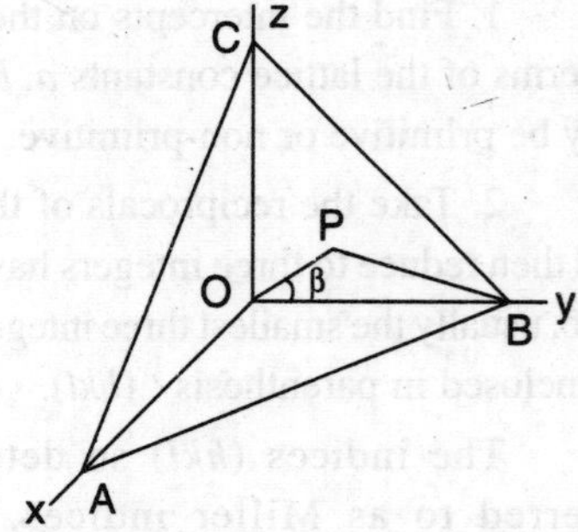

Fig. 20.13. Intersection of (*h k l*) plane with the three axis at *A*, *B* and *C* respectively such that *OA* = *a*/*h*, *OB* = *b*/*k* and *OC* = *c*/*l*. *a*, *b* and *c* denote lattice constants.

If $OP = d$, $\angle POB = \alpha$ and $OB = b/k$.

$$d = OB \cos(\beta) = b/k \cos(\beta) \text{ i.e., } \cos\beta \propto k/b.$$

Similarly, $\cos\alpha \propto h/a$ and $\cos\gamma \propto l/c$. These are the direction cosines of the normal to (*h k l*) plane along *OP*. Origin *O* lies on the adjoining parallel plane.

Thus,
$$\cos(\alpha) = (h/a)/[(h/a)^2 + (k/b)^2 + (l/c)^2]^{1/2}$$
$$\cos(\beta) = (k/b)/[(h/a)^2 + (k/b)^2 + (l/c)^2]^{1/2}$$
and
$$\cos(\gamma) = (l/c)/[(h/a)^2 + (k/b)^2 + (l/c)^2]^{1/2}$$

Hence, $d = b/k \cos(\beta) = 1/[(h/a)^2 + (k/b)^2 + (l/c)^2]^{1/2}$

For a cubic crystal $a = b = c$. So,

$$d = a/[h^2 + k^2 + l^2]^{1/2}$$

20.10 RECIPROCAL LATTICE

A lattice constructed with primitive lattice vectors **A**, **B**, **C** that are related to the primitive vectors of crystal lattice **a**, **b**, **c** through the following relation is known as reciprocal lattice to the crystal lattice.

$$\mathbf{A} = 2\pi\,[(\mathbf{b} \times \mathbf{c})/(\mathbf{a} \cdot \mathbf{b} \times \mathbf{c})]$$
$$\mathbf{B} = 2\pi\,[(\mathbf{c} \times \mathbf{a})/(\mathbf{a} \cdot \mathbf{b} \times \mathbf{c})] \qquad \text{...(20.5)}$$
$$\mathbf{C} = 2\pi\,[(\mathbf{a} \times \mathbf{b})/(\mathbf{a} \cdot \mathbf{b} \times \mathbf{c})]$$

Using various simple properties of cross and dot products of vectors, it is quite easy to deduce that

$$\mathbf{A}.\mathbf{a}=2\pi \quad \mathbf{B}.\mathbf{a}=0 \quad \mathbf{C}.\mathbf{a}=0$$
$$\mathbf{A}.\mathbf{b}=0 \quad \mathbf{B}.\mathbf{b}=2\pi \quad \mathbf{C}.\mathbf{b}=0$$
$$\mathbf{A}.\mathbf{c}=0 \quad \mathbf{B}.\mathbf{c}=0 \quad \mathbf{C}.\mathbf{c}=2\pi$$

A vector $\mathbf{G}=v_1\mathbf{A}+v_2\mathbf{B}+v_3\mathbf{C}$, with v_1, v_2, v_3 as integers is called as a reciprocal lattice vector. The importance of a reciprocal lattice lies in the fact that it is a map of diffraction pattern of a crystal. The dimensions of vectors in crystal lattice are that of length while in reciprocal lattice these dimensions are that of length^{-1}. Wigner-Seitz cell *i.e.*, cell of minimum area (of two dimensional lattice) or volume (of three dimensional lattice), in reciprocal lattice is known as Brillouin zone. The first Brillouin zone is the smallest volume enclosed by planes that are the perpendicular bisectors of the reciprocal lattice vectors drawn from the origin. The Brillouin zone gives geometrical interpretation to diffraction conditions.

Reciprocal Lattice to Simple Cubic (*sc*) Lattice

The primitive translation vectors of a *sc* lattice are given by eq. (20.2)

$$\mathbf{a}=a\mathbf{x},\ \mathbf{b}=a\mathbf{y},\ \mathbf{c}=a\mathbf{z}$$

Now, the corresponding translation vectors of reciprocal lattice **A, B, C** can be determined by using eqn.(20.5)

$$\mathbf{A}=2\pi\,[(\mathbf{b}\times\mathbf{c})/(\mathbf{a}.\mathbf{b}\times\mathbf{c})] = 2\pi\,(a^2/a^3)\,\mathbf{x}=(2\pi/a)\,\mathbf{x}$$
$$\mathbf{B}=2\pi\,[(\mathbf{c}\times\mathbf{a})/(\mathbf{a}.\mathbf{b}\times\mathbf{c})] = 2\pi\,(a^2/a^3)\,\mathbf{y}=(2\pi/a)\,\mathbf{y}$$
$$\mathbf{C}=2\pi\,[(\mathbf{a}\times\mathbf{b})/(\mathbf{a}.\mathbf{b}\times\mathbf{c})] = 2\pi\,(a^2/a^3)\,\mathbf{z}=(2\pi/a)\,\mathbf{z}$$

Hence the reciprocal lattice of *sc* lattice is itself a *sc* lattice but of lattice constant $(2\pi/a)$.

Reciprocal Lattice to Body Centered Cubic (*bcc*) Lattice

The primitive translation vectors of the *bcc* lattice are [eqn. (20.3)]

$$\mathbf{a}=a/2\,(\mathbf{x}+\mathbf{y}-\mathbf{z}),\ \mathbf{b}=a/2\,(-\mathbf{x}+\mathbf{y}+\mathbf{z}),\ \mathbf{c}=a/2\,(\mathbf{x}-\mathbf{y}+\mathbf{z})$$

Now, the corresponding translation vectors of reciprocal lattice **A, B, C** may be obtained as follows

$$\mathbf{A}=2\pi\,[(\mathbf{b}\times\mathbf{c})/(\mathbf{a}.\mathbf{b}\times\mathbf{c})]$$

$$\mathbf{b}\times\mathbf{c}=(a^2/4)\begin{vmatrix}\mathbf{x} & \mathbf{y} & \mathbf{z}\\ -1 & 1 & 1\\ 1 & -1 & 1\end{vmatrix}=(a^2/4)\,[2x+2y]=(a^2/2)\,[x+y]$$

$$\mathbf{a}.\mathbf{b}\times\mathbf{c}=(a^3/4)[(\mathbf{x}+\mathbf{y}-\mathbf{z}).(\mathbf{x}+\mathbf{y}+0\mathbf{z})]=(a^3/4)\,(1+1+0)=a^3/2$$

$$\therefore \quad \mathbf{A}=2\pi\,[\{(a^2/2)(x+y)\}/(a^3/2)]=(2\pi/a)\,(\mathbf{x}+\mathbf{y})$$

Similarly,

$$\mathbf{B}=(2\pi/a)\,(y+z) \quad \text{and} \quad \mathbf{C}=(2\pi/a)\,(z+x)$$

The vectors **A, B, C** are the same as that of the *fcc* lattice. Hence the reciprocal lattice to *bcc* lattice is a *fcc* lattice.

Reciprocal Lattice to Face Centered Cubic (*fcc*) Lattice

The primitive translation vectors of the *fcc* lattice are [eqn. (20.4)]

$$\mathbf{a}=a/2\,(x+y),\ \mathbf{b}=a/2\,(y+z),\ \mathbf{c}=a/2\,(z+x)$$

Now, the corresponding translation vectors of reciprocal lattice **A, B, C** may be obtained as follows

$$\mathbf{A}=2\pi\,[(\mathbf{b}\times\mathbf{c})/(\mathbf{a}.\mathbf{b}\times\mathbf{c})]$$

$$\mathbf{b} \times \mathbf{c} = (a^2/4)\begin{vmatrix} x & y & z \\ 0 & 1 & 1 \\ 1 & 0 & 1 \end{vmatrix} = (a^2/4)\,[x + y - z]$$

$$\mathbf{a} \cdot \mathbf{b} \times \mathbf{c} = (a^3/8)[(x + y + 0z) \cdot (x + y - z)] = (a^3/8)\,(1 + 1 + 0) = a^3/4$$

$$\therefore\ \mathbf{A} = 2\pi\,[\{(a^2/4)(x + y - z)\}/(a^3/4)] = (2\pi/a)\,(x + y - z)$$

Similarly,

$$\mathbf{B} = (2\pi/a)\,(-x + y + z) \quad \text{and} \quad \mathbf{C} = (2\pi/a)\,(x - y + z)$$

The vectors **A**, **B**, **C** are the same as that of the body centered cubic (*bcc*) lattice. Hence the reciprocal lattice to *fcc* lattice is a *bcc* lattice.

20.11 SIMPLE CRYSTAL STRUCTURES

Here, we present some simple crystal structures like sodium chloride, cesium chloride, hexagonal closed packed, diamond, cubic zinc sulphide and hexagonal zinc sulphide. These structures are presented one by one in the following sections.

(*i*) **Sodium Chloride Structure.** The basis of sodium chloride, NaCl, structure consists of one sodium atom and one chlorine atom. The Na^+ ion and Cl^- ion are arranged alternately at the lattice points of a simple cubic lattice resulting in the structure shown in Fig. 20.14 for NaCl crystal. The Bravais lattice of NaCl crystal is face centered cubic with one Cl^- ion at 000 and one Na^+ ion at 1/2 1/2 1/2. Every unit cube contains four units of NaCl having the positions

Cl	0	0	0	1/2	1/2	0	1/2	0	1/2	0	1/2	1/2
Na	1/2	1/2	1/2	0	0	1/2	0	1/2	0	1/2	0	0

Each atom has as nearest neighbors six atoms of the opposite kind. The cube edge a of unit cube is 5.63 Å (5.63×10^{-10} m).

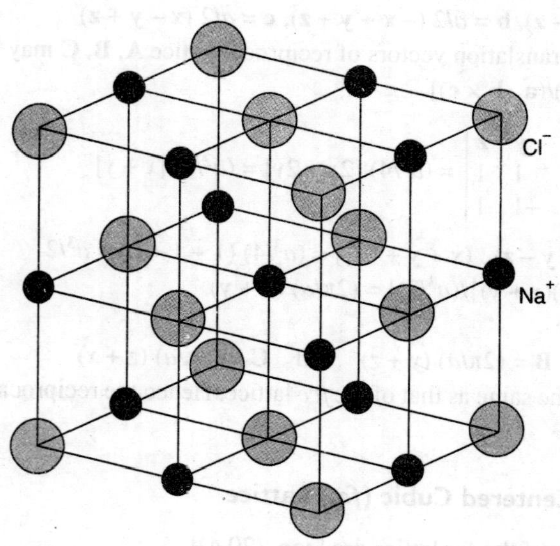

FIGURE 20.14 **Structure of NaCl crystal.**

(*ii*) **Cesium Chloride Structure.** The basis of cesium chloride, CsCl, structure consists of one Cs atom and one Cl atom. Its space lattice is simple cubic with one Cs^+ ion at 000 and one Cl^- ion at 1/2 1/2 1/2.

This crystal structure is shown in Fig. 20.15. The Bravais lattice is body centered cubic and cube edge '*a*' of unit cube is 4.11 Å (4.11×10^{-10} m).

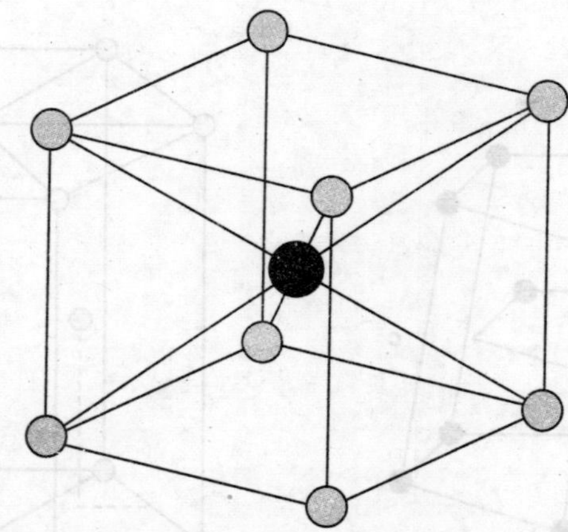

FIGURE 20.15 **The crystal structure of CsCl.**

(*iii*) **Hexagonal Closed Packed (*hcp*) Structure.** When one tries to arrange equivalent spheres in a regular array to minimize the interstitial volume, he comes across two different ways to do so. One layer of spheres is arranged touching each other as shown in Fig. 20.16 with centers at points marked *A*. The second layer of spheres is placed over points marked *B*. Now, there are two different choices for the arrangement of third layer of spheres. It may be placed over points marked *A* or marked *B*. If third layer is arranged over *A*, then the sequence is *ABABAB*... and if it is placed over points marked *C*, then the sequence is *ABCABCABC* ... The structure obtained by the sequence *ABABAB* ... is hexagonal closed packed (*hcp*) and that obtained by the sequence *ABCABCABC* ... is face centered cubic (*fcc*).

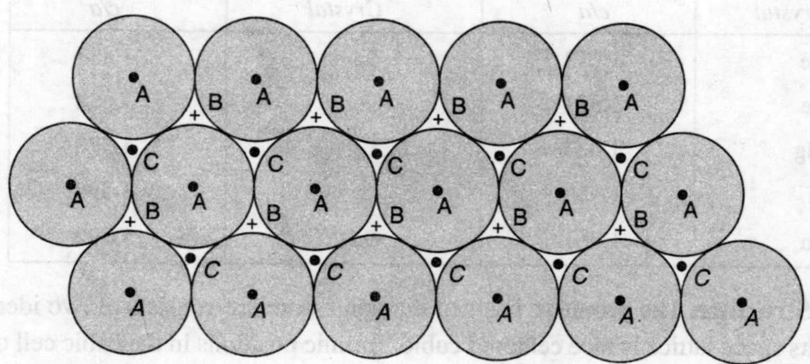

FIGURE 20.16 **A closed-packed layer of spheres, with centers at points marked *A*. A second and identical layer of spheres can be placed over this, with centers over the points marked *B*(or, equivalently, over the points marked *C*).**

The hexagonal closed packed structure has hexagonal symmetry [Fig. 20.17(*a*)]. It has a hexagonal primitive cell ; the basis consists of two atoms as shown in Fig. 20.17. The primitive cell has $a = b$, with an included angle of 120°. The *c* axis is normal to the plane containing *a* and *b*. The *c*/*a* ratio for an ideal *hcp* structure is $(8/3)^{1/2} = 1.633$. [see Fig. 20.17(*b*)] However, certain crystal having a slightly different value of *c*/*a* may also be referred to as *hcp* crystals *e.g.*, zinc with $c/a = 1.86$ and magnesium with $c/a = 1.623$. Many metals easily transform from *fcc* to *hcp* structures and vice-versa at appropriate temperatures. The co-ordination number [number of nearest neighbor atoms] is 12 for both the structures.

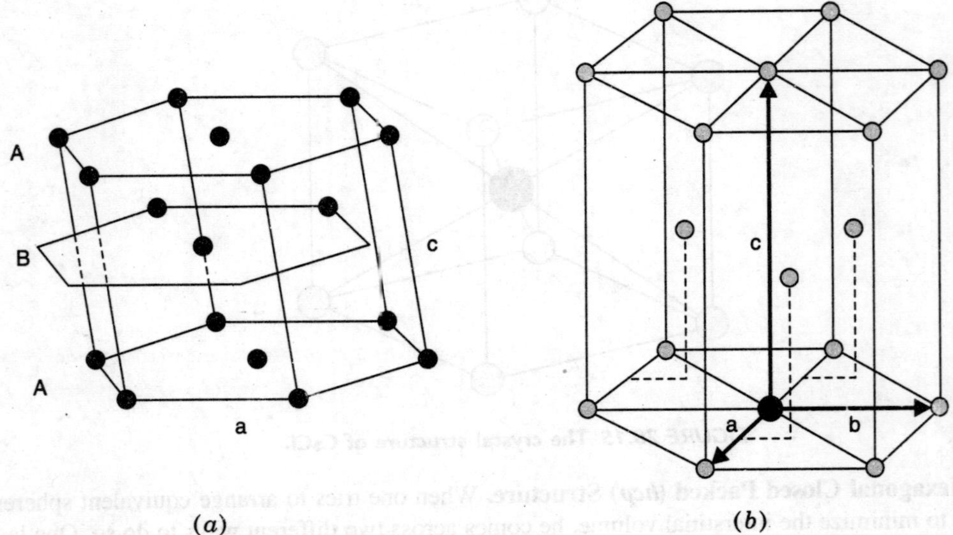

FIGURE 20.17 **The hexagonal closed-packed structure.**

Some examples of crystals having *hcp* structure are given in Table 20.5.

TABLE 20.5

Crystal	*c/a*	*Crystal*	*c/a*
He	1/633	Cd	1.886
Be	1.581	Co	1.622
Mg	1.623	Y	1.570
Ti	1.586	Zr	1.594
Zn	1.861	Gd	1.592

(*iv*) **Diamond Structure.** The primitive basis of diamond structure consists of two identical atoms at 0 0 0 and 1/4 1/4 1/4. Its space lattice is face centered cubic. Atomic positions in the cubic cell of the diamond structure projected on a cube face is shown in Fig. 20.18. In this figure fractions denote height above the base in units of cube edge. The points at 0 and $\frac{1}{2}$ are on the *fcc* lattice and those at 1/4 and 3/4 are on a similar lattice displaced along the body diagonal by one fourth of its length. The diamond structure have tetrahedral bonding arrangement as shown in Fig. 20.19. Each atom has four nearest neighbors and 12 next nearest neighbors. A unit cube contains 8 atoms. Carbon, Silicon, Germanium and Gray Tin crystallize in the diamond structure with lattice constants a = 3.56, 5.43, 5.65 and 6.46 Å, respectively. This structure is the result of directional covalent bonding.

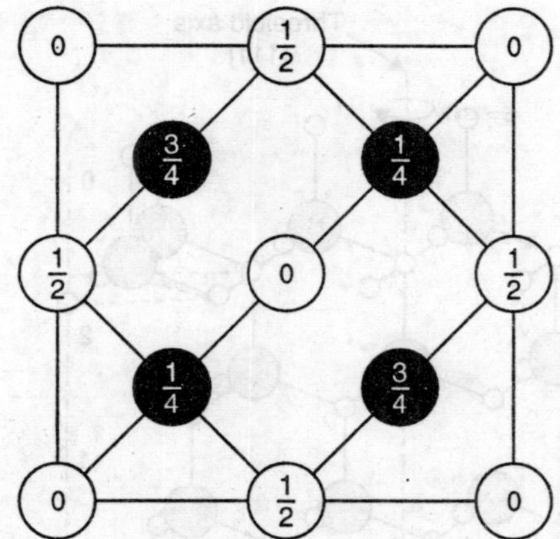

FIGURE 20.18 Atomic positions in the cubic cell of the diamond structure projected on a cube face.

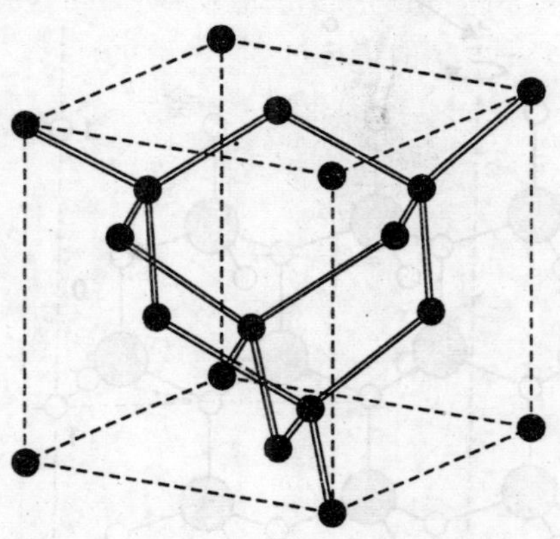

FIGURE 20.19 Crystal structure of diamond showing tetrahedral bond arrangement.

(*v*) **Cubic Zinc Sulphide Structure.** The diamond structure is an arrangement of two identical atoms on two *fcc* structures displaced from each other by one quarter of a body diagonal. If zinc atoms are placed on one *fcc* lattice and sulphur atoms on the other, instead of two identical atoms, then one get cubic zinc sulphide crystal as shown in Fig. 20.20. The conventional cell is a cube. The coordinates of the zinc atoms are 000, 0 1/2 1/2 , 1/2 0 1/2 , 1/2 1/2 0 ; the coordinates of the sulphur atoms are 1/4 1/4 1/4, 1/4 3/4 3/4 , 3/4 1/4 3/4 , 3/4 3/4 1/4. The space lattice is *fcc.* There are four molecules of ZnS per conventional cell. About each atom there are four equally distant atoms of the opposite kind arranged at the corners of a regular tetrahedron. The cube edge of the unit cell is 5.41 Å.

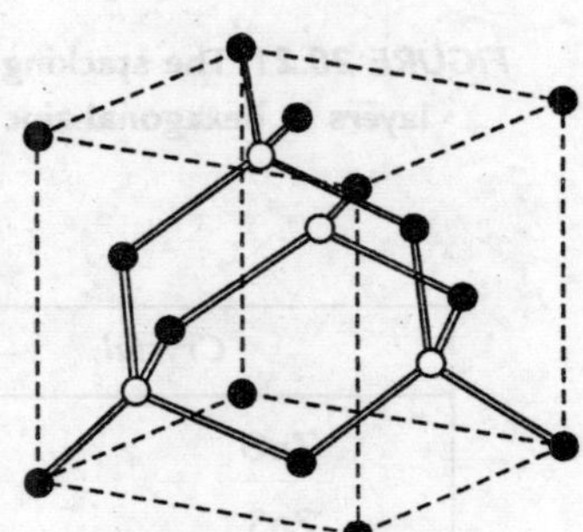

FIGURE 20.20 The cubic zinc sulphide crystal structure.

(*vi*) **Hexagonal Zinc Sulphide (Wurtzite) Structure.** Like cubic zinc sulphide structure, it is also a special form of diamond structure. Fig. 20.21 gives a view of hexagonal ZnS with a set of tetrahedral bonds in the upright position. It is important to compare this structure with cubic ZnS structure shown in Fig. 20.22. In both the figures the large atoms are sulphur, while the smaller one are zinc. Both cubic and hexagonal structures are made up of pairs of layers one plane Zn and the other plane S. The layer pairs in both the structures are stacked one above the other, and successive layer pairs are displaced horizontally. This displacement preserves the tetrahedral bonding scheme, otherwise the sequence S-Zn-S would be a straight line. The stacking sequence for cubic zinc sulphide is 012012012 … while that for hexagonal zinc sulphide is 010101 … as shown in Fig. 20.21 and Fig. 21.22 respectively. In order to distinguish the hexagonal ZnS structure from cubic ZnS structure, one must look beyond to the second nearest neighboring atoms. Some of the crystals having hexagonal zinc sulphide structures are given in Table 20.6.

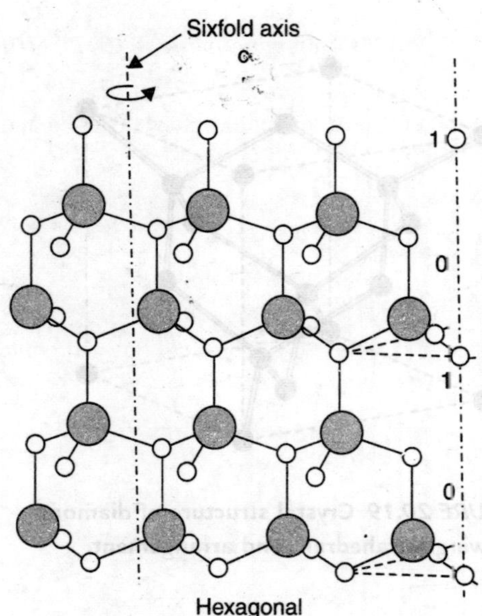

FIGURE 20.21 **The stacking of tetrahedral layers in hexagonal zinc sulphide.**

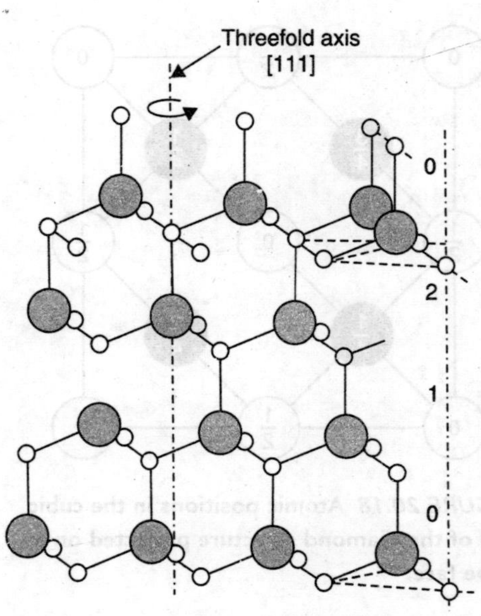

FIGURE 20.22 **The stacking of tetrahedral layers in cubic zinc sulphide.**

TABLE 20.6

Crystal	*A (Å)*	*C (Å)*
ZnO	3.25	5.12
ZnS	3.81	6.93
ZnSe	3.98	6.53
ZnTe	4.27	6.99
SiC	3.25	5.21
Hexagonal diamond	2.52	4.12
CdS	4.13	6.75
CdSe	4.30	7.02

SOLVED NUMERICAL PROBLEMS

Problem 1. (*a*) *How many atoms are there in the primitive cell of diamond ? (b) What is the length of a primitive translation vector ? Given a = 3.56 Å.*

Solution. (*a*) Diamond has *fcc* Bravais lattice with 2 atoms in a primitive cell and 8 atoms in a conventional cube.

(*b*) If '*a*' is the cube edge then $a/\sqrt{2}$ is the length of primitive translation vector

$$a/\sqrt{2} = 3.56/1.41 = 2.52 \text{ Å}.$$

Problem 2. *The lattice constant of a cubic lattice is 'a'. Find the spacing between (011), (101), and (110) planes.*

Solution. Using $d_{hkl} = a/[h^2 + k^2 + l^2]^{1/2}$ with a = lattice constant, d = interplaner distance and hkl = Miller indices,

We have, for plane (011), $h = 0, k = 1, l = 1 \quad \therefore \quad d_{011} = a/\sqrt{2}$

Similarly for plane (101), $h = 1, k = 0, l = 1 \quad \therefore \quad d_{101} = a/\sqrt{2}$

and for plane (110), $h = 1, k = 1, l = 0 \quad \therefore \quad d_{110} = a/\sqrt{2}$

Problem 3. *In a triclinic crystal, a cleavage plane makes intercepts 2.93, 4.47 and 2.35 mm along the three crystallographic axes with the corresponding primitives being 3.05, 6.99, and 4.90 Å. Calculate the Miller indices of the cleavage plane.*

Solution. It is given that

$$u : v : w = 2.93/3.05 : 4.47/6.99 : 2.35/4.90 = 0.961 : 0.640 : 0.480 = 6 : 4 : 3$$

$$\therefore \quad u^{-1} : v^{-1} : w^{-1} = 1/6 : 1/4 : 1/3 = 2 : 3 : 4$$

Hence the Miller indices of the given plane are (234).

APPENDIX

Atomic Packing Fraction

The packing fraction or relative packing density is defined as the ratio of the volume of the atoms occupying the unit cell to the volume of the unit cell relating to that structure.

$\therefore$ Packing Fraction = Volume of the atoms per unit cell (v)/Volume of the unit cell (V)

The packing fraction is directly related to co-ordination number (number of equidistant neighbours). Greater the coordination number, more closely packed will be the structure. Before calculating the packing fraction of different cubic structure, we calculate atomic radius in terms of cube edge 'a'. Here, we present the evaluation of the packing fraction of some of the simple crystal structures.

(*a*) **Simple Cubic Lattice.** A two dimensional projection of *sc* lattice is shown in Fig. (A-1). There are eight corner atoms and each of these eight atoms is a member of eight cells. So, there is one atom per unit cell. The atoms are to be treated as hard spheres each of radius r.

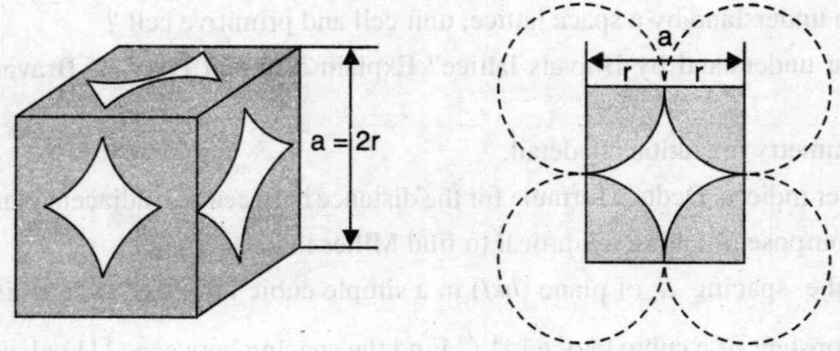

FIGURE A.1 **Simple cubic structure and two dimensional projection.**

Figure (A-1). 'a' is the size of the cubic egde. $r = a/2$.

It is obvious from the figure that

$$2r = a \quad \text{or} \quad r = a/2$$

Now packing fraction = $v/V = (4/3)\pi r^3/a^3 = (4/3)\pi(a/2)^3/a^3 = \pi/6 = 0.52$

(*b*) **Face Centered Cubic Lattice.** It has 8 corner atoms and each of these atoms is a member of 8 cells. There are 6 face centered atoms, each is a member of two cells on its either side. So, each unit cell will have 3 + 1 = 4 atoms. Let r be the radius of each hard sphere. The translation vectors of *fcc* crystal are

$$\mathbf{a} = a/2(\mathbf{x} + \mathbf{y}),\ \mathbf{b} = a/2(\mathbf{y} + \mathbf{z}),\ \mathbf{c} = a/2(\mathbf{x} + \mathbf{z})$$

So, $$2r = (\mathbf{a}.\mathbf{a})^{1/2} = (\mathbf{b}.\mathbf{b})^{1/2} = (\mathbf{c}.\mathbf{c})^{1/2} = a/\sqrt{2}.$$

or $$r = a/2\sqrt{2}$$

Now packing fraction = v/V

v(volume of 4 atoms) = 4 × volume of one atom = 4 × $(4/3)\pi r^3$

V (volume of unit cell) = a^3

Packing Fraction = $(16/3)\pi r^3/a^3 = 16\pi(\sqrt{2}\,a/4)^3/3a^3 = (\pi\sqrt{2})/6 = (3.14 \times 1.414)/3 = 0.74.$

(*c*) **Body Centered Cubic Lattice.** In a *bcc* unit cell there are 8 atoms at corners and each one of these atom is a member of 8 cells. There is one atom per cell at the body center. So, total no. of atoms per unit cell is 1 + 1 = 2. Now, the primitive translation vectors of *bcc* are

$$\mathbf{a} = a/2(\mathbf{x} + \mathbf{y} - \mathbf{z}),\ \mathbf{b} = a/2(-\mathbf{x} + \mathbf{y} + \mathbf{z}),\ \mathbf{c} = a/2(\mathbf{x} - \mathbf{y} + \mathbf{z})$$

As a result $(2r)^2 = \mathbf{a}.\mathbf{a} = 3a^2/4$ or $r = \sqrt{3}\,a/4$

Now packing fraction = v/V

v (volume of 2 atoms) = 2 × volume of one atom = 2 × $(4/3)\pi r^3$

V (volume of unit cell) = a^3

∴ Packing Fraction = $(8/3)\pi r^3/a^3 = 8\pi(\sqrt{3}\,a/4)^3/3a^3 = (\pi\sqrt{3})/8 = (3.14 \times 1.732)/8 = 0.68.$

(*d*) **Hexagonal Closed Packed System (hcp).** In this system there are two atoms per unit cell and the distance between nearest neighbours is $a/2$ while the volume of the unit cell is $\sqrt{2}\ a^3$. Packing fraction of *hcp* is, therefore, given by

$$f = (2\,.\,(4\pi/3)\,.\,r^3)/\sqrt{2}\,a^3$$
$$= (\pi/3)\,(1/\sqrt{2}) = 0.74 \text{ same as of } fcc \text{ structure.}$$

EXERCISE

1. Differentiate between crystalline and amorphous solids.
2. What do you understand by a space lattice, unit cell and primitive cell ?
3. What do you understand by Bravais lattice? Explain different types of Bravais lattices in two and three dimensions.
4. Describe symmetry operations in detail.
5. Explain Miller indices. Deduce formula for the distance between two adjacent planes of a simple cubic lattice.
6. What is the purpose of taking reciprocal to find Miller indices ?
7. Show that the spacing d of plane (*hkl*) in a simple cubic lattice of side 'a' is $d_{hkl} = a/[h^2 + k^2 + l^2]^{1/2}$.
8. The lattice constant of a cubic lattice is 'a'. Find the spacing between (111) planes. **[Ans.** $a/\sqrt{2}$ **]**
9. In an orthorhombic crystal whose primitives are 1.21, 1.84 and 1.97 Å, a cleavage plane whose Miller indices are (231) cuts off intercept 1.68 mm along x-axis. Find the lengths of intercepts on y-axis and z-axis.

   [Ans. 1.70 mm, – 5.46 mm**]**
10. Prove that in a crystal of cubic symmetry a direction [*hkl*] is perpendicular to the plane (*hkl*) with same indices.

11. The primitive translation vectors of the hexagonal space lattice may be taken as $\mathbf{a} = (\sqrt{3}/2a)i + (a/2)j$, $\mathbf{b} = -(\sqrt{3}/2a)i + (a/2)j$ and $\mathbf{c} = ck$. Show that the volume of the primitive cell is $\sqrt{3}/2\, a^2c$.
12. What do you mean by reciprocal lattice ? Show that the reciprocal lattice of a simple cubic lattice is itself a simple cubic lattice.
13. Show that the reciprocal lattice of *fcc* lattice is *bcc* and vice-versa.
14. Explain the following simple crystal structures (*i*) Sodium chloride structure (*ii*) Cesium chloride structure (*iii*) Hexagonal closed-packed structure.
15. Show that the *c*/*a* ratio for an ideal *hcp* structure is $(8/3)^{1/2} = 1.633$ and hence the packing fraction is 0.74.
16. Explain the following terms :
    (*a*) Space lattice
    (*b*) Translation vector
    (*c*) Unit cell
    (*d*) Packing efficiency.
17. Explain how would you arrange identical spheres in a hexagonal closed packed structure ? Also calculate its *c*/*s* ratio.
18. How do you assign Miller indices to represent crystallographic planes and directions ?
19. Show that in a cubic lattice, the distance between successive planes of Miller indices (*hkl*) is given by $d = a/(h^2 + k^2 + l^2)^{1/2}$.
20. Calculate the distance between adjacent planes of the type (100), (110) and (111) in a *sc* lattice of lattice constant '*a*'.
21. What is packing efficiency ? Calculate the packing efficiency in case of a *hcp* crystal.
22. (1) Describe diamond structure and find its packing efficiency.
    (2) Discuss sodium chloride and zinc blend structures. What are the coordination number in these cells ?
23. Show that in a crystal of cubic symmetry, a direction [*hkl*] is perpendicular to the planes having the same indices.

Chapter 21 BINDINGS IN SOLIDS

21.1 INTRODUCTION

The binding of constituent atoms of solids remained a subject of great interest to solid state physicists. The nature of bindings in solids depends on the type of constituent atoms *i.e.*, whether these are inert gas atoms,

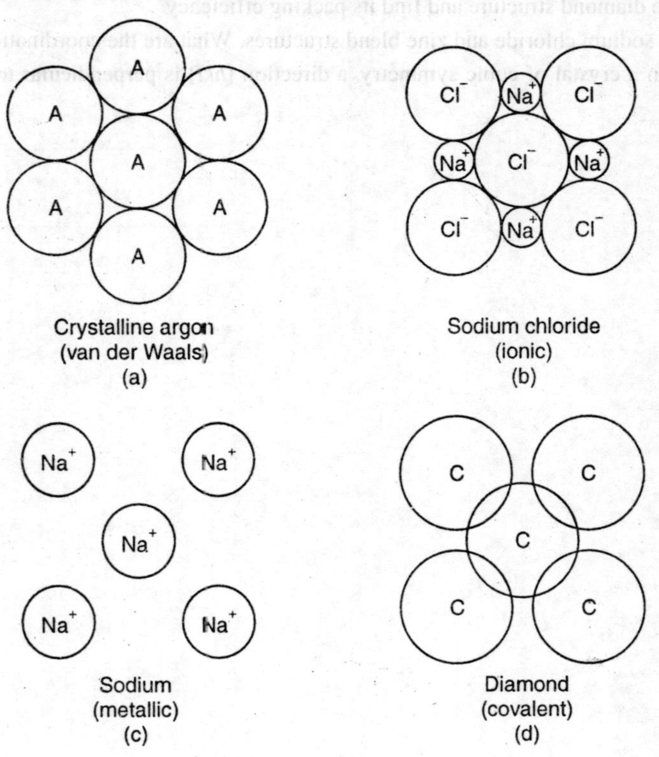

FIGURE 21.1 **Various type of bindings in solids.**

alkali halides, alkali metals or of atoms having four valence electrons. In case of crystalline solids comprising inert gas atoms and alkali halides the bonds have been well understood through simple models. The results of these models explained the experimental results quite accurately. The reason being the orderly and regular arrangement of constituent atoms in the solids. The variation of binding energy per atom, melting point etc. of solids have been predicted fairly well through these models.

Accordingly the bindings in solids are of following types (Fig. 21.1)

(*i*) van der Waals

(*ii*) Ionic

(*iii*) Covalent

(*iv*) Metal

(*v*) Hydrogen bonds

Table 21.1 shows the relative strengths of these bindings. For instance one finds that van der Waal bonds are the weakest while covalent are the strongest.

TABLE 21.1 Cohesive Energies in eV/atom at 0 K and 1 atm. pressure

Ne 0.02	Li 1.63	C 7.37
Ar 0.08	Na 1.113	Si 4.63
Kr 0.116	K 0.934	Ge 3.85
Xe 0.16	Rb 0.852	Sn 3.14
Rn 0.202	Cs 0.804	Pb 2.03

The trend of melting point as well as of bulk modulii of these solids is same as that of binding energy/atom (or cohesive energy/atom). We shall now discuss these bindings one by one.

21.2 VAN DER WAALS BINDINGS

The bindings in crystal of inert gas atoms is very simple and weak. It being small the electron charge distribution around the atoms in solids would not be very much different from spherically symmetric as in free inert gas atoms. The forces with which these atoms are bound in solids are termed as van der Waals forces, of course are attractive in nature. It is assumed that as atoms come closer to each other, the spherically symmetric charge distribution gets distorted and dipole moments in each other is induced. Therefore, for simplicity we consider the dipole-dipole interactions. Let x_1 and x_2 be the co-ordinates of the two dipoles (or the identical oscillators) and p_1 and p_2 be their conjugate momenta. (See Fig. 21.2)

The centers of positive and negative charges oscillate in the two oscillators such that non interaction Hamiltonian would be

$$H_0 = p_1^2/2m + cx_1^2/2 + p_2^2/2m + cx_2^2/2 \qquad ...(21.1)$$

where c denotes the force constants and m the masses. These will further interact through electrostatic energies given by

$$H_I = e^2/R + e^2/(R + x_1 - x_2) - e^2/(R + x_1) - e^2/(R - x_2)$$

where e denotes the positive as well as negative charge. By assuming x_1 and x_2 as small in comparison with R one can simplify the above expression using binomial theorem. The result is

$$H_I = e^2/R + e^2/R[1 - (x_1 - x_2)/R + ((x_1 - x_2)/R)^2] - e^2/R[1 - x_1/R + x_1^2/R^2]$$
$$- e^2/R[1 + x_2/R + x_2^2/R^2]$$
$$H_I = e^2/R^3[(x_1 - x_2)^2 - x_1^2 - x_2^2]$$
$$H_I = -2\,(e^2/R^3)\, x_1 x_2 \qquad \text{...(21.2)}$$

(the van der Waals attractive interaction)

Let us now transform x_1, x_2 and also p_1, p_2 through following relations

$$x_1 = (x_s + x_a)/\sqrt{2} \quad ; \qquad p_1 = (p_s + p_a)/\sqrt{2}$$
$$x_2 = (x_s - x_a)/\sqrt{2} \quad ; \qquad p_2 = (p_s - p_a)/\sqrt{2}$$

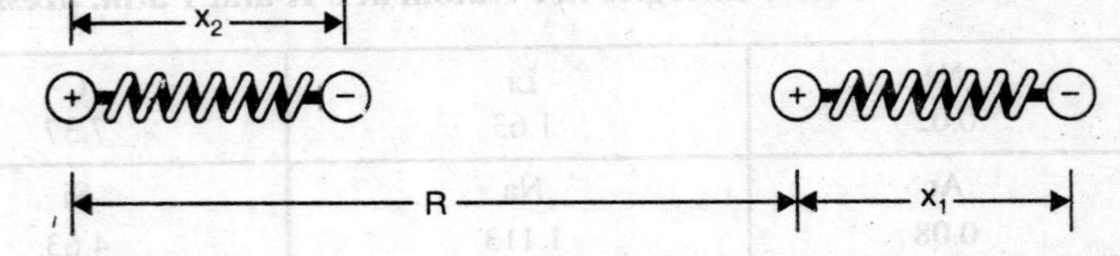

FIGURE 21.2 **The interaction between two dipoles ; x_1, x_2 denote their coordinates.**

As a result

$$H = H_0 + H_I = (1/2m)(p_s^2 + p_a^2) + (c/2)(x_s^2 + x_a^2) - (e^2/R^3)(x_s^2 - x_a^2)$$
$$H = p_s^2/2m + (½)(c - 2e^2/R^3)x_s^2 + p_a^2/2m + (½)(c + 2e^2/R^3)x_a^2$$
$$H = p_s^2/2m + (½)c_s x_s^2 + p_a^2/2m + (½)c_a x_a^2 \qquad \text{...(21.3)}$$

This hamiltonian represents a combination of two non-interacting oscillators having frequencies

$$\omega_s^2 = c_s/m \text{ and } \omega_a^2 = c_a/m$$

and having zero point energies as $\hbar\omega_s/2$ and $\hbar\omega_a/2$ respectively.

Here, $c_s = c - 2e^2/R^3$ and $c_a = c + 2e^2/R^3$

Thus, the lowest energy of the two systems would be

$$(\hbar/2)(\omega_s + \omega_a) = (\hbar/2)\sqrt{(c/m)}\ [(1 - 2e^2/cR^3)^{1/2} + (1 + 2e^2/cR^3)^{1/2}]$$
$$= \hbar\omega_0 - (\hbar\omega_0)(1/8)(2e^2/cR^3)^2$$

where $\omega_0 = \sqrt{(c/m)}$

Therefore, the change in the lowest energies of the two atoms would be

$$\Delta U = -(\hbar\omega_0)(1/8)(2e^2/cR^3)^2$$

So, the energy gets reduced and this accounts for the binding of two inert gas atoms in the solid. This is usually written as

$$A/R^6 \text{ where } A = (\hbar\omega_0/8)(4e^4/c^2)$$

In other words van der Waals interaction varies inversely as the sixth power of R, the closest distance between atoms.

Repulsive Interaction

As the electron charge distributions of neighboring atoms start overlapping a repulsive interaction comes into play because of Pauli Exclusion Principle. This is because no two electrons can occupy same state [see Fig. 21.3]

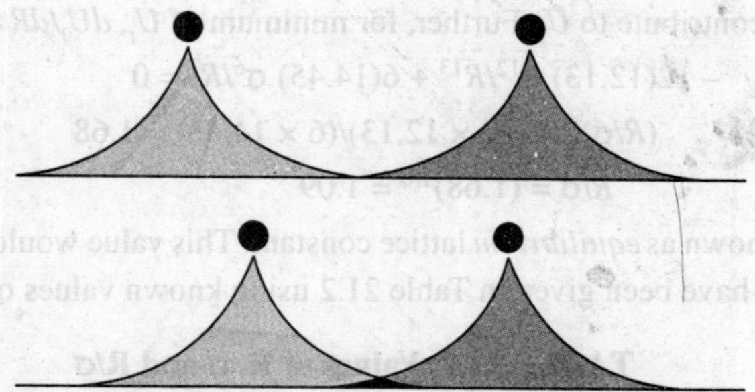

FIGURE 21.3 **Display of overlap of electronic charge distributions as atoms approach. Solid circles denote inert gas atoms.**

This repulsive energy varies as B/R^{12}, so that total energy of each atom in the solid becomes

$$U(R) = -A/R^6 + B/R^{12}$$

It is usual to write

$$A = 4\varepsilon\sigma^6 \text{ and } B = 4\varepsilon\sigma^{12}$$

such that $U(R) = 4\varepsilon[-(\sigma/R)^6 + (\sigma/R)^{12}]$ ...(21.4)

This is called Lennard-Jones interaction energy per atom. Its variation is being shown in Fig. 21.4.

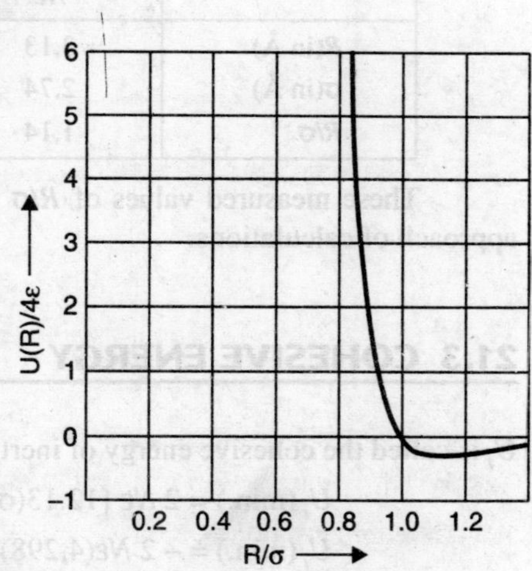

FIGURE 21.4 **Plot of Lennard-Jones interaction energy vs *R*.**

One notes the following

(*i*) $U(R) = 0$ at $R = \sigma$ or $R/\sigma = 1$

(*ii*) $U(R)$ becomes minimum at $dU/dR = 0$ *i.e.*, when $2\sigma^{12}/R^{13} = \sigma^6/R^7$ or $(R/\sigma) = 2^{1/6} = 1.12$

(*iii*) $U(R)$ has a steep rise between $R/\sigma = 1$ and 0.8 while its variation is flat for $R/\sigma > 1.12$ and starts approaching zero.

Lattice Constants

If there are N atoms in the given crystal and the kinetic energy of each is neglected, the total energy would be

$$U_t = \frac{1}{2} N 4\varepsilon [\Sigma_j'(\sigma/r_{ij})^{12} - \Sigma_j'(\sigma/r_{ij})^6] \quad ...(21.5)$$

where r_{ij} denotes the distance of jth neighboring atom from ith atom and prime on summation means that $j \neq i$. The factor $\frac{1}{2}$ has been taken to avoid the doubling of the pairs i, j while calculating U_t. Let us take r_{ij} in terms of R *i.e.*, $r_{ij} = P_{ij}$, R, R being the distance of nearest neighbors. Thus, we get for total energy

$$U_t = \frac{1}{2} N 4\varepsilon [(\sigma/R)^{12}\Sigma_j'(P_{ij})^{-12} - (\sigma/R)^6\Sigma_j'(P_{ij})^{-6}]$$

It has been seen that for inert gas atom crystals

$$\Sigma_j'(P_{ij})^{-12} = 12.132 \quad \text{and} \quad \Sigma_j'(P_{ij})^{-6} = 14.454$$

Thus $\qquad U_t = 2N\varepsilon\,[12.132(\sigma/R)^{12} - 14.454(\sigma/R)^6]$

Since these values are not far from 12 and that there are 12 nearest neighbors in *fcc* structural crystals, only nearest neighbors contribute to U_t. Further, for minimum of U_t, $dU_t/dR = 0$

That is, $\qquad -12(12.13)\,\sigma^{12}/R^{13} + 6(14.45)\,\sigma^6/R^7 = 0$

or $\qquad (R/\sigma)^6 = (12 \times 12.13)/(6 \times 14.45) = 1.68$

and $\qquad R/\sigma = (1.68)^{1/6} = 1.09$

This ratio R/σ is known as *equilibrium* lattice constant. This value would be same for all types of inert gas crystals. Some values have been given in Table 21.2 using known values of R and σ

TABLE 21.2 Values of R, σ and R/σ

Quantity	*Inert Gas Atoms*			
	Ne	*Ar*	*Kr*	*Xe*
R(in Å)	3.13	3.76	4.01	4.35
σ(in Å)	2.74	3.40	3.65	3.98
R/σ	1.14	1.106	1.10	1.09

These measured values of R/σ are very close to the theoretical value 1.09, thereby justifying the approach of calculations.

21.3 COHESIVE ENERGY

U_t is called the cohesive energy of inert gas crystals. Let us write this for equilibrium state *i.e.*,

$$U_t(\text{min.}) = 2\,N\varepsilon\,[12.13(\sigma/R)^{12} - 14.45(\sigma/R)^6] = 2\,N\varepsilon\,[4.318 - 8.616] \text{ for } R/\sigma = 1.09$$

$$U_t(\text{min.}) = -2\,N\varepsilon(4.298) = -4\,N\varepsilon(2.15) \qquad \text{...(21.6)}$$

This is the cohesive energy when the atoms are at rest. However, the existence of kinetic energy will reduce this value. The lowest kinetic energy can be calculated quantum mechanically by the formula

$K.E. = p^2/2m = (h/\lambda)^2(1/2m)$; λ being the de Broglie wavelength and h and m denote Plank's constant and the atomic mass. That is heavier the atom smaller the correction due to kinetic energy. This trend is confirmed by experimental values.

21.4 IONIC BINDING

In alkali halides the bindings are caused by electrostatic force between +ve and –ve ions of the molecules. Such crystalline solids are called ionic crystals. Examples are LiF, NaCl, KCl, RbF etc. The constituent atoms of the alkali halides have either one electron in surplus or deficit before forming the molecules. For instance, Na atom has the configuration $1s^2\,2s^2\,2p^6\,3s^1$ *i.e.*, one valence electron while Cl has one electron less in the last shell ($1s^2\,2s^2\,2p^6\,3s^2\,3p^5$). As a result when these atoms come closer Na parts with one electron and becomes Na^+ while Cl captures one electron to form Cl^-. Now, these ions Na^+ and Cl^- have completed outermost shells and condense to form NaCl molecules of the solid. This way formation of ionic crystal is explained. The whole process of condensation is shown in Fig. 21.5.

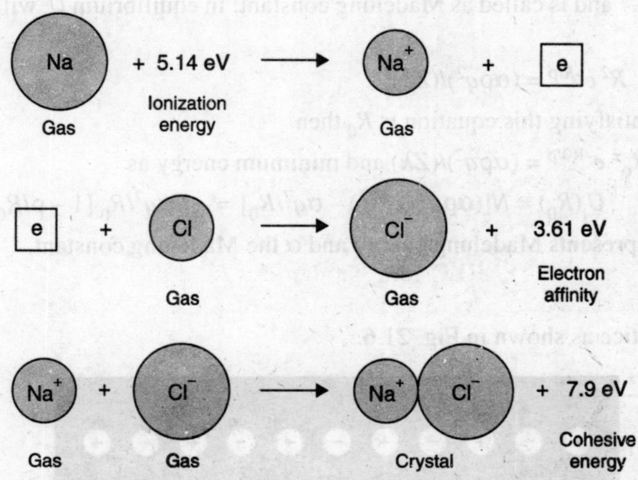

FIGURE 21.5 **The steps through which NaCl crystalline solid is formed.**

In such crystals the main forces are electrostatic. For instance in NaCl, inter ionic distance is about 2.81×10^{-8} cm. Then the coulomb interaction in these will be given by

$$q^2/(2.81 \times 10^{-8}) \text{ ergs} ; [q = 4.8 \times 10^{-10} \text{ e.s.u. of charge in CGS system}]$$
$$= 1.44/0.281 \text{ eV} = 5.12 \text{ eV}$$

This value is comparable with the experimental value of 7.9 eV per molecule.

Electrostatic or Madelung Energy

Let us next calculate the binding energy per molecule of ionic crystals consisting of alkali halides. It is mainly electrostatic and is called Madelung energy. Let *i*th ion be surrounded by other ions named as *j*. Then electrostatic energy on *i*th ion would be

$$U_i = \Sigma_j' U_{ij} \text{ ; prime means } i \neq j$$

Now, $$U_{ij} = \lambda \text{ Exp}(-r_{ij}/\rho) \pm q^2/r_{ij}$$

where first term represents repulsive energy due to +ve charge in the ionic cores and 2nd term represents electrostatic energy between the +ve and –ve ion pairs their charges taken as point charges. (+) sign is taken for similar charges while (–) sign is taken for opposite charges. r_{ij} represents distance and λ and p the parameters called strength and range of repulsive energy term. Note that we have used $\lambda e^{-R/\rho}$ instead of (B/R^{12}) form. Then total energy of N molecules would be

$$U_t = NU_i = N[\Sigma_j' \lambda \text{ Exp}(-r_{ij}/\rho) \pm \Sigma_j' q^2/r_{ij}]$$

In first term main contribution comes from nearest neighbors at distance R each. Let these be Z in number. Hence

$$U_t = N[Z\lambda \text{ Exp}(-R/\rho) - Zq^2/R + \Sigma_j' \pm q^2/r_{ij}]$$

as nearest ions are having opposite charges to the *i*th one and summation excludes such ions. It is also usual to write $r_{ij} = p_{ij} R$, so that

$$U_t(R) = N[Z\lambda \, \text{Exp}(-R/\rho) - \alpha q^2/R]$$

where $\alpha = -Z + \sum_j{}' \pm (p_{ij})^{-1}$ and is called as Madelung constant. In equilibrium U_t will be lowest and $dU_t/dR = 0$ or R will satisfy

$$R^2 e^{-R/\rho} = (\alpha\rho q^2)/(Z\lambda)$$

If the value of R satisfying this equation is R_0 then

$R_0^2 e^{-R0/\rho} = (\alpha\rho q^2)/(Z\lambda)$ and minimum energy as

$$U_t(R_0) = N[(\alpha\rho q^2)/(R_0^2) - \alpha q^2/R_0] = -N\alpha q^2/R_0 \, [1 - \rho/R_0] \text{ in CGS system.}$$

Here $-N\alpha q^2/R_0$ represents Madelung energy and α the Madelung constant.

Evaluation of α

Take the case of linear lattice as shown in Fig. 21.6.

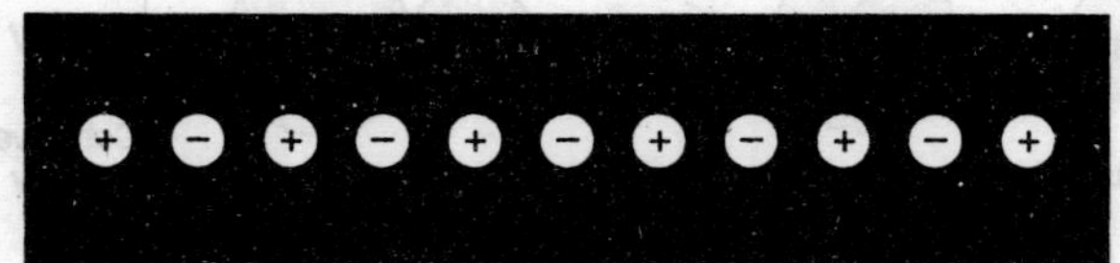

FIGURE 21.6 **Linear arrangement of +ve and −ve ions.**

From Fig. 21.6 it is obvious that

$$\alpha = \sum_j \pm (p_{ij})^{-1}$$

or
$$\alpha/R = \sum_j \pm 1/Rp_{ij} = 2[1/R - 1/2R + 1/3R - 1/4R + \ldots]$$

Thus,
$$\alpha = 2[1 - 1/2 + 1/3 - 1/4 + \ldots]$$

since
$$\log(1 + x) = 1 - x/2 + x^3/3 - x^4/4 + \ldots$$

$$\alpha = 2 \log(1 + 1) = 2 \log 2 = 1.39$$

This is not far off from typical values known, as shown in Table 21.3

TABLE 21.3

Crystalline Solids	α
NaCl	1.75
CsCl	1.76
ZnS	1.64

These values as expected are larger than 1.39 calculated on the basis of linear lattice. In reality, lattice is 3-dimensional.

For KCl crystal,

$$U_t = (2.4 \times 10^4 \, \text{Exp}(-R/0.30) - 25.2/R) \text{ eV}$$

and are plotted in Fig. 21.7 for more insight and understanding.

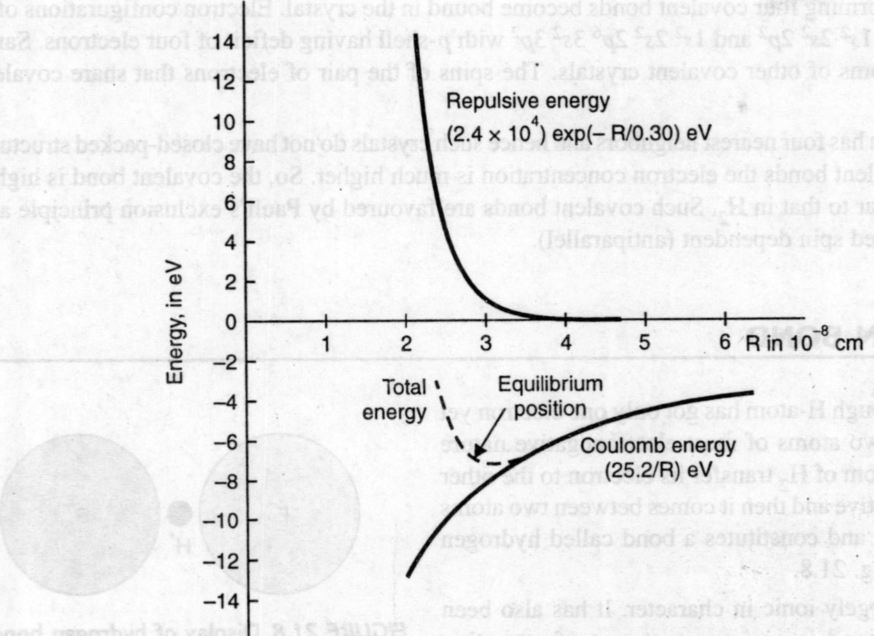

FIGURE 21.7 **Plot of repulsive energy, Madelung energy and total energy for the KCl crystal.**

We next give in Table 21.4 properties of some alkali halides with NaCl structure in order to have a feeling about how much predictions (calculated) agree with experimental values of lattice energies.

TABLE 21.4 Some properties of alkali halides

Alkali Halide	*Nearest Neighbor Separation*	*Repulsive Energy Parameters*		*Lattice Energy U, in kcal/mol*	
	R_0 *in Å*	$Z\lambda$*(in* 10^{-8} *ergs)*	ρ *in Å*	*Expt.*	*Theor.*
LiF	2.01	0.296	0.291	242.3	242.2
NaF	2.32	0.64	0.29	214.41	215.2
NaCl	2.82	1.05	0.32	182.6	178.6
KF	2.674	1.31	0.298	189.8	189.1
KCl	3.147	2.05	0.326	165.8	161.6
RbF	2.815	1.78	0.301	181.4	180.4
RbCl	3.291	3.19	0.323	159.3	155.4

21.5 COVALENT BINDING

The crystals of C, Si, Ge, Sn posses covalent bonds. In these the last shell of the constituent atoms runs short of four electrons. Therefore, these electrons form bonds with those of neighboring four atoms. These bonds are called covalent and the crystal as covalent crystals.

Each atom by forming four covalent bonds become bound in the crystal. Electron configurations of C and Si respectively are $1s^2\ 2s^2\ 2p^2$ and $1s^2\ 2s^2\ 2p^6\ 3s^2\ 3p^2$ with *p*-shell having deficit of four electrons. Same is the case with the atoms of other covalent crystals. The spins of the pair of electrons that share covalent bonds are antiparallel.

Thus every atom has four nearest neighbors and hence such crystals do not have closed-packed structure. Further, along the covalent bonds the electron concentration is much higher. So, the covalent bond is highly directional and is similar to that in H_2. Such covalent bonds are favoured by Pauli's exclusion principle and energy per bond is called spin dependent (antiparallel).

21.6 HYDROGEN BOND

It has been seen that though H-atom has got only one electron yet it forms a bond with two atoms of most electronegative nature like F, O and N. One atom of H_2 transfer its electron to the other and become electropositive and then it comes between two atoms of electronegative type and constitutes a bond called hydrogen bond as displayed in Fig. 21.8.

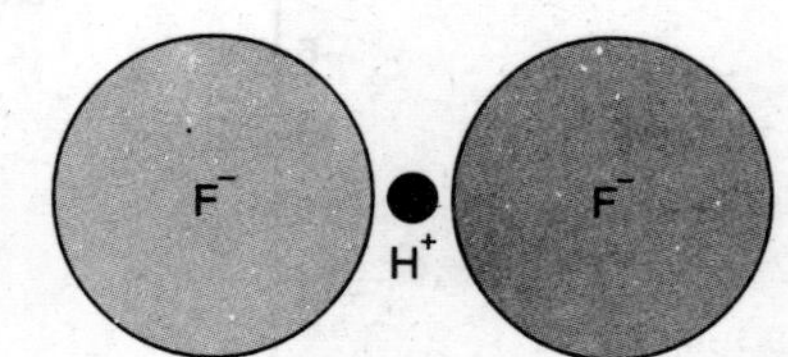

FIGURE 21.8 **Display of hydrogen bond, an extreme model of this bond.**

This bond is largely ionic in character. It has also been found that hydrogen bond is an important part of attraction between H_2O molecules and this leads to striking physical properties of water and ice.

21.7 METALLIC BOND

The elements placed in first three groups of periodic table along with certain other elements are usually referred to as metals. The metals are mainly characterized by their high electrical and thermal conductivities, opacity, surface luster and some other properties. These properties of metals are chiefly attributed to the availability of large number of free electrons in metal crystals.

The valence electron (or electrons) in metal atoms are loosely bound. When these atoms come closer to form a crystal all the valence electrons become free from their parent atoms. The free electrons, which can move around inside the crystal, act like electron gas. Hence, the metal crystal can be supposed to be ions embedded in the sea of electrons and the electron gas acts so as to hold the ions together. The free electrons in metals are also called as conduction electrons. Since the free electrons in metal crystals are not assigned to particular atoms therefore two metals having atoms of nearly same size can be combined to form alloys.

The total energy of metallic crystals, like all other solid crystals, is found to be lower than that of separated metal atoms. The reduction in energy is due to the following two factors. The valence electron in the metal crystal is nearer to one ion or another in comparison to isolated atom. Hence potential energy of electron is less in the crystal than in the isolated atom. Secondly, the kinetic energy of electron in the crystal increases which causes the lowering of potential energy. Metallic bonding occurs when the reduction in electron potential energy is balanced by the increase in electron kinetic energy. If atoms having more valence electrons tend to crystallize then increase in average kinetic energy of electrons is rather large causing an imbalance between drop in potential energy and increase in kinetic energy hence avoiding metallic bond. It is because of this reason the metallic elements are found mainly in first three groups of periodic table.

It may be mentioned here that the transition metals may have additional binding from inner electron shells. Transition metals and the metals immediately following them in the periodic table have large *d*-electron shells and are characterized by high binding energy. This may be caused in part by covalent bonding and in part by the van der Waals interaction of the cores. The typical examples are iron and tungsten, in these metals *d*-electrons make a substantial contribution to the binding energy.

Usually, metals tend to crystallize in relatively closed-packed structure like *hcp*, *fcc*, *bcc* and some other closely related structures and not in loosely packed structures such as diamond.

SOLVED NUMERICAL PROBLEMS

Problem 1. *If the potential energy function is expressed as*

$$U = -A/R^6 + B/R^{12} \text{ show that}$$

(*a*) *The intermolecular distance R_0 for which the potential energy is minimum is given by $(2B/A)^{1/6}$.*

(*b*) *The minimum potential energy is given by $U_{min.} = -A^2/4B$.*

Solution. (*a*) The condition for the minima in potential energy is

$$(dU/dR) = 0 \quad (\text{at } R = R_0)$$

$$\therefore \quad (6A/R^7 - 12B/R^{13}) = 0 \quad (\text{at } R = R_0)$$

or $$(6A/R_0^{\,7} - 12B/R_0^{\,13}) = 0$$

which gives $R_0 = (2B/A)^{1/6}$.

(*b*) The minimum potential energy can be obtained by inserting this value of R_0 in

$$U = -A/R^6 + B/R^{12}$$

$$\therefore \quad U_{min.} = -A(A/2B)[1 - (B/A)(A/2B)] = -(A^2/2B)[1 - \tfrac{1}{2}] = -(A^2/4B)$$

Problem 2. *Consider a line of 2N ions of alternating charge ± q with a repulsive potential energy A/R^n between nearest neighbors. Show that (a) at the equilibrium separation i.e., at $R = R_0$, $U(R_0) = -(2Nq^2 \log 2/R_0)(1 - 1/n)$.*

Solution. Let the cohesive energy be the sum of the repulsive and attractive potential as follows

$$U(R) = N(A/R^n - \alpha q^2/R) \qquad ...(1)$$

where $\alpha = 2 \log 2$ is the Madelung constant and n is the repulsive exponent. In equilibrium $R = R_0$ and $dU/dR = 0$

$$\therefore \quad dU/dR = N(-nA/R^{n+1} + \alpha q^2/R^2)$$

or $$N(-nA/R_0^{\,n+1} + \alpha q^2/R_0^{\,2}) = 0$$

or $$R_0^{\,n-1} = nA/\alpha q^2 \qquad ...(2)$$

Putting in eqn. (1) the value of A, we get

$$U(R_0) = -\alpha Nq^2/R_0(1 - 1/n) = -(2Nq^2 \log 2/R_0)(1 - 1/n).$$

Problem 3. *Calculate the cohesive energy per atom of Ne crystal given that R = 3.13 Å, σ = 2.74 Å and $\varepsilon = 50 \times 10^{-16}$ ergs.*

Solution. The cohesive energy of inert gas crystal is given by

$$U(R) = 2N\varepsilon[12.132(\sigma/R)^{12} - 14.454(\sigma/R)^6]$$

and per atom it would be

$$U(R)/N = 2\varepsilon[12.132(\sigma/R)^{12} - 14.454(\sigma/R)^6]$$

Here, R denotes the distance of nearest neighbors.

Now, $\sigma/R = 2.74/3.13$ and $(\sigma/R)^6 = 0.451$ and $(\sigma/R)^{12} = 0.203$

Next, $\varepsilon = 50 \times 10^{-16}$ ergs $= (50 \times 10^{-16})/(1.6 \times 10^{-12}) = (50/1.6) \times 10^{-4}$ eV

Therefore,

$$U(R)/N = 2 \times (50/1.6) \times 10^{-4}[12.132 \times 0.203 - 14.454 \times 0.451] = 0.025 \text{ eV}.$$

Problem 4. *Calculate cohesive energy per mole of Kr crystal from the data $R = 4.01$ Å, $\sigma = 3.65$ Å and $\varepsilon = 225 \times 10^{-16}$ ergs.*

Solution. $\sigma/R = 0.9102$; $(\sigma/R)^6 = 0.569$ and $(\sigma/R)^{12} = 0.323$

Therefore total energy per mol

$$U_{tot.}(R) = 2N\varepsilon[12.132 \times 0.323 - 14.454 \times 0.569]$$
$$= 2 \times 6.022 \times 10^{23} \times 225 \times 10^{-23}[3.923 - 8.224] \text{ Joule/mol}$$
$$= - 11.6 \text{ kJ/mol.}$$

So, cohesive energy/mol $= - U_{tot.}(R) = 11.6$ kJ/mol

Problem 5. *Calculate lattice energy of ionic crystal per mol from the data $R = 2.82$ Å, $Z\lambda = 1.05 \times 10^{-8}$ erg, and $\rho = 0.321$Å.*

Solution. Lattice energy is given by

$$U_{tot.}(R) = N[Z\lambda\, e^{-R/\rho} - \alpha q^2/R] \text{ where}$$

N = Avogadro's number

α = Madelung Constant = 1.748

and q = ionic charge = 1 esu

Now, $q^2 = 14.4$ eVÅ

Therefore, $\alpha q^2/R = 14.4 \times 1.748/2.82 \text{ eV} = 8.926$ eV

Also, $Z\lambda = 1.05 \times 10^{-8} \text{ erg} = (1.05/1.6) \times 10^4$ eV

and $e^{-R/\rho} = 1.53 \times 10^{-4}$

So, $Z\lambda\, e^{-R/\rho} = 1.05 \times 1.53/1.6 = 1.00$

Finally, $U_{tot.}(R) = N[1.00 - 8.926]$ eV

$$= - (6.022 \times 1.6 \times 10^4/4.186)\, [7.926] \text{ J/mol}$$
$$= - 6.022 \times 1.6 \times 10 \times 7.926/4.186 \text{ kJ/mol}$$
$$= - 182.4 \text{ kcal/mol}$$

Lattice energy $= - U_{tot.}(R) = 182.4$ kcal/mol.

EXERCISE

1. Explain the difference between
   (*a*) ionic and covalent type of binding
   (*b*) van der Waals and ionic type of binding.
2. Obtain the binding energy of an ionic crystal and derive the expression for Madelung constant. Calculate the value of this constant for a linear ionic lattice.
3. Show that the Madelung constant for a one-dimensional array of ions of alternating sign with a distance r between successive ions is equal to 2 log 2.
4. Explain in detail the hydrogen bond and the metallic bond.
5. Calculate cohesive energy per mol of Ar crystal given that $R = 3.76$ Å, $\sigma = 3.40$ Å and $\varepsilon = 167 \times 10^{-16}$ ergs.
   [Ans. – 8.6 kJ/mol]
6. Calculate the lattice energy per mol of KF crystal. Given that $R = 2.674$ Å, $Z\lambda = 1.31 \times 10^{-8}$ erg and $\rho = 0.298$ Å.
   [Ans. 192.7 kcal/mol**]**
7. Classify crystals on the basis of energy of bonding.
8. Discuss the main features of ionic and covalent solids.
9. Explain different types of bonding in solids.

Chapter 22

THERMAL PROPERTIES OF CRYSTALLINE SOLIDS

22.1 INTRODUCTION

In crystalline solids three types of vibrations are possible. These are longitudinal (one) and transverse (two) vibrations. The energy of these vibrations are quantized and propagates as phonons. The direction of propagation is expressed through **k**, the wave vectors. When the displacement of the atoms is along **k**, the vibrations are called longitudinal and when the displacement is perpendicular to **k** it is called transverse. Thus degree of freedom for vibrations of atoms is three, one of longitudinal and two of transverse. According to classical approach each degree of freedom is assigned thermal energy k_BT where $k_B = 1.38062 \times 10^{-23}$ J/K and T the temperature. Thus, if there are N atoms ($= 6.0222 \times 10^{23}$) per mol, the total thermal energy would be

$$U = 3Nk_BT$$

and heat capacity due to lattice vibrations at constant volume

$$C_{\text{Lat.}} = dU/dT = 3Nk_B = 3 \times 6.0222 \times 1.38062 = 24.943 \text{ J/mol/K}$$

This value is called the classical limit. However, the observed value is found as proportional to T^3(low temperature) which approaches classical limit at high temperature. Therefore, classical approach fails to explain the observed heat capacity of crystalline solids that varies with temperature as T^3.

22.2 QUANTUM MECHANICAL APPROACH

To understand quantum mechanical approach it is convenient to take the case of crystals with monoatomic basis. That is there is one atom in each primitive cell of the crystal. Also we shall assume that atom in each plane of the crystal vibrates in same phase, longitudinally or transversely (Figs. 22.1 and 22.2) and that the vibrations can be described by a single displacement say u_s, s denoting a particular plane.

We further assume that force on plane s due to a plane $s + p$ is proportional to the difference between their displacements *i.e.*, u_{s+p} and u_s. That is

$$F_s = c(u_{s+p} - u_s)$$

where c is the constant of proportionality. Since the main contribution would come from the immediate neighboring planes the force on s will be

$$F_s = c[(u_{s+1} - u_s) + (u_{s-1} - u_s)] = c[u_{s+1} + u_{s-1} - 2u_s]$$

Thus if M denote the atomic mass in that plane

$$M\, d^2u_s/dt^2 = c[u_{s+1} + u_{s-1} - 2u_s]$$

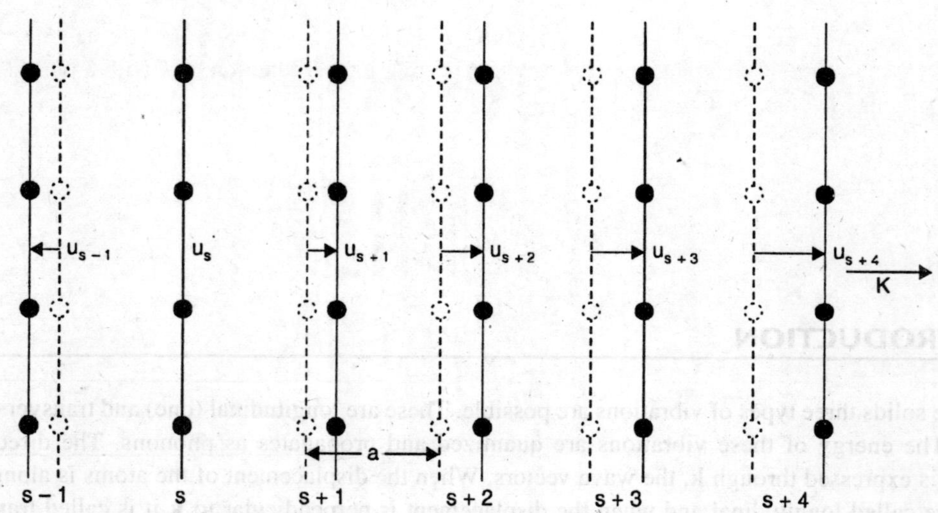

FIGURE 22.1 (Dashed lines) Planes of atoms when in equilibrium. (Solid lines) Planes of atoms when displaced as for a longitudinal wave. The coordinate *u* measures the displacement of the planes.

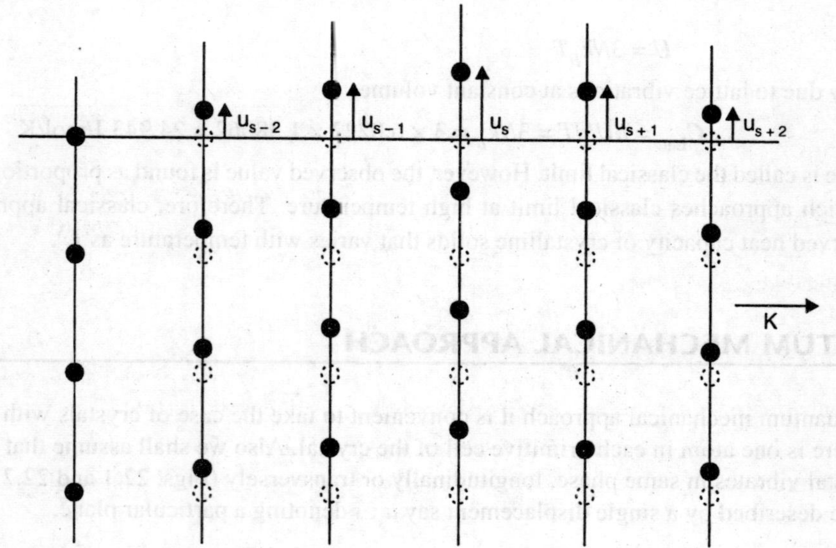

FIGURE 22.2 Planes of atoms as displaced during passage of a transverse wave.

Taking the time dependence of the wave as $e^{-i\omega t}$ such that $d^2u_s/dt^2 \longrightarrow -\omega^2 u_s$, the above equation becomes

$$-M\omega^2 u_s = c[u_{s+1} + u_{s-1} - 2u_s]$$

Now we look for a solution of the type $u_s = ue^{iksa}$ where k is the magnitude of the wave vector and '$\vec{a}$' the lattice constant (interatomic distance). So, we get

$$-M\omega^2 ue^{iksa} = cu[e^{ik(s+1)a} + e^{ik(s-1)a} - 2e^{iksa}]$$

On simplification we obtain

$$-M\omega^2 = c[e^{ika} + e^{-ika} - 2]$$

or $$\omega^2 = 2c/M\,[1 - \cos(ka)] \qquad ...(22.1)$$

Eqn. (22.1) is called the dispersion relation expressing ω as a function of k with a period $ka = 2\pi$. Thus 1st Brillouin zone has the boundaries at

$$k = \pm\,\pi/a$$

Eqn. (22.1) also gives us

$$\omega = (4c/M)^{1/2}\sin(ka/2)\,;$$

for long wavelength it comes out as $ka(c/M)^{1/2}$. That is, $\omega \propto k$.

22.3 GROUP VELOCITY

The waves mentioned above are called phonon elastic waves and k is the phonon wave vectors. The group velocity v_g is given by

$$v_g = d\omega/dk = (a/2)(4c/M)^{1/2}\cos(ka/2)$$
$$= (a^2c/M)^{1/2}\cos(ka/2)$$

This has been plotted in Fig. 22.3

Thus at the 1st Brillouin zone boundaries $k = \pm\,\pi/a$, group velocity vanishes and that $\omega/k = (a^2c/M)^{1/2}$, a constant quantity, when $ka << 1$ (for long wavelenghts)

So, long wave length limit is equivalent to the statement that velocity is constant *i.e.*, independent of frequency, ω. This statement is always valid whenever $\omega \propto k$.

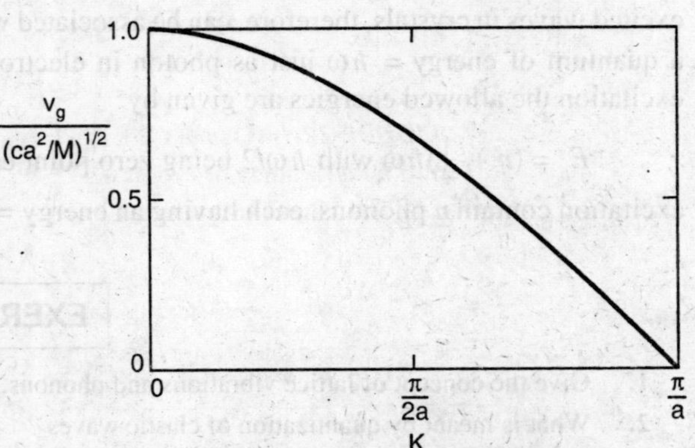

FIGURE 22.3 **Plot of group velocity $v_g/(a^2c/M)^{1/2}$ *vs k*.**

Significant Range of Wave Vector, k

The displacements of atoms in two successive planes of the crystal has the ratio

$$u_{s+1}/u_s = \text{Exp}(ika)$$

i.e., the phase difference is ka. Since, beyond the range $-\pi \le ka \le \pi$, Exp(ika) values will get repeated, the independent values of k are therefore

$$-\pi/a \le k \le \pi/a$$

Generalization of Dispersion Relation

In metals, however, the forces may extend beyond immediate neighboring planes. As a result,

$$F_s = \Sigma_{p>0}(u_{s+p} + u_{s-p} - 2u_s)$$

and proceeding in the same fashion as earlier on and replacing c by c_p we get

$$-M\omega^2 = \Sigma_p\, c_p[e^{ikpa} + e^{-ikpa} - 2]$$
$$\omega^2 = (2/M)\, \Sigma_p\, c_p[1 - \cos(kpa)]$$

As ω^2 depends on k we write

$$\omega_k^{\,2} = (2/M)\, \Sigma_p\, c_p[1 - \cos(kpa)]$$

Multiplying both sides by $\cos(kra)$, r is the integer, and integrating over k from $-\pi/a$ to π/a we obtain

$$\int_{-\pi/a}^{\pi/a} dk\omega_k^{\,2} \cos(kra) = (2/M)\, \Sigma_p\, c_p \int_{-\pi/a}^{\pi/a} dk(1 - \cos(kpa)) \cos(kra)$$
$$= (2/M)\, \Sigma_p\, c_p(-\pi/a)\, \delta_{pr} = (-2/M)\, c_r\, (\pi/a)$$

Thus, $$c_r = -(Ma/2\pi) \int_{-\pi/a}^{\pi/a} dk\omega_k^{\,2} \cos(kra)$$

This gives us a method to calculate force constants from experiments.

22.4 QUANTIZATION OF ELASTIC WAVES

We have seen earlier that lattice planes execute vibrations just like simple harmonic oscillators. These thermally excited waves in crystals, therefore, can be associated with quantum of energies called phonons each carrying a quantum of energy = $\hbar\omega$ just as photon in electromagnetic waves. Further, for each mode of thermal excitation the allowed energies are given by

$E_n = (n + \frac{1}{2})\hbar\omega$ with $\hbar\omega/2$ being zero point energy and n as integer. We also say that this mode of excitation contain n phonons, each having an energy = $\hbar\omega$.

EXERCISE

1. Give the concept of lattice vibrations and phonons.
2. What is meant by quantization of elastic waves ?
3. What do you mean by group velocity ? Derive its expression for lattice vibrations and deduce its value for long wavelengths.

Chapter 23 HEAT CAPACITY

23.1 PHONONS CONTRIBUTION TO HEAT CAPACITY

Thermal vibrations are set in the solids. These vibrations are in longitudinal (one) and transverse (two). So, these vibrations have three degree of freedom which are assigned to the associated phonons. In other words the energy of the thermal vibrations is propagated through phonons each carrying an amount equal to $\hbar\,\omega_{k,p}$, p denoting the polarization and k the wave vector. If we know the number of phonons for each (k, p) we can write total energy of the material and then its specific heat at constant volume.

The classical estimate is calculated as follows

N = Number of atoms

T = Temperature

k_B = Boltzmann constant

since there are three types of waves, one longitudinal and two types of transverse *i.e.*, 3 degree of freedom and the energy with each is k_BT per atom the total internal energy would be

$$U = 3Nk_BT$$

As a result the heat capacity

$$C_{\text{Lat}} = dU/dT = 3Nk_B = 3 \times 6.0222 \times 10^{23} \times 1.38062 \times 10^{-23} \text{ J/mol/K.}$$

$$= 24.943 \text{ J/mol/K.}$$

However, observed value is seen as dependent on temperature and varies as T^3 at low temperature while it approaches the above mentioned classical limit = 24.943 J/mol/K at high temperature. This is also known as the Dulong and Petit law.

23.2 QUANTUM MECHANICAL APPROACH

Let there be $n_{k,p}$ number of occupancy (phonons) of energy $\hbar\omega_{k,p}$ at temperature T K. The total energy would be

$$U = \Sigma_{k,p}\, U_{k,p} = \Sigma_{k,p}\, n_{k,p}\hbar\omega_{k,p}\ ;$$

k and p denote the wave vector and polarization respectively. The phonon number $n_{k,p}$ follows Einstein distribution law as

$$n_{k,p} = 1/[\text{Exp}(\hbar\omega_{k,p}/k_B T) - 1]\ ;\ \hbar = 1.0546 \times 10^{-34}\ Js$$

and $k_B = 1.3806 \times 10^{-23}\ J/K$

Therefore, $U = \Sigma_{k,p}\, \hbar\omega_{k,p}/[\text{Exp}(\hbar\omega_{k,p}/\tau) - 1]$ where $\tau = k_B T$.

It is usual to replace summation over k-values by integration by introducing a density function $D_p(\omega)$ such that $D_p(\omega)d\omega$ denote the no. of modes in the range ω to $\omega + d\omega$ of given polarization. As a result

$$U = \Sigma_p \int D_p(\omega)d\omega\ \hbar\omega/[\text{Exp}(\hbar\omega/\tau) - 1]$$

and

$$C_{\text{Lat}} = dU/dT = \Sigma_p \int D_p(\omega)d\omega\ \hbar\omega/[\text{Exp}(\hbar\omega/\tau) - 1]^2\ (\hbar\omega/k_B T^2)\ \text{Exp}(\hbar\omega/\tau)$$

$$= \Sigma_p\, k_B \int D_p(\omega)d\omega\ \{(\hbar\omega/\tau)^2\ \text{Exp}(\hbar\omega/\tau)\}/[\text{Exp}(\hbar\omega/\tau) - 1]^2$$

C_{Lat} denotes the lattice heat capacity of the material.

Putting $\hbar\omega/\tau = x$, we get for lattice heat capacity

$$C_{\text{Lat}} = \Sigma_p\, k_B \int D_p(\omega)d\omega\ \{(x)^2\ \text{Exp}(x)\}/[\text{Exp}(x) - 1]^2 \quad ...(23.1)$$

Thus, the calculation of C_{Lat} involves the knowledge of density function $D(\omega)$ for each polarization. We now discuss the method to obtain $D(\omega)$.

One Dimensional Case

Let us suppose that atoms are arranged in a line with 'a' distance apart as

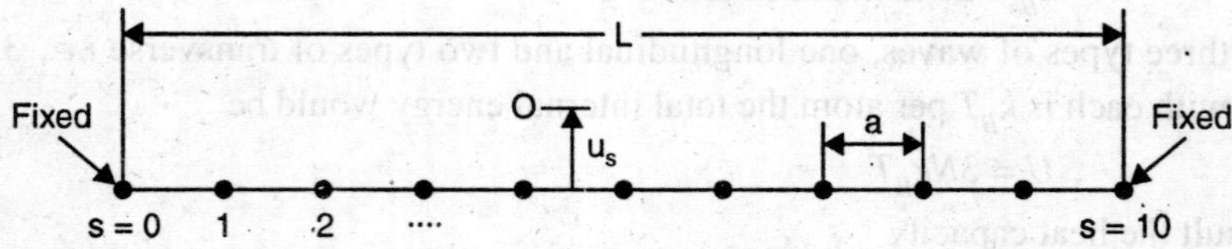

FIGURE 23.1 **Linear lattice with atoms at the sites s = 0, 1, 2, 3...N = 10**

If we set up standing waves along this lattice with wave vectors k, the form of this wave would be

$u_s = u \sin(ska)\ \text{Exp}(-i\omega_k t)$ assuming same frequency ω_k for each polarization

This function has to satisfy

$$u_s = 0 \text{ for } s = 0 \text{ as well as } s = N$$

if the first and the last atoms are supposed as fixed. This requires that

$$\sin(Nka) = 0$$

i.e., $Nka = n\pi$ where n is a positive integer.

or $\quad k = n\pi/Na = n\pi/L \qquad [\because \; L = Na$, the extension of the lattice]

So, possible values of k are 0, π/L, $2\pi/L$, $3\pi/L \ldots (N-1)\,\pi/L$, $N\pi/L$

But, for $k = 0$ and $N\pi/L$ the atoms would not have any wave motion (or non-zero wave function). Hence, ignoring these two values we are left with the following values for k,

$$k = \pi/L,\ 2\pi/L,\ 3\pi/L \ldots (N-1)\,\pi/L \text{ a total of } (N-1) \text{ values.}$$

This is equal to the number of mobile atoms as $s = 0$ and N atoms are being kept as fixed. This can also be put as that there is one mode per unit cell of the linear lattice.

The above mentioned solution $u_s = u \sin(ska)\,\text{Exp}(-i\omega_k t)$ represent standing waves. However, we can work equally well with travelling periodic waves of the form

$$u_s = u\,\text{Exp}\,[+\,i(ska - \omega_k t)]$$

Let us suppose that these are periodic with a period L such that

$$u_s = u_{s+N}, \text{ where } L = Na$$

i.e., $\qquad e^{iska} = e^{i(s+N)ka}$

or $\quad Nka = 2\pi n$ with n an integer positive or negative

Thus, possible values of k are

$$k = 0,\ \pm\,2\pi/L,\ \pm\,4\pi/L,\ \ldots\ \pm\,N\pi/L$$

assuming N as even. We have terminated the values at $\pm\,N\pi/L = \pm\,\pi/a$ as k is to satisfy the inequality

$$-\,\pi/a \le k \le \pi/a \text{ (Explained earlier)}$$

Again these $N + 1$ values equal to the number of mobile atoms. So, in each case there is one mode of k per atom or unit cell.

Three Dimensional Case

The above treatment can be easily extended for the case of actual crystal *i.e.*, to three dimensional lattice. Then the travelling periodic waves will be of the form

$$u_k(\mathbf{r}) = u\,\text{Exp}[i(\mathbf{k}.\mathbf{r} - \omega_k t)]$$

such that Exp $[i(\mathbf{k}.\mathbf{r})]$ is periodic in $\mathbf{r}$

We have considered a cube of sides L containing N atoms in all.

This requires that

$$k_x, k_y, k_z = 0,\ \pm\,2\pi/L,\ \pm\,4\pi/L,\ \ldots\ \pm\,N\pi/L$$

Again we have $Na = L$, 'a' being the lattice constant. It is quite obvious that in **k**-space there is only one distinct set of (k_x, k_y, k_z) in the volume $(2\pi/L)^3$. That is there are $(L/2\pi)^3$ number of possible modes of **k** per unit volume of **k**-space. If we have a sphere of radius say k in this space, it will contain

$(L/2\pi)^3\, 4\pi/3\, k^3$ number of possible modes. As in the case of linear lattice this number should be equal to the number of mobile atoms say N so that

$$N = (L/2\pi)^3\, 4\pi/3\, k^3 = (V/6\pi^2)k^3$$

or $\qquad k = (6\pi^2\, N/V)^{1/3}\ ;\ V = L^3 \qquad$ ...(23.2)

Further, by definition

$$D(\omega) = dN/d\omega = (dN/dk)(dk/d\omega) = (V/2\pi^2)\ k^2(dk/d\omega)$$

In other words if we know the dispersion relation of ω in terms of k one can calculate $d\omega/ak$ and hence the density of states of function $D(\omega)$ from above. Two approaches are available to obtain $d\omega/dk$. These are Debye model and Einstein model which we discuss next.

23.3 DEBYE MODEL FOR THE HEAT CAPACITY

In Debye's approach the velocity of elastic waves in solid is to be taken as constant and also same for each polarization. That is

$\omega = kv$ which gives us

$d\omega/dk = v$; $\omega_D = k_D v$ where ω_D and k_D denote the upper limits,

and $D(\omega) = (V/2\pi^2)\, k^2\, 1/v = (V/2\pi^2)\, \omega^2/v^3$; v = velocity

Hence, $C_{\text{Lat}} = \Sigma_p\, k_B \int D(\omega)d\omega\, \{(x)^2 \text{Exp}(x)\}/[\text{Exp}(x) - 1]^2$ [from eqn. (23.1)]

where $x = \hbar\omega/k_BT$

Putting the value of $D(\omega)$ we get

$$C_{\text{Lat}} = \Sigma_p\, k_B (V/2\pi^2)\, 1/v^3 \int_0^{\omega_D} d\omega\, \omega^2\, \{(x)^2 \text{Exp}(x)\}/[\text{Exp}(x) - 1]^2$$

As there are three polarization and v is same for each we get

$$C_{\text{Lat}} = 3k_B (V/2\pi^2)\, (k_BT/\hbar v)^3 \int_0^{x_D} dx\, \{(x)^4 \text{Exp}(x)\}/[\text{Exp}(x) - 1]^2$$

where $x_D = \hbar\omega_D/k_BT = \theta/T$. The constant $\theta = \hbar\omega_D/k_B = \dfrac{\hbar v}{k_B k_D}$ is known as Debye temperature.

We further obtain

$$C_{\text{Lat}} = 3k_B (V/2\pi^2)\, (T/\theta)^3\, k_D{}^3 \int_0^{x_D} dx\, \{(x)^4 \text{Exp}(x)\}/[\text{Exp}(x) - 1]^2$$

[From eqn. (23.2)]

However, $k_D = (6\pi^2 N/V)^{1/3}$

Therefore, $C_{\text{Lat}} = 3k_B\, (V/2\pi^2)\, (T/\theta)^3\, (6\pi^2 N/V) \int_0^{x_D} dx\, \{(x)^4 \text{Exp}(x)\}/[\text{Exp}(x) - 1]^2$

$$= 9Nk_B\, (T/\theta)^3 \int_0^{x_D} dx\, \{(x)^4 \text{Exp}(x)\}/[\text{Exp}(x) - 1]^2 \quad ...(23.3)$$

Special Cases

(*a*) **High Temperature Limit.** As T is large, x will be small and

$e^x \to 1 + x$. As a result

$$\int_0^{x_D} dx\, \{(x)^4 \text{Exp}(x)\}/[\text{Exp}(x) - 1]^2 \;... \to \int_0^{x_D} x^2 dx\, ;\, x_D = (\theta/T)$$

$$= x_D{}^3/3 = 1/3\, (\theta/T)^3$$

Hence, $C_{\text{Lat}} = 3Nk_B = 24.943$ J/mol/K. [using eqn. (23.3)]

So, heat capacity approaches classical value.

(*b*) **Low Temperature Limit.** At low temperature x $(= \hbar\omega/k_BT)$ will be large and the upper limit of the integration x_D could be taken as ∞. That is

$\int_0^{x_D} dx\, \{(x)^4 \text{Exp}(x)\}/[\text{Exp}(x) - 1]^2 \;... \to \int_0^{\infty} x^4 \text{Exp}(-x)dx = 24$, obtained through integration by parts.

Hence, $C_{\text{Lat}} \to 216\, Nk_B(T/\theta)^3 \propto T^3$

[**Note.** More accurate calculations give $C_{\text{Lat}} = 234\, Nk_B(T/\theta)^3$]

This is well known Debye's T^3 law. In Fig. 23.2, we also show the plot of C_{Lat} *vs* T/θ.

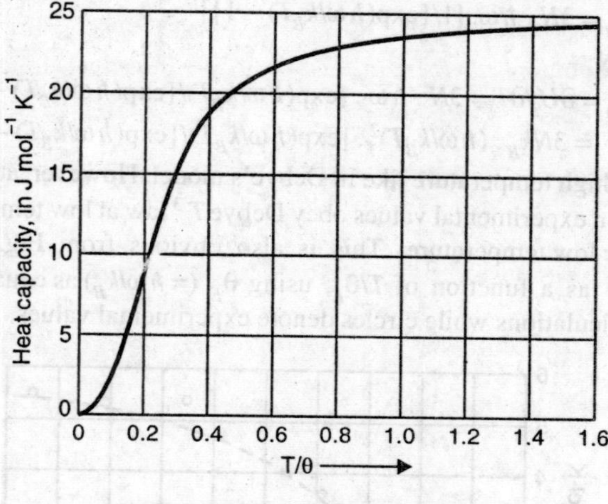

FIGURE 23.2 **Plot of C_{Lat} *vs* (T/θ).**

The observed values (experimental) of C_{Lat} for Ge and Si follow the same trend as predicted theoretically above and shown in Fig. 23.2. However, at low temperature the agreement between observed values and the Debye T^3 law has been found as excellent and has been shown in Fig. 23.3.

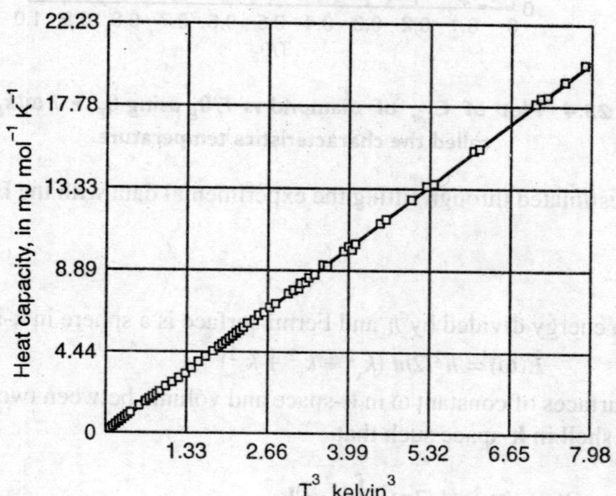

FIGURE 23.3 **Variation of heat capacity with T^3 at low temperature ; soild line is theoretical while circles denote experimental values.**

23.4 EINSTEIN MODEL FOR HEAT CAPACITY

In Einstein model one assumes N identical oscillators each oscillating with well defined frequency ω. Then total energy of these oscillating atoms at temperature T would be given by

$$U = N \,.\, \hbar\omega.\, [1/\{\exp(\hbar\omega/k_BT) - 1\}]$$

as $$D(\omega) = N\,\delta(\omega' - \omega) \text{ and } \int \delta(x-a)\, f(x)\, dx = f(a)$$

When there are 3 modes of vibrations this energy would become 3 times. That is

$$U_t = 3N \,.\, \hbar\omega.\, [1/\{\exp(\hbar\omega/k_BT) - 1\}]$$

and the lattice heat capacity

$$C_{\text{Lat}} = \partial U_t/\partial T = 3N \,.\, \hbar\omega \,.\, [\exp(\hbar\omega/k_BT)/\{\exp(\hbar\omega/k_BT) - 1\}^2]\ \hbar\omega/k_BT^2$$
$$= 3Nk_B \,.\, (\hbar\omega/k_BT)^2 \,.\, [\exp(\hbar\omega/k_BT)/\{\exp(\hbar\omega/k_BT) - 1\}^2]$$

which approaches $3Nk_B$ at high temperature like in Debye's model. However, at low temperature it will vary as $\exp(-\hbar\omega/k_BT)$. However, experimental values obey Debye T^3 law at low temperature. Therefore, Einstein model is not applicable for low temperature. This is also cbvious from Fig. 23.4 where heat capacity of diamond has been plotted as a function of T/θ_E, using θ_E $(= \hbar\omega/k_B)$ as equal to 1320 K. The dashed line curve is Einstein model calculations while circles denote experimental values.

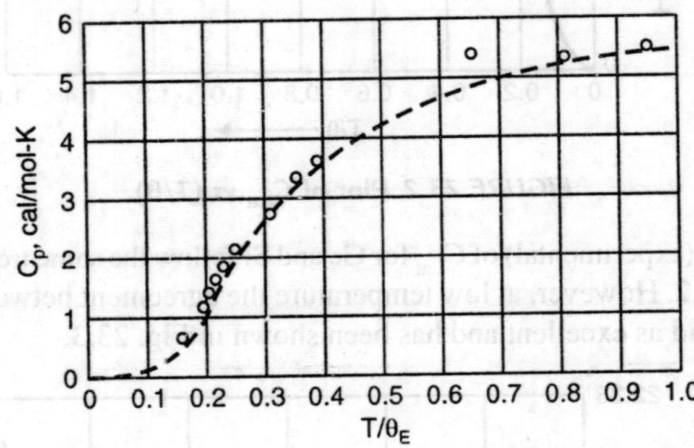

FIGURE 23.4 **Plot of C_{Lat} of diamond *vs* T/θ_E using $\theta_E = \hbar\omega/k_B$ = 1320 K, called the characteristics temperature.**

The value θ_E is estimated through fitting the experimental data with the Einstein expression of C_{Lat}.

General Case

Frequency ω is equal to energy divided by $\hbar$ and Fermi surface is a sphere in **k**-space as

$$E(\omega) = \hbar^2/2m\,(k_x^2 + k_y^2 + k_z^2)$$

Thus we have surfaces of constant ω in **k**-space and volume between two surfaces having frequency ω and $\omega + d\omega$ will be a shell in **k**-space such that

$$D(\omega)\, d\omega = (L/2\pi)^3 \int_{\text{shell}} d^3\mathbf{k}$$

because the number of possible states of **k** in unit volume of **k**-space is $(L/2\pi)^3$. The integral gives the volume of the shell. If we know that dispersion relation $\omega(\mathbf{k})$, we can obtain the value of integral and hence of $D(\omega)$ the density of states. For example, if $dk_\perp$ is normal to two surfaces ω and $\omega + d\omega$ in **k**-space

$$\int_{\text{shell}} d^3\mathbf{k} = \int dS_\omega dk_\perp$$

and $$D(\omega)\, d\omega = (L/2\pi)^3 \int dS_\omega dk_\perp$$

$$= V/(2\pi)^3 \int dS_\omega dk_\perp ; \quad L^3 = V, \text{ volume of the crystal.}$$

On the surface with $\omega(\mathbf{k})$ = constant

$$d\omega(\mathbf{k}) = 0 = \nabla_k\omega \cdot d\mathbf{k}$$

That is $\nabla_k\omega$ is perpendicular to the surface. Hence for two surfaces ω and $\omega + d\omega$, we have

$$d\omega = |\nabla_k\omega|\, dk_\perp$$

As a result $\quad D(\omega)\, d\omega = V/(2\pi)^3 \int dS_\omega\, d\omega/|\nabla_k\omega|$

cancelling $d\omega$ from both sides as integration is over the area of shell in **k**-space with ω constant.

$$D(\omega) = V/(2\pi)^3 \int dS_\omega/|\nabla_k\omega| = V/(2\pi)^3 \int dS_\omega/v_g$$

where v_g is the group velocity by definition.

APPENDIX

Here, we calculate the average energy of the oscillator, which can have energies equal to integral multiple of $\hbar\omega$ where ω is the characteristics frequency. We also assume Boltzmann distribution function $\exp(-E/k_BT)$ for the energy E at temperature T, $k_B = 1.38062 \times 10^{-23}$ J/K being the Boltzmann constant. Thus, the average energy of the oscillator would be

$$E = [\Sigma_n E_n \exp(-E_n/k_BT)]/[\Sigma_n \exp(-E_n/k_BT)]$$
$$= [\Sigma_n n\hbar\omega \exp(-n\hbar\omega/k_BT)]/[\Sigma_n \exp(-n\hbar\omega/k_BT)] \;;\; E_n = n\hbar\omega$$
$$= k_BT\,[\Sigma_n nx \exp(-nx)]/[\Sigma_n \exp(-nx)] \;;\; x = \hbar\omega/k_BT$$
$$= k_BT\,x\,[(e^{-x} + 2e^{-2x} + 3e^{-3x} + \ldots)/(1 + e^{-x} + e^{-2x} + e^{-3x} + \ldots)]$$
$$= -k_BT \cdot x \cdot d[\ln(1 + e^{-x} + e^{-2x} + e^{-3x} + \ldots)]/dx$$
$$= -k_BT \cdot x \cdot d[\ln(1/(1 - e^{-x}))]/dx$$

using the formula for summing the G.P. series.

$$= k_BT \cdot x \cdot d[\ln(1 - e^{-x})]/dx$$
$$= k_BT \cdot x \cdot [e^{-x}/(1 - e^{-x})]$$
$$= k_BT \cdot x \cdot [1/(e^{-x} - 1)]$$
$$E = \hbar\omega/[\exp(\hbar\omega/k_BT) - 1]$$

For the case of three-dimensional oscillators, the average energy would be

$$U = 3N\hbar\omega/[\exp(\hbar\omega/k_BT) - 1]$$
$$= 3N\,k_BT\;(\hbar\omega/k_BT)/[\exp(\hbar\omega/k_BT) - 1]$$

Heat capacity,

$$C_{\text{Lat}} = \partial U/\partial T = 3Nk_B \exp\left(\frac{\hbar\omega}{k_BT}\right)\;(\hbar\omega/k_BT)^2/[\exp(\hbar\omega/k_BT) - 1]^2$$

→ $3Nk_B$ for high temperature T

→ varies as $\exp(-\hbar\omega/k_BT)$ at low temperature T.

EXERCISE

1. **Explain how lattice vibrations in solid give rise to specific heat of solids.**
2. **Discuss Debye model for the density of phonon states and calculate the expression of heat capacity.**
3. **What is Debye – T^3 law ? Derive this law.**
4. **Discuss Einstein model for the calculation of heat capacity of solids. What is Dulong-Petit value ?**
5. **Distinguish between Debye's model and Einstein's model.**

Chapter 24

FREE ELECTRON THEORY OF METALS

24.1 INTRODUCTION

It was known long back that the free electron theory could explain many physical properties of metals. The classical approach used in such a model though had initial success but it could not explain a large mean free path of conduction electrons. It also could not explain the specific heat as well as the magnetic susceptibility as observed in metals. Therefore, quantum mechanical approach was developed for the motion of almost free electrons moving around in the metals. In fact, one can easily understand the undeflected motion of electrons provided these have periodic wave motion. Also, using Fermi-Dirac statistics and Pauli exclusion principle for many electrons system one can easily explain specific heat as well as magnetic susceptibility. We now discuss this free electron theory by first assuming one dimensional motion and then 3-dimensional motion of electrons.

24.2 ONE-DIMENSIONAL FREE ELECTRON MOTION

As the metal (or solid) is formed, the valence electrons of atoms becomes free. That is N monovalent atoms will provide N electrons. These not only provide stability to the positive ions but also are responsible for the physical properties of metals. In alkali metals every atom has one valence electron as their outermost orbit has ns^1 configuration where $n = 2, 3, 4, \ldots$. For example, electronic configuration of Li is $1s^2\ 2s^1$ of Na is $1s^2\ 2s^2\ 2p^6\ 3s^1$ and of K is $1s^2\ 2s^2\ 2p^6\ 3s^2\ 3p^6\ 4s^1$. These valence electrons form a sea in which the ions arrange themselves as regular and periodic arrays as shown in Fig. 24.1.

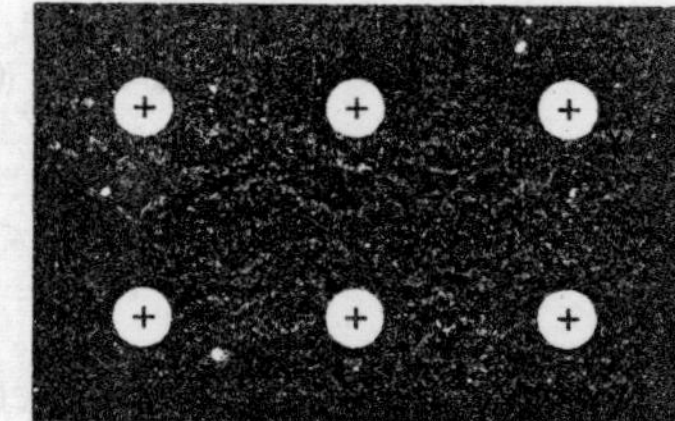

FIGURE 24.1. **Schematic model of a crystal of sodium metal.**

The sea of electrons, obeying Pauli exclusion principle constitute a gas called electron fermi gas. Since any quantum state of the electron is designated by a set of quantum numbers, the Pauli exclusion principle would require that no two electrons be in the same state *i.e.*, the sets of quantum numbers of states of pair of electrons would not be same or equal. At least one of the sets of quantum numbers will be different. For example, if the energy quantum number of two electrons is same their spin quantum number m_s will be different. If one has $m_s = +1/2$, the other electron will be in the state of $m_s = -1/2$.

We now discuss the system of free electrons moving in one-dimensional space, that is say along x-axis. Quantum mechanically this motion is described as

$$H\psi(x) = E\psi(x) \quad ...(24.1)$$

where H is the Hamiltonian $= p^2/2m = -(\hbar^2/2m)d^2/dx^2$

So,
$$-(\hbar^2/2m)d^2\{\psi(x)\}/dx^2 = E\psi(x) \quad ...(24.2)$$
$$d^2\{\psi(x)\}/dx^2 + k^2\psi(x) = 0 \quad ...(24.3)$$

where $k^2 = (2mE/\hbar^2)$ ...(24.4)

m = mass of electron, $\hbar = h/2\pi$, k is also called the magnitude of wave vector such that $k = 2\pi/\lambda$

Equation (24.3) has solutions of the type sin(kx) and cos(kx).

A general solution of eqn.(24.3) will be

$$\psi(x) = A\sin(kx) + B\cos(kx)$$

It has to satisfy boundary conditions such as

$\psi(x) = 0$ for $x = 0$ as well as L, the linear size of the crystal

which means $B = 0$ and that

$$A\sin(kL) = 0$$

i.e. $kL = \pi n$, where n is any integer.

More explicitly we write

$$k_n L = \pi n \quad \text{or} \quad k_n = \pi n/L \quad ...(24.5)$$

Corresponding to these k-values, the solutions will be

$$\psi_n(x) = A\sin(k_n x) = A\sin(\pi n x/L) \quad ...(24.6)$$

If $\psi_n(x)$ is to be normalized, we must have

$$\int_0^L |\psi_n(x)|^2\, dx = 1$$

i.e.
$$A^2 \int_0^L dx\, \sin^2(\pi n x/L) = 1$$

or
$$A^2/2 \int_0^L (1 - \cos(2\pi n x/L)dx = 1$$
$$A^2(L/2) = 1$$

So,
$$A = \sqrt{(2/L)} \quad \text{and} \quad \psi_n(x) = \sqrt{(2/L)}\, \sin(\pi n x/L)$$

We are mainly concerned with the electron energy levels and filling of these levels with electrons. Obviously possible energies are (see eqn. (24.4))

$$k_n^2 = (2mE_n/\hbar^2)$$

i.e.
$$E_n = (\hbar^2 k_n^2)/(2m) = \hbar^2/(2m)\,(\pi n/L)^2 \quad \text{[From eqn. (24.5)]}$$

or
$$E_n/\{(\pi^2\hbar^2)/(2mL^2)\} = n^2$$

By Pauli exclusion principle we can fill two electrons in each level, one with $m_s = +1/2$ (spin up) and with $m_s = -1/2$ (spin down). This way all energy levels E_n may be filled by such pairs. In other words N electrons would require $N/2$ energy levels for the ground state of the system of N-electrons. Fig. 24.2 show energy levels E_n in the unit $(\pi^2\hbar^2)/(2mL^2)$ and the wave function $\psi_n(x)$ in arbitrary units. The wavelength of the functions $\psi_n(x)$ in terms of L has also been shown there.

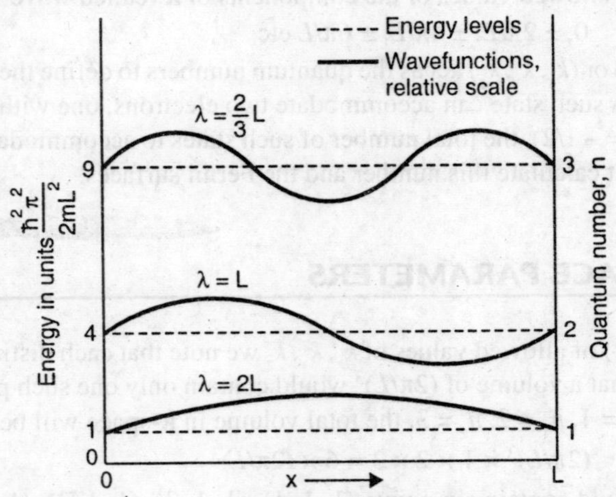

FIGURE 24.2 **First three energy levels and wave function of a free electron.**

When N is even, the number of lowest levels to accommodate these electrons will be $N/2 = n$. The topmost level is usually written as n_F, the Fermi level upto which the electrons get filled at absolute zero temperature. The energy of this Fermi level will be given by

$$E_F = \hbar^2/(2m)\,(\pi n_F/L)^2 = \hbar^2/(2m)\,(\pi N/2L)^2 \qquad ...(24.7)$$

This is known as Fermi energy of the system of N-electrons.

24.3 THREE-DIMENSIONAL FREE ELECTRON MOTION

In actual case, the conduction electrons in metals have got three dimensional motion and these constitute free electron gas called Fermi gas. The three dimensional time independent Schrodinger equation is

$$H\psi(\mathbf{r}) = E\psi(\mathbf{r})$$

$$(1/2m)(p_x^2 + p_y^2 + p_z^2)\,\psi(\mathbf{r}) = E\psi(\mathbf{r})$$

As $\quad p_x = -i\hbar\,\partial/\partial x,\ p_y = -i\hbar\,\partial/\partial y \quad$ and $\quad p_z = -i\hbar\,\partial/\partial z$

and $\quad \nabla^2 = \partial^2/\partial x^2 + \partial^2/\partial y^2 + \partial^2/\partial z^2$

we have $\quad -(\hbar^2/2m)\,\nabla^2\psi(\mathbf{r}) = E\psi(\mathbf{r}) \qquad ...(24.8)$

or $\quad \nabla^2\psi(\mathbf{r}) = -k^2\psi(\mathbf{r}) \qquad ...(24.9)$

where $\quad k^2 = (2mE/\hbar^2) = k_x^2 + k_y^2 + k_z^2 \qquad ...(24.10)$

If we require periodic travelling waves to be associated with these free electrons, the obvious solution would of the form

$$\psi(\mathbf{r}) = A\,e^{i\mathbf{k}.\mathbf{r}} \qquad ...(24.11)$$

This satisfies eqn. (24.8) as can be seen by substituting this in the equation. For this to be periodic in x, y, z co-ordinates with a period L, we must have

$$\psi(x + L, y, z) = \psi(x, y, z)$$

i.e. $e^{ik_x(x+L)} = e^{ik_x x}$

or $k_x L = 2\pi n_x$

where n_x is any integer. Similarly

$$k_y L = 2\pi n_y \quad \text{and} \quad k_z L = 2\pi n_z$$

In other words the allowed values of the components of **k** (called wave vector) would be

$$0, \pm 2\pi/L, \pm 4\pi/L, \pm 6\pi/L \text{ etc.} \qquad ...(24.12)$$

The sets (n_x, n_y, n_z) or (k_x, k_y, k_z) act as the quantum numbers to define the states $\psi(\mathbf{r})$ *i.e* distinct states are $\psi n_x n_y n_z(\mathbf{r})$. Since each such state can accommodate two electrons, one with spin up ($m_s = +1/2$) and the other with spin down ($m_s = -1/2$), the total number of such states to accommodate N electrons would be half of this number. Let us next calculate this number and the Fermi surfaces.

24.4 FERMI SURFACE PARAMETERS

From the condition (24.12) of allowed values of k_x, k_y, k_z we note that each distinct set (k_x, k_y, k_z) would be a point in the **k**-space and that a volume of $(2\pi/L)^3$ would contain only one such point and hence one state (or orbital). Further, when $n_x = 1$, $n_y = 2$, $n_z = 3$, the total volume in **k**-space will be

$$(2\pi/L)^3 \times 1 \times 2 \times 3 = 6 \times (2\pi/L)^3$$

Such a volume would contain six points (1, 1, 1), (1, 1, 2), (1, 1, 3), (1, 2, 1), (1, 2, 2) and (1, 2, 3) proving thereby that in a volume of $(2\pi/L)^3$ there will be only one state.

Next, the energy is given by

$$E = (\hbar^2/2m)\,(k_x^2 + k_y^2 + k_z^2) = (\hbar^2 k^2/2m)$$

i.e. $k^2 = (2m/\hbar^2)E.$

So, we can draw a sphere of radius k in **k**-space whose volume will be

$$(4\pi/3)k^3 = (4\pi/3)\,(2mE/\hbar^2)^{3/2}$$

The number of states in this volume will be $(4\pi/3)k^3/(2\pi/L)^3$

Since each state accommodates two electrons, we must have

$$N = 2 \times (4\pi/3)k^3/(2\pi/L)^3 = (L^3k^3)/(3\pi^2) = (Vk^3)/(3\pi^2)$$

as total number of electronic states

and $k = (3\pi^2 N/V)^{1/3}$, where $V = L^3$ denote the volume of the periodic cell.

At absolute zero temperature, all states are filled by N electrons and the radius of the sphere in **k**-space just enclosing these states is called the radius of the Fermi sphere. Let it be k_F. Then

$$k_F = (3\pi^2 N/V)^{1/3} \qquad ...(24.13)$$

and corresponding energy

$$E_F = \text{Fermi energy} = (\hbar^2/2m)\,(3\pi^2 N/V)^{2/3} \qquad ...(24.14)$$

and the velocity

$$v_F = \hbar k_F/m \qquad ...(24.15)$$

Usually V is taken as unity and $\frac{N}{V} = c_e$ as the electron concentration in the metals. As a result, we can easily calculate k_F, v_F, and E_F through eqns.(24.13) – (24.15) *i.e.* using

$$k_F = (3\pi^2 c_e)^{1/3}$$

$$E_F = (\hbar^2/2m)k_F^2 \quad \text{and} \quad v_F = \hbar k_F/m = (10.546/9.1)k_F \text{ cm s}^{-1}$$

(with $m = 9.1 \times 10^{-31}$ kg and $\hbar = 1.0546 \times 10^{-34}$ J-sec)

These calculated values for certain metals have been given in Table 24.1.

TABLE 24.1 Calculated values of Fermi surface parameters

Metals	c_e *(in 10^{22} cm^{-3})*	k_F *(in 10^8 cm^{-1})*	v_F *(in 10^8 cm/s)*	E_F *(in eV)*
Li	4.70	1.116	1.29	4.75
Na	2.65	0.922	1.07	3.24
K	1.40	0.746	0.86	2.12
Cu	8.45	1.358	1.57	7.03
Au	5.90	1.204	1.395	5.53
Ca	4.60	1.108	1.28	4.68
Cd	9.28	1.401	1.62	7.49
Al	18.06	1.749	2.027	11.67

[**Note :** 1 eV = 1.602×10^{-12} erg]

Lastly, let us find the density of electron states. Consider eqn.(24.14) again to write

$$N = \frac{V}{3\pi^2}\,(2mE/\hbar^2)^{3/2}$$

That is, $\ln(N) = 3/2 \ln(E) + \text{constant}$

and $dN/N = 3/2\ dE/E$ or $dN/dE = 3/2(N/E) = D(E)$, ...(24.16)

which shows that density of electron states is equal to 3/2 times the ratio of number of electrons to energy E. This is expected as upto energy level E there were N electronic states.

24.5 HEAT CAPACITY OF METALS

The alkali metals like Li, Na, K etc. consist of monovalent atoms. Each atom contributes one valence electron to what we call the free electron gas in metals. If there are N-atoms the free electron gas would contain N-electrons. Thus at room temperature T, the thermal energy of these would be

$$U_{th} = 3/2\ N\, k_B T \quad \text{(from kinetic theory of gases)}$$

and the heat capacity(specific heat) will be given by

$$C_{el} = dU_{th}/dT = 3/2\ N\, k_B$$

This result leads to a large value as compared to the observed value. In fact, the observed value is even less than 0.01 of this calculated value. So, there is the need of a drastic correction to this classical approach. The correction was first suggested by Fermi himself.

According to Fermi, the electrons should not contribute to heat capacity at absolute zero temperature as the specific heat vanishes as $T \to 0$ and at low temperature it is proportional to T. Also, as specimen is heated the electrons get excited to above Fermi energy as per Fermi-Dirac distribution function

$$f(E,T) = 1/[1 + \text{Exp}((E - \mu)/(k_B T))]$$

where μ denotes chemical potential of the specimen (or metal) under consideration. At low value of $k_B T$, $\mu = E_F$, the Fermi energy. So, for low temperature one may use

$$f(E,T) = 1/[1 + \text{Exp}((E - E_F)/(k_B T))]$$

The plots of this $f(E,T)$ as a function of E have been shown in Fig. 24.3 for $E_F/k_B = 5 \times 10^4$ K and T = 500 K, 5000K and 10^4 K. Thus when temperature is raised above absolute zero, only some of the electrons would get excited above the Fermi energy and that only this fraction of electrons should contribute to the heat capacity of the metals. It is seen that the number of such electrons lies within k_BT. *e.g.* see Fig. 24.4 where density of states, $D(E)$, has been plotted as a function of energy, E.

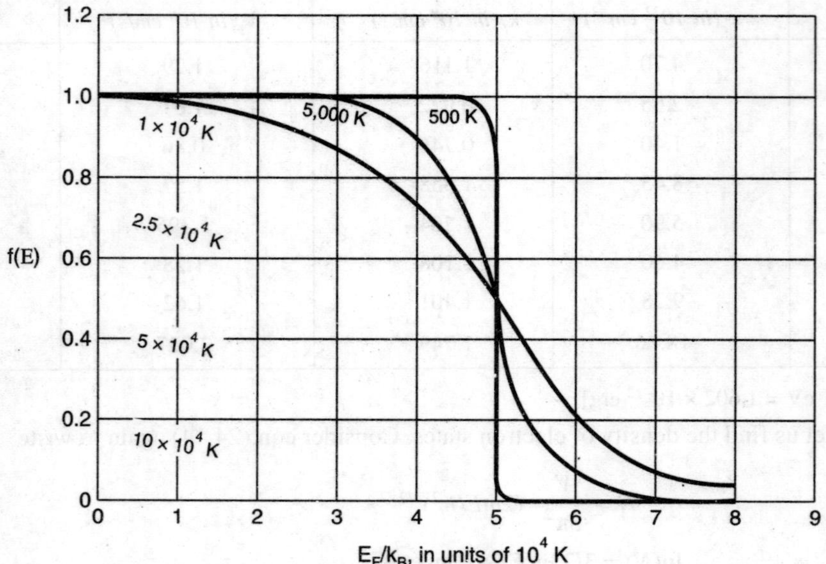

FIGURE 24.3 **Fermi-Dirac distribution function at various temperatures, for $T_F \equiv E_F/k_B$ = 50,000 K. The results apply to a gas in three dimensions. The total number of particles is constant, independent of temperature. The chemical potential at each temperature may be read off the graph as the energy at which f = 0.5.**

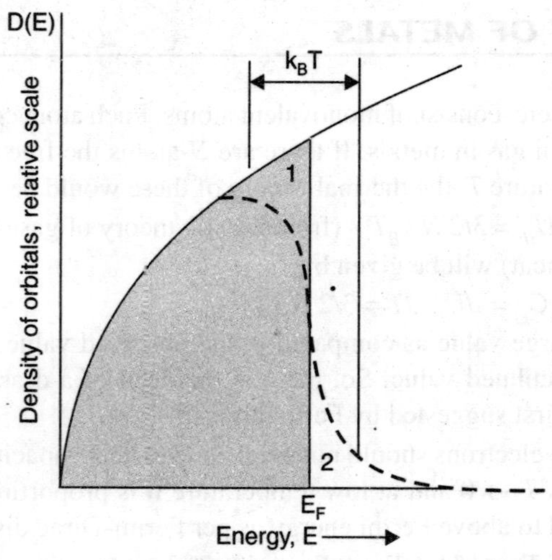

FIGURE 24.4 **Density of single-particle states as a function of energy, for a free electron gas in three dimensions. The dashed curve represents the density $f(E, T)D(E)$ of filled orbitals at a finite temperature, but such that k_BT is small in comparison with E_F. The shaded area represents the filled orbitals at absolute zero. The average energy is increased when the temperature is increased from 0 to T, for electrons are thermally excited from region 1 to region 2.**

That is, about $dE\,D(E)$ electrons leave the Fermi surface and go to higher states. The energy of these electrons lies between E_F and $E_F + k_BT$. That is, $dE \approx k_BT$. Thus, the corresponding thermal energy of the material would be given by

$$U_{th} = dN\,k_BT = D(E_F)\,k_BT \times dE\;;\; dN = D(E_F)\,dE$$
$$= (3/2)(N/E_F)\,k_B{}^2T^2\;;\; N \text{ denotes the total number of conduction electrons}$$
$$= (3/2)(Nk_B)T^2/T_F \text{ as } E_F = k_BT_F$$

Thus, heat capacity is given as

$$C_{el} = dU_{th}/dT = 3Nk_B(T/T_F) \quad \{\text{due to conduction electrons}\}$$

While the classical value calculated earlier is 3/2 Nk_B. Thus, this is about (T/T_F) times the classical value. Taking $T = 300$ K and $T_F = 5 \times 10^4$ K it comes out as $2 \times (3/5) \times 10^{-2}$ or 0.014. But, this rough estimate of heat capacity comes very close to the observed value and also shows that specific heat is proportional to T. So, one may now proceed to do more accurate calculations.

24.6 QUANTITATIVE EXPRESSION OF C_{el}

When temperature of the material increases, the electrons tend to cross over the Fermi energy. Let us suppose an electron gas consisting of N conduction electrons at temperature T. Then, change in energy of N-electrons system will be

$$\Delta U = U_{th}(T) - U_{th}(0) = \int_0^\infty ED(E)\,f(E,T)\,dE - \int_0^{E_F} D(E)dE \qquad \text{...(24.17)}$$

while

$$N = \int_0^\infty D(E)\,f(E,T)dE = \int_0^{E_F} D(E)dE \qquad \text{...(24.18)}$$

because at absolute zero $f(E,T) = 1$ irrespective of E, and all levels are filled completely. Multiplying eqn.(24.18) by E_F to get

$$\int_0^\infty E_FD(E)\,f(E,T)dE - \int_0^{E_F} E_FD(E)dE = 0 \qquad \text{...(24.19)}$$

Subtracting eqn.(24.19) from eqn.(24.17)

$$\Delta U_{th} = \int_0^\infty (E - E_F)D(E)f(E,T)dE - \int_0^{E_F} (E - E_F)D(E)dE$$
$$= \int_0^{E_F} (E - E_F)D(E)f(E,T)dE + \int_{E_F}^\infty (E - E_F)D(E)f(E,T)dE - \int_0^{E_F} (E - E_F)D(E)dE$$
$$= \int_{E_F}^\infty (E - E_F)D(E)f(E,T)dE + \int_0^{E_F} (E_F - E)(1 - f(E,T))D(E)dE \qquad \text{...(24.20)}$$

Both terms in the R.H.S. are positive. First term measures the energy required to lift the electrons from Fermi energy surface to higher energies while second term gives the energy required to raise the electrons from below Fermi surface to the Fermi surface.

Now, differentiate both sides with respect to T, we get

$$C_{el} = dU_{th}(T)/dT = d(\Delta U_{th})/dT$$
$$= \int_{E_F}^\infty (E - E_F)D(E)[df(E,T)/dT]dE - \int_0^{E_F} (E_F - E)\,[df(E,T)/dT]D(E)dE$$

$$= \int_0^\infty (E - E_F)D(E)[df(E, T)/dT]dE \qquad ...(24.21)$$

Next, at temperatures of interest *i.e.* when $k_BT << E_F$,

$$D(E) \to D(E_F) = (3/2)(N/E_F) \quad \text{and} \quad \mu \to E_F. \text{ Hence,}$$

$$C_{el} \approx (3/2)(N/E_F) \int_0^\infty dE(E - E_F)[\text{Exp}((E - E_F)/k_BT)/\{1 + \text{Exp}((E - E_F)/k_BT)\}]^2 \times (E - E_F)/k_BT^2$$

$$= (3/2)(N/E_F)(k_B{}^2T) \int_{-E_F/k_BT}^{\infty} dx\, x^2e^x/(1 + e^x)^2 \text{ where } x = (E - E_F)/k_BT$$

As E_F/k_BT is very large, the lower limit of the integration can safely be taken as $-\infty$ *i.e.*

$$C_{el} = (3/2)(N/E_F)(k_B{}^2T) \int_{-\infty}^{\infty} dx\, x^2e^x/(1 + e^x)^2$$

The value of the integral is found to be $\pi^2/3$. So,

$$C_{el} = (\pi^2/2)\,(N/E_F)(k_B{}^2T) = (\pi^2/2)\,(Nk_B)(T/T_F) \text{ as } E_F = k_BT_F$$

In case of potassium metal, $N = 6.0222 \times 10^{23}$/mol and $T_F = 2.46 \times 10^4$K, the specific heat at room temperature (300 K) will come out as equal to 1.67×10^{-3} T. That is $C_{el}/T = 1.67$ mJ/mol-K^2 which may be compared with the observed value equal to 2.08 (see Fig. 24.5).

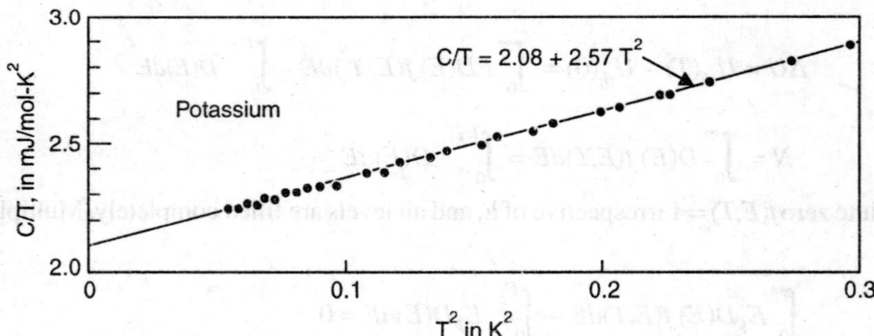

FIGURE 24.5 **Experimental heat capacity values for potassium, plotted as C/T versus T^2. T^2 term denotes the phonons contribution to heat capacity.**

It is a common feature to express the ratio of the observed value to the above mentioned calculated value as a ratio of thermal effective mass to the free electron mass. That is

$$[\text{Observed } C_{el}/T]/[\text{Calculated } C_{el}/T] = m_{th}/m = 2.08/1.67 = 1.246.$$

24.7 EFFECT OF TEMPERATURE ON FERMI ENERGY

Fermi energy in a metal is equal to the top most energy occupied by the conduction electrons at absolute zero temperature. The calculations have shown that this energy is given by

$$E_F = (\hbar^2/2m)(3\pi^2\, N/V)^{3/2}$$

where N/V ($= c_e$) denotes the number of electrons per unit volume of the metal or its concentration (for metals comprising of monovalent atoms). However, at other temperatures the electron distribution in various orbits is given by Fermi-Dirac distribution function

$f(E) = [1 + \text{Exp}(E - \mu)/k_BT]^{-1}$, μ being the chemical potential of the given metal which depends on T. For low temperatures μ is almost equal to Fermi energy E_F. Even at room temperature $T = 300$ K, μ does not differ significantly from E_F. How exactly it varies with T may be found in the following fashion.

Denoting μ by E_f we write

$$f(E) = [1 + \text{Exp}(E - E_f)/k_BT]^{-1}$$

Since this function gives the probability of occupancy of an orbital with energy E, we have

$$N = \int_0^\infty D(E) f(E)\, dE,$$

where $D(E)$ is the density of states equal to $(V/2\pi^2)(2m/\hbar^2)^{3/2}E^{1/2}$ in the energy range E to $E + dE$.

Suppose $g(E)$ is a function such that

$$dg(E)/dE = D(E)$$

i.e.
$$g(E) = \int_0^E D(E)\, dE = 0 \quad \text{at} \quad E = 0.$$

Then
$$N = \int_0^\infty dg(E)/dE\, f(E)\, dE$$

Integrating by parts we obtain

$$= |\, g(E)\, f(E)\, |_0^\infty - \int_0^\infty g(E)\, f'(E)\, dE$$

$$= -\int_0^\infty g(E)\, f'(E)\, dE \qquad \text{because } g(0) = 0 \text{ and } f(\infty) = 0.$$

Now expand $g(E)$ by Taylor theorem around E_f as

$$g(E) = g(E_f) + (E - E_f)g'(E_f) + (E - E_f)^2/2\, g''(E_f) + \ldots$$

so that
$$N = -g(E_f)\int_0^\infty f'(E)\, dE - g'(E_f)\int_0^\infty (E - E_f)\, f'(E)\, dE$$

$$- g''(E_f)/2 \int_0^\infty (E - E_f)^2 f'(E)\, dE$$

and neglecting other terms.

Next, as $f(E) = [1 + \text{Exp}(E - E_F)/k_BT]^{-1}$ and $f'(E)$, the derivative of $f(E)$ has a sharp peak at $E = E_F$. We may extend the lower limit of the integration to $-\infty$ without causing any error.

Then
$$N = -g(E_f)\,|\, f(E)\,|_{-\infty}^{\infty} - g'(E_f)\int_{-\infty}^{\infty} (E - E_f)\, f'(E)\, dE$$

$$- g''(E_f)/2 \int_{-\infty}^{\infty} (E - E_f)^2 f'(E)\, dE$$

As $f(-\infty) = 1$ and $f(\infty) = 0$ we have by putting the value of $f'(E)$

$$N = g(E_f) + g'(E_f)\int_{-\infty}^{\infty} (E - E_f)/k_BT\; e^{(E - E_f)/k_BT}/[1 + e^{(E - E_f/k_BT}]^2\, dE$$

$$+ g''(E_f)/2 \int_{-\infty}^{\infty} (E - E_f)^2/k_BT\; e^{(E - E_f)/k_BT}/[1 + e^{(E - E_f/k_BT}]^2\, dE$$

The first integral may be rewritten as ($x = (E - E_f)/k_BT$)

$$g'(E_f)\,(k_BT)\int_{-\infty}^{\infty} x\,e^x/[1+e^x]^2\,dx = 0 \text{ as } x\,e^x/[1+e^x]^2 \text{ is an odd function of } x.$$

The second integral is similarly rewritten as

$$g''(E_f)/2\,(k_BT)^2\int_{-\infty}^{\infty} x^2\,e^x/[1+e^x]^2\,dx = g''(E_f)/2\,(k_BT)^2\,\pi^2/3$$ as $\pi^2/3$ is the value of the integral. Thus

$$N = g(E_f) + g''(E_f)\,\pi^2/6\,(k_BT)^2$$

Now, N is also given by $\int_0^{E_F} D(E)\,dE = \int_0^{E_F} d(g(E))/dE\,dE = g(E_F)$ because all states are filled up to E_F, and that $g(0) = 0$

Therefore, $$g(E_F) = g(E_f) + g''(E_f)(k_BT)^2\,\pi^2/6$$

or $$g(E_f) - g(E_F) = -\,g''(E_f)(k_BT)^2\,\pi^2/6$$

when E_f is close to E_F as is the case at low temperatures including room temperature, we may write

$$[g(E_f) - g(E_F)]/(E_f - E_F) = g'(E_F) = D(E_F)$$

Thus, $$(E_f - E_F)\,D(E_F) = -\,g''(E_f)(k_BT)^2\,\pi^2/6$$

or $$E_f = E_F - g''(E_f)/D(E_F)(k_BT)^2\,\pi^2/6$$

Once again assuming

$$g''(E_f) = g''(E_F) = D'(E_F) \text{ and since } D'(E_F)/D(E_F) = 1/2\,(1/E_F)$$

$$[\because\ D(E) = CE^{1/2} \text{ and } D'(E)/D(E) = 1/2\ 1/E]$$

$$E_f = E_F\,[1 - \pi^2/12\,(k_BT/E_F)^2] = E_F\,[1 - \pi^2/12\,(T/T_F)^2]$$

which is the required variation of Fermi energy with temperature provided temperature is not high. Now, $T \sim 10^2$ and $T_F \sim 10^4$, $(T/T_F)^2 \sim 10^{-4}$. As a result, $\pi^2/12\,(T/T_F)^2 \sim 10^{-4}$. Thus Fermi energy, more exactly μ, varies slowly with temperature.

Variation of Average Energy with Temperature

At absolute zero temperature, the average energy of electrons is $3/5\,E_F$. At any other temperature, T, the average energy would be given by

$$E_T = (1/N)\int_0^{\infty} E\,D(E)\,f(E)\,dE$$

Proceeding in the same way as above we find that

$$E_T = E_0[1 + 5\pi^2/12\,(k_BT/E_F)^2]$$
$$= E_0[1 + 5\pi^2/12\,(T/T_F)^2] \qquad [\because\ E_F = k_BT_F]$$

where $E_0 = 3/5\,E_F$. Thus, average energy of electron increases but slightly with temperature. This is very much expected as some of the electrons get excited to higher energies when temperature is increased.

24.8 THERMIONIC EMISSION

When the metal is heated the electrons are emitted from the surface of the metal. This is termed as thermionic emission. The flux of these electrons coming out from unit area of metal surface per second constitutes the electric current which would depend on

(*i*) Nature of metal and
(*ii*) Temperature.

The expression of such a current density as a function of temperature for a particular metal is known Richardson-Dushman equation which we derive in the following.

Richardson-Dushman Equation

Let the metal surface be a *Y-Z* plane and p_x be the momentum of electrons along *X*-axis which strikes the surface before coming out. If $n(p_x)dp_x$ is the density of electrons which strike the surface per second the flux of such electrons will be

$$p_x/m\ n(p_x)dp_x \qquad (\because \quad v_x = p_x/m)$$

We are concerned with the current due to these electrons. Not all these will succeed in coming out due to the surface barrier offered by the metal. If *r* denotes the probability of reflection of these elctrons from the surface (quantum mechanical effect), only $(1 - r)$ fraction of electrons having kinetic energy $\geq E_s$ will come out. That is

$$p_x^2/2m \geq E_s$$

and $\quad$ let $E_s = p_s^2/2m \qquad$ (Barrier height)

Then the total current density would be given by

$$J = e\int_{p_s}^{\infty} (1 - r)\ p_x/m\ n(p_x)dx\ ;$$

e denotes the charge of electron.

Assuming *r* to be independent of p_x, one gets

$$J = e(1 - r)/m \int_{p_s}^{\infty} p_x\ n(p_x)dx$$

Thus, knowing $n(p_x)$ the expression of *J* will be known by integration. So, let us now find $n(p_x)$ with the help of Fermi-Dirac distribution law, and the informations of density of states.

Density of states $D(E)$ lying between E and $E + dE$ is given by

$$D(E)dE = V/2\pi^2\ (2m/\hbar^2)^{3/2}\ E^{1/2}\ dE$$

where *V* is the volume of the given metal. Since,

$$p^2/2m = E \quad \text{and} \quad pdp = mdE,$$

$$D(E)dE = D(p)dp = V/\pi^2\ (p^2/\hbar^2)\ dp.$$

In these states the electrons are distributed as per Fermi-Dirac distribution function

$f(E) = 1/\{\text{Exp}[(E - E_F)/k_BT] + 1\}$; E_F being the Fermi energy (provided $k_BT << E_F$) writing $E = (p_x^2 + p_y^2 + p_z^2)/2m$, we get

$$f(p_x, p_y, p_z) = 1/\{\text{Exp}[((p_x^2 + p_y^2 + p_z^2)/2m - E_F)/k_BT] + 1\}$$

Therefore, number of electrons per unit volume having momentum in the range

$$p_x \text{ to } p_x + dp_x,$$

$$p_y \text{ to } p_y + dp_y \quad \text{and} \quad p_z \text{ to } p_z + dp_z$$

will be given by

$$n(p_x, p_y, p_z)dp_x dp_y dp_z = D(p)dp/V \times (dp_x dp_y dp_z)/4\pi p^2 dp \times f(p_x, p_y, p_z)$$

$$= 1/4\pi^3 \times 1/\hbar^3 \times f(\mathbf{p})\ dp_x dp_y dp_z$$

That is, $\quad n(p_x)dp_x = dp_x/4\pi^3\hbar^3 \int_{-\infty}^{\infty} dp_y dp_z\ f(\mathbf{p})$

$$[f(\mathbf{p}) = f(p_x, p_y, p_z)]$$

Usually, $E > E_F$,

As a result $f(\mathbf{p}) = \text{Exp}[-(E - E_F)/k_BT]$

by ignoring unity from denomenator

$$= \text{Exp}[-(\Sigma p_x^2/2m - E_F)/k_BT]$$

Also, $\int_{-\infty}^{\infty} dp_y \text{Exp}[-p_y^2/2mk_BT]$

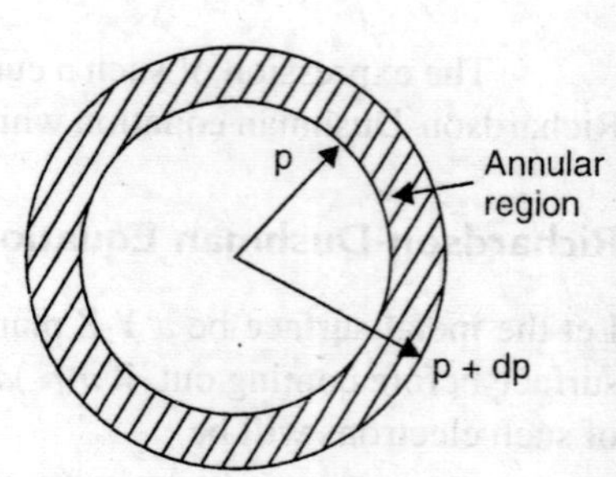

$$= \int_{-\infty}^{\infty} dp_z \text{Exp}[-p_z^2/2mk_BT]$$

$$= (2\pi mk_BT)^{1/2}$$

FIGURE 24.6 Volume of annular region $= 4\pi p^2 dp$; $dp_x dp_y dp_z$ is the incremental volume within this region.

With these results we finally obtain

$$n(p_x)dp_x = dp_x/4\pi^3\hbar^3(2\pi mk_BT)\text{ Exp}[-(p_x^2/2m - E_F)/k_BT]$$

and hence $J = (1 - r)\, ek_BT/2\pi^2\hbar^3 \text{ Exp}[E_F/k_BT] \int_{p_s}^{\infty} p_x \text{ Exp}[-p_x^2/2mk_BT]\, dp_x$

It is easy to see that

$$\int_{p_s}^{\infty} p_x \text{ Exp}[-p_x^2/2mk_BT]\, dp_x$$

$$= m(k_BT)\text{ Exp}[-E_s/k_BT]$$

$$[\because \quad E_s = p_s^2/2m]$$

Hence, $J = (1 - r)\, em(k_BT)^2/2\pi^2\hbar^3$

$$\text{Exp}[-(E_s - E_F)/k_BT]$$

$$= (1 - r)\, em(k_BT)^2/(2\pi^2\hbar^3)$$

$$\text{Exp}[-\phi/k_BT]$$

$$\phi = E_s - E_F,$$

the metal work function by definition

We can also write,

$$J = AT^2 \text{ Exp}[-\phi/k_BT] \text{ where}$$

$$A = (1 - r)\, emk_B^2/(2\pi^2\hbar^3)$$

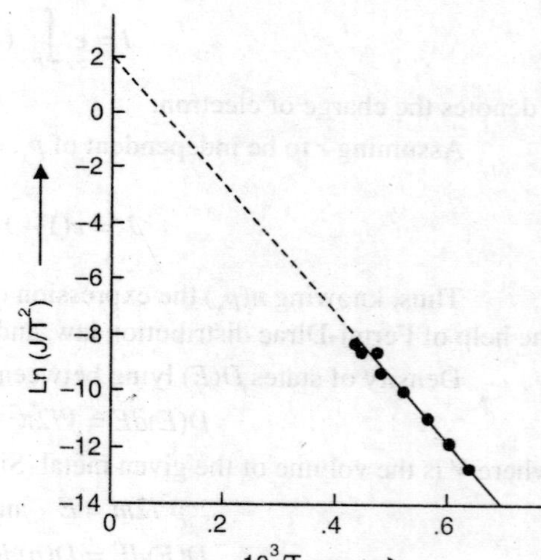

FIGURE 24.7 Plot of $Ln(J/T^2)$ vs $1/T$ called Richardson line. Points are experimental data for Tungsten metal.

Taking ; $e = 1.6 \times 10^{-19}\ C$

$$m = 9.1 \times 10^{-28}\ gm$$

$$\hbar = 1.0546 \times 10^{-27}\ erg\text{-}sec$$

and $k_B = 1.38 \times 10^{-16}\ erg/K.$

$$A = (1 - r)\ 119.76 \approx 120\ (1 - r)$$

Also, $Ln(J/T^2) = Ln(A) - \phi/k_BT$

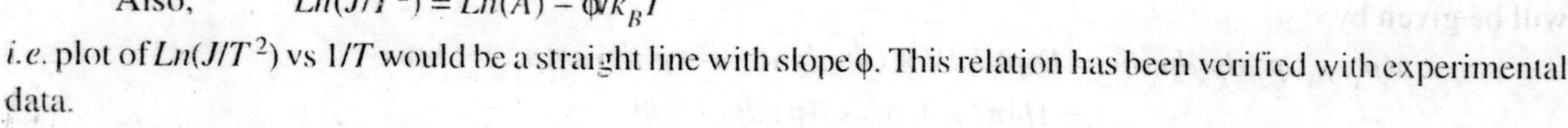

i.e. plot of $Ln(J/T^2)$ vs $1/T$ would be a straight line with slope ϕ. This relation has been verified with experimental data.

The agreement between theory and experimental value may be off the mark because of

(*a*) Surface may not be perfectly plane.

(*b*) ϕ may not be same along the entire surface emitting the electrons.

(*c*) ϕ might change because of experimental conditions, and

(*d*) Variation of r as well as ϕ with temperature of the emitter.

24.9 OHM'S LAW

Whenever an electric field is applied along a conductor, it begins to conduct current. The current is found to be proportional to the applied electric field or the potential difference between the ends of the conductor. If this P.D. is maintained fixed, the current in the conductor will be constant. This is Ohm's law

Let **E** be the electric field applied across the conductor (say Cu). The momentum of the conduction electrons, given by $m\mathbf{v} = \hbar\mathbf{k}$ will change because a force $-e\mathbf{E}$ will act on it. Thus by law of force

$$\hbar\, d\mathbf{k}/dt = -\, e\mathbf{E} \quad i.e., \quad m\, d\mathbf{v}/dt = -\, e\mathbf{E}$$

or

$$\Delta\mathbf{v} = -\,(e\mathbf{E}/m)\,\Delta t$$

However, this increment in velocity will be resisted by the collision between electrons and impurities as well as imperfections and phonons in the conductor. As a result, the conduction electrons tend to maintain a constant increment in velocity at a value

$$\Delta\mathbf{v} \equiv \mathbf{v} = -\,(e\mathbf{E}/m)\,\tau$$

where $\Delta t = \tau$, the collision time, so, if there are n electrons per unit volume of the conductor, the current density would be given by

$$\mathbf{j} = n(-\,e)\mathbf{v} = ne^2\tau\mathbf{E}/m = \sigma\mathbf{E} \qquad \text{(Ohm's law)}$$

where we have set

$\sigma = ne^2\tau/m$ (ohm^{-1}cm^{-1}) called the conductivity of the conductor. It therefore depends on the electron concentration and the collision time. Further, the collision time can be increased by cooling the conductor. For instance, the conductivity of Cu at 4 *K* is 10^5 times the conductivity at 300 *K*.

The resistivity, ρ, is equal to the reciprocal of conductivity. That is

$$\rho = 1/\sigma = m/ne^2\tau \text{ (ohm-cm)}$$

m being the mass and e the charge of the electron.

[**Note:** In ground state, the filled orbitals with momentum vector +**k** and –**k** are equal in number. So, there is no current in the absence of external electric field. For the same reason we have denoted the incremental velocity $\Delta\mathbf{v}$ by **v** while calculating the current.]

If the collision of electrons with phonons and imperfections are independent which is so in most of the conductors, the resistivity may be split as

$$\rho = \rho_L + \rho_i$$

ρ_L being the resistivity due to collisions with phonons and ρ_i due to collisions with imperfections. As collisions with phonons are temperature dependent and collisions with imperfections are temperature independent we may write

$$\rho_L(T) = \rho - \rho_i(0)$$

In fact, $\rho_L(0)$ is almost zero. This approach is expressed as *Matthiessen's rule.*

24.10 HALL EFFECT

Whenever a conductor carrying current is placed in an external magnetic field, an electric field is developed at right angle to both the directions of current as well as magnetic field. This phenomenon is called as Hall effect. The value of this electric field per unit current per unit applied magnetic field is called as Hall co-efficient. If the magnetic field B_z is applied along say z-axis and current in the conductor is flowing along x-axis, then the electric field E_y (called Hall field) is developed along y-axis and is related to Hall co-efficient, R_H, as

$$R_H = E_y/J_x B_z$$

The following figures display the phenomenon of Hall effect

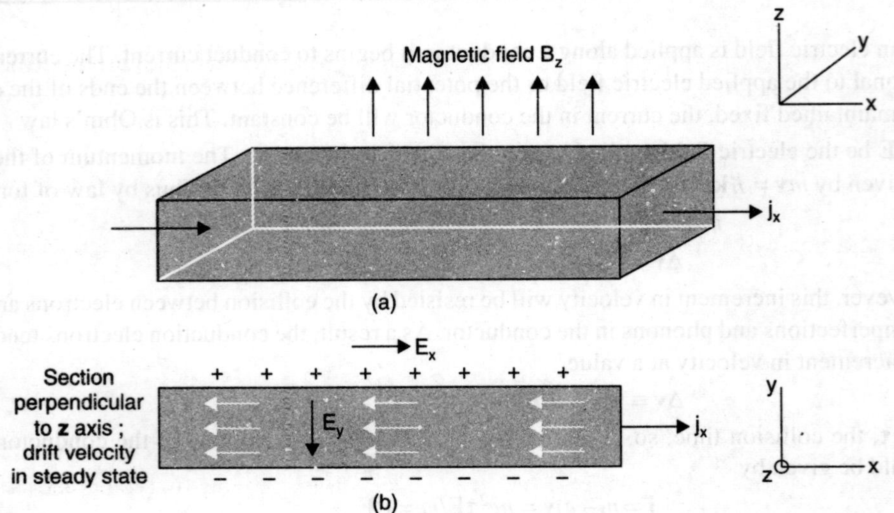

***FIGURE 24.8** (a) B_z is the uniform magnetic field, j_x the current and E_x the external electric field applied across the rectangular conductor ; (b) Shows the steady state of flow of electrons and the transverse electric field E_y (Hall field).*

Calculations of Hall Co-efficient

Because of electric field, E_x, maintained along x-axis and uniform magnetic field $B_z (= B)$, the Lorentz force, F, will act on the conduction electrons. The equation of motion of these electrons will be

$$\mathbf{F} = m\, d\mathbf{v}/dt + m\mathbf{v}/\tau = m(d/dt + 1/\tau)\mathbf{v}$$

τ is the collision time and $m\mathbf{v}/\tau$ is the frictional force due to electrons collisions with impurities and imperfections.

So, $\quad -e(\mathbf{E} + \mathbf{v} \times \mathbf{B}/c) = m(d/dt + 1/\tau)\mathbf{v}$.

Taking individual components of both sides we get

$$-e(E_x + v_y \times B_z/c) = m(d/dt + 1/\tau)v_x \quad \text{(since } B_x = B_y = 0\text{)}$$

$$-e(E_y - v_x \times B_z/c) = m(d/dt + 1/\tau)v_y$$

and

$$-eE_z = m(d/dt + 1/\tau)v_z.$$

Now, motion of electrons along y-axis is not possible as current is drawn along x-axis, $v_y = 0$ and j_x is constant *i.e.*

$$-eE_x = mv_x/\tau \quad \text{and} \quad E_y = v_x B_z/c = v_x B/c \qquad \text{...(24.22)}$$

These two equations yield

$$E_y = -(B/c)(e\tau/m)\, E_x$$

But E_x is related to j_x through

$$j_x = -nev_x = -ne^2\tau/m\, E_x \qquad \text{...(24.23)}$$

Now, from eqns.(24.22) and (24.23) we get

$$E_y/j_xB = -\frac{1}{nec} = R_H,$$

Hall co-efficient in C.G.S. units

or $$R_H = -\frac{1}{ne} \text{ in SI units } i.e.\ c^{-1}m^3.$$

It is quite obvious that the Hall Co-efficient R_H becomes positive when the current in the conductor is due to the holes.

For such a case

$$R_H = +\frac{1}{ne} \text{ [S.I.]}, = +\frac{1}{nec} \text{ [C.G.S.]}$$

In connection with certain problems one talks about the Hall Resistance, ρ_H defined as

$$\rho_H = BR_H = B\left(\frac{E_y}{j_xB}\right) = \frac{E_y}{j_x}$$

$$= \frac{\text{Hall field}}{\text{Current density}} \quad (\text{V–m}^2/\text{Amp})$$

Conductivity and carrier Mobility : It can be shown that conductivity and carrier mobility are connected mutually through Hall co-efficient. In fact,

$$\mu = \left|\frac{v_x}{E_x}\right| \quad \text{and} \quad j_x = -ne\, v_x = \sigma E_x \quad \text{(for electrons)}$$

$$\therefore \qquad \mu = \left|-\frac{\sigma}{ne}\right| = \sigma R_H$$

So, by measuring R_H one can calculate μ or σ as the case may be.

We, next, give the table of Hall co-efficients for some metals in order to have the knowledge of the order of this co-efficient.

TABLE 24.2 Experimental values of R_H for some metals

Metal	*R_H in $10^{-10}\ m^3c^{-1}$ [SI]*
Li	– 1.7
Na	– 2.36
K	– 4.45
Cu	– 0.54
Au	– 0.72
Be	+ 2.43
Al	+ 1.02
As	+ 45.0

SOLVED NUMERICAL PROBLEMS

Problem 1. *Show that density of orbitals is proportional to the square root of energy.*

Solution. $N = (V/3\pi^2)\, k^3$; V = volume and k = wave vector

Since $\hbar^2k^2/2m = E$, energy

$$N = (V/3\pi^2)\,(2m/\hbar^2)^{3/2}\, E^{3/2}$$

Density of orbitals or states, $D(E)$, is defined as

$$D(E) = dN/dE = (V/2\pi^2)\,(2m/\hbar^2)^{3/2}\, E^{1/2} \propto \sqrt{E}\,.$$

Problem 2. *Calculate Fermi wave vector and energy from the data $N/V = 2.65 \times 10^{22}\ cm^{-3}$.*

Solution. Since $N = (V/3\pi^2)\, k_F^3$ at absolute zero temperature.

$$k_F = [(N/V) \times 3\pi^2]^{1/3} = (2.65 \times 30\pi^2)^{1/3} \times 10^7\ \text{cm}^{-1} = 0.922 \times 10^8\ \text{cm}^{-1}.$$

Also, Fermi energy $= E_F = \hbar^2k_F^2/2m = (1.0546 \times 0.922/1.6)^2 \times (9/2 \times 0.51)$ eV

$[\because \ mc^2 = 0.51 \times 10^6\ \text{eV},\ c^2 = 9 \times 10^{20}$ and $\hbar = (1.0546/1.6) \times 10^{-15}\ \text{eV}]$

Finally, $E_F = (1.0546 \times 0.922 \times 3/1.6)^2 \times (1/2 \times 0.51) = 3.26$ eV.

Problem 3. *Calculate Fermi temperature in the above problem.*

Solution. Fermi temperature, T_F, is given by

$$T_F = E_F/k_B$$

where k_B = Boltzmann constant = $1.38 \times 10^{-16}\ erg/K = (1.38/1.6) \times 10^{-4}\ eV/K$.

$$T_F = (3.26 \times 1.6/1.38) \times 10^4\ K = 3.78 \times 10^4\ K.$$

Problem 4. *Calculate the conductivity of the material having charge carrier concentration $8.5 \times 10^{28}\ m^{-3}$ and collision time 2.7×10^{-14} s. Given that the effective mass of the carrier is $1.2\ m_e$.*

Solution. Conductivity σ is given as

$$\sigma = \frac{ne^2\tau}{m^*}$$

where $m^* = 1.2\ m_e = 1.2 \times 9.1 \times 10^{-31}$ kg

$e = 1.6 \times 10^{-19}$ C

$\tau = 2.7 \times 10^{-14}$ s and $n = 8.5 \times 10^{28}\ \text{m}^{-3}$

Therefore, $$\sigma = \frac{8.5 \times (1.6)^2 \times 2.7}{1.2 \times 9.1} \times 10^7 = 5.38 \times 10^7\ \text{Ohm}^{-1}\ \text{m}^{-1}$$

Problem 5. *Find the resistivity of K from the data :*

$n = 1.3 \times 10^{28}\ m^{-3}$

$e = 1.6 \times 10^{-19}\ C$

$\tau = 4.3 \times 10^{-14}\ s$ and $m^* = 9.1 \times 10^{-31}\ kg.$

Solution. Now, $$\sigma = \frac{ne^2\tau}{m^*}$$

and $$\rho = \frac{1}{\sigma}\ \text{(ohm-m)}$$

$$= \frac{9.1}{1.3 \times (1.6)^2 \times 4.3} \times 10^{-7} = 6.36 \times 10^{-8}\ \text{ohm-m}.$$

Problem 6. *Calculate the mean free path of electron in Cu when $\tau = 2.4 \times 10^{-14}$ s and Fermi velocity $v_F = 1.55 \times 10^6$ ms^{-1}.*

Solution. Mean free path λ_F is defined as

$$\lambda_F = v_F \times \tau = 1.55 \times 10^6 \times 2.4 \times 10^{-14}$$
$$= 1.55 \times 2.4 \times 10^{-8} \text{ m} = 37.2 \times 10^{-9} \text{ m} = 37.2 \text{ nm.}$$

Problem 7. *The conductivity of Cu increases by a factor 10^5 when its temperature increases from 4K to 300K. Find the ratio of mean free path of the electrons at the two temperature. If mean free path at 4K is 0.3 cm. Calculate its value at 300 K.*

Solution. Since the conductivity is proportion to the collision time,

$$\tau(4\text{K}) = 10^5\, \tau(300\text{K})$$

Now as $\lambda_F = v_F \tau$,

$$\frac{\lambda_F\,(4\text{K})}{\lambda_F\,(300\text{K})} = \frac{\tau\,(4\text{K})}{\tau\,(300\text{K})} = 10^5 \quad \text{as } v_F \text{ remains same.}$$

Thus $\lambda_F(300\text{K}) = \lambda_F(4\text{K}) \times 10^{-5} = 3 \times 10^{-6}$ cm.

Problem 8. *Density of Silver is 10.5 gm cm^{-3} and its atomic w.t. 107.87. Assuming it a monovalent metal calculate the Fermi energy.*

Solution. If 6×10^{23} is the Avogadro number the electron concentration $\frac{N}{V}$ will be given by

$$\frac{N}{V} = \frac{6 \times 10^{23} \times 10.5}{107.87} = \frac{1}{3\pi^2} k_F^3$$

$$\therefore \qquad k_F = \left(\frac{3 \times \pi^2 \times 6 \times 10.5}{107.87} \times 100 \right)^{1/3} \times 10^7 = 12 \times 10^7 \text{ cm}^{-1}$$

and the Fermi energy

$$E_F = \frac{\hbar^2}{2m} k_F^2 = \frac{(1.0546 \times 12)^2}{2 \times 9.1 \times 10^{-28}} \times 10^{-54} \times 10^{14} \text{ ergs}$$

$$= \frac{(1.0546 \times 12)^2}{2 \times 9.1 \times 1.6} \text{ eV} = 5.5 \text{ eV}$$

EXERCISE

1. Explain classical free electron theory of metals and give its limitations. Derive an expression for electrical conductivity of free electron gas using collision time concept.
2. Explain the quantum theory of free electrons in three dimensions and show that $\overline{E} = (3/5)E_F$.
3. (*a*) What is Fermi energy? Derive Richardson-Dushman equation for thermionic emission from a metal surface.

   (*b*) The atomic weight of aluminium is 27 and its density is 2.7 *g/cc*. Calculate the value of Fermi energy at $T = 0K$. **[Ans.** 5.6 eV]
4. (*a*) What is Fermi level ? What does it signifies? Discuss how it changes with temperature and concentration of electrons.

(*b*) Silver has *fcc* crystal structure and its atomic radius is 1.44 Å assuming silver to be monovalent metal calculate the value of Fermi energy, Fermi temperature and Fermi velocity.

5. Using quantum theory of free electrons, calculate the density of energy states and Fermi level energy.
6. Write down the expression of Fermi-Dirac distribution law mentioning the symbols used.
7. Define thermionic emission. Derive Richardson-Dushman equation.
8. Discuss free electron Fermi gas model and explain what are various Fermi surface parameters.
9. Obtain the expression $D(E) = 3/2\ (N/E)$ for the density of orbitals where letters have got usual meanings.
10. Calculate Fermi wave vector, energy and temperature for *Rb* metal. Given that $N/V = 1.15 \times 1022$ electrons per unit volume. [Ans. $k_F = 0.698 \times 10^8\ \text{cm}^{-1}$; $E_F = 1.86$ eV ; $T_F = 2.16 \times 10^4$K]
11. Explain Hall effect, Hall co-efficient and prove that carrier mobility, μ = Hall co-efficient × conductivity.
12. Calculate the Hall co-efficient of Na having *bcc* structure of a cell of side 4.28 Å.

[Ans. $2.45 \times 10^{-10}\ \text{m}^3\ \text{C}^{-1}$.]

Chapter 25

BAND THEORY OF SOLIDS

25.1 NEARLY FREE ELECTRON MODEL

Motion of a valence electron in metals is described by a plane wave $e^{i\mathbf{k}.\mathbf{r}}$ with energy, E, given by

$$E = (\hbar^2k^2)/2m,\ k^2 = k_x^2 + k_y^2 + k_z^2 \quad \text{and} \quad k_x, k_y, k_z = 0, \pm 2\pi/L, \pm 4\pi/L, \ldots$$

In other words, E is almost continuous function of k having values from 0 to infinite. Such electrons have linear momentum given by $\mathbf{p} = \hbar\mathbf{k}$. However, the lattice atoms do exert some influence on the motion of the electrons. As a result, for certain values of k such as $k = n\pi/a$ with n as integer, motion is no longer describable by plane waves, $e^{i\mathbf{k}.\mathbf{r}}$, but only by standing waves like sin ($\mathbf{k}.\mathbf{r}$) and cos ($\mathbf{k}.\mathbf{r}$). The values of $k = \pm\pi/a$ define the boundary of 1st Brillouin zone for a linear lattice. Similarly, $k = \pm 2\pi/a$, $\pm 3\pi/a$ define the boundaries of 2nd and 3rd Brillouin zones respectively. This is because at these values of k, there are Bragg diffractions and the energy of electrons is no longer continuous but has gaps called energy gaps. That is, 1st gap occurs at $k = \pm\pi/a$. The extent of gap depends on the magnitude of interaction between electrons and the lattice atoms or ion cores.

The above mentioned approach is called nearly free electron model in which electrons are assumed as interacting weakly with ion cores. For one dimensional case the motion of electrons for $k = \pm\pi/a$ can be described in two ways as

$$\psi^{(+)}(x) = 1/\sqrt{2}\ (e^{i\pi x/a} + e^{-i\pi x/a}) = \sqrt{2}\ \cos(\pi x/a)$$

and

$$\psi^{(-)}(x) = 1/\sqrt{2}\ (e^{i\pi x/a} - e^{-i\pi x/a}) = \sqrt{2}\ i \sin(\pi x/a)$$

The probability density of these waves would be

$$|\psi^{(+)}|^2 = \rho^{(+)} \propto \cos^2(\pi x/a)$$

and

$$|\psi^{(-)}|^2 = \rho^{(-)} \propto \sin^2(\pi x/a) \text{ respectively.}$$

The first wave function, therefore, implies higher concentration of electrons at $x = na$ with n as integer *i.e.* at the sites of ion cores (a is the lattice constant). However, 2nd wave function with $\rho(-) \propto \sin^2(\pi x/a)$

implies zero concentration at the lattice ion cores. As a result, electrons describable by cos $(\pi x/a)$ will have less potential energy while those describable by sin $(\pi x/a)$ will have more potential energy as compared to electrons described by e^{ikx} for which $\rho = |\psi|^2 = 1$, constant. Thus, there will be difference in the total energy of these two types of electrons *i.e.*, those described by $\psi^{(-)}(x)$ will have higher energy while those with $\psi^{(+)}(x)$ will have lower energy at $|k| = \pi/a$. The difference will be equal to the magnitude of the periodic potential which these electrons experience during motion in the solid.

This situation is qualitatively explained in the following *E-k* diagrams :

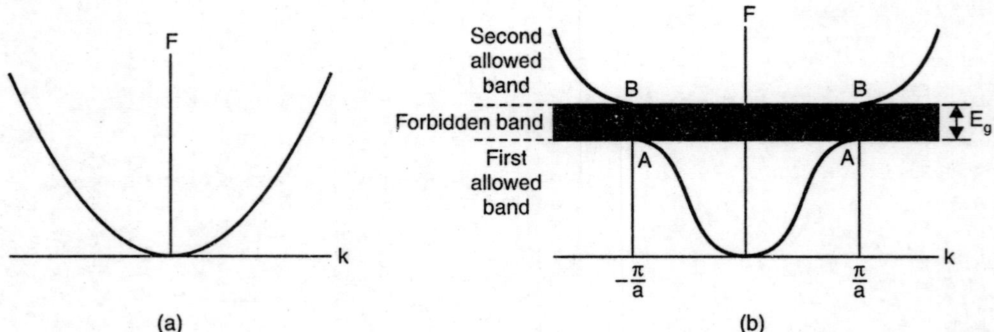

FIGURE 25.1 (*a*) Plot of *E* vs *k* for free electrons (*b*) plot of *E* vs *k* for nearly free electrons moving in a weak periodic potential in a linear lattice of lattice constant, *a*. E_g denotes the energy gap.

Just at *B* for $k = \pm\pi/a$ state of electrons is given by $\psi^{(-)}(x)$ while just at *A* state is given by $\psi^{(+)}(x)$. The energy gap, E_g, shown in Fig. 25.1, in the energy distribution occurs at $k = \pm\pi/a$. Such gaps also occur at $\pm 2\pi/a$, $\pm 3\pi/a$ etc. This is how one can explain the existence of energy bands and energy gaps and hence the behaviour of insulators, conductors and semiconductors.

Calculation of Energy Gap

When the electrons interact weakly with lattice atoms through a periodic potential, say,

$U(x) = U\cos(2\pi x/a)$ such that $U(x + a) = U(x)$, the change in total energy is given by (1st order perturbation theory)

$$\Delta E = \int_0^1 |\psi(x)|^2\, U(x)\, dx$$

$$= 2U\int_0^1 \cos^2(\pi x/a)\cos(2\pi x/a)\, dx \qquad [\text{for } |\psi(x)|^2 = |\psi^{(+)}|^2 = 2\cos^2(\pi x/a)]$$

$$= U/2$$

and

$$= 2U\int_0^1 \sin^2(\pi x/a)\cos(2\pi x/a)\, dx \qquad [\text{for } |\psi(x)|^2 = |\psi^{(-)}|^2 = 2\sin^2(\pi x/a)$$

$$= -\,U/2$$

Hence, the energy of electrons in the state $\psi^{(+)}(x) = \sqrt{2}\,\cos(\pi x/a)$ gets raised by an amount $U/2$ (when U is positive) while the energy of those described by $\psi^{(-)}(x) = \sqrt{2}\,\sin(\pi x/a)$ gets lowered by the same amount $= U/2$. As a result, the gap would be equal to U(see the figure above). This discussion, therefore, explained the origin of energy gaps and energy bands. More precise calculations regarding bands and gaps are explained in the Kronig-Penney model.

25.2 KRONIG-PENNEY MODEL

Kronig-Penney model explains the existence of energy bands and gaps when a periodic potential acts on the conduction electrons. It also assumes that if $\psi_k(x)$ is the solution (Bloch function), then

$$\psi_k(x + a) = e^{ika}\psi_k(x)$$

where 'a' is the lattice constant and k is the wave vector.

Consider the one dimensional periodic potential $U(x)$ having constant strength U and range b and periodicity $(a + b)$. That is

$$U(x) = U(x + a + b)$$

It is being shown in Fig. 25.2. This also means that for any solution $\psi(x)$ we must have

$$\psi_k(x + a + b) = e^{ik(a+b)}\ \psi_k(x)$$

the wave vector k distinguishing a particular solution. This equation also defines Bloch theorem.

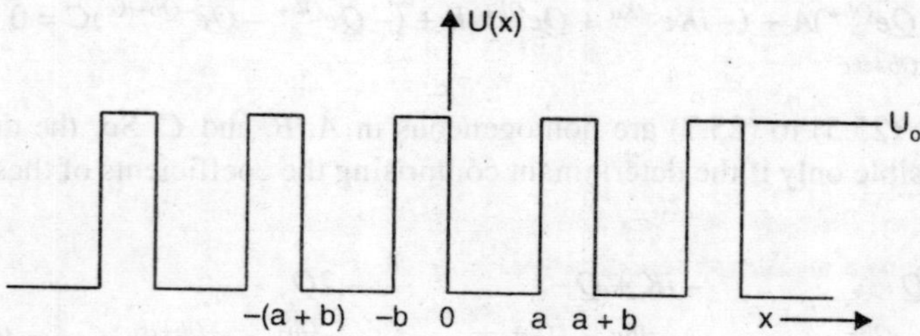

FIGURE 25.2 **Square-well periodic potential as introduced by Kronig and Penney.**

Let us write Schrodinger equation for the motion of the electron in such a periodic square-well potential. The equation in the region where $U(x) = 0$ would be

$$-(\hbar^2/2m)\, d^2\psi(x)/dx^2 = E\psi(x) \qquad \text{[Time independent equation]}$$

or

$$d^2\psi(x)/dx^2 + K^2\psi(x) = 0 \qquad \text{where } (\hbar^2K^2)/2m = E$$

Its solution (general) will be

$$\psi_1(x) = Ae^{iKx} + Be^{-iKx}$$

The Schrodinger equation in the region of square-well potential would be

$$-(\hbar^2/2m)\, d^2\psi(x)/dx^2 + U\psi(x) = E\psi(x)$$

or

$$d^2\psi(x)/dx^2 - Q^2\psi(x) = 0 \quad \text{where } (\hbar^2Q^2)/2m = (U - E)$$

Its general solution would be

$$\psi_2(x) = Ce^{Qx} + De^{-Qx}$$

The set of four constant A, B, C, D are determined from various boundary conditions. For instance, the continuity of ψ requires that

$$\psi_1(x) = \psi_2(x) \quad \text{and} \quad d\psi_1(x)/dx = d\psi_2(x)/dx$$

at the boundaries joining the two regions of zero potential and of non zero potential. Let us apply these at $x = 0$ to obtain

$$A + B = C + D \qquad \text{...(25.1)}$$

and

$$iK(A - B) = Q(C - D) \qquad \text{...(25.2)}$$

We also apply these conditions for solutions at $x = a$ where

$$\psi_2(x) = e^{ik(a+b)}\,\psi_2(x = -b) \quad \text{(Using Bloch Theorem)}$$

because of translational properties of the wave function mentioned in the beginning. Thus, we must have as per requirements of continuity condition

$$\psi_1(x = a) = \psi_2(x = -b)e^{ik(a+b)}$$

and $$(d\psi_1/dx)_{x=a} = e^{ik(a+b)}\,(d\psi_2/dx)_{x=-b}$$

That is $$Ae^{iKa} + Be^{-iKa} = (Ce^{-Qb} + De^{Qb})e^{ikc} \qquad ...(25.3)$$

and $$iKAe^{iKa} - iKBe^{-iKa} = Q(Ce^{-Qb} - De^{Qb})e^{ikc} \qquad ...(25.4)$$

where $c = a + b$.

Next, eliminate D from equations (25.2), (25.3) and (25.4) with the help of eqn. (25.1), we obtain

$$(iK + Q)\,A + (Q - iK)B - 2QC = 0 \qquad ...(25.5)$$

$$(e^{iKa} - e^{Qb+})A + (e^{-iKa} - e^{Qb+})B + (e^{Qb+} - e^{-Qb+ikc})C = 0 \qquad ...(25.6)$$

and $$(iKe^{iKa} + Qe^{Qb+})A + (-\,iKe^{-iKa} + Qe^{Qb+})B + (-\,Qe^{Qb+} - Qe^{-Qb+ikc})C = 0 \qquad ...(25.7)$$

Here, $e^{Qb+} = e^{Qb+ikc}$

These equations (25.5) to (25.7) are homogeneous in A, B, and C. So, the determination of these constants would be possible only if the determinant comprising the coefficients of these constants vanishes. That is, if

$$\begin{vmatrix} iK + Q & -\,iK + Q & -\,2Q \\ e^{iKa} - e^{Qb+} & e^{-iKa} - e^{Qb+} & e^{Qb+} - e^{-Qb+ikc} \\ iKe^{iKa} + Qe^{Qb+} & -\,iKe^{-iKa} + Qe^{Qb+} & -\,Qe^{Qb+} - Qe^{-Qb+ikc} \end{vmatrix} = 0$$

OR

$$\begin{vmatrix} iK + Q & -\,iK + Q & -\,2Q \\ e^{iKa} - e^{Qb+} & e^{-iKa} - e^{Qb+} & 2\sinh(Qb)\,e^{ikc} \\ iKe^{iKa} + Qe^{Qb+} & -\,iKe^{-iKa} + Qe^{Qb+} & -\,2Q\cosh(Qb)\,e^{ikc} \end{vmatrix} = 0; \quad (\because \; Qb+ = Qb + ikc)$$

Cancelling $-\,2e^{ikc}$ from the last column and solving the determinant (see appendix) one obtains

$$[(Q^2 - K^2)/2KQ]\sinh(Qb)\sin(Ka) + \cosh(Qb)\cos(Ka) = \cos(ka) \qquad ...(25.8)$$

Eqn. (25.8) is the important relation in the Kronig-Penney Model. We, next, discuss the implications of this result.

It can be assumed that the periodic potential is very large and $b \to 0$ such that $Qb \to$ small but finite. Within this limit, we neglect K^2 in comparison with Q^2, take $\sinh(Qb) = Qb$ while $\cosh(Qb) = 1$ and $\mathbf{Q^2ab/2}$ is denoted by P. Hence, eqn. (25.8) becomes

$$P(\sin(Ka)/Ka) + \cos(Ka) = \cos(ka) \qquad ...(25.9)$$

Which is the required condition to explain the energy bands and band gaps for a given solid. For example, let us plot the left hand side as ordinate($= y$) and Ka as abscissa as in Fig. 25.3. We have also drawn boundary lines at $y = \pm 1$ because of the R.H.S. of equation (25.9) as $\cos(ka)$ cannot exceed these limits. So, all values of y which lie outside these lines are not allowed. One can also find the limits of Ka within which these values are to be rejected. Since energy $E = (\hbar^2K^2)/2m = \hbar^2(Ka^2)/2ma^2$, one can find the energy gaps which are not allowed for the conduction electrons. These gaps are also called band gaps. The shaded regions in the figure correspond to these values. A graph has also been plotted between energy, rather $\frac{E}{\left(\frac{\hbar^2\pi^2}{2ma^2}\right)}$, and ka showing gaps at $ka = \pi, 2\pi, 3\pi$ etc. (see Fig. 25.3 *b*). More details may be seen in the appendix.

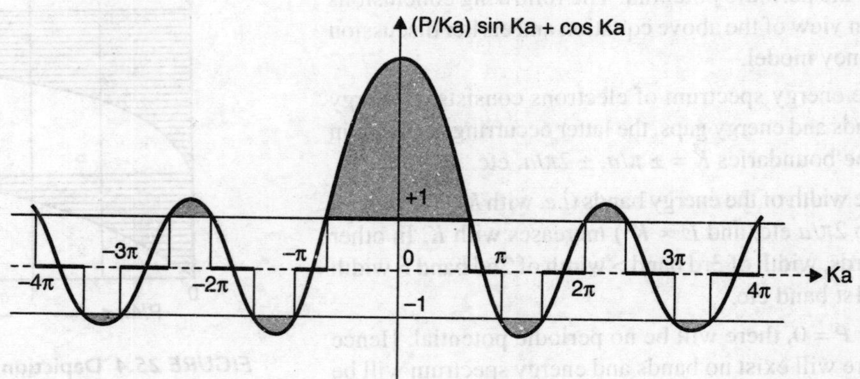

FIGURE 25.3 **(*a*) Plot of the function (*P*/*Ka*) sin *Ka* + cos *Ka*, for *P* = 3π/2. The allowed values of the energy *E* are given by those ranges of $Ka = (2mE/\hbar^2)^{1/2}\,a$ for which the function lies between ± 1. For other values of the energy there are no travelling waves or Bloch-like solutions to the wave equation, so that forbidden gaps in the energy spectrum are formed.**

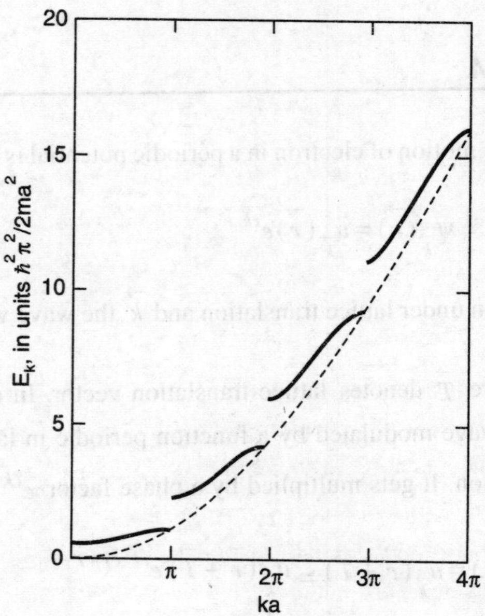

FIGURE 25.3 **(*b*) Plot of energy *vs.* wave number for the Kronig-Penney potential, with *P* = 3π/2. Notice the energy gaps at *ka* = π, 2π, 3π...**

Conclusions of Kronig-Penney Model

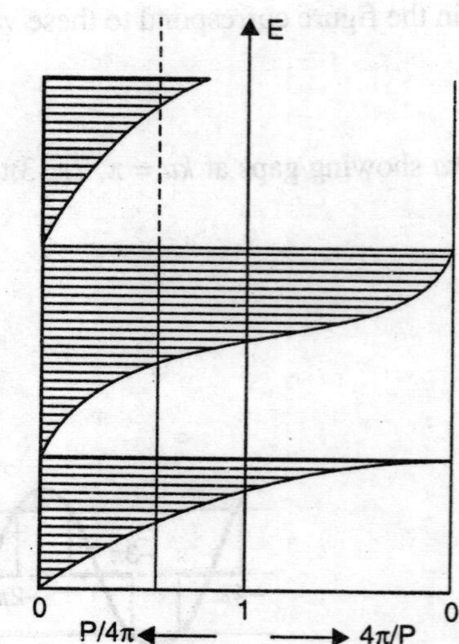

FIGURE 25.4 **Depiction of energy spectrum, *E*, as *P* varies from 0 to ∞.**

The key equation in Kronig-Penney model is

$$P(\sin (Ka)/Ka) + \cos (Ka) = \cos (ka)$$

where energy of the electron is $E = (\hbar^2 K^2)/2m$ and P depends on the strength of the periodic potential. The following conclusions can be drawn in view of the above equation and earlier discussion of Kronig-Penney model.

1. The energy spectrum of electrons consists of energy bands and energy gaps, the latter occurring at Brillouin zone boundaries $K = \pm \pi/a, \pm 2\pi/a$, etc.
2. The width of the energy bands (*i.e.* with $K = 0$ to π/a, π/a to $2\pi/a$ etc. and $E \propto K^2$) increases with K. In other words, width of 3rd band > width of 2nd band > width of 1st band etc.
3. For $P = 0$, there will be no periodic potential. Hence there will exist no bands and energy spectrum will be continuous as in case of free electron gas model.
4. For $P = \infty$, the lattice atoms will show infinite barriers to their respective electrons and hence the electrons will not leave their parent atoms. So, the electron spectrum will be a line spectrum as in isolated atoms as shown in Fig. 25.4.

It is also obvious that the energy gaps would increase with increasing P or strength of periodic potential while with increasing E or K, the gaps would decrease.

25.3 BLOCH THEOREM

The Bloch theorem states that the motion of electron in a periodic potential is described by a function $\psi_{\vec{k}}(\vec{r})$ such that

$$\psi_{\vec{k}}(\vec{r}) = u_{\vec{k}}(\vec{r})e^{i\vec{k}.\vec{r}}$$

where $u_{\vec{k}}(\vec{r})$ is invariant function under lattice translation and $\vec{k}$ the wave vector. This means that

$u_{\vec{k}}(\vec{r}+\vec{T}) = u_{\vec{k}}(\vec{r})$ where $\vec{T}$ denotes lattice translation vector. In other words, the motion of an electron in a crystal is a plane wave modulated by a function periodic in lattice translation. The function $\psi_k(\vec{r})$ is known as Bloch function. It gets multiplied by a phase factor $e^{i\vec{k}.\vec{T}}$ under the lattice translation operation. In fact,

$$T\psi_k(\vec{r}) = u_{\vec{k}}(\vec{r}+\vec{T}) = u_{\vec{k}}(\vec{r}+\vec{T})\,e^{i\vec{k}.(\vec{r}+\vec{T})}$$

$$= (u_{\vec{k}}(\vec{r})\,e^{i\vec{k}.\vec{r}})\,e^{i\vec{k}.\vec{T}} = e^{i\vec{k}.\vec{T}}\psi_{\vec{k}}(\vec{r}); \quad \because u_{\vec{k}}(\vec{r}+\vec{T}) = u_{\vec{k}}(\vec{r})$$

Bloch function is, therefore, an eigen function of $\vec{T}$ with an eigen value $e^{i\vec{k}.\vec{T}}$.

Proof. Let us assume a linear lattice for which $\psi_k(x) = u_k(x)\, e^{ikx}$ is the Bloch function.

For a periodic potential, $U(x) = U(x + a)$, we can use Fourier expansion.

$$U(x) = \sum_G U_G\, e^{iGx} \quad \text{where } G \text{ is the reciprocal lattice vector defined as } G = \frac{2\pi n}{a}$$

with n any integer and a the lattice constant.

Next, let us write the schrodinger equation for the conduction electron

$$\frac{-\hbar^2}{2m}\frac{d^2}{dx^2}\psi + U(x)\psi = E\psi$$

Using Fourier expansion of $U(x)$ and

$$\psi(x) = \sum_k C(k)\, e^{ikx}$$

we get $\sum_k (\lambda_k - E)\, C(k)\, e^{ikx} + \sum_k \sum_G U_G\, C(k)\, e^{i(k+G)x} = 0$

where $\lambda_k = \dfrac{\hbar^2 k^2}{2m}$ and $C(k)$ depending upon the wave vector k (Allowed values of k are 0, $\pm\dfrac{2\pi}{L} \pm \dfrac{4\pi}{L}$, ... for a crystal of length L ; $L = Na$, N being the number of unit cells or lattice points).

The above equation may be rewritten as

$$\sum_k \left[(\lambda_k - E)\, C(k) + \sum_G U_G\, C(k - G)\right] e^{ikx} = 0$$

Hence, the algebraic equation

$$(\lambda_k - E)\, C(k) + \sum_G U_G\, C(k - G) = 0$$

There will be as many equation as is the number of co-efficients $C(k)$.

Usually, k is chosen within Ist Brillouin zone *i.e.* $\dfrac{-\pi}{a} \le k \le \dfrac{\pi}{a}$. For a lattice of N unit cells, there will be N distinct values of k which are 0, $\pm\dfrac{2\pi}{L}, \pm\dfrac{4\pi}{L}, \ldots \dfrac{N\pi}{L} = \dfrac{\pi}{a}$. For the case $N = 20$, there will be 20 values of k within $\dfrac{-\pi}{a}$ to $\dfrac{\pi}{a}$ and we may use $k = -8 \cdot \dfrac{2\pi}{L}$ with $L = 20a$ so that the list of $C(k)$ co-efficients could be taken as

$$C\left(-8\left(\frac{2\pi}{L}\right) \pm \frac{2\pi n}{a}\right), \; n \text{ being any integer}$$

or

$$C\left(-8\left(\frac{2\pi}{L}\right) \pm 20\left(\frac{2\pi}{L}\right)n\right)$$

Finally, when these co-efficients $C(k - G)$ have been determined, the wave function for the electron may be written as :

$$\psi_k(x) = \sum_G C(k - G)\, e^{i(k-G)x}$$

$$= \left[\sum_G C(k - G)\, e^{-iGx}\right] e^{ikx} = u_k(x)\, e^{ikx}$$

where we have used $\quad u_k(x) = \sum_G C(k - G)\, e^{-iGx}$

which is periodic with respect to lattice displacement as G is the reciprocal lattice vector. In three dimensional space the solution takes the form

$$\psi_{\vec{k}}(\vec{r}) = u_{\vec{k}}(\vec{r})\, e^{i\vec{k}.\vec{r}}$$

which proves the Bloch theorem for the motion of electron in a periodic potential of the crystal.

Lastly, for the case of zero potential, the algebraic equation in the co-efficients reduces to

$$(\lambda_k - E)\, C(k) = 0$$

i.e. all C's vanish except the one, $C(k)$, for which $E = \lambda_k = \dfrac{\hbar^2 k^2}{2m}$ and the wave function becomes

$$\psi_k = e^{i\vec{k}.\vec{r}},$$

the plane wave of a free electron motion as expected.

Further, the solution of the central equation

$$(\lambda_k - E)\, C(k) + \sum_G U_G\, C(k - G) = 0$$

will yield various values of E for a particular k. These values of E for same k (in the Ist Brillouin zone) will belong to different energy bands. The corresponding values of the co-efficients $C(k-G)$ will also be different. So, for a particular band, say nth, the wave functions will be

$$\psi_{nk}(\vec{r}) = \sum_G C_n(\vec{k} - \vec{G})\, e^{i(\vec{k}-\vec{G}).\vec{r}}$$

and the energy eigen values as E_n.

25.4 ZONE SCHEMES

The construction of Fermi surfaces one requires the knowledge of various zone schemes. There are three schemes namely extended zone scheme, reduced zone scheme and periodic zone scheme.

Extended Zone Scheme

About this extended zone scheme we have already talked about in the discussion of Kronig-Penney model. There we found that the Ist band occurred within $\dfrac{-\pi}{a} \le k \le \dfrac{\pi}{a}$, while 2nd at k from $\dfrac{\pi}{a}$ to $\dfrac{2\pi}{a}$ and from $\dfrac{-\pi}{a}$ to $\dfrac{-2\pi}{a}$ and so on. Different energy bands were separated by gaps called band gaps.

Such a representation of energy bands is termed as extended zone scheme.

Reduced Zone Scheme

It is possible to map the higher energy bands back into the Ist Brillouin zone $\left(\dfrac{-\pi}{a} \le k \le \dfrac{\pi}{a}\right)$ by suitably translating the wave vector of higher zones with the reciprocal lattice vector such as

$$\vec{k} = \vec{k}' + \vec{G}$$

where $\vec{k}'$ lies outside the first Brillouin zone while k lies within $\frac{-\pi}{a}$ to $\frac{\pi}{a}$, so that energy $E_{k'}$ of higher band is equal to E_k in the first zone. The result of this mapping has been shown in Fig. 25.5 along with other zone schemes.

Periodic Zone Scheme

We can also translate different energy bands to cover the entire range of k values by using suitable reciprocal lattice vector $\vec{G}$ (*e.g.* $G = \pm\frac{2\pi}{a}, \pm\frac{3\pi}{a}$ etc. in one dimensional case). In such a scheme energy eigen values E_{nk} becomes periodic in $\vec{k}$-space. Such a representation of energy bands is known as periodic zone scheme. This has also been shown in Fig. 25.5.

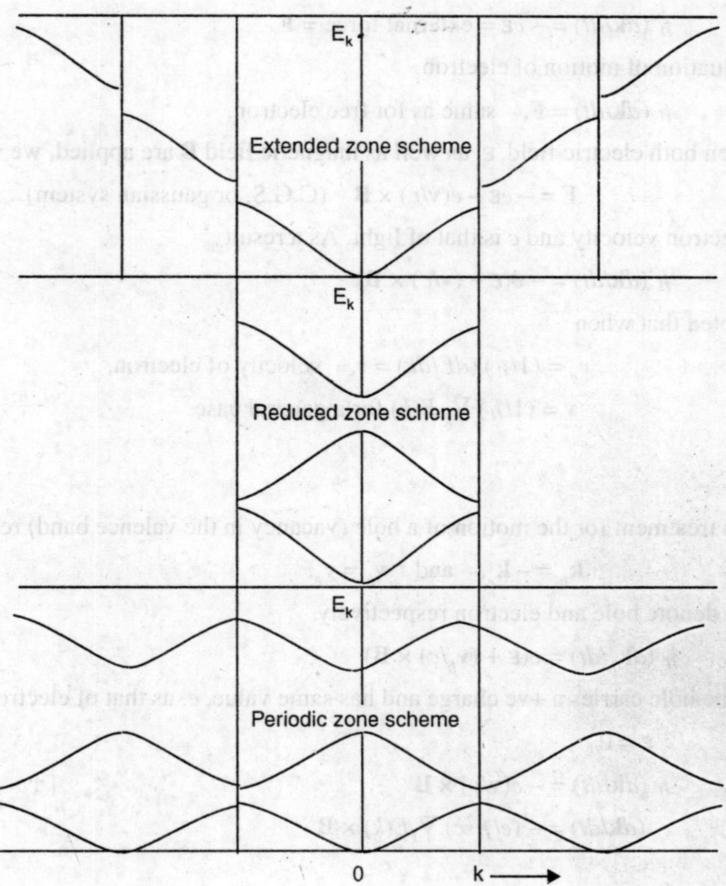

FIGURE 25.5 **Three energy bands of a linear lattice plotted in the extended (Brillouin), reduced, and periodic zone schemes.**

25.5 EQUATION OF MOTION

Now, we discuss the motion of conduction electrons in solids. We associate a wave packet with the motion of an electron having a velocity v. Let the wave vectors of the waves forming the wave packet to be centered around k such that the group velocity is given by

$$v_g = (1/\hbar)\, dE/dk \text{ (By definition)}$$

where $E = \hbar\omega$.

When a constant electric field ε is applied, the change in the energy of electron will be

$$dE = -e\varepsilon dx \quad \text{(Field is along } x\text{-axis)}$$

$$= -e\varepsilon v_g dt = -e\varepsilon(1/\hbar)\,(dE/dk)dt$$

Also, $\qquad dE = (dE/dk)dk$

Hence $\quad (dE/dk)dk = -e\varepsilon(1/\hbar)\,(dE/dk)dt \quad i.e. \quad \hbar(dk/dt) = -e\varepsilon$

For, a general case this equation becomes

$$\hbar\,(d\mathbf{k}/dt) = -e\varepsilon = \text{external force} = \mathbf{F}$$

which means the equation of motion of electron

$$\hbar\,(d\mathbf{k}/dt) = \mathbf{F}, \quad \text{same as for free electron.}$$

Further, when both electric field, $\boldsymbol{\varepsilon}$, as well as magnetic field $\mathbf{B}$ are applied, we will have

$$\mathbf{F} = -e\boldsymbol{\varepsilon} - e(\mathbf{v}/c) \times \mathbf{B} \quad \text{(C.G.S. or gaussian system)}$$

where $\mathbf{v}$ denotes electron velocity and c is that of light. As a result

$$\hbar\,(d\mathbf{k}/dt) = -e(\boldsymbol{\varepsilon} + (\mathbf{v}/c) \times \mathbf{B})$$

It is to be noted that when

$$v_g = (1/\hbar)\,(dE/dk) = v, \quad \text{velocity of electron,}$$

$$\mathbf{v} = (1/\hbar)\,\nabla_k E(k) \text{ for a general case.}$$

Motion of Holes

We may extend this treatment for the motion of a hole (vacancy in the valence band) remembering that

$$\mathbf{k}_h = -\mathbf{k}_e, \quad \text{and} \quad \mathbf{v}_h = \mathbf{v}_e,$$

where label h and e denote hole and electron respectively.

That is, $\qquad \hbar\,(d\mathbf{k}_h/dt) = e(\boldsymbol{\varepsilon} + (\mathbf{v}_h/c) \times \mathbf{B})$

which shows as if the hole carries a +ve charge and has same value, e, as that of electron.

When $\qquad \varepsilon = 0$

$$\hbar\,(d\mathbf{k}/dt) = -e(\mathbf{v}/c) \times \mathbf{B} \qquad [\because \; v = +(1/\hbar)\,\nabla_k E(k)]$$

or

$$(d\mathbf{k}/dt) = -(e/\hbar^2 c)\,\nabla_k E(k) \times \mathbf{B}.$$

25.6 EFFECTIVE MASS

In an energy band the energy of electron does not vary as k^2. Not only this, the variation of $E(k)$ is slow as k increases from zero and becomes faster as k approaches the middle of Brillouin zone till the point of inflection of $E(k)$ vs k graph and starts slowing down as k approaches the boundary of the Brillouin zone, $\pm \pi/a$.

However, by definition mass is given by $1/m^* = (1/\hbar^2)\, d^2E(k)/dk^2$. That is the curvature of $E(k)$ vs k graph determines the effective mass, m^*, of the electron during its motion in a conduction band. In fact the group velocity is given as

$$v_g = (1/\hbar)\,(dE/dk)$$

$$dv_g/dt = \text{acceleration} = (1/\hbar)\,(d^2E/dk^2)(dk/dt)$$

$$= (1/\hbar^2)\,(d^2E/dk^2)F = F/\text{mass}$$

$\therefore$ $1/m^* = (1/\hbar^2)\, d^2E/dk^2$; m^* being the effective mass of conduction electron.

Now, since this curvature varies, with k, m is not constant during motion. Greater the curvature, smaller the effective mass. Thus around $k = 0$ and $\pm\pi/a$ where the $E - k$ plot has large curvature the effective mass of electron will be small, while near the points of inflection, the effective mass will be very large. Its variation with k is of the type as shown in Fig. 25.6.

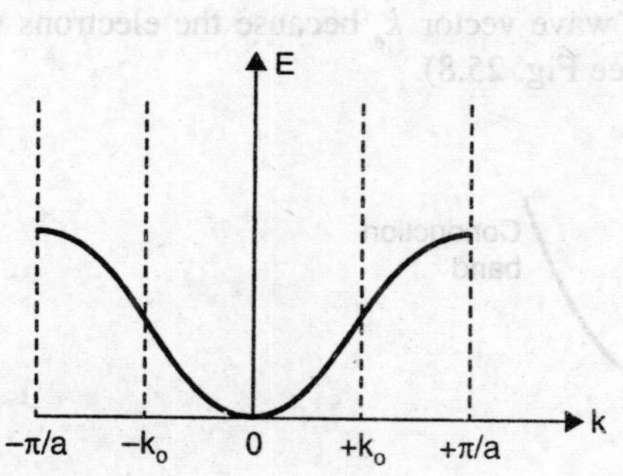

FIGURE 25.6 **(*a*) Plot of *E*(*k*) vs *k* within the Ist Brillouin zone. ± k_0 indicate points of inflection.**

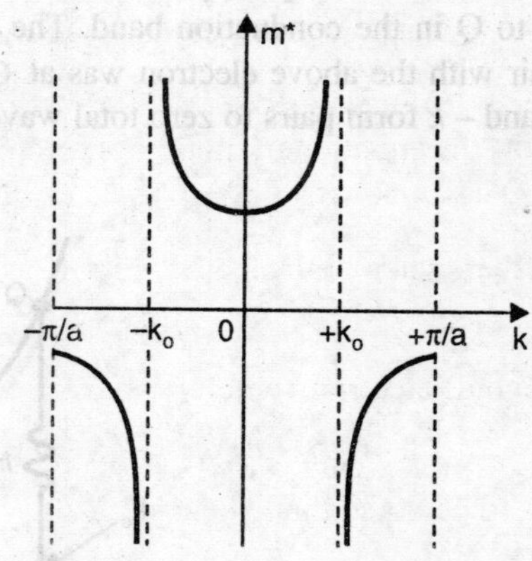

FIGURE 25.6 **(*b*) Plot of effective mass *m** vs *k* in the Ist Brillouin zone. ± k_0 denote points of inflection.**

Velocity of Electron

Velocity of electron is the same as the group velocity given by $(1/\hbar)\,(dE/dk)$ which is zero at $k = 0$, and $\pm\pi/a$. It also increases as k approaches $\pm k_0$ and then decreases to zero as $k \to \pm\pi/a$. So, the electron velocity varies as shown in Fig. 25.7.

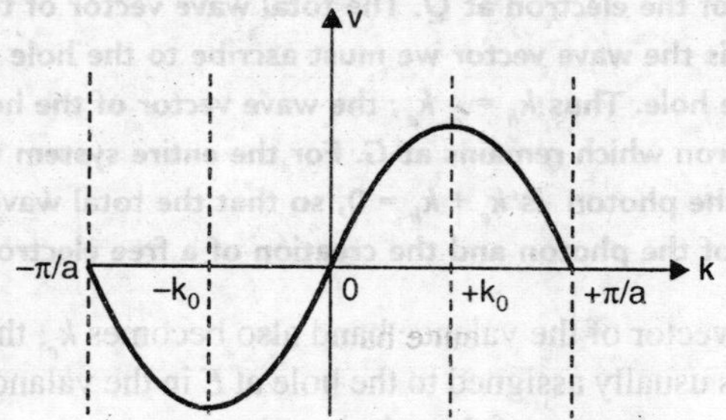

FIGURE 25.7 **Variation of electron velocity with *k* in the energy band $-\pi/a \le k \le \pi/a$ (Ist Brillouin zone), k_0 denotes the point of inflection.**

25.7 DYNAMICS OF A HOLE

A vacancy is left in the energy band when any electron from valence band is pushed to some state in the conduction band. This vacancy is termed as 'hole'. It can be shown that this hole behaves like a particle such that

$$m^*_{\hbar} = -m^*_e$$

$$\mathbf{k}_h = -\mathbf{k}_e \; ; \; dk_h/dk_e = -1$$

and $$E_h(k_h) = -E_e(k_e) = -E_e(-k_e)$$

here suffix h denotes hole and e denotes electron.

Further, hole behaves as if it carries a positive charge equal to e, the electronic charge. Let us suppose that the valence band is completely filled and an electron from E after absorbing a photon of negligible wave vector goes to Q in the conduction band. The wave vector of this electron was $-k_e$. The electron which formed a pair with the above electron was at G having a wave vector k_e because the electrons with wave vectors $+k$ and $-k$ form pairs to zero total wave vector (see Fig. 25.8).

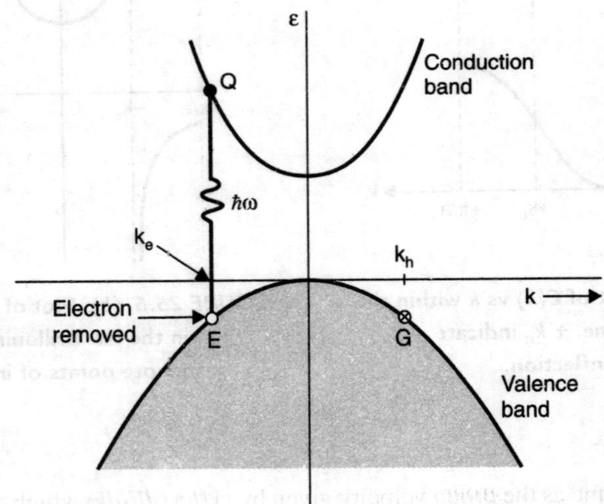

FIGURE 25.8 Absorption of a photon of energy $\hbar\omega$ and negligible wave vector takes an electron from E in the filled valence band to Q in the conduction band. If k_e was the wave vector of the electron at E, it becomes the wave vector of the electron at Q. The total wave vector of the valence band after the absorption is $-k_e$, and this is the wave vector we must ascribe to the hole if we describe the valence band as occupied by one hole. Thus $k_h = -k_e$; the wave vector of the hole is the same as the wave vector of the electron which remains at G. For the entire system the total wave vector after the absorption of the photon is $k_e + k_h = 0$, so that the total wave vector is unchanged by the absorption of the photon and the creation of a free electron and free hole.

However, the total wave vector of the valence band also becomes k_e, the wave vector of the unpaired electron at G. This wave vector is usually assigned to the hole at E in the valance band. That is, the hole wave vector $k_h = -k_e$, opposite to the wave vector of the missing electron.

Further, since the top of the valence band is at zero energy, lower the missing electron, higher the energy of the system in which it is left. That is if the energy of the missing electron was $-E_e(k_e)$, that of remaining system would increase by $E_e(-k_e)$. When this energy is assigned to the hole as E_h, we write

$$E_h(\mathbf{k}_h) = -\text{ the energy of missing electron from the valence band.}$$

In general, therefore,

$$E_h(\mathbf{k}_h) = -E_e(-\mathbf{k}_e) = -E_e(+\mathbf{k}_e)$$

In the last step we used that E_e depends on k_e^2. On this basis we can draw hole band as in Fig. 25.9.

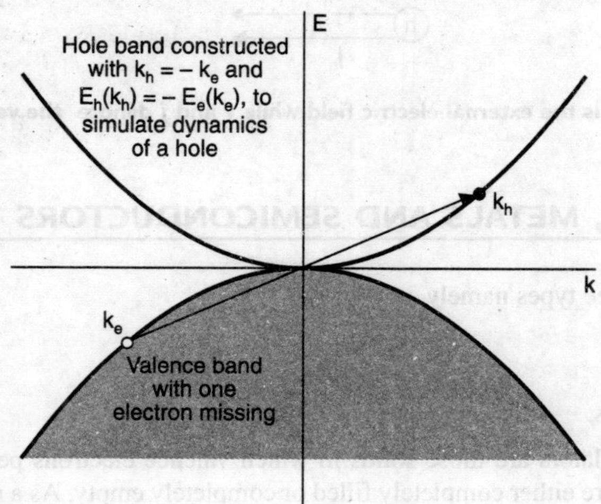

FIGURE 25.9 **Description of valence band and the corresponding hole band. Missing electron from k_e in valence band corresponds to a hole at k_h in the hole band such that $E_h(k_h) = -E_e(\pm k_e)$; and $k_h = -k_e$.**

It is also obvious from Fig. 25.9 that curvature of hole band is opposite to that of valence band. So,

$$m_h^* = -m_e^*.$$

That is effective mass of hole is minus times the effective mass of electron in a band.

The velocity of hole may be similarly related as

$$\mathbf{v}_h = (1/\hbar)\,\nabla^{kh} E_h(\mathbf{k}_h) \qquad \text{(definition of velocity)}$$

$$= -(1/\hbar)\,\nabla^{kh} E_e(-\mathbf{k}_e) \qquad \text{Using } E_h = -E_e$$

$$= (1/\hbar)\,\nabla^{ke} E_e(\mathbf{k}_e) \qquad \because\ dk_h/dk_e = -1 \text{ and } E_h(\mathbf{k}_h) = -E_e(\pm\mathbf{k}_e)$$

$$= \mathbf{v}_e$$

Lastly, we may write the equation of motion for the hole as explained below.

For electron $\quad \hbar\,(d\mathbf{k}_e/dt) = -e(\boldsymbol{\varepsilon} + (\mathbf{v}_e/c) \times \mathbf{B})$

or $\quad \hbar\,(d\mathbf{k}_h/dt) = e(\boldsymbol{\varepsilon} + (\mathbf{v}_h/c) \times \mathbf{B}) \qquad \because\ \mathbf{k}_e = -\mathbf{k}_h \text{ and } \mathbf{v}_e = \mathbf{v}_h$

which is a motion of a +ve charge e in electromagnetic field. Thus, hole is assigned a positive charge = e.

In summary, $\quad \mathbf{k}_h = -\mathbf{k}_e$, when $\mathbf{k}_e$ is the wave vector of missing electron.

$$E_h(\mathbf{k}_h) = -E_e(\pm\mathbf{k}_e),$$

$$\mathbf{v}_h = \mathbf{v}_e,$$

$$m_h^* = -m_e^* \quad \text{and} \quad e_h = e$$

This is all about the dynamics of the hole or vacancy in the valence band.

Direction of current may also be depicted as in Fig. 25.10.

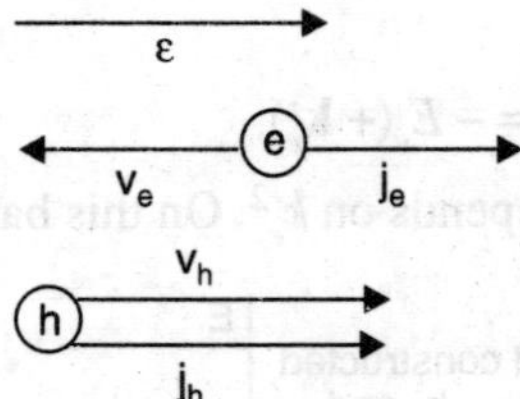

FIGURE 25.10 **E is the external electric field while *v* and *j* denote the velocity and current.**

25.8 INSULATORS, METALS AND SEMICONDUCTORS

Solids are divided into three types namely

1. Insulators
2. Metals
3. Semiconductors.

1. Insulators. Insulators are those solids in which valence electrons per primitive cell are even in number and all the bands are either completely filled or completely empty. As a result, there is no movement of electrons when external field, ε, is applied. The resistivity can be as high as 10^{22} ohm-cm. The energy gaps are also large [see Fig. 25.11].

2. Metals. In metals, however, one of the energy band is partially filled (say from 10 to 90%). As a result, the electrons do move when external electric field is applied. This partially filled band is known as conduction band. As a result, metals conduct electricity and are said to be good conductors. The resistivity is as low as 10^{-10} ohm-cm. Alkali and noble metals are the examples. In certain cases such as Bismuth one of the bands is almost empty [see Fig. 25.11] at absolute zero because of overlapping of bands. These solids are called semimetals.

3. Semiconductors. Pure semiconductors are insulators at absolute zero because these have eight valence electrons per primitive cell. However, at finite temperature some of the electrons get excited thermally to the next higher empty band, as shown in left of right in Fig. 25.11. Silicon is an example. In other cases due to the presence of impurities there becomes deficiency or surplus of electrons. As a result, one of the bands remain slightly empty or becomes slightly filled. These semiconductors are called *p*-type or *n*-type respectively. So, pure crystals such as Diamond, Silicon, Germanium are insulators at absolute zero because bands do not overlap and there being eight valence electrons per primitive cell.

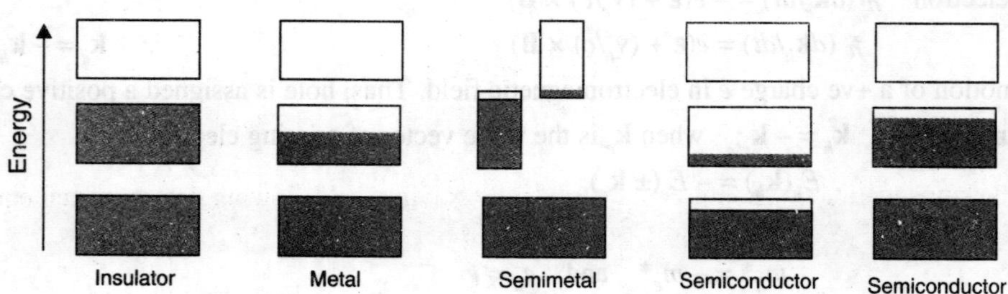

FIGURE 25.11 **Display of occupancy of energy bands in insulator, metal, semimetal, and semiconductor.**

Semiconductor compounds are also available. Such compounds have symbolic formula AB where A and B are elements forming the compound. When A is trivalent and B is pentavalent, the semiconductor compound is designated as III–IV. Similarly, other types of semiconductors are named as IV–IV, II–VI. Table 25.1 shows the list of such semiconductors besides the pure ones along with their energy gaps at room temperature.

TABLE 25.1 List of semiconductors along with the energy gaps

Crystal	*Energy Gap E_g at 300 K. (eV)*	*Crystal*	*Energy Gap E_g at 300 K.(eV)*
Si	1.11	ZnS	3.6
Ge	0.66	ZnO	3.2
InSb	0.17	HgTe	– 0.30
InAs	0.36	PbS	0.34 – 0.37
GaSb	0.68	CdS	2.42
GaAs	1.43	αSn	0.00

Note. HgTe is semimetal just as Bi and HgSe. Bands overlap in semimetals.

These three types of crystals may also be distinguished through $E - k$ diagrams as in Fig. 25.12.

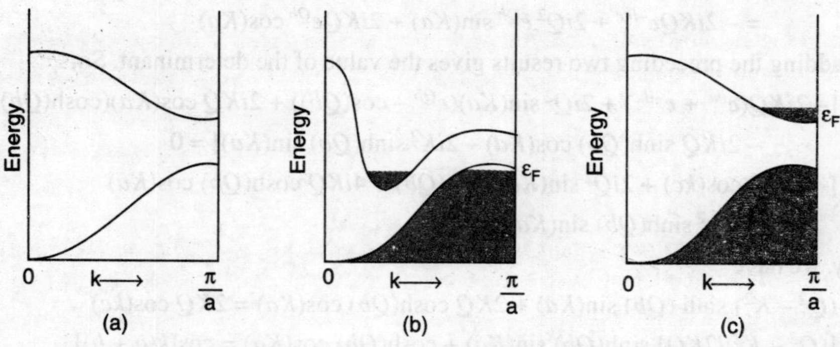

FIGURE 25.12 **Occupied states and band structures giving (*a*) an insulator ; (*b*) a metal or a semimetal because of band overlap ; and (*c*) a metal because of electron concentration.**

APPENDIX

Kronig-Penney Model

Let us derive the key equation of Kronig-Penney model by solving the following determinantal equation

$$\begin{vmatrix} iK+Q & -iK+Q & -2Q \\ e^{iKa} - e^{Qb+} & e^{-iKa} - e^{Qb+} & 2\sinh(Qb)\, e^{ikc} \\ iKe^{iKa} + Qe^{Qb+} & -iKe^{-iKa} + Qe^{Qb+} & -2Q\cosh(Qb)\, e^{ikc} \end{vmatrix} = 0$$

In the last column let us cancel the common factor $-2e^{ikc}$ and write

$$\begin{vmatrix} iK+Q & -iK+Q & Qe^{-ikc} \\ e^{iKa}-e^{Qb+} & e^{-iKa}-e^{Qb+} & -\sinh(Qb) \\ iKe^{iKa}+Qe^{Qb+} & -iKe^{-iKa}+Qe^{Qb+} & Q\cosh(Qb) \end{vmatrix} = 0$$

The first term of determinant is

$$(iK+Q)[Q(e^{-iKa}-e^{Qb+})\cosh(Qb) + (-iKe^{-iKa}+Qe^{Qb+})\sinh(Qb)]$$

and the second term would be

$$(-iK+Q)[(iKe^{iKa}+Qe^{Qb+})(-\sinh(Qb)) - Q(e^{iKa}-e^{Qb+})\cosh(Qb)]$$

Sum of these terms

$$\begin{aligned} Q[-2iQ\cosh(Qb)\sin(Ka) - 2iK\sinh(Qb)\cos(Ka)] \\ + iK[2Q\cosh(Qb)\cos(Ka) - 2K\sinh(Qb)\sin(Ka) \\ - 2Q\cosh(Qb)e^{Qb+} + 2Q\sinh(Qb)e^{Qb+}] \\ = [-2iQ^2\cosh(Qb)\sin(Ka) - 2iKQ\sinh(Qb)\cos(Ka) \\ + 2iKQ\cosh(Qb)\cos(Ka) - 2iK^2\sinh(Qb)\sin(Ka) - 2iKQe^{ikc}] \end{aligned}$$

$$[\because \quad \cosh(Qb) - \sinh(Qb) = e^{-Qb} \text{ and } e^{-Qb}e^{Qb+} = e^{ikc}]$$

Now, the third term of the determinant is

$$\begin{aligned} Qe^{-ikc}[(e^{iKa}-e^{Qb+})(-iKe^{-iKa}+Qe^{Qb+}) - (e^{-iKa}-e^{Qb+})(iKe^{iKa}+Qe^{Qb+})] \\ = Qe^{-ikc}[-2iK + 2iQe^{Qb+}\sin(Ka) + 2iKe^{Qb+}\cos(Ka)] \\ = -2iKQe^{-ikc} + 2iQ^2e^{Qb}\sin(Ka) + 2iKQe^{Qb}\cos(Ka) \end{aligned}$$

Next, adding the preceding two results gives the value of the determinant. So,

$$\begin{aligned} [-2iKQ(e^{ikc}+e^{-ikc}) + 2iQ^2\sin(Ka)(e^{Qb} - \cos(Qb)) + 2iKQ\cos(Ka)(\cosh(Qb) + e^{Qb}) \\ - 2iKQ\sinh(Qb)\cos(Ka) - 2iK^2\sinh(Qb)\sin(Ka)] = 0 \end{aligned}$$

$$\begin{aligned} [-4iKQ\cos(kc) + 2iQ^2\sin(Ka)\sinh(Qb) + 4iKQ\cosh(Qb)\cos(Ka) \\ - 2ik^2\sinh(Qb)\sin(Ka)] = 0 \end{aligned}$$

Finally, we have

$$(Q^2 - K^2)\sinh(Qb)\sin(Ka) + 2KQ\cosh(Qb)\cos(Ka) = 2KQ\cos(kc)$$

or

$$[(Q^2 - K^2)/2KQ]\sinh(Qb)\sin(Ka) + \cosh(Qb)\cos(Ka) = \cos[k(a+b)]$$

$$[\because \quad c = a + b \text{ and } e^{Qb} = \cosh(Qb) + \sinh(Qb)]$$

This is the standard result of Kronig–Penney model. With the assumption that as $b \to 0$, $Q \to \infty$ such that Qb remains finite but small and $Q^2 >> K^2$, we obtain

$$(Q^2ab/2)\sin(Ka)/Ka + \cos(Ka) = \cos(ka)$$

$$[\because \text{ for small } Qb, \cosh(Qb) \to 1, \sinh(Qb) \to Qb]$$

Putting $P = Q^2ab/2$, we get

$$P\sin(Ka)/Ka + \cos(Ka) = \cos(ka)$$

Which equation is used to find the energy bands and energy gaps.

Energy Bands and Energy Gaps

We may explain the formation of energy bands and energy gaps graphically by taking $P = 5\pi/2$ and varying Ka in steps in the above equation. The left hand side is plotted along y-axis while Ka along x-axis. We can also draw two parallel lines to x-axis at $y = +1$ and -1 as the R.H.S. of the equation is a cos function whose values lie within these limits. Only those values of Ka are permitted for which y would lie between -1 to $+1$ while other values of Ka are not allowed for which y becomes $> +1$ and < -1. Since energy is given by $E = (\hbar^2K^2)/2m$, energies corresponding to the above mentioned regions of Ka would give rise to energy bands and energy gaps. In the graph, the shaded regions corresponding to gaps while the remaining regions to energy bands. We have also shown the $E - k$ graph with gaps at $k = \pm \frac{\pi}{a}, \pm \frac{2\pi}{a}$ and $\pm \frac{3\pi}{a}$, in Fig. 25.13.

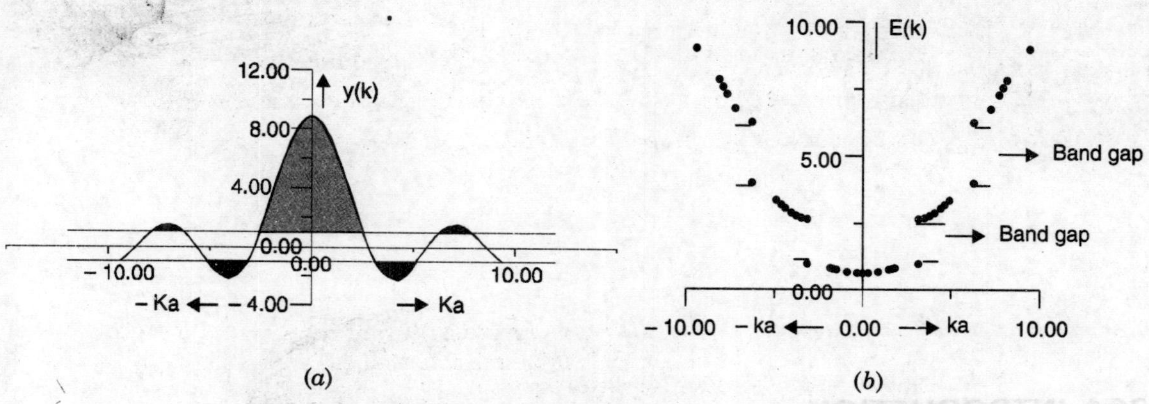

FIGURE 25.13 **(a) Plot of** $Y(K) = \frac{P \sin Ka}{Ka} + \cos Ka$ **vs** ka **with** $p = \frac{5\pi}{2}$.
(b) E – k graph depicting energy bands and band gaps at $ka = \pm \pi, \pm 2\pi$, a **being the lattice constant.**

EXERCISE

1. Discuss the origin of energy bands in solids. How does it explain the behaviour of metals, semiconductors and insulators?
2. State and explain Kronig-Penny model of solids. *(K.U. June, 2007)*
3. (*a*) What is meant by "effective mass" of an electron ? Show that the effective mass of an electron can become positive, negative and infinity.
   (*b*) Show that the total number of possible states in an energy band of a finite crystal is equal to the number of primitive cells in it.
   (*c*) What is Hall effect ? Derive an expression for Hall co-efficient. Discuss some important application of this effect.
   (*d*) Calculate the value of Hall co-efficient and charge carrier density if the observed Hall voltage is 10 mV for a current of 20 mA flowing through a specimen 1 mm thick placed in a magnetic field of 2×10^3 G.
4. (*a*) What are the conclusions of Kronig-Penney model ?
   (*b*) Explain the concept of Brillouin zones.
5. What is the Bloch theorem ? Derive Bloch theorem by a simple model.
6. What are zone schemes to represent the energy bands ? Explain them.

Chapter **26** *SUPERCONDUCTIVITY*

26.1 INTRODUCTION

The phenomenon in which electrical resistivity of many metals, alloys, and certain chemical compounds drop to zero when they are cooled to a sufficiently low temperature (often a temperature in the liquid helium range) is known as superconductivity. The superconducting specimen are referred to as superconductors. The sufficient low temperature at which a particular specimen undergoes a phase transition from normal conducting state to superconducting state is known as critical temperature or transition temperature of that specimen. Presently, various modern technologies have a direct bearing on the phenomenon of superconductivity. This chapter is largely devoted to the study of the phenomenon of superconductivity and its applications in modern technologies.

26.2 DISCOVERY AND THE EXPERIMENTAL SURVEY

Superconductivity was discovered by the Dutch physicist Heike Kamerlingh Onnes in Leiden in 1911, three years after he first liquefied helium. He studied the variation of the electrical resistivity of mercury with temperature. The value of electrical resistance of mercury used was 172.7 ohms in the liquid condition at 0°C. In the solid state, the resistance at 0°C was extrapolated by using the temperature co-efficient of solid mercury and predicted to be 39.7 ohms. At 4.3 K this had sunk to 0.084 ohms, *i.e.*, to 0.0021 times the resistance which the solid mercury would have at 0°C. At 3 K the resistance was found to have fallen below 3×10^{-6} ohms, *i.e.*, to one ten millionth of the value which it would have at 0°C. As the temperature sank further to 1.5 K this value remained the upper limit of the resistance. This behavior of the resistance of mercury with respect to temperature is shown in Fig. 26.1. This observation marked the phenomenon of superconductivity.

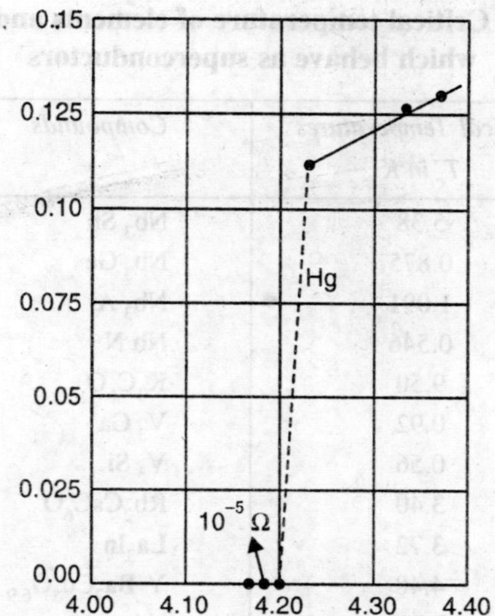

FIGURE 26.1 **Variation of resistivity of mercury (*Hg*) with temperature in K. *R* (ohms) drops to negligible value at $T \cong 4.2$ K.**

26.3 EXPERIMENTAL FACTS ABOUT SUPERCONDUCTIVITY

After the discovery of the phenomenon of superconductivity, a lot of experiments were performed to explore various important properties of superconductors. Certain important facts about superconductors, revealed through these experiments, are presented below.

In the superconducting state the *d.c.* electrical resistivity is zero or very close to zero so that one must expect persistent electrical current within such a material for long time. This fact has actually been observed by File and Mills (1963) who concluded that the decay time of supercurrent in a solenoid of NbZr was not less than 10^5 years. Thus existence of persistent current is a proof of the occurrence of superconducting state.

Besides their unusual electrical properties, the magnetic properties of superconductors were also observed to be equally dramatic. It is an experimental fact that a bulk superconductor in a weak magnetic field will act as a perfect diamagnetic with zero magnetic induction in the interior.

In superconducting state, it was further observed that the conduction electrons were in ordered state. That is, the electrons are ordered at temperatures below the transition temperature, and they are disordered above the transition temperature. Bardeen, Cooper, and Schrieffor explained the nature and origin of the superconducting state of the material through such ordering of electrons in pairs.

It has been also found that properties of superconductors can be changed by varying temperature, magnetic field, size and isotopic mass of the specimen. These variations in the properties of superconductors have been thoroughly studied. In fact, material behaves like superconductors at $T \leq T_c$ and a normal conductor for $T > T_c$. We give the value of this temperature for different elements and compounds in Table 26.1.

TABLE 26.1 Critical temperature of elements and compounds which behave as superconductors

Element	*Critical Temperatures* T_c *in K*	*Compounds*	*Critical Temperatures* T_c *in K*
V	5.38	$Nb_3\,Sn$	18.05
Zn	0.875	$Nb_3\,Ge$	23.2
Ga	1.091	$Nb_3\,Al$	17.5
Zr	0.546	Nb N	16.0
Nb	9.50	K_3C_6O	19.2
Mo	0.92	$V_3\,Ga$	16.5
Cd	0.56	$V_3\,Si$	1.71
In	3.40	Rb_2CsC_6O	31.3
Sn	3.72	La_3In	10.4
Ta	4.48	$Y\,Ba_2Cu_3O_{6.9}$	90.0
Hg	4.153		
Pb	7.193		

26.4 PERFECT DIAMAGNETISM OR THE MEISSNER EFFECT

Meissner, in 1933 found that if a superconductor is cooled in a magnetic field down to the transition temperature, then at the transition, the lines of induction B are pushed out in Fig. 26.2. This phenomenon is called Meissner effect. This shows that, in an applied external field, a bulk superconductor behave as if inside the specimen $B = 0$.

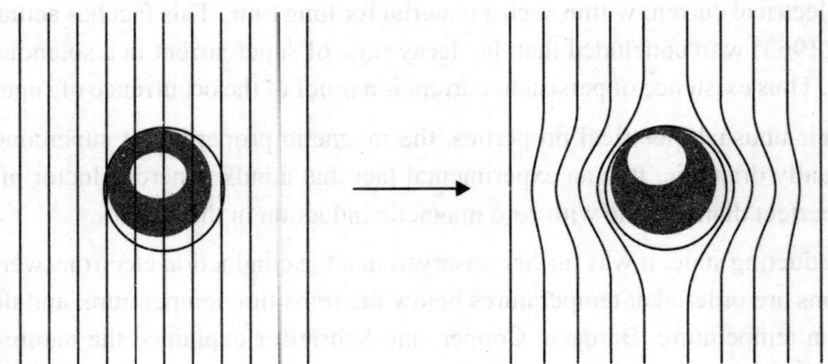

FIGURE 26.2 The pushing of magnetic flux out of the specimen on cooling below its critical temperature, T_c.

For example, Fig. 26.2 shows that magnetic flux is ejected out as the temperature of the material goes below its critical temperature T_c.

Consequently, from the relation

$$B = H + 4\pi M \text{ (in CGS system)} \qquad \text{...(26.1)}$$

and when $B = 0$ inside the specimen

$$H = -4\pi M$$

Then, susceptibility $\chi = M/H = -1/4\pi$

or $\chi = -1/4\pi$ (in CGS system)

which is negative and shows that superconductor exhibit perfect diamagnetism. However, it is much higher than that of a normal diamagnetic substance.

Earlier, it was considered that Maxwell equations are sufficient to explain superconductivity. From Maxwell's equations we have

$$\nabla \times \mathbf{E} = (-1/c)\, \partial \mathbf{B}/\partial t \qquad \text{...(26.2)}$$

From Ohm's law $\mathbf{E} = \rho \mathbf{J}$ for superconductor $\rho = 0$ while for finite **J**, **E** must be zero. Then from eqn. (26.2) $\partial \mathbf{B}/\partial t = 0$ or $\mathbf{B}$ = constant, which predicts that the magnetic flux through the specimen cannot change on cooling through the transition temperature. But Meissner showed that a change does occur as the material passes over to the superconducting phase. From this we conclude that both Meissner effect and Maxwell's equation contradicts to each other. However, this magnetic property is used to characterize the material into Type I and Type II superconductors.

Effect of Strong Magnetic Fields. At low applied magnetic fields the superconductors are diamagnetic (Meissner effect). But when this magnetic field exceeds a particular limit the superconducting state gets destroyed and the superconductor becomes a normal conductor. This limiting value of the external magnetic field is called the critical value say B_c (= H_c in CGS) and this value depends on the low temperature of the superconductor. Lower is this temperature higher the value of magnetic field required to destroy the superconductivity. This effect of magnetic field was extensively studied experimentally and the results are being shown in Fig. 26.3 as plot of $H_c(T) = B_c(T)$ (CGS) as a function of temperature for different elements. Clearly the left lower part of each curve indicates the superconducting state while the right upper part the normal state with the curve dividing the two regions.

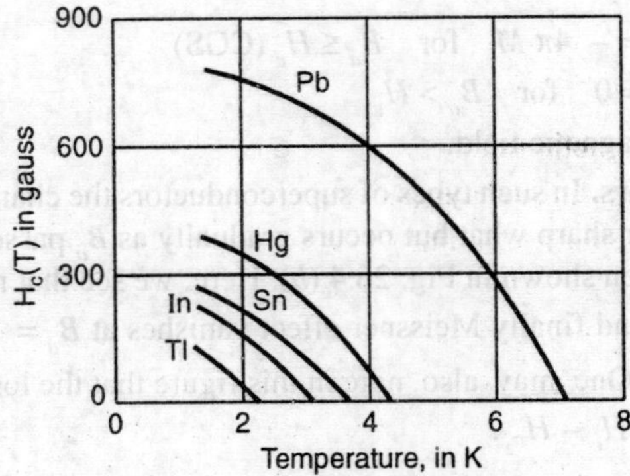

FIGURE 26.3 **Experimental threshold curves of the critical field $H_c(T)$ versus temperature for several superconductors. A specimen is superconducting below the curve and normal above the curve.**

26.5 TYPE I AND TYPE II SUPERCONDUCTORS

Zero resistivity and the Meissner effect have been found as the essential properties of a superconductor. We have also seen that strong magnetic field when applied on the superconductor destroys its superconductivity.

These properties of the superconductors lead to the characterization of the superconductors into two categories as—Type I Superconductors and Type II Superconductors.

Type I Superconductors. In such type of superconductors the diamagnetic behaviour is of the nature as shown in Fig. 26.4 (*a*). This trend shows that type I superconductors are completely diamagnetic upto $B_a = H_c$ (CGS) and becomes non magnetic thereafter as B_a is increased. Besides this, the change from magnetic to non magnetic state is abrupt. In other words,

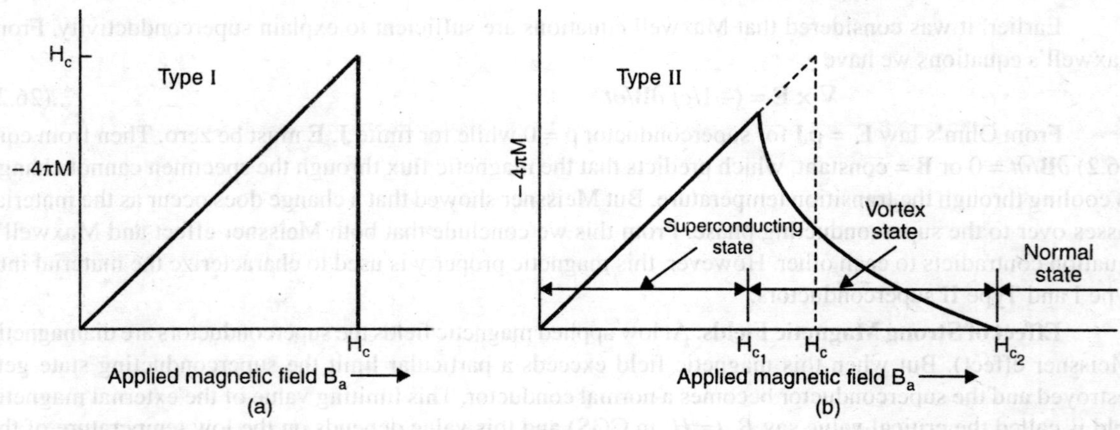

FIGURE 26.4 **(*a*) Magnetization versus applied magnetic filed for a bulk superconductor exhibiting a complete Meissner effect (perfect diamagnetism). A superconductor with this behaviour is called a type I superconductor ; (*b*) Superconducting magnetization curve of a type II superconductor. The flux starts to penetrate the specimen at a field H_{c_1} and H_{c_2} lower than the thermodyanamic critical field H_{c_1} and continue to do so till $B_a = H_{c_2}$.**

$$B_a = -\ 4\pi M \quad \text{for} \quad B_a \leq H_c \text{ (CGS)}$$

and
$$M = 0 \quad \text{for} \quad B_a > H_c$$

where B_a denotes the applied magnetic field.

Type II Superconductors. In such types of superconductors the change over from diamagnetic state to non magnetic is not abrupt or sharp what but occurs gradually as B_a passes across H_c. This behaviour of Type II superconductors has been shown in Fig. 26.4 (*b*). Here, we see that magnetic flux starts penetrating the material at $B_a = H_{c_1} < H_c$ and finally Meissner effect vanishes at $B_a = H_{c_2}$ *i.e.* thereafter the material becomes a normal conductor. One may also note in this figure that the loss of area by the curve between $H_{c_1} - H_c$ gets restored between $H_c - H_{c_2}$.

In Fig. 26.5, we show this behaviour of lead and various lead-indium alloys. The dots represent experimental points. Here, we find that a pure lead is Type I superconductor while its alloys are Type II superconductors of increasing extent as the proportion of indium increases in the alloy.

Another point to note is that in the range $H_{c_1} - H_{c_2}$ the state of superconductor is called vortex state and that the value of H_{c_2} is much higher than that of H_c, it could be even 100 times higher. Such a property of superconductors alloys has assumed lot of importance in the superconductor technology. Because of this variety of alloys of *Nb*, *Al* and *Ge* have been made available.

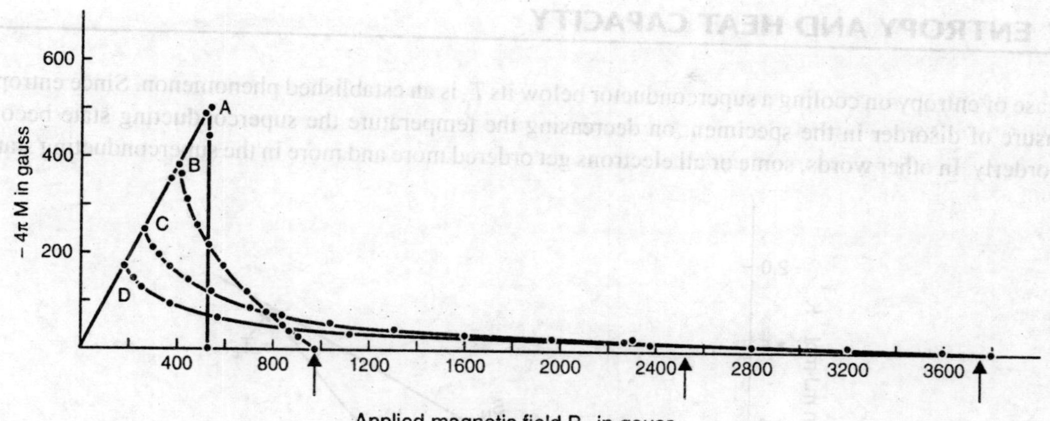

FIGURE 26.5 **Superconducting magnetization curves of annealed polycrystalline lead and lead-indium alloys at 4.2 K. (A) lead ; (B) lead – 2.08 wt. percent indium ; (C) lead – 8.23 wt. percent indium ; (D) lead–20.4 wt. percent indium.**

26.6 ISOTOPIC EFFECT

It has also been observed experimentally that the critical temperature, T_c, (transition temperature) of a particular element varies from isotope to isotope. That is T_c depends on the mass of the isotope of the element. For instance T_c of H_g varies from 4.185 K to 4.146 K as its atomic mass, M, varies from 199.5 to 203.4 a.m.u. T_c, even changes when different isotopes are mixed. This phenomenon is termed as isotopic effect. The experimental values of T_c usually follow

$$M^{\alpha} T_c = \text{constant}$$

where parameter α is less than unity. It is about $\frac{1}{2}$. So, greater the atomic mass M, smaller the critical temperature T_c. In Table 26.2 has been listed values of α for certain elements obtained through fitting the data.

TABLE 26.2 Value of the constant α which fits $M^{\alpha}T_c$ = constant relation

Substance	α	*Substance*	α
Zn	0.45 ± 0.05	Os	0.15 ± 0.05
Cd	0.32 ± 0.07	Mo	0.33
Sn	0.47 ± 0.02	Nb_3Sn	0.08 ± 0.02
Hg	0.50 ± 0.03	Zr	0.00 ± 0.05

The BCS theory has also predicted that

$$T_c \propto M^{-\frac{1}{2}}$$

but without taking into consideration the coulomb interaction between electrons. From the table it is clear that in Zr isotopic effect is negligible. Likewise there are other cases also.

26.7 ENTROPY AND HEAT CAPACITY

Decrease of entropy on cooling a superconductor below its T_c is an established phenomenon. Since entropy is a measure of disorder in the specimen, on decreasing the temperature the superconducting state becomes more orderly. In other words, some or all electrons get ordered more and more in the superconducting state as

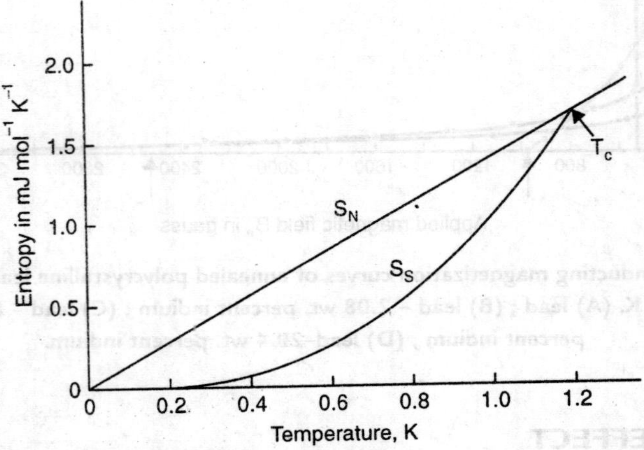

FIGURE 26.6 **(*a*) Plot of entropy *vs* temperature ; S_N for normal state of Al and S_s for its superconducting state. T_c denotes the critical temperature.**

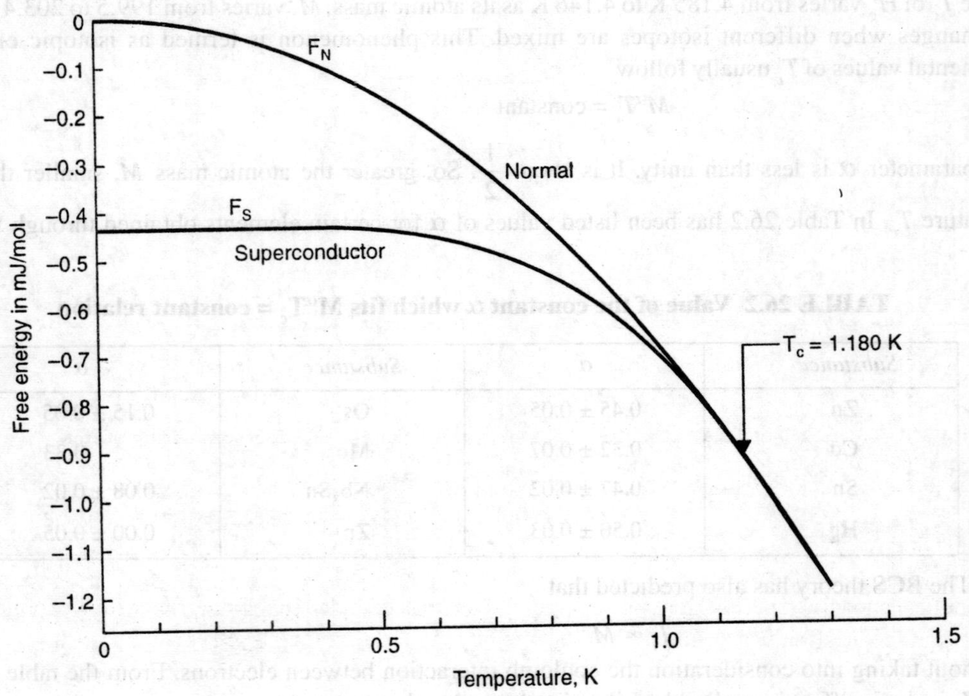

FIGURE 26.6 **(*b*) Plot of free energy *vs* temperature for normal state (F_N) and superconducting state (F_S).**

the temperature is lowered more and more and hence do not contribute to entropy of the system. Further, as the change in entropy is small, only a small fraction of total conduction electrons gets ordered and contribute to the superconducting state. In Frg. 26.6(*a*) we show the trend of variation of entropy *O, S,* for normal state as well as superconducting state of Al as a function of temperature. In Fig. 26.6(*b*) we also show the variation of free energy, *F*, as a function of temperature where one notes that free energy in superconducting state is lower than in normal state at temperatures below T_c. Clearly the two curves in both cases merge into each other at the critical (transition) temperature, T_c as expected.

The specific heat (or heat capacity) of gallium (Ga) was also measured at various temperatures below its critical temperature $T_c = 1.091$ K. The results were plotted as function of temperature and is being shown in Fig. 26.7. The normal state was maintained by applying a magnetic field of 200 G. From the graph one finds that while for normal state the curve of *C/T* vs T^2 follows the usual relation $\frac{C}{T} = \gamma + AT^2$ the curve for superconducting state does not. It has also been seen that the electronic part of heat capacity varies as $\frac{C_{el}}{\gamma T_c} = 7.46 \; exp \; [-1.39 \; T_c/T]$ with $\gamma = 0.60$ mJ mol^{-1} deg^{-2}. This trend suggests that there is excitation of electrons across the energy gap in the superconducting state.

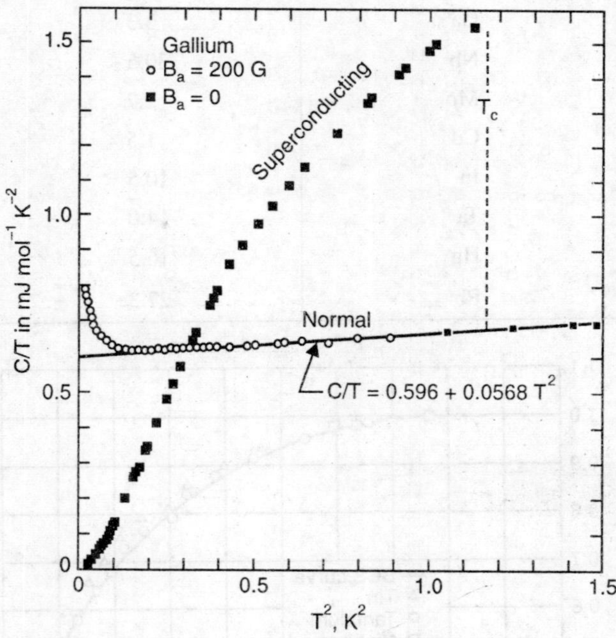

FIGURE 26.7 **The heat capacity of gallium in the normal and superconducting states.**

26.8 ENERGY GAP

The study of heat capacity (specific heat) of superconductors has revealed that this varies with *T* exponentially. In fact, for Ga the variation was found as

$$\frac{C_{el}}{\gamma T_c} = 7.46 \; exp \; [-1.39 \; T_c/T]$$

with $\gamma = 0.60$ mJ mol^{-1} k^{-2}

This, therefore, indicates the existence of energy gap in superconductors. However, the energy gap, E_g, was found to appear in the exponential as $exp\left[-\frac{E_g}{2k_BT}\right]$ where k_B denotes Boltzmann constant. In the above case of Ga, therefore, $\frac{E_g}{2} = 1.39\ k_BT_c$

or
$$E_g = \frac{2 \times 1.39 \times 1.38}{1.6} \times 10^{-4}\ T_c\ \text{eV} = 2.40 \times 10^{-4}\ T_c\ \text{eV}.$$

Taking $T_c = 1.1$ K, E_g comes out as 2.64×10^{-4} eV. In Table 26.3 we list the value of energy gap E_g at $T = 0$ K for different superconductors.

TABLE 26.3 Values of energy gap in superconductor at T = 0 K

Element	$E_g(0)$ *in* 10^{-4} *eV*
V	16
Zn	2.4
Ga	3.3
Nb	30.5
Mo	2.7
Cd	1.5
In	10.5
Ta	14.0
Hg	16.5
Pb	27.3

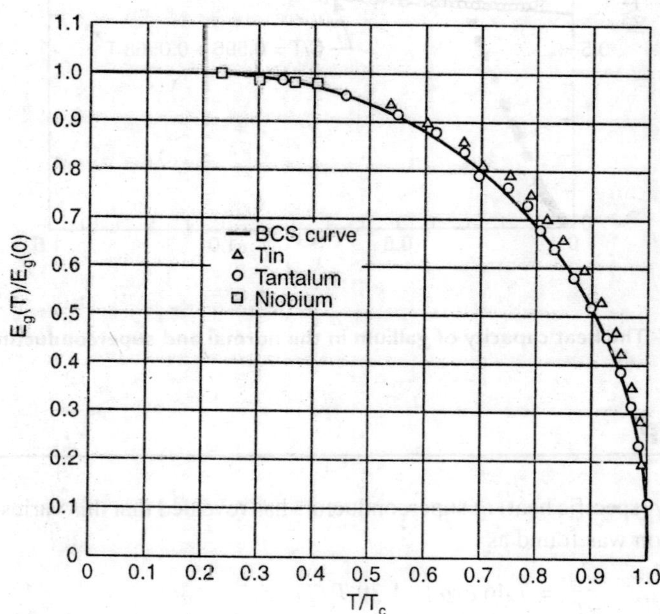

FIGURE 26.8 **Reduced values of the observed energy gap $E_g(T)/E_g(0)$ as a function of the reduced temperature T/T_c. The solid curve is drawn for the BCS theory.**

In Fig. 26.8, the measured values of $\frac{E_g(T)}{E_g(0)}$, the reduced value of energy gap, has been plotted as a function of T/T_c which agrees very well with BCS theory shown as solid curve.

The energy gap in superconductor, thus, is of different type than those in insulators and semiconductors. In superconductors the gap arises from the electron-electron interaction ordering the electrons in $\vec{k}$-space while in normal solids energy gap arises from electron-lattice interaction. Further, we note from Fig. 26.8 that energy gap reduces to zero as T approaches T_c but continuously instead of suddenly as happens during the transition from superconducting state to normal state. Since this transition of normal state into superconducting state occurs without latent heat, it is called second order transition.

26.9 APPLICATIONS OF SUPERCONDUCTIVITY

High Field Magnets

Magnetic field greater than about 30 kilo-oersted is produced conventionally by passing a large electric current through a water-cooled copper or aluminium solenoid. But the power of a megawatt is required to produce 100 kilo-oersted magnetic field in one inch diameter bore. Most of this power input is released as joule heating in the coil and some precautions must made for its removal if the magnet is to run continuously. Neither the large power supply nor the means of removing the heat are needed when the resistance of the coil is zero, *i.e.*, if it is wound with superconducting wire. The only power required by a superconducting magnet is to keep the magnet cool (so that the wire remains superconducting), to overcome the small amount of dissipation in the normal leads and to set up the magnetic field. Once the required field has been established the two ends of the wire in the coil can be joined with a superconducting link (the magnet is then in the persistent mode) and the field will remain even when the power supply is removed. In this case the only power required to maintain the field is that needed to cool the solenoid and for high fields this is several orders of magnitude less than that required to maintain the same field using a conventional resistive solenoid. The low power requirement and simple cooling have lead to a large reduction in the cost of producing high magnetic fields and high magnetic field facilities need no longer be restricted to few laboratories.

Flux Trapping and Shielding

When a Type II superconducting cylinder is cooled in a magnetic field and if the field is large enough to penetrate into the cylinder then persistent currents are induced in the cylinder walls and a magnetic field is trapped in the ring. The field inside the cylinder will remain effectively constant as long as it is superconducting and therefore it may used as a very high field permanent magnet.

Magnetic Bearings

The possibility of flux trapping and flux exclusion suggests that a superconducting cylinder and disc be used as magnetic bearing. The bearing is restricted in its lateral movement but can spin on a cushion of flux. If the material of the moving part of the bearing is chosen so that no flux will penetrate into it the only friction in the bearing will come from the viscosity of the surrounding medium and this may be made very small.

Energy Storage

Another possible use of superconducting coil is an energy store. A trapped field of H oersted store an energy of ($H^2/8\pi$) ergs/cm^3 so that a trapped field of 100 kilo-oersted correspond to a stored energy of 10^8 erg/cm^3. A conventional capacitor bank stores about 10^6 ergs/cm^3. There is, however, a difficulty in regaining the electrical energy quickly from the superconducting store since the inductance of the coil may be high and this will lead to long time constant in the withdrawal circuit.

D.C. Transformer

In the presence of a transport current lines if magnetic flux are moved as if under the action of Lorentz force. If a magnetic field is established around the device and the input voltage applied to the wider strip, then the flux lines in the primary circuit will move cutting the secondary and, therefore, inducing a voltage at the output. The efficiency of the transformer is only about 10% but the voltage output may be made larger by increasing the number of thin film units to which the primary voltage is applied.

D.C. Power Transmission

Since there is no joule heat loss in a superconducting cable. So we prefer to use superconducting cables for d.c. power transmission. To use such cables, however, large amounts of power would be lost in rectifying the alternating current coming from the generators and also in keeping the cables cool. The power loss in conventional cable is fairly small, so that with the present materials and cost of rectification and cooling is reducing and in applications where direct current is already used, *e.g.* in parts of chemical industry, the time may soon come when the costs of using superconducting and conventional cables may break even.

Electrical Switching Element

Because of the infinite ratio between resistance of a metal in the normal state and superconducting state, a superconductor can be used to switch electrical cuurent between different possible paths. Either temperature or magnetic field can be used to control the change of state of such a switch. A temperature controlled switch uses a film of lead as the switching element. The current carrying film is operated below the transition temperature so as to be superconducting. The current can be interrupted by means of a heating coil made of non-superconducting metal. As the lead film becomes resistive, the current flowing through it is divided through a parallel path of low resistance.

Superconducting Fuse and Breaker

Thin insulated superconductor film gets its normal state after the current of more than critical density has passed through it. Such film can be used instead of a fuse. Selecting the necessary material for a film it is possible to regulate very precisely the admissible intensity of current. If a long film is taken and at the moment of its transition into normal state its resistance is rather high, then it can be used as a breaker.

26.10 LONDON EQUATION

It has been found that the Meissner effect in bulk superconductors led to the equation

$$\vec{B} = \vec{B}_a + 4\pi\vec{M} \text{ (CGS)} = 0 \quad i.e. \quad (\vec{B}_a + 4\pi\chi\,\vec{B}_a) = 0$$

or
$$\chi = -\frac{1}{4\pi} \qquad (\because \ \vec{B}_a = \vec{H}_a \text{ in CGS units} = \text{Applied field})$$

and
$$-4\pi\vec{M} = \vec{B}_a$$

But when the superconductor is taken as a thin film there is a penetration of magnetic flux into the superconduction. This was later explained through London equation which is being given below.

Let us postulate that current density, $\vec{J}$, in superconductor is proportional to the vector potential $\vec{A}$ so that

$$\vec{J} = -\frac{c}{4\pi\lambda_L^2}\vec{A} \quad \text{and} \quad \vec{B} = \vec{\nabla}\times\vec{A} \text{ at any point } \vec{r}.$$

$$-\frac{c}{4\pi\lambda_L^2} \text{ being the constant of proportionality}$$

Then, $\vec{\nabla}\times\vec{J} = -\frac{c}{4\pi\lambda_L^2}\vec{\nabla}\times\vec{A} = -\frac{c}{4\pi\lambda_L^2}\vec{B}$, c is the velocity of light.

This relation between $\vec{J}$ and $\vec{A}$ is termed as London equation. The boundary conditions $\vec{\nabla}\cdot\vec{A} = \vec{\nabla}\cdot\vec{J} = 0$ and $A_n = J_n = 0$ are implied in the London equation.

Now, we use the Maxwell's equation

$$\vec{\nabla}\times\vec{B} - \frac{1}{c}\frac{\partial\vec{D}}{\partial t} = \frac{4\pi\vec{J}}{c} \text{ (CGS)}$$

For the static case $\frac{\partial}{\partial t}D = 0$. Hence.

$$\vec{\nabla}\times\vec{B} = \frac{4\pi\vec{J}}{c} \quad \text{and} \quad \nabla\times(\nabla\times\vec{B}) = \frac{4\pi}{c}\vec{\nabla}\times\vec{J}$$

That is $-\nabla^2\vec{B} = -\frac{1}{\lambda_L^2}\vec{B}$, obtained by using the earlier value of $\vec{\nabla}\times\vec{J}$ and the fact that

$$\vec{\nabla}\cdot\vec{B} = 0$$

Thus, we have
$$\nabla^2\vec{B} = \frac{1}{\lambda_L^2}\vec{B}$$

which result clearly suggests that $\vec{B}$ is not a constant quantity, rather it would decrease with depth. For example, taking $\vec{B}$ along parallel to the face of the semi infinite thin superconductor which is placed along positive side of x-axis (see Fig. 26.9), we get

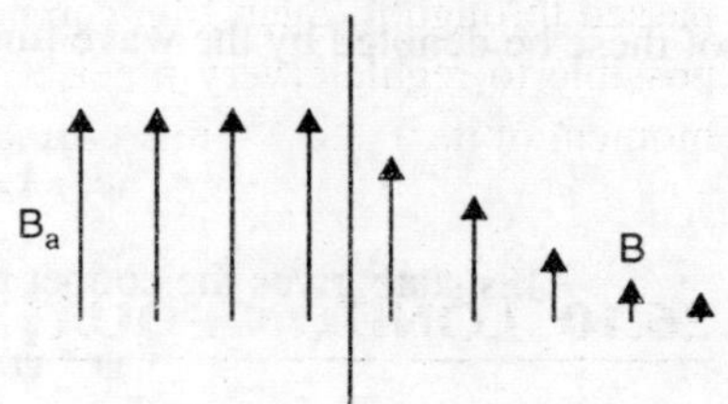

FIGURE 26.9 **Penetration of applied magnetic field in the superconductor.**

$$\frac{d^2}{dx^2}B = \frac{1}{\lambda_L^2}B$$

On integration one gets

$$B(x) = B(0)\, e^{-\frac{x}{\lambda_L}},$$

$B(0)$ being the field at $x = 0$ (face of the superconductor). Thus the applied magnetic field, B, gets damped with increasing penetration and falls to e^{-1} times $B(0)$ just after $x = \lambda_L$; λ_L is called London penetration depth. The semi quantum mechanical calculations yield the value of λ_L as

$$\lambda_L = \left(\frac{mc^2}{4\pi nq^2}\right)^{1/2}$$

where m, n, and q denote the mass, concentration and charge of the carrier in the superconductor. The calculated values of λ_L for some material have been listed in Table 26.4.

TABLE 26.4 Calculated values of λ_L in 10^{-6} cm

Metal	*λ_L in 10^{-6} cm (CGS)*	*SI Units*
Sn	23	$\lambda_L = \left(\frac{\varepsilon_0 mc^2}{nq^2}\right)^{1/2}$
Al	160	because
Pb	8.3	$\frac{1}{4\pi} = \varepsilon_0$
Cd	76	
Nb	3.8	

Note that the above relation $\nabla^2 \vec{B} = \frac{\vec{B}}{\lambda_L^2}$ ensures that if $\vec{B}$ has a constant value it must be zero and thus accounts for the Meissner effect in superconductors and lastly $\vec{J} = 0$ in a region where $\vec{B} = 0$.

26.11 DERIVATION OF LONDON EQUATION

Let us suppose that there are n electron-pairs (cooper pairs) per unit volume in a superconductor and the state of these be denoted by the wave function

$$\psi(\vec{r}) = n^{\frac{1}{2}}\, e^{i\theta(\vec{r})}$$

This state gives the cooper pair density as

$$\psi^* \psi = n \quad \text{as expected}$$

Further, since the velocity operator is (in electromagnetic field)

$$\vec{v} = \frac{1}{m}\left(\vec{p} - \frac{q}{c}\vec{A}\right)$$

where various letters have usual meanings, the current density will be given by

$$\vec{J} = q\psi^* \vec{v}\, \psi^-$$

where q denotes the charge on the electron-pairs which cause the super current. By substituting the value of $\vec{v}$ one gets

$$\vec{J} = \frac{nq}{m}\left(-i\hbar\vec{\nabla}\theta - \frac{q}{c}\vec{A}\right) \quad \text{because} \quad \vec{p} = -i\hbar\vec{\nabla}$$

Taking curl of both sides, we get

$$\vec{\nabla}\times\vec{J} = \frac{-nq^2}{mc}\nabla\times A$$

But $\vec{\nabla}\times\vec{A} = \vec{B}$,

therefore,

$$\vec{\nabla}\times\vec{J} = \frac{-nq^2}{mc}\vec{B} \quad \text{in C.G.S. units}$$

$$= \frac{nq^2}{m}\vec{B} \text{ in SI units.}$$

This equation is known as London equation in the literature. It takes the form

$$\vec{\nabla}\times\vec{J} = -\frac{c}{4\pi\lambda_L^2}\vec{B} \text{ (C.G.S.)}$$

such that

$$\lambda_L = \sqrt{\frac{mc^2}{4\pi nq^2}}$$

where m, n and q denote the mass, concentration and the charge of the electron-pair. c being the velocity of light. This London penetration depth for the given superconductor may be calculated knowing the value of n as

$$q = 2e$$

$m = 2m_e$, m_e being the mass of electron. It is of the order of 10^{-6} cm.

Also, this equation reveals that when no current passes through the superconductor while placed in magnetic field, $\vec{B}$ will be zero within this superconductor. This explains Meissner effect.

26.12 SQUID—A BRIEF NOTE

SQUID is a device abbreviated from superconducting quantum interference device. It is constructed using superconducting Josephson junctions. Josephson junction consists of a very thin layer of insulation (~ 10 Å) say oxidized layer sandwitched between two superconducting films say of tin or lead. This junction is usually known as weak link. The important characteristic of such a junction is that it becomes resistive when the superconducting current exceeds a certain limit called critical current. The device which uses such junctions are called SQUIDS.

SQUID circuits are of two types, *d.c.* and *a.c.* SQUID circuits. In *d.c.* type, two weak links are used in a ring while in *a.c.* circuit only one weak link (Josephson junction) is used. These weak links of three types shown in Fig. 26.10. In Fig. 26.10(*a*) a thin oxidized strip is sandwitched between two superconducting films.

In Fig. 26.10(*b*) the two superconducting films A and B are separated by a thin insulating film all are deposited on the same substrate of glass. In Fig. 26.10(*c*) there is a superconducting screw with an oxidized point pressing against a superconducting film.

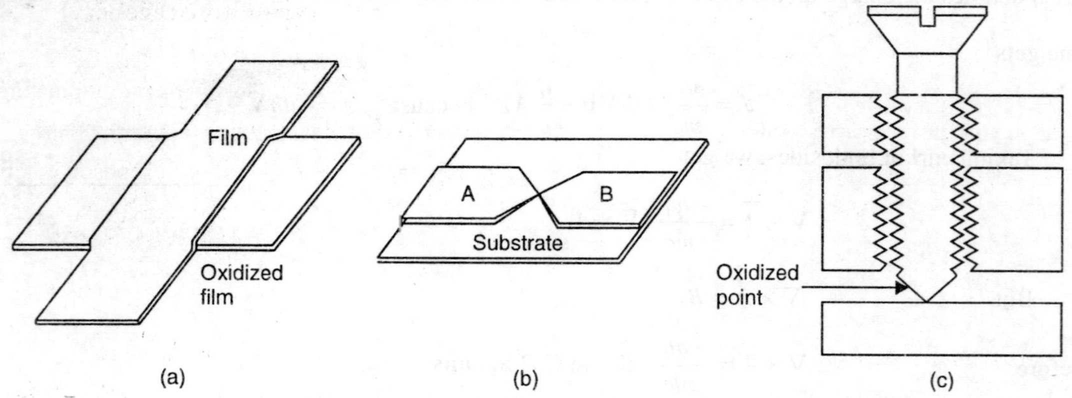

FIGURE 26.10 **Display of different types of Josephson junction in use in SQUID devices.**

As has been mentioned, a SQUID device is a circuit containing one or two junctions (also called as weak links). This device works on the following principles which have been experimentally established and explained through BCS (Bardeen, Cooper, Schrieffer) theory of superconductivity.

1. Quantization of Magnetic Flux through a Superconducting Ring. This states that a superconducting ring always encloses a magnetic flux equal to an integral multiple of $\frac{2\pi\hbar}{q}$ SI units with q as charge on electron pairs (= 2e). These electron-pairs called cooper pairs conduct super currents in a superconductor. The value of this quantum of flux will be equal to

$$\frac{\pi \times \hbar}{e} = \frac{3.142 \times 1.0546}{1.6} \times 10^{-15} \text{ Tesla m}^2$$

$$= 2.071 \times 10^{-15} \text{ Tesla}$$

This unit of flux is called fluxoid or fluxon. Thus, a SQUID device can measure a magnetic flux of this order.

2. D.C. Josephson Effect. A constant *d.c.* current always flows through a junction or weak link without applying potential difference or magnetic field. This is called *d.c.* Josephson effect which is the result of electron pair tunneling through the junction. The maximum value of this *d.c.* current depends on the square root of the electron pairs concentrations on the two sides of the junction. This is easily explained by BCS theory and accounts for the presence *d.c.* current through the junctions.

3. A.C. Josephson Effect. When a *d.c.* potential difference is applied across the junction an oscillator current is set up in the circuit along with the emission or absorption of radiation. The frequency of such a current or for that matter of the radiation is given by (BCS theory)

$$\omega = \frac{2eV}{\hbar}, \quad V = \text{applied p.d.}$$

or

$$2\pi f = \frac{2eV}{\hbar}$$

Hence, $$f = \frac{eV}{\pi\hbar}$$

which for 1 μV potential difference comes out as

$$\frac{1.6}{3.142 \times 1.0546} \times 10^9 \text{ Hz} = 483 \text{ MHz}$$

Therefore, by measuring the frequency of light emitted one can either calculate the important constant $\hbar/e$ or V, very accurately. So, SQUID device using such a junction may be used as voltage standard.

Therefore, the SQUID circuits in which one weak link or two weak links (called Josephson junctions) are employed may be used as

(*a*) Magnetocardiographs

(*b*) Magnetoexcephalographs

(*c*) Voltage standard device and

(*d*) Small magnetic flux detector.

These are some of many applications in which SQUID circuits are put to use.

EXERCISE

1. What is superconductivity ? Outline the experimental facts about this phenomenon.
2. Explain Meissner effect.
3. Bring out the salient experimental features of superconductivity.
4. Mention the five important applications of superconducting materials.
5. Define critical magnetic field and how does it varies with temperature in Type I and Type II superconductors.
6. Explain the effect of isotopes on superconductors.
7. Explain the behavior of Type I and Type II superconductors.
8. Explain the variation of entropy and thermal conductivity when a specimen changes from normal to superconducting state.
9. What does the equation $dB/dt = 0$ tells ? Is this equation adequate to explain superconductivity ? If not, why ?
10. Describe Meissner effect in superconducting materials and explain main feature of Type I and Type II superconductors.
11. Explain how Meissner effect contradict Maxwell's equation ?
12. Define superconductivity and give its features.
13. Give important features and applications of superconductivity.
14. What are persistent currents explain.
15. Write a short note on 'SQUIDS'.
16. What is the effect of magnetic field and temperature on superconductivity ?
17. Write a short note on : (1) Isotope effect, (2) Variation of entropy with temperature, (3) Variation of specific heat with temperature, and (4) Variation of thermal conductivity with temperature.
18. Write detailed note on superconductivity. (*K.U. June, 2007*)

Chapter 27 PHOTOCONDUCTIVITY

27.1 INTRODUCTION

Photoconductivity is the phenomenon of increasing the electrical conductivity of a crystal by irradiating it with photons of suitable energy. We know that when photons of energy greater than the energy gap are absorbed in the crystal, free electron-hole pairs are formed. The absorbed photon raises the electron from valence band to conduction band. Consequently, a hole is created in the valence band; and both, hole in valence band and electron in conduction band, may move independently through the crystal and thus enhance the conductivity.

Once electron and hole have been produced by the absorption of photon of sufficient energy, they will remain free until they are captured at an imperfection. These capturing imperfections are called as capturing centers and are of the following two types

(*i*) Trapping center

(*ii*) Recombination center

If the captured charge carrier has greater probability of being thermally reexcited to free state than that of combining with a charge carrier of opposite sign at the imperfection then the capturing center is called as trapping center. Obviously trapping center removes the charge carrier temporarily. On the other hand, if the captured charge carrier has a greater probability of combining with a charge carrier of opposite sign at the imperfection than that of being reexcited to the free state then the capturing center is called as recombination center. Thus recombination results in the permanent loss of charge carrier.

It may be mentioned here that the holes and electrons will eventually reunite with permanent loss of charge carriers either by direct recombination or through recombination centers. But before reuniting, each may spend different amounts of time trapped on the impurities and imperfections in the crystal. Since traps have different influence on the holes and electrons, it is difficult to conclude that both types of charge carriers make equal contribution to the photoconductivity in a given specimen.

27.2 PHOTOCONDUCTING MATERIALS

In 1873 Smith observed a photoconductive effect in Selenium. Later, it was found that the Selenium photocell depended not on photoconductivity in Selenium but depended on the photovoltaic effect *i.e.* the generation of a voltage because of absorption of light.

In 1916 Pfund announced that he discovered photoconductivity in cuprous oxide. A few years later in 1920, case disclosed another new photosensitive material, namely thallous sulphide. Some other photoconductive materials reported are bismuth sesquisulphide, molybdenum sulphide, silver sulphide, lead sulphide, silver oxide and mercury sulphide.

27.3 SIMPLE MODEL OF A PHOTOCONDUCTOR

In the simplest model we suppose that electron-hole pairs are produced uniformly throughout the volume of the crystal by radiation incident upon it ; and in recombination process, electrons directly combine with holes. We further suppose that the electrons, equal to number of electrons leaving the crystal at one electrode, enter into the crystal from the opposite electrode. Since the mobility of holes is small compared to that of electrons, the former can be neglected. With all these supposition, we can write the rate of change of electron co-ncentration 'n' as

$$dn/dt = L - Anp$$

where L is the number of photons absorbed per unit volume per unit time, and Anp is bimolecular recombination rate, which is proportional to the product of hole and electron concentration and A is called recombination coefficient.

Using $n = p$, we get

$$dn/dt = L - An^2 \qquad \text{...(27.1)}$$

In steady state, $dn/dt = 0$

so that from eqn.(27.1), we get

$$L = An_o^2, \quad \text{where } n_o \text{ is the steady state electron concentration}$$

or

$$n_o = (L/A)^{1/2} \qquad \text{...(27.2)}$$

Since the corresponding conductivity is given as

$$\sigma = n_o e\mu, \quad \mu \text{ being the mobility and } e \text{ the electronic charge}$$

therefore, using eqn.(27.2), we have

$$\sigma = (L/A)^{1/2} e\mu. \qquad \text{...(27.3)}$$

Hence, the photo current density will be

$$J = (L/A)^{1/2} e\mu(V/d) \qquad (\because \quad J = \sigma E \quad \text{and} \quad E = V/d) \qquad \text{...(27.4)}$$

Where V is the voltage across the specimen and d is the thickness of the specimen.

Eqn.(27.4) shows that photo current varies as $(L)^{1/2}$. However, the actual exponents observed range between $\frac{1}{2}$ and 1 or even higher.

If the light is switched off, then $L = 0$ and eqn.(27.1) provides

$$dn/dt = -An^2 \qquad \text{...(27.5)}$$

representing the decay of charge carriers.

Now, eqn.(27.5) can be written in the following form

$$dn/n^2 = -Adt$$

integrating it we obtain

$$1/n = At + C \qquad ...(27.6)$$

where C is the constant of integration.

Since, at $t = 0$, $n = n_o$(electron concentration at $t = 0$)

therefore, $C = 1/n_o$

Thus, $$1/n = At + 1/n_o \quad \text{or} \quad 1/n = (1 + Atn_o)/n_o \qquad ...(27.7)$$

or $$n = n_o/(i - Atn_0)$$

Obviously, if $$t = t_o = 1/An_o,$$

then $n = n_o/2$. Therefore time in which charge carrier concentration drops to half of its steady state value is

$$t_0 = 1/An_o = (LA)^{-1/2} \qquad \text{[using eqn.(27.2)]}$$

$$= n_o/L \qquad ...(27.8)$$

or $$t_o = \sigma/Le\mu \qquad \text{[using eqn.(27.3)]}$$

It predicts that response time to should be directly proportional to the photoconductivity at a given illumination level L. In other words, sensitive photoconductors should have long response time and is found to be of several seconds in case of cadmium sulphide light meter.

The photoconductivity G (also called gain factor) is defined as the number of charge carriers crossing the specimen (J/e) to the photon absorbed in the specimen (Ld). Thus using eqn.(27.4), we have

$$G = (J/e)/Ld = [(L/A)^{1/2}\mu(V/d)]/Ld = V\mu/d^2\,(LA)^{-1/2} \qquad ...(27.9)$$

obtained using eqn. (27.4)

However, the value of G calculated through this relation is found to be much larger (by a factor of 10^8 in some cases) than that observed experimentally.

27.4 EFFECTS OF TRAPS ON PHOTOCONDUCTIVITY

A trap is an impurity atom or imperfection in the crystal capable of capturing an electron or hole. In other words, it is an energy level that can capture either electron or hole easily but not both. The captured charge carriers may be reemitted at a subsequent time and consequently, becomes free to move until captured by a recombination center or by another trap.

Generally, two kinds of traps are observed. One kind of trap acts principally to promote electron hole recombination by capturing electrons from the conduction band and subsequently transferring them to valence band whenever a hole appears near the trap to recombine with the trapped electron. The pair thus lost and the trapping level returns to its original state. In this process of annihilation no or little electromagnetic energy is emitted and since the trapping center is tightly coupled to the lattice, energy and momentum initially belonging to the electron and hole is converted into lattice vibrations. The four fundamental processes involved in recombination through traps are illustrated in Fig. 27.1.

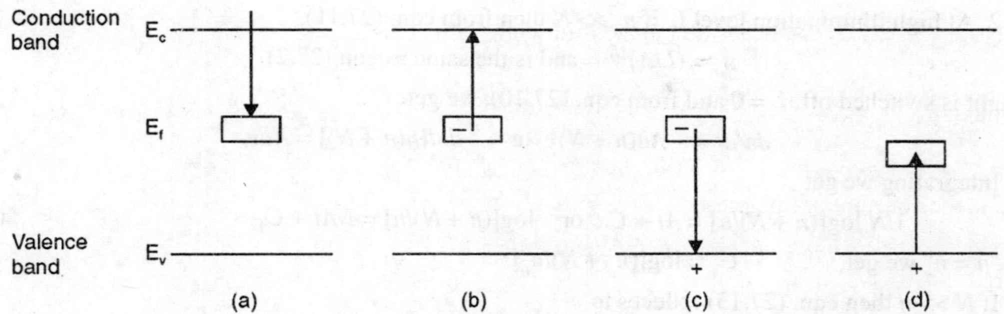

FIGURE 27.1 ***(a)*** **The capture of an electron from the conduction band by an initially neutral empty trap ; *(b)* The emission of an electron from the trap to the conduction band ; *(c)* The capture of a hole from the valence band by a trap containing an electron ; *(d)* The promotion of a valence electron into an initially empty trap.**

Such type of traps assists in restoring thermal equilibrium. Under such circumstances, the rate at which electrons from the conduction band are captured by traps will clearly be proportional to the number of electrons in the conduction band and to the number of empty traps available to receive them.

Another type of traps, with which we are concerned here, do not directly help in recombination process but restrict the freedom of motion of holes or electrons. Let us consider a crystal with N electron trap levels per unit volume (Fig. 27.2).

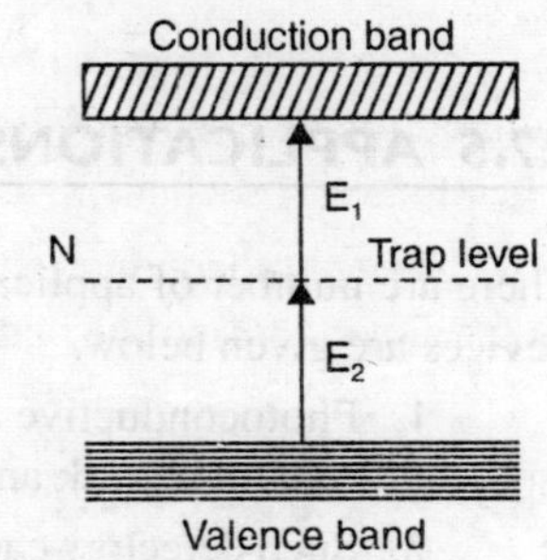

FIGURE 27.2 **Model for photoconductivity with electron traps in concentration N.**

We suppose that the temperature is quite low compared to ionization energies so as to neglect the concentration of thermal charge carriers. If the recombination coefficient A is same for both the electron hole recombination and the electron trap capture, then the rate of change of electron concentration is given by

$$dn/dt = L - A(n + N)n \qquad ...(27.10)$$

where n is the electron concentration in the conduction band. In the above equation, we have neglected the term representing the rate of thermal evaporation of trapped charge carrier back into the conduction band.

In steady state $\quad dn/dt = 0,$

giving

$$L/A = (n_o + N)n_o, \qquad ...(27.11)$$

where n_o is the steady state electron concentration. Since to grow the crystal with trap concentration (N) much less than 10^{14} cm^{-3} is difficult, so we can take the least value of N to be of the order of 10^{14} cm^{-3}. Two cases arise

1. At low current level the charge carrier concentration n_o will be of the order of 10^8 or 10^{10} cm^{-3}. We can then neglect n_o compared to N so that eqn.(27.11) gives

$$n_o N = L/A \quad \text{or} \quad n_o = L/NA$$

then photo current density will be

$$J = n_o e\mu\, V/d$$

$$e\mu\, LV/(NAd) \qquad ...(27.12)$$

This equation indicates that photocurrent is directly proportional to illumination and presence of traps reduces the conductivity.

2. At high illumination level L, if $n_o >> N$ then from eqn. (27.11),

$$n_o = (L/A)^{1/2} \quad \text{and is the same as eqn.(27.2)}$$

when light is switched off, $L = 0$ and from eqn. (27.10), we get

$$dn/dt = -An(n + N) \quad \text{or} \quad -dn/[n(n + N)] = Adt$$

Integrating we get

$$1/N \log[(n + N)/n] = At + C \quad \text{or} \quad \log[(n + N)/n] = NAt + C_1 \qquad ...(27.13)$$

at $t = 0$, $n = n_o$ we get, $\quad C_1 = \log[(n_o + N)/n_o]$

If $N >> n$ then eqn. (27.13) reduces to

$$\log(N/n) - \log(N/n_o) = NAt$$

or

$$\log(n/n_o) = -NAt \quad \text{or} \quad n = n_o e^{-NAt} \qquad ...(27.14)$$

For

$$t = t_o = 1/NA, \qquad ...(27.15)$$

$$n = n_o/e$$

So the time in which charge carrier concentration falls to $1/e$ of its steady state value is $1/NA$. Which is entirely different from the values of t_o when traps are absent. From eqns. (27.12) and (27.15) one conclude that conductivity and response time reduces in the presence of traps.

27.5 APPLICATIONS OF PHOTOCONDUCTIVITY

There are number of applications of photoconductive devices. Some of the important applications of these devices are given below.

1. Photoconductive light detectors can be used in the home to control automatic night lamps which turn on at dusk and turn off at dawn.
2. Such detectors can also be used to measure illumination level, as in exposure meters for cameras.
3. Detectors are used in optical signaling system in which information is transmitted by a light beam and is received at photoconductive cell.
4. Many systems include a light beam aimed at the photoconductor, to signal the presence of an object between the source and detector.
5. Such systems are also useful in moving object counters etc.
6. These are used for burglar alarms and many other applications.

27.6 PHOTOVOLTAICS AND PHOTOVOLTAIC CELL

The direct conversion of solar energy into electrical energy by means of the photovoltaic effect, that is, the conversion of light or other electromagnetic radiation into electricity is called photovoltaics. The photovoltaic effect is defined as the generation of an electromotive force as a result of the absorption of the ionizing radiation. Energy conversion devices that are used to convert sunlight into electricity by the use of the photovoltaic effect are called solar cell. A single coverter cell is called a solar cell or, more generally, a photovoltaic cell, and combination of such cells, designed to increase the electric power output is called a solar module or solar array.

Photovoltaic cells are made up of semiconductors that generate electricity when they absorb light. As photons are received, free electrical charges are generated that can be collected on contacts applied to the surfaces of the semiconductors. Generally, these are made to operate in the presence of the weak energy flux of solar radiation, at room temperature. These devices have theoretical efficiencies of the order of 25%. Actual operating efficiencies are less than half of this value, and decreases fairly rapidly with increasing temperature.

27.7 PRINCIPLE OF A PHOTOVOLTAIC CELL

The photovoltaic effect can be observed in nature in a variety of materials, but the materials that have shown the best performance in sunlight are the semiconductors as mentioned earlier. When photons from the sun are absorbed in a semiconductor, they create free electrons with higher energies than the electrons which provide the bonding in the bare crystal. Once these electrons are created, there must be an electric field to induce these higher electrons to flow out of the semiconductor to do useful work. The electric field in most solar cells is provided by a junction of materials, which have different electrical properties.

To obtain a useful power output from photon interaction in a semiconductor three processes are required.

1. The photons have to be absorbed in the active part of the material resulting in the excitation of electrons to a higher energy state.

2. The electron-hole charge carrier created by the absorption must be physically separated and moved to the edge of the cell.

3. The charge carriers must be removed from the cell and delivered to a useful load before they loose their extra potential.

In order to complete the above processes, a solar cell consists of

(*a*) Semiconductor in which electron-hole pairs are created by absorption of incident solar radiation.

(*b*) Region containing a drift field for charge separation, and

(*c*) Charge collecting front and back electrodes.

The photovoltaic effect can be described easily for *p-n* junction in a semiconductor. The most normal configuration for a solar cell to make a *p-n* junction is shown in Fig. 27.3.

The junction of the *p*-type and *n*-type materials provide an inherent electric field which separates the charge created by the absorption of sunlight. This *p-n* junction is usually obtained by putting a *p*-type based material into a diffusion furnace containing a gaseous *n*-type dopent such as phosphorus and allowing the *n*-dopent to diffuse into the surface about 0.2 μm. The junction is thus formed slightly below the planar surface of the cell and light impinges perpendicular to the junction.

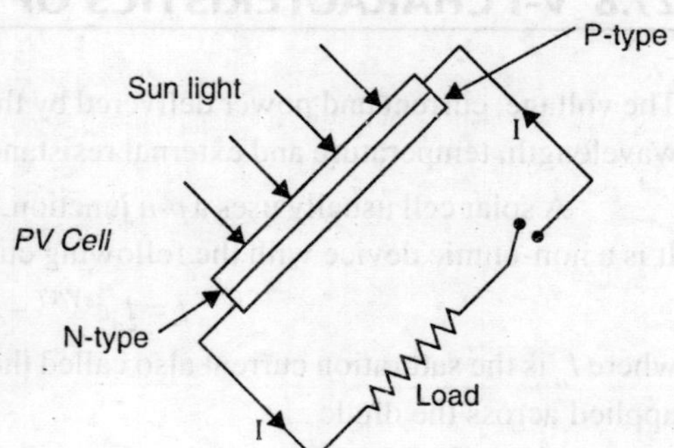

FIGURE 27.3 **Schematic diagram of a PV-cell (solar cell).**

The positive and negative charges created by the absorption of photons are thus encouraged to drift to the front and back of the solar cell. The back is completely covered by a metallic contact to remove the charges to the electric load. The collection of charges from the front of the cell is aided by a fine grid of narrow metallic fingers. An antireflective coating is applied on the top of the cell.

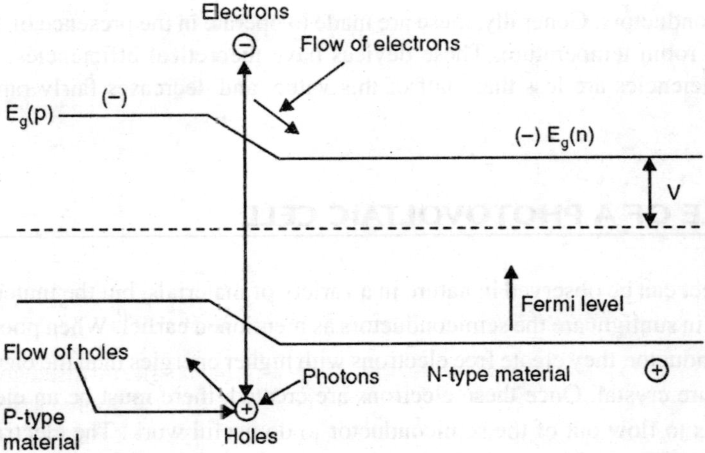

FIGURE 27.4 ***p-n* junction electric fields.**

Fig. 27.4 demonstrates how this *p-n* junction provides an electric field that sweeps the electrons in one direction and the positive holes in the other. If the junction is in thermodynamic equilibrium, then the Fermi energy must be uniform throughout. Since the Fermi level is near the top of the gap of an *n*-doped material and near the bottom of the *p*-doped side, an electric field must exist at the junction providing the charge separation function of the cell. Important characteristics of the Fermi level is that, in thermodynamic equilibrium, it is always continuous across the contacts between the two materials.

Each of the individual solar cell in a solar module will produce power at about 0.5 V with the current directly proportional to the cell's area. The individual cells are connected in series-parallel combination to meet the voltage, power and reliability requirements of the particular applications.

27.8 V-I CHARACTERISTICS OF A SOLAR CELL

The voltage, current and power delivered by the solar cell are influenced by conditions of sunlight, intensity, wavelength, temperature and external resistance (R).

A solar cell usually uses a *p-n* junction, its physical configuration is shown schematically in Fig. 27.3. It is a non-ohmic device with the following current and voltage relationship

$$I = I_o e^{eV/kT} - 1 \quad \text{...(27.16)}$$

where I_o is the saturation current also called the dark current and is obtained when a large negative voltage is applied across the diode

V = voltage across junction

e = electronic charge

k = Boltzman constant

T = absolute temperature.

When light impinges on the junction, electron-hole pairs are created at a constant rate providing an electrical current flow across the junction. The net current is thus the difference between the normal diode current and light generated current I_L. Simplified equivalent circuit for the cell is shown in Fig. 27.5.

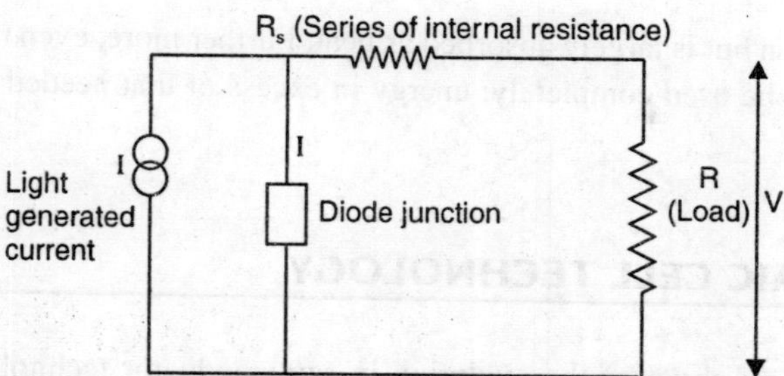

FIGURE 27.5 **The equivalent circuit of solar cell.**

The internal series resistance R_s is mostly due to high resistance of the diffused layer which is in series with the junction. The light generated current, I_L, acts as a constant current source supplying the current to either the junction or a useful load depending on the junction characteristics and the value of external load resistance. The net current I_n is given by

$$I_n = I_L - I = I_L - I_o[e^{eV/kT} - 1] \quad ...(27.17)$$

The internal voltage drop in a cell can usually be minimized, and for ideal cell R_s may be assumed equal to zero *i.e.*, $R_s = 0$, with these the corresponding *I-V* plot is given in Fig. 27.6.

The open circuit voltage V_{OC} for ideal cell is given by

$$V_{OC} = (kT/e) \log[(I_L/I_o) + 1] \quad ...(27.18)$$

If $I_L >> I_o$ then $I_L/I_o + 1 = I_L/I_o$

The open circuit voltage

$$V_{OC} = (kT/e) \log(I_L/I_o) \quad ...(27.19)$$

In practice the open circuit voltage of the cell decreases with increasing temperature. The maximum power that can be derived from the device is given by

$$P_{max} = V_m.I_m \quad ...(27.20)$$

where V_m and I_m are the voltage and current at maximum power point as shown in Fig. 27.6. The maximum efficiency, η, of the cell is obtained by dividing $V_m.I_m$ by the total power density of the sunlight P_s

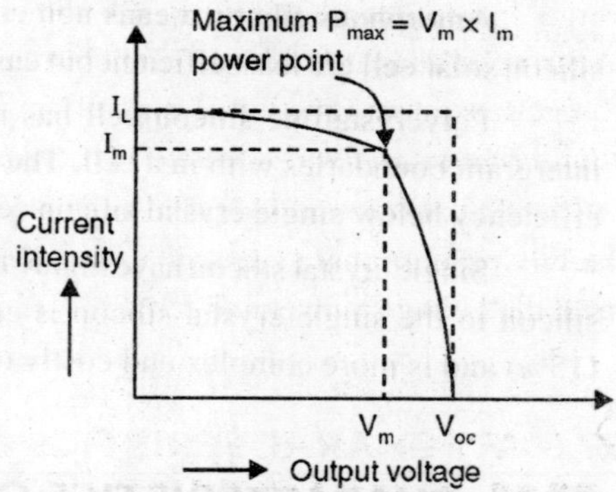

FIGURE 27.6 **A typical *I-V* plot for ideal solar cell.**

Thus $\eta = V_m I_m / P_s = (E_g I_L / eP_s)(V_m I_m / I_L V_{OC})(eV_{OC}/E_g)$

The fill factor (*FF*) for a solar cell is defined as the ratio of two areas shown in Fig. 27.6

or
$$FF = V_m I_m / I_L V_{OC} \quad ...(27.21)$$

The solar cell designer strive to increase the fill factor values, to minimize internal losses. Maximum power can be defined in terms of V_{OC} and I_L and is given by

$$P_{max} = I_L.V_{OC}.FF \quad ...(27.22)$$

A typical value of the fill factor for a good silicon cell is about 0.8. The voltage factor eV_{OC}/E_g is determined by the basic properties of the materials in the cell and is typically about 0.5 for silicon cell.

For both theoretical and practical reasons, not all of the solar radiation energy falling on a solar cell can be converted into electrical energy. A specific amount of energy is required to produce a free electron and a hole in the semiconductor material. In silicon for example, minimum energy is 1.1 eV and this is available in radiation having a wavelength of 1.1 cm. Consequently infrared radiation of longer wave length has no photovoltaic effect in silicon but is largely absorbed as heat. Further more, even the energy of radiation with shorter wavelength cannot be used completely; energy in excess of that needed to free a bound electron is simply converted into heat.

27.9 PHOTOVOLTAIC CELL TECHNOLOGY

The *PV* technology is an off shoot of the evolution in semiconductor technology during 1980s. Several different base materials and doping materials were tried during 1980s. By mid 1980s the silicon has been adopted as base material for producing *PV* cells by almost all the PV cell manufacturers. PV cells which have only silicon as the base for *p-n* junction are called homojunction cells and those which have two base materials (*e.g.*, CdS/Cu_2S) are called heterojunction cells.

Homojunction cells with silicon base are most successful and have following three types

1. Amorphous silicon
2. Polycrystalline silicon
3. Single crystal silicon

Amorphous silicon means non crystalline silicon. Such material is used in filmprocess. Amorphous silicon solar cell are least efficient but easy to manufacture. Efficiency of amorphous cells do not exceed 5%.

Polycrystalline silicon cell has many crystal in a single silicon. The polycrystalline material has intergrain boundaries with in a cell. The electrons are inhibited at these boundaries resulting in reduction in efficiency below single crystal silicon cells. Efficiency of polycrystalline solar cell is about 7%.

Single crystal silicon have highly ordered molecular structure. The process of converting polycrystalline silicon to the single crystal silicon is called 'Crystal growing'. Single crystal silicon has high efficiency (15%) and is more complex and costly to produce.

27.10 CHARACTERISTICS OF PHOTOVOLTAIC CELLS

Following are some characteristics of Photovoltaic Cells.

1. By these cells conversion of light into electricity is possible even at room temperature.
2. Modular nature in which desired currents, voltage and power levels can be mere integration.
3. These are easy to fabricate and have low maintenance cost, long life and high reliability.
4. These do not have moving and pollution creating parts.
5. No fuel is required to operate.
6. These have wide power handling capabilities from microwatts to kilowatts or even megawatts when modules are combined into large area arrays. Solar cells can be used in combination with proper power conditioning circuitry to feed power into utility grid.
7. These have high power to weight ratio. This characteristics is more important for space applications than terrestrial.
8. These can be used with or without sun tracking, making a wide range of applications feasible.

EXERCISE

1. What is photoconductivity ? Show how does photoconductivity depend on intensity of incident light in the presence of traps. Derive an expression for photoconductive gain and give some use of photoconductivity.
2. Define photoconductivity. What are traps ?
3. Discuss the effect of traps on photoconductivity of a photoconductor.
4. Describe principle of solar photovoltaic energy conversion. *(K.U. June, 2007)*
5. What are the advantages and disadvantages of photovoltaic solar energy conversion ?
6. Discuss the *V-I* characteristics of a solar cell and find expression for the efficiency and Fill Factor (*FF*) of a solar cell.
7. A S_i solar cell has an area of 5 cm^2. Under AMI illumination it develops $V_{OC} = 0.7$ V. Under these conditions $V_m/V_{OC} = 0.85$ and the fill factor is 0.75. Estimate the value of I_m. If the incident power is 50 mW/cm^2, calculate the cell efficiency.
8. Describe the simple model which would explain the effect of variation of illumination and effect of traps. *(K.U. June, 2007)*

Chapter 28

MAGNETIC PROPERTIES OF MATERIALS

28.1 INTRODUCTION

In order to understand the behaviour of magnetic materials such as dia-, para-, and ferro-magnetic and their role in the magnetic circuits one requires the background of magnetic effects due to current circuits (current loops). For example, a current loop acts as magnetic dipole and hence a source of magnetic field. Also one current circuit exerts a force on the other as had been seen experimentally. Therefore, first of all we discuss the various laws such as Lorentz force, Biot-Savart law, Ampere's circuital law etc. and then how these are used to understand the properties of magnetic materials.

28.2 LORENTZ FORCE AND MAGNETIC INDUCTION

A moving charge experiences a force in a magnetic field according to

$$\mathbf{F} = q\,\mathbf{v} \times \mathbf{B}$$

where q is the quantity of charge, **v**, the velocity and the vector **B** known as magnetic induction. This result is well known as Lorentz force. Thus, the unit of **B** will be newton/Ampere-meter. If we have a wire carrying a steady current, I, and is placed in a magnetic field, it will also experience a force. Suppose the wire contains N number of charge carrier per unit volume and has a cross-sectional area A, the steady current will be given by

$$\mathbf{I} = NAq\mathbf{v}$$

and force on dl length will be

$$dF = NA\,dl\,q\,\mathbf{v} \times \mathbf{B} = I\,\mathbf{dl} \times \mathbf{B}$$

we have taken **dl** as a vector along the direction of current or **v** and of magnitude dl. The total force on the wire will be

$$F = I\int_c \mathbf{dl} \times \mathbf{B}$$

where c represents the closed contour of the circuit carrying a current I.

28.3 MAGNETIC MOMENT OF THE CIRCUIT

If A denotes the area within the loop of current, the magnetic moment of the loop is given by

$$\mathbf{m} = I\mathbf{A} \quad \text{where } I \text{ is the current}$$

$$= I/2 \int_c \mathbf{r} \times \mathbf{dl}$$

since $1/2 \int_c \mathbf{r} \times \mathbf{dl}$ = area of the closed contour, $\mathbf{r}$ being the position of $\mathbf{dl}$ from the reference point within the contour.

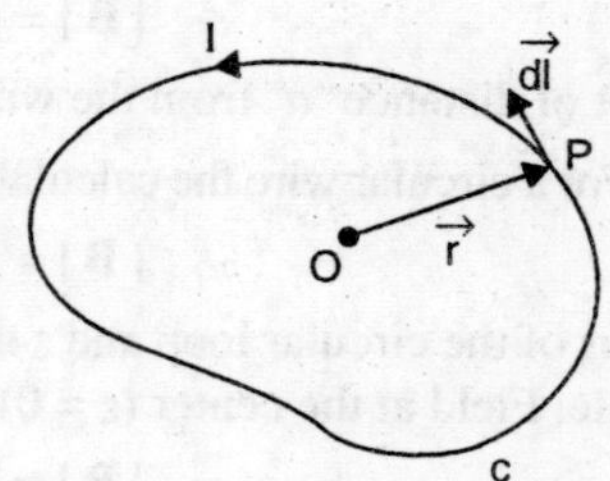

FIGURE 28.1 **c is the contour of the loop carrying current I and O the reference point ; OP = r.**

28.4 BIOT-SAVART LAW

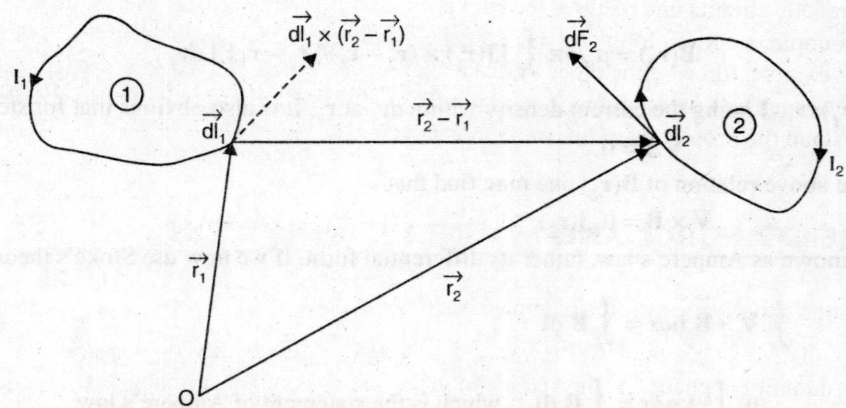

FIGURE 28.2 **Interaction between two current circuits.**

According to this law $\mathbf{dl}_1$ element of the circuit(1) will exert a force $\mathbf{dF}_2$ on $\mathbf{dl}_2$ element of the circuit (2) by the formula

$$\mathbf{dF}_2 = \mu_0/4\pi\, I_1 I_2\, \mathbf{dl}_2 \times [\mathbf{dl}_1 \times (\mathbf{r}_2 - \mathbf{r}_1)/|\, \mathbf{r}_2 - \mathbf{r}_1 |^3]$$

and on integration one gets

$$\mathbf{F}_2 = \mu_0/4\pi\, I_1 I_2 \int_1 \int_2 \mathbf{dl}_2 \times [\mathbf{dl}_1 \times (\mathbf{r}_2 - \mathbf{r}_1)/|\, \mathbf{r}_2 - \mathbf{r}_1 |^3]$$

The directions of various vectors have been well explained in the figure above. The factor $\mu_0/4\pi$ is a constant required for the law to be compatible with a set of units. Its value is given as $\mu_0/4\pi = 10^{-7}$ n/Amp^2.

From the above relation of $\mathbf{F}_2$ it is clear that

$$\mathbf{B}(\mathbf{r}_2) = \mu_0/4\pi\, I_1 \int_1 [\mathbf{dl}_1 \times (\mathbf{r}_2 - \mathbf{r}_1)/|\,\mathbf{r}_2 - \mathbf{r}_1|^3],$$

the magnetic induction due to the first circuit at $\mathbf{r}_1$. This relation has been used to find **B** for simple and symmetrical circuits [This formula is also called Biot-Savart law]. For instance, the lines of magnetic induction due to a long straight wire are found as circular with centers on the wire and planes with the wire as normal. The value of **B** comes out as

$$|\,\mathbf{B}\,| = \mu_0 I/2\pi a,$$

at a point of distance '*a*' from the wire carrying current *I*.

For a circular wire the calculations show that the magnetic induction on any point on its axis is given by

$$|\,\mathbf{B}\,| = \mu_0 I/2a^2/(z^2 + a^2)^{3/2},$$

a = radius of the circular loop and z the distance of the point on the axis. Its direction obeys the right-handed screw rule. Field at the center ($z = 0$) will be

$$|\,\mathbf{B}\,| = \mu_0 I/2a \qquad \text{[SI]}$$

28.5 AMPERE'S CIRCUITAL LAW

From the magnetic induction due to a circuit carrying current *I* given by

$$\mathbf{B}(\mathbf{r}_2) = \mu_0 I_1/4\pi \int_1 [\mathbf{dl}_1 \times (\mathbf{r}_2 - \mathbf{r}_1)/|\,\mathbf{r}_2 - \mathbf{r}_1|^3],$$

one may write

$$\mathbf{B}(\mathbf{r}_2) = \mu_0/4\pi \int_{v_1} [\mathbf{J}(\mathbf{r}_1) \times (\mathbf{r}_2 - \mathbf{r}_1)/|\,\mathbf{r}_2 - \mathbf{r}_1|^3]\, dv_1$$

since $I_1\mathbf{dl}_1 = \mathbf{J}(r_1)dv_1$, **J** being the current density within dv_1 at $\mathbf{r}_1$. It is also obvious that for steady currents

$$\nabla.\mathbf{J} = 0$$

From the above relation of $\mathbf{B}(\mathbf{r}_2)$ one may find that

$$\nabla \times \mathbf{B} = \mu_0\mathbf{J}(\mathbf{r}_2)$$

This is known as Ampere's law, rather its differential form. If we now use Stoke's theorem, we get

$$\int_s \nabla \times \mathbf{B}.\mathbf{n}ds = \int_c \mathbf{B}.\mathbf{dl}$$

i.e.

$$\mu_0 \int_s \mathbf{J}.\mathbf{n}ds = \int_c \mathbf{B}.\mathbf{dl},$$ which is the statement of Ampere's law.

28.6 MAGNETIC VECTOR POTENTIAL

It can be proved that

$$\mathbf{B} = \text{curl } \mathbf{A}$$

in which case **A** is called magnetic vector potential. Let us consider

$$\mathbf{B}(\mathbf{r}_2) = \mu_0/4\pi \int_{v_1} [\mathbf{J}(\mathbf{r}_1) \times (\mathbf{r}_2 - \mathbf{r}_1)/|\,\mathbf{r}_2 - \mathbf{r}_1|^3]\, dv_1$$

$$= \mu_0/4\pi \int_{v_1} [\mathbf{J}(\mathbf{r}_1) \times \nabla_1 1/|\mathbf{r}_2 - \mathbf{r}_1|] \, dv_1$$

$$[\because \quad (\mathbf{r}_2 - \mathbf{r}_1)/|\mathbf{r}_2 - \mathbf{r}_1|^3 = \nabla_1 1/|\mathbf{r}_2 - \mathbf{r}_1|]$$

As $\nabla_1 = -\nabla_2$

$$\mathbf{B}(\mathbf{r}_2) = -\mu_0/4\pi \int_{v_1} [\mathbf{J}(\mathbf{r}_1) \times \nabla_2 1/|\mathbf{r}_2 - \mathbf{r}_1|] \, dv_1$$

Use the following identity

$$\nabla \times (\mathbf{A}f) = f\nabla \times \mathbf{A} - \mathbf{A} \times \nabla f$$

we obtain $\mathbf{B}(\mathbf{r}_2) = \mu_0/4\pi \int_{v_1} \nabla_2 \times [\mathbf{J}(\mathbf{r}_1)/|\mathbf{r}_2 - \mathbf{r}_1|] \, dv_1$

because $\nabla_2 \times \mathbf{J}(\mathbf{r}_1) = 0$ as **J** is independent of $\mathbf{r}_2$.

Now, since integration is over $\mathbf{r}_1$ variables while $\nabla_2 \times$ is w.r.t. $\mathbf{r}_2$, ∇_2 may be taken out of the integration so that

$$\mathbf{B}(\mathbf{r}_2) = \nabla_2 \times [\mu_0/4\pi \int_{v_1} [\mathbf{J}(\mathbf{r}_1)/|\mathbf{r}_2 - \mathbf{r}_1|] \, dv_1]$$

$$= \nabla_2 \times \mathbf{A}(\mathbf{r}_2) \quad \text{where } \mathbf{A}(\mathbf{r}_2) = \mu_0/4\pi \int_{v_1} [\mathbf{J}(\mathbf{r}_1)/|\mathbf{r}_2 - \mathbf{r}_1|] \, dv_1$$

Next, as $\mathbf{B} = \nabla \times \mathbf{A}$

$\nabla.\mathbf{B} = 0$ showing thereby that magnetic induction does not have isolated magnetic poles.

Further, $\nabla \times \mathbf{B} = \nabla \times (\nabla \times \mathbf{A}) = \mu_0 \mathbf{J}(\mathbf{r}_2)$ $\quad [\because \nabla \times \mathbf{B} = \mu_0 \mathbf{J}(\mathbf{r})]$

i.e. $-\nabla^2\mathbf{A} + \nabla(\nabla.\mathbf{A}) = \mu_0 \mathbf{J}(\mathbf{r}_2)$

or $\nabla^2\mathbf{A} = -\mu_0 \mathbf{J}$,

Poisson equation to be satisfied by each component of **A** whenever $\nabla.\mathbf{A} = 0$ is satisfied. This is also called coulomb or solenoidal gauge in the electromagnetic theory. The solution of this equation is well known and hence we again get

$$\mathbf{A}(\mathbf{r}_2) = \mu_0/4\pi \int_{v_1} [\mathbf{J}(\mathbf{r}_1)/|\mathbf{r}_2 - \mathbf{r}_1|] \, dv_1 \quad \text{[SI]}$$

28.7 MAGNETIC FIELD OF A REMOTE CIRCUIT

For larger distances $\mathbf{r}_2$, the denominator of the integrand in

$$\mathbf{A}(\mathbf{r}_2) = \mu_0/4\pi \int_{v_1} [\mathbf{J}(\mathbf{r}_1)/|\mathbf{r}_2 - \mathbf{r}_1|] \, dv_1$$

can be expanded. In fact, taking $\mathbf{J}(\mathbf{r}_1)\, dv_1 = I\, d\mathbf{r}_1$ and expanding the denominator we can easily obtain

$$\mathbf{A}(\mathbf{r}_2) = \mu_0/4\pi \, [I/2 \int \mathbf{r}_1 \times d\mathbf{r}_1] \times \mathbf{r}_2/r_2^{\,3} = \mu_0/4\pi \, \mathbf{m} \times \mathbf{r}_2/r_2^{\,3}$$

where $\mathbf{m} = I/2 \int \mathbf{r}_1 \times d\mathbf{r}_1$ is called magnetic dipole moment of the distant current circuit. We may rewrite this result as

$$\mathbf{A}(\mathbf{r}) = \mu_0/4\pi \, \mathbf{m} \times \mathbf{r}/r^3$$

where **r** is the position vector w.r.t. the distant circuit or magnetic dipole of moment **m**.

28.8 MAGNETIC FLUX

The magnetic flux, ϕ, is defined as $\phi = \int_s \mathbf{B.n}\, ds$. Its unit is weber. As a result, weber/m^2 is also the unit of **B**.

Since by divergence theorem

$$\oint_s \mathbf{B.n}\, ds = \int_v \nabla.\mathbf{B}\, dv = 0$$

So, magnetic flux through closed surface is always zero.

28.9 MAGNETIC PROPERTIES

Now we are well prepared to discuss the magnetic properties of the matter comprising the medium. So far we have been doing our calculations in vacuum.

Magnetization

The material medium consists of atoms (or molecules) and in each atom electrons have circulatory motion around its center. This circulatory motion of electrons gives rise to effective current loops, of course very tiny loops of atomic dimensions. This atomic currents would act as distant circuits or magnetic dipoles since we have seen that distant circuits are equivalent to magnetic dipoles with magnetic moment

$$\mathbf{m} = I\mathbf{A} = I/2 \int \mathbf{r} \times d\mathbf{r}$$

Taking $\mathbf{m}_i$ as the magnetic moment of ith atom and N as the number of atoms in a volume Δv of the material medium, we define magnetic moment per unit volume by

$$\mathbf{M} = Lt_{\Delta v \to 0}\ 1/\Delta v\ \Sigma_{i=1}^{N}\ \mathbf{m}_i$$

This quantity is called magnetization of the material and in a continuous material it is assumed as a function of position of that elemental volume Δv denoted as $\mathbf{M}(x,y,z)$ or $\mathbf{M}(\mathbf{r})$. When m_i are distributed randomly within the elemental volume, their sum and hence **M** will turn out to be zero. The material medium would be in unmagnetized state. However, in the presence of external magnetic field these atomic dipoles may be aligned along this field making **M** non-zero function and the medium becomes magnetized. Such medium is called magnetic material.

Once again situation is very tricky in the sense that external field **B** causes the existence of **M** while this **M** tends to change **B**. However, using the formalism of magnetic effects of current circuits one may find the contribution of **M** to the magnetic induction **B**.

In general, therefore, we have two types of magnetic fields, one due to the current caused by transportation of free electrons and the second due to the atomic currents of the magnetic substances. The latter is attributed to the magnetization, **M**, of the material medium.

For a uniform magnetization of the magnetic material one does not get flow of any current in the material and hence these cause no magnetic field. However, when **M** changes from region to region in the material, one gets the flow of net current in a particular direction and hence there is magnetic effects. It has been found that such a current density is related to the material magnetization through the equation $\mathbf{J}_M = \nabla \times \mathbf{M}$, where $\mathbf{J}_M$ denotes the magnetization current density due to atomic currents.

Magnetic Field Produced by Magnetized Material

If we consider an elemental volume dv' of the material of magnetization **M**, its magnetic dipole moment (by definition) will be

$$\Delta \mathbf{m} = \mathbf{M}\, dv'$$

and the vector potential at a distant point **r**,

$$\mathbf{A}(\mathbf{r}) = \mu_0/4\pi \int_{v'} \mathbf{M} \times (\mathbf{r} - \mathbf{r}')/|\,\mathbf{r} - \mathbf{r}'\,|^3\, dv'\ ;\ \mu_0/4\pi = 10^{-7}\ \mathrm{n/Amp^2}.$$

The following figure explains the various quantities

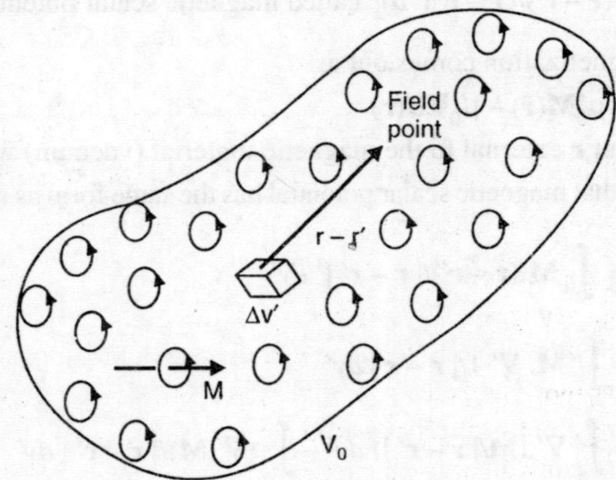

FIGURE 28.3 **M is the magnetization, r′ the location of Δv′ and r the point *P* where field is being obtained.**

$$\mathbf{A}(\mathbf{r}) = \mu_0/4\pi \int_{v'} \mathbf{M} \times \nabla'\, 1/|\,\mathbf{r} - \mathbf{r}'\,|\, dv'$$

$$= \mu_0/4\pi \left[\int_{v'} \mathrm{curl}\mathbf{M}/|\,\mathbf{r} - \mathbf{r}'\,|\, dv' - \int_{v'} \mathrm{curl}(\mathbf{M}/|\,\mathbf{r} - \mathbf{r}'\,|)\, dv' \right]$$

obtained by using the vector identity

$$\nabla \times (\mathbf{A}f) = -\,\mathbf{A} \times \nabla f + f\, \nabla \times \mathbf{A}$$

prime indicates differentiation w.r.t. **r′**. Now use the definition of curl

$$\int_{v'} \mathrm{curl}\mathbf{A}\, dv' = \int_{s'} \mathbf{n} \times \mathbf{A}\, ds', \quad (s' \text{ is the surface bounding } v')$$

to obtain

$$\mathbf{A}(\mathbf{r}) = \mu_0/4\pi \int_{v'} \mathrm{curl}\mathbf{M}/\,|\,\mathbf{r} - \mathbf{r}'\,|\, dv' + \mu_0/4\pi \int \mathbf{M} \times \mathbf{n}/|\,\mathbf{r} - \mathbf{r}'\,|\, ds'$$

$$= \mu_0/4\pi \int_{v'} \mathbf{J}_M/|\mathbf{r} - \mathbf{r}'|\, dv' + \mu_0/4\pi \int \mathbf{j}_M/|\,\mathbf{r} - \mathbf{r}'\,|\, ds'$$

where $\mathbf{J}_M = \mathbf{M} \times \mathbf{n}$, the magnetization current per unit length flowing in the surface layer and $\nabla \times M = \mathbf{J}_M$.

B is now found from

$$\mathbf{B}(\mathbf{r}) = \nabla \times \mathbf{A}(\mathbf{r})$$

It is a tough task to compute this magnetic field. However, it is enough to know how one finds this contribution of the magnetized material to the magnetic field. There exists an alternative but easier approach to compute **B**. Let us consider expression of vector potential

$$\mathbf{A}(\mathbf{r}) = \mu_0/4\pi \int_{v'} \mathbf{M} \times (\mathbf{r} - \mathbf{r}')/|\,\mathbf{r} - \mathbf{r}'\,|^3 \, dv'$$

So that, $\mathbf{B}(\mathbf{r}) = \text{curl}\mathbf{A}(\mathbf{r}) = \mu_0/4\pi \int \text{curl}[\mathbf{M} \times (\mathbf{r} - \mathbf{r}')/|\,\mathbf{r} - \mathbf{r}'\,|^3]dv'$

By using vector identity and $\nabla^2 \dfrac{1}{|\,\mathbf{r} - \mathbf{r}'\,|} = -4\pi\delta(\mathbf{r} - \mathbf{r}')$ the right hand side gets simplified to

$$= \mu_0\mathbf{M}(\mathbf{r}) - \mu_0\nabla U(\mathbf{r})$$

where $U(\mathbf{r}) = 1/4\pi \int_{v'} \mathbf{M}(\mathbf{r}').\,(\mathbf{r} - \mathbf{r}')/|\,\mathbf{r} - \mathbf{r}'\,|^3 \, dv'$ called magnetic scalar potential due to magnetic material.

So, magnetic field due to magnetization comes out as

$$\mathbf{B}(\mathbf{r}) = \mu_0\mathbf{M}(\mathbf{r}) - \mu_0\nabla U(\mathbf{r})$$

$\Rightarrow \; -\mu_0\nabla U(\mathbf{r})$ at point $\mathbf{r}$ external to the magnetic material (vacuum) where $\mathbf{M} = 0$.

It is interesting to note that magnetic scalar potential has the same form as its counter part in electrostatic.

In fact, $4\pi\, U(\mathbf{r}) = \int_{v'} \mathbf{M}.(\mathbf{r} - \mathbf{r}')/|\,\mathbf{r} - \mathbf{r}'\,|^3 \, dv'$

$$= \int \mathbf{M}.\,\nabla'\, 1/|\,\mathbf{r} - \mathbf{r}'|\, dv'$$

$$= \int \nabla'.(\mathbf{M}/|\,\mathbf{r} - \mathbf{r}'\,|)\, dv' - \int (\nabla'.\mathbf{M})/|\,\mathbf{r} - \mathbf{r}'\,|\, dv'$$

Obtained by using vector identity

$$\nabla.(\mathbf{A}f) = f\,\nabla.\mathbf{A} + \mathbf{A}.\nabla f$$

Now using divergence theorem we get

$$4\pi\, U(\mathbf{r}) = \int_{s'} (\mathbf{M}.\mathbf{n})/|\,\mathbf{r} - \mathbf{r}'\,|\, ds' + \int (-\nabla'.\mathbf{M})/|\,\mathbf{r} - \mathbf{r}'\,|\, dv'$$

$$= \int \sigma_M ds'/|\,\mathbf{r} - \mathbf{r}'\,| + \int \rho_M dv'/|\,\mathbf{r} - \mathbf{r}'\,|$$

where $\sigma_M = (\mathbf{M}.\mathbf{n})$, surface density of magnetic poles and $\rho_M = -\nabla'.\mathbf{M}$, the volume density of magnetic pole distribution. The magnetic field is now obtained through

$$\mathbf{B}(\mathbf{r}) = \mu_0\mathbf{M}(\mathbf{r}) - \mu_0\nabla U(\mathbf{r})$$

$$= \mu_0/4\pi \int \sigma_M(\mathbf{r} - \mathbf{r}')/|\,\mathbf{r} - \mathbf{r}'\,|^3\, ds' + \mu_0/4\pi \int \rho_M(\mathbf{r} - \mathbf{r}')/|\,\mathbf{r} - \mathbf{r}'\,|^3\, dv'$$

$$+ \mu_0\mathbf{M}(\mathbf{r}) \quad \left(\text{since } \nabla \frac{1}{|\,\mathbf{r} - \mathbf{r}'\,|} = -\frac{\mathbf{r} - \mathbf{r}'}{|\,\mathbf{r} - \mathbf{r}'\,|^3}\right)$$

= contribution of the magnetic material to the magnetic field at $\mathbf{r} = (x, y, z)$.

Now, we can write the magnetic induction due to both true currents $\mathbf{J}$ and as well as the magnetisation current $\mathbf{J}_\mathbf{M}$.

For example, $B(r) = \dfrac{\mu_0}{4\pi} \displaystyle\int_{v'} J(r') \times \frac{\mathbf{r} - \mathbf{r}'}{|\,\mathbf{r} - \mathbf{r}'\,|^3}\, dv' + \mu_0\mathbf{M}(r) - \mu_0\nabla U(r)$

and hence $\mathbf{H}(r) = \left(\dfrac{\mathbf{B}}{\mu_0} - \mathbf{M}\right) = \left[\dfrac{1}{4\pi} \displaystyle\int_{v'} J(r') \times \frac{\mathbf{r} - \mathbf{r}'}{|\,\mathbf{r} - \mathbf{r}'\,|^3}\, dv' - \nabla U(r)\right]$

The Field Relations

A field vector, **H**, is introduced such that

$$\mathbf{H} = \mathbf{B}/\mu_0 - \mathbf{M}$$

where **B** and **M** are respectively magnetic induction and magnetization of the medium. This vector field is called magnetic intensity and has the same unit as **M** *i.e.* Amp/m. In vacuum as the medium **M** = 0 and **B** = $\mu_0\mathbf{H}$ where $\mu_0 = 4\pi \times 10^{-7}$ n/Amp^2.

We still have $\nabla.\mathbf{B} = 0$

since **B** can always be derived from a vector potential **A** through

$$\mathbf{B} = \nabla \times \mathbf{A}$$

whatever may be the source of current loop. However, now we must write

$$\nabla \times \mathbf{B} = \mu_0(\mathbf{J} + \mathbf{J}_m)$$

as there are two types of current densities, true, **J**, and atomic, $\mathbf{J}_m$ (or magnetization). Also,

$$\mathbf{J}_m = \nabla \times \mathbf{M}.$$

Therefore $\nabla \times (\mathbf{B}/\mu_0 - \mathbf{M}) = \mathbf{J}$

or

$$\nabla \times \mathbf{H} = \mathbf{J}$$

Thus, the curl of magnetic intensity is given in term of true current density **J**.

Next, using Stokes' theorem we have

$$\int_s \nabla \times \mathbf{H} \,.\, \mathbf{n}\, ds = \int_c \mathbf{H}.d\mathbf{l},\ c \text{ is the closed curve enclosing } s.$$

i.e.

$$\int_c \mathbf{H}.d\mathbf{l} = \int_s \mathbf{J} \,.\, \mathbf{n}\, ds = I$$

So, the line integral of magnetic intensity along the closed path is equal to the true current through the surface enclosed by the closed path.

28.10 MAGNETIC SUSCEPTIBILITY AND PERMEABILITY

The study of magnetic properties of the material requires interdependence of **M** and one of **H** and **B**. Such a relation has been experimentally found as dependent on the nature of the magnetic material. We have materials for which

$$\mathbf{M} = \chi_m \mathbf{H}$$

where the constant χ_m is called magnetic susceptibility of the material.

When $\chi_m > 0$ the material is called as paramagnetic. In such medium **B** gets strengthen because

$$\mathbf{B} = \mu_0(\mathbf{H} + \mathbf{M})$$

and = $\mu_0\mathbf{H}$ when there is no magnetic medium, **M** = 0.

when $\chi_m < 0$ the material (or medium) is called diamagnetic and magnetic induction gets weakened in such materials. Further, for these para- and dia-magnetic materials, χ_m remains small. χ_m is also temperature dependent. Values of χ_m for some materials have been listed in Table 28.1.

TABLE 28.1 Magnetic susceptibilities of some para- and dia-magnetic materials

Material	$\chi_m \times 10^{-5}$
Al	2.3
Bi	– 1.66
Cu	– 0.98
Au	– 3.6
Hg	– 3.2
Ag	– 2.6
$GdCl_3$	276.0
Ti	7.06
W	6.8
N_2	– 0.0005
O_2	0.21

[**Note** : when magnetic moment per unit mass is to be calculated, we write $\mathbf{M} = \chi_{m,mass}\mathbf{H}$ and for per mole $\mathbf{M} = \chi_{m,\text{molar}}\mathbf{H}$]

Also, as **M** is proportion to **H** we may also write

$$\mathbf{B}/\mu_0 = (\mathbf{H} + \mathbf{M}) = (1 + \chi_m)\mathbf{H} \quad \text{or} \quad \mathbf{B} = \mu\mathbf{H}$$

where $\mu/\mu_0 = (1 + \chi_m)$

and μ is called permeability of the material. Sometimes one uses $\mu/\mu_0 = K_m$, called relative permeability.

28.11 FERROMAGNETIC MATERIALS AND HYSTERESIS

The ferromagnetic materials require separate discussion because of their markedly different behavior and important role in the applications

Firstly, the relation $\mathbf{B} = \mu\mathbf{H}$

is not linear in case of ferromagnetic materials. The value of μ, the permeability depends on **H** itself and is written as $\mu(H)$. It takes wide range of values lying between 0 and ∞ and may be positive as well as negative. The measured values of μ for the case of commercial iron has been shown in Fig. 28.4. Along with this the values of magnetic induction **B** has also been plotted.

Such a *B-H* plot is known as magnetization curve of ferromagnetic substance. Since $\mathbf{B} = \mu_0\mathbf{H} + \mathbf{M}$ and **M** gets maximized (up to $H = 100$ Amp/m) to its saturated value, B increases with H more steadily as H increases beyond this value because now **B** increases due to $\mu_0\mathbf{H}$ term only. Because of the same reason, μ/μ_0 first increases to its maximum value of about 6000 at the knee of *B-H* curve before starting decreasing as H continues to increase.

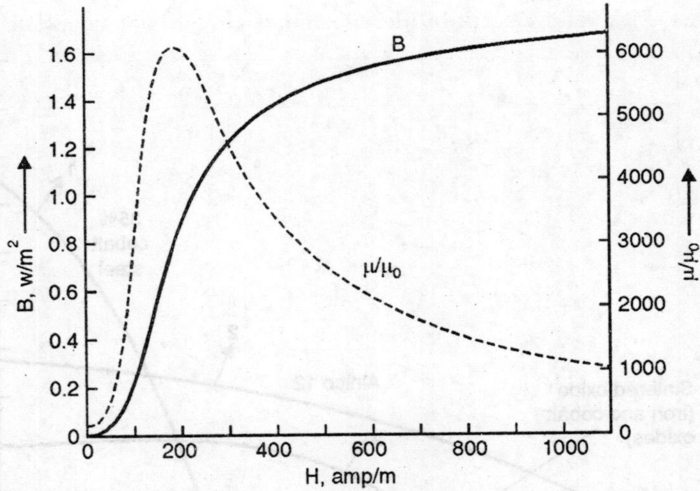

FIGURE 28.4 **Plot of μ/μ_0 and B vs H of commercial iron. μ/μ_0 is maximum at the knee of B and H plot.**

Second, the variation of **B** as H decreases does not follow the same path which it followed with increasing H. This behavior is called Hysteresis and is shown by only ferromagnetic materials. Typical of which is being shown in Fig. 28.5. These plots of B vs H are called Hysteresis loop of ferromagnetic materials.

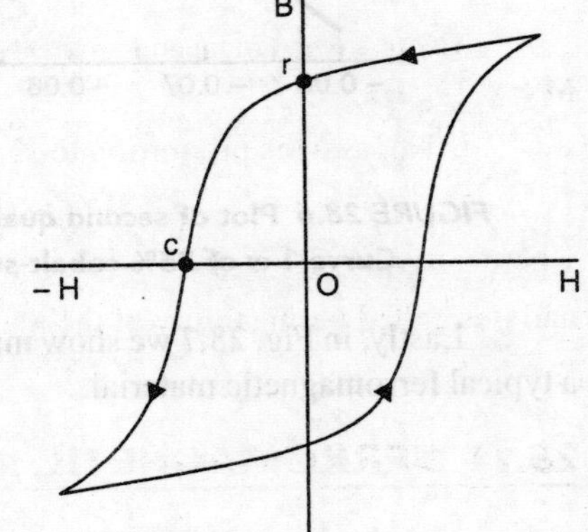

FIGURE 28.5 **B-H plot of ferromagnetic material a typical hysteresis curve.**

In this graph we find that once the material is magnetized it does not follow the B-H plot of Fig. 28.4 with decreasing H but rather intersects the B-axis at r for $H = 0$ and then on increasing H in the reverse direction the curve cuts the H-axis at c there after it continues to the third quadrant of the loop in a symmetrical fashion. After this if H is increased the magnetic induction follows the lower part of the loop till it gets closed to the initial value.

The study of the hysteresis loop makes it clear that B retains finite value even when H becomes zero while it was decreasing because of permanent magnetization (**M**). This value of B at r is known as retentivity or remanence. Further, B becomes zero along this route at c with H having some negative value $= -M$. The magnitude of this value of H is called coercive force or coercivity of the material.

Thirdly, the value of μ (or μ/μ_0) is negative in the second and fourth quadrants [see Fig. 28.5 and field relation $\mathbf{B} = \mu\mathbf{H}$].

Fourth, in certain applications of ferromagnetic material one requires the knowledge of incremental permeability μ_{in} given by

$$\mu_{in} = \Delta B/\Delta H.$$

This is approximately equal to the slope of the hysteresis loop at a particular operating point. The most important use of ferromagnetic material is found in increasing the magnetic flux in a circuit and where it becomes a core of that device. Another use of ferromagnetic materials is in the production of permanent magnets. Thus, the part of hysteresis loop in the second quadrant assumes a lot of importance. This part of the loop (2nd quadrant) has been shown in Fig. 28.6 for various permanent magnet materials.

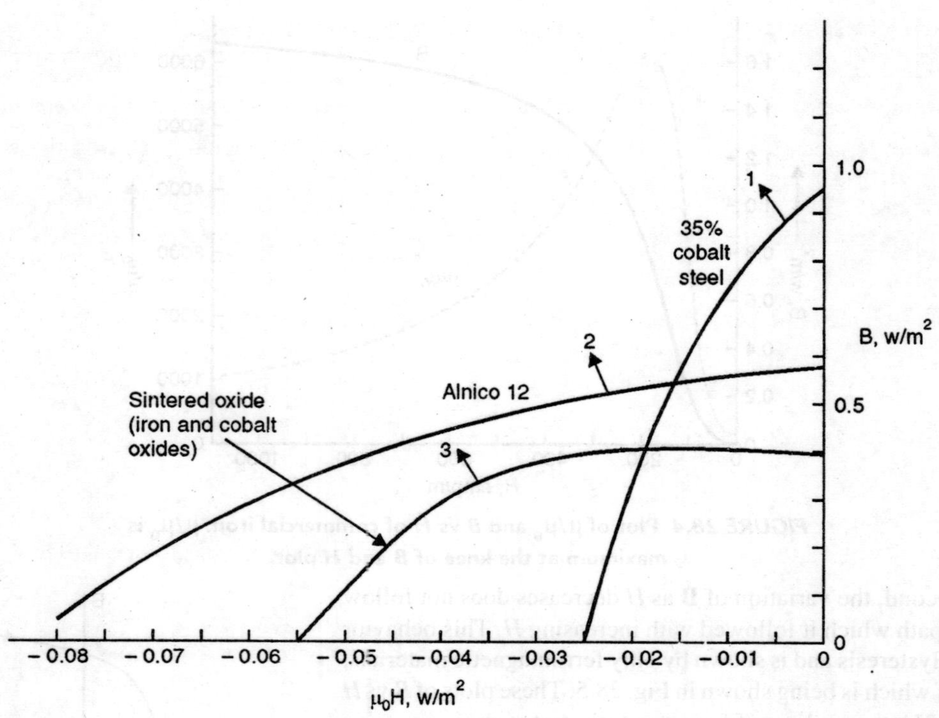

FIGURE 28.6 **Plot of second quadrant of hysteresis loop of various permanent magnetic materials. Curve-1 is of 35% cobalt-steel, curve-2 of Alnico12 and curve-3 of iron-cobalt oxides.**

Lastly, in Fig. 28.7 we show major hysteresis loop (outermost) and several minor hysteresis loops for a typical ferromagnetic material.

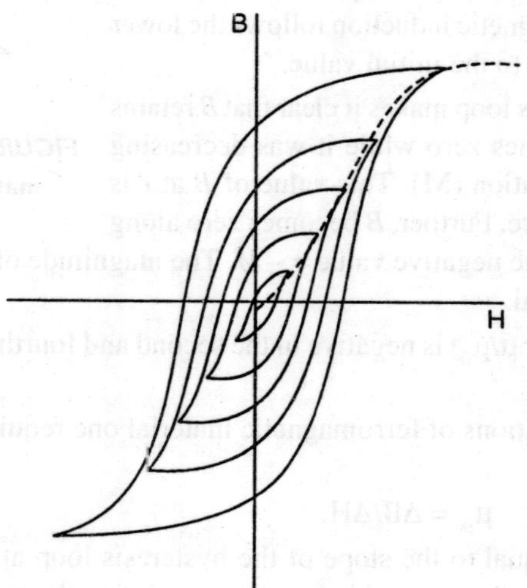

FIGURE 28.7 **Display of major hysteresis loop (outermost) and several minor hysteresis loops.**

The shape of hysteresis loop depends not only on the nature of the magnetic material but also on the maximum value of H to which the material is subjected (*e.g.* the major and minor hysteresis loops).

28.12 THEORY OF DIA-, PARA- AND FERROMAGNETISM

Origin of diamagnetism. The orbiting of atomic electrons in a molecule is equivalent to some effective currents and hence magnetic moments. Due to the application of external magnetic field $\vec{B}$, these atomic electrons speed up or slow down due to the force $-e\,\vec{v}\times\vec{B}$ and this leads to change in the atomic currents and hence the effective magnetic moments of the molecules. By Lenz's law this change of magnetic moment will be opposite to $\vec{B}$.

By simple calculations, this change in magnetic moment is found as

$$\Delta\vec{M} = \frac{-e^2}{4m} R^2 \vec{B}$$

where e = electronic charge,

R = radius of the circular orbit of electron

and m = electron mass

In case of N molecules per unit volume of the magnetic material, the magnetisation will be

$$\vec{M} = -\frac{Ne^2}{4m}\vec{B}\sum_i R_i^2$$

where summation in over the number of electrons in the molecule. Writing $\vec{B} = \mu_0\vec{H}$

$$\vec{M} = -\left[\frac{Ne^2\mu_0}{4m}\sum_i R_i^2\right]\times\vec{H}$$

and hence the magnetic susceptibility,

$$\chi_m = -\frac{Ne^2\mu_0}{4m}\sum_i R_i^2$$

This is a –ve quantity and explains the property of diamagnetic materials. It is to be noted that in the absence of external magnetic field the effective magnetic moments of molecules are generally zero, in diamagnetic materials.

So, diamagnetic behaviour is more prominent in materials comprising of atoms having closed electronic shells. In case of materials with molecules having non-zero magnetic moments, the para- and ferromagnetic effects over shadow the diamagnetic effects.

Origin of paramagnetism. In certain magnetic materials the molecules have effective magnetic moments due to the orbital and spin motion of the atomic electrons comprising each molecule. The magnetization of such a material vanishes because of random orientation of the molecular dipoles. That is,

$\mathbf{M} = \frac{1}{\Delta v}\sum_i \mathbf{m}_i = 0$, $\mathbf{m}_i$ being the molecular moments and sum is over molecules within the volume dv.

However, when this material is placed in a magnetic field say $\mathbf{B} = \mu_0\mathbf{H}$, each of the molecules of same type gains potential energy given by

$$W_p = -m_0 B\cos\theta = -m_0\mu_0 H\cos\theta$$

where θ denotes the inclination of molecular moment, $\mathbf{m}_0$, with **B** or **H.** Because of the thermal contacts of the molecules the probability distribution of molecules will be proportional to Maxwell's factor

$$e^{-W_p/kT} = e^{m_0\mu_0 H\cos\theta/kT}$$

for inclination θ w.r.t. **H**, where k is the Boltzmann constant and T is the absolute temperature. Thus the average value of the molecular magnetic moments along **H** would be

$$<m_0 \cos\theta> = \frac{\int m_0 \cos\theta \, e^{m_0\mu_0 H\cos\theta/kT}\, d\Omega}{\int e^{m_0\mu_0 H\cos\theta/kT}\, d\Omega} = \frac{\int m_0 \cos\theta \, e^{y\cos\theta} \sin\theta\, d\theta}{\int e^{y\cos\theta}\sin\theta\, d\theta}$$

where $y = \dfrac{m_0\,\mu_0\,H}{kT}$

On integration one gets

$$<m_0\cos\theta> = m_0\left[\coth y - \frac{1}{y}\right] = m_0\, L(y),$$

which is known as Langevin formula, $L(y)$ being the Langevin function (see Fig. 28.8). Since y remains small, at normal temperatures,

$$L(y) \to \frac{y}{3}$$

and the average molecular magnetic moment tends to

$$\frac{m_0 y}{3} = \frac{m_0^2\,\mu_0 H}{3kT}$$

Thus, magnetisation will be

$$\vec{M} = \left(\frac{N\, m_0^2\,\mu_0}{3kT}\right)\vec{H}$$

and magnetic susceptibility,

$$\chi_m = \frac{N\, m_0^2\,\mu_0}{3kT} > 0.$$

(+ve quantity) and varies as $\dfrac{1}{T}$ called Curie-law.

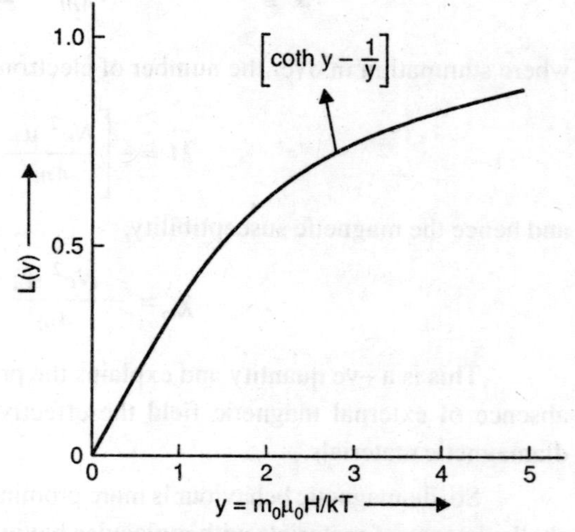

FIGURE 28.8 **Plot of the Langevin function. The asymptotic value of $L(y)$ as $y \to \infty$ is one.**

Further, this value of **M** is far less than the saturation value of $|\mathbf{M}| = Nm_0$ which is the value when there is complete alignment of molecular moments along the direction of H. However, the above value of χ_m agrees with observations.

Theory of ferromagnetism. In the Langevin theory of paramagnetism it is assumed that the magnetic field which acts on the molecules (called molecular) of the material is same as the external field. This is because the effect of magnetisation is negligible. The simple calculations show that the molecular field $\mathbf{H}_m$ differs from **H** by $\dfrac{\mathbf{M}}{3}$. That is

$$\mathbf{H_m} = \mathbf{H} + \frac{\mathbf{M}}{3} = \left(\mathbf{H} + \frac{\chi_m}{3}\mathbf{H}\right) \simeq \mathbf{H} \text{ as } \frac{\chi_m}{3}$$ is usually of the order of 10^{-5} in case of paramagnetic materials.

But in case of ferromagnetic materials the situation seems to be different as the molecular dipoles are aligned substantially even in the absence of external magnetic field **H**.

So, if we assume that $\mathbf{H_m} = \mathbf{H} + \gamma\mathbf{M}$,

this field will be $\gamma\mathbf{M}$ when $\mathbf{H} = 0$.

Now, the question arises whether this molecular field $\gamma\mathbf{M}$ maintains the given value of magnetisation of the ferromagnetic material. For example, in ferromagnetic materials

$$M > 0.7\, Nm_0$$

By Langevin theory, since

$$M = Nm_0\left[\coth y - \frac{1}{y}\right],$$

$$\coth y - \frac{1}{y} > 0.7 \quad \text{or} \quad y > 3$$ (see the graph of $L(y)$ vs y)

i.e. $$\frac{m_0\mu_0 H_m}{kT} > 3.$$

when $H_m = \gamma M$, this condition leads to

$$\frac{m_0\mu_0\gamma M}{kT} > 3 \quad \text{or} \quad 0.7\,\frac{\gamma N m_0^2 \mu_0}{kT} > 3$$ (obtained by putting $M > 0.7\, Nm_0$)

This condition gives that $\gamma \simeq 10^3$.

But molecular theory predicts the value of $\gamma = \frac{1}{3}$. So, ferromagnetism is not as simple as paramagnetic behaviour.

Weiss theory of ferromagnetism. Weiss started his approach by assuming that $\vec{H}_m = \gamma\vec{M}$ is a fact and did not bother to explain the co-efficient r.

According to Langevin theory, the magnetisation is given by

$$M = Nm_0\left[\coth y - \frac{1}{y}\right]$$

where $$y = \frac{m_0\mu_0 H_m}{kT} = \gamma\frac{m_0\mu_0 M}{kT}$$

as $$H_m = \gamma M \text{ when } \vec{H} = 0$$

So, for the consistency of the theory

$$M = \frac{ykT}{\gamma m_0\mu_0}$$

and at a given temperature M should satisfy both

$$M = Nm_0\left[\coth y - \frac{1}{y}\right]$$

as well as $$M = \frac{ykT}{\gamma m_0 \mu_0}$$

Thus, the spontaneous magnetisation (for $H = 0$) of the ferromagnetic material will be the intersection of curves $Nm_0\left(\coth y - \frac{1}{y}\right)$ vs y and $\frac{y\,kT}{\gamma m_0 \mu_0}$ vs y as in Fig. 28.9.

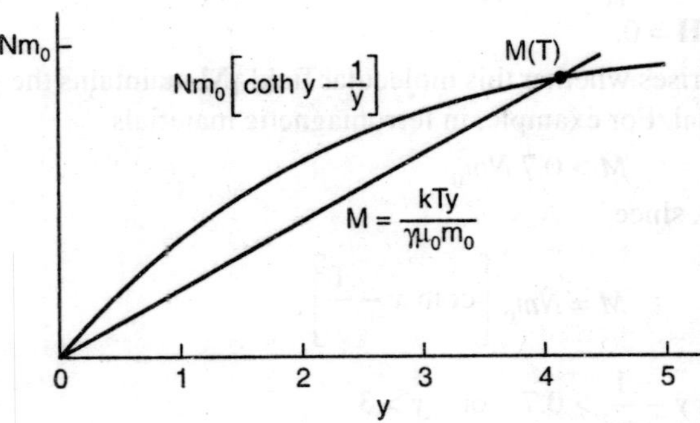

FIGURE 28.9 **Determination of the spontaneous magnetization *M*(*T*) with the aid of the Langevin function.**

As per this approach it is obvious that as T increases the 2nd curve $M = \frac{k\,Ty}{\gamma m_0 \mu_0}$ becomes steeper and resulting a lower value of spontaneous magnetisation, $M(T)$. As T is increased further this line becomes tangent to the Ist curve where $M \to 0$. This temperature is called Curie temperature T_c above which spontaneous (or permanent) magnetization vanishes and paramagnetic behaviour results. This trend of magnetisation of ferromagnetic material is displayed in Fig. 28.10 which agrees well with experimental values provided quantum corrections are applied.

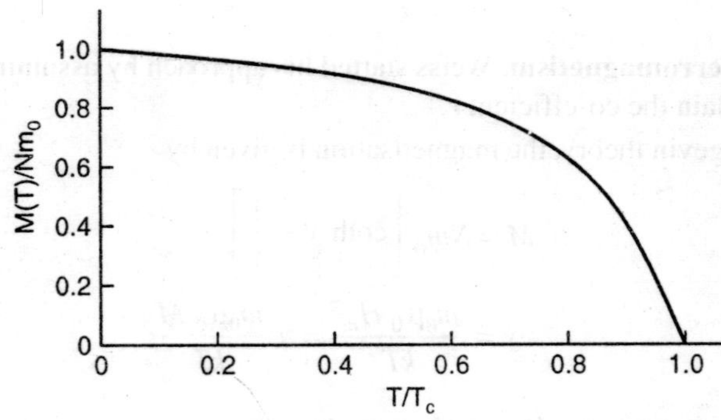

FIGURE 28.10 **The magnetization of a ferromagnetic material as a function of temperature. T_c is called the Curie temperature.**

28.13 FERROMAGNETIC DOMAINS

On the basis of classical Langevin theory and concept of Curie temperature one finds that ferromagnetic material should possess saturation magnetisation at temperatures below Curie temperature. This does not agree with observations. In fact, a piece of iron can exist both in unmagnetized as well as magnetized state. This paradox led to the formulation of domain theory of ferromagnetism. According to domain theory, the ferromagnetic material consists of regions called domains and each of these are completely magnetised. However, when the magnetization of these domains are randomly oriented, the material will behave as unmagnetised.

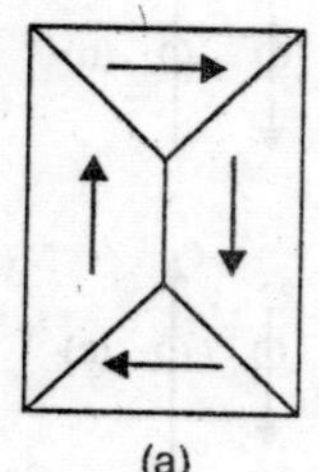
(a)

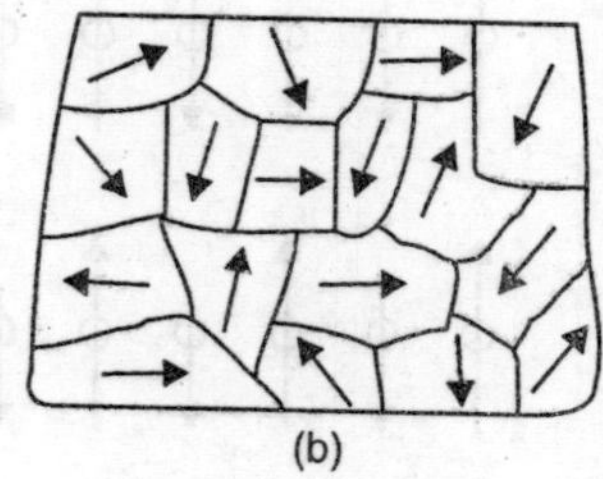
(b)

FIGURE 28.11 Ferromagnetic domain structures : (*a*) Single crystal ; (*b*) Polycrystalline specimen. Arrows represent the direction of magnetization.

In the regions between the adjacent domains the molecular moments rotate gradually towards the new direction and such regions act as 'domain wall'. It is obvious that smaller the number of domains, higher must be the external field and higher the number of domains, smaller the exciting field. Thus, multidomains structure having smaller magnetic energy is usually favoured energetically.

The increase in the magnetisation with the application of external field is attributed to two types of processes. First, the size of a particular domain with moment parallel to $\vec{H}$ expands at the cost of other domains. This process is termed as wall motion [see Fig. 28.12 (*b*)]. Second, the spin moments of the individual domains rotate towards the field direction [see Fig. 28.12(*c*)]. In weak fields the reversible domain wall motion is more probable while in stronger fields the magnetization increases by irreversible wall motion followed by domain rotation. These phenomena have been observed experimentally.

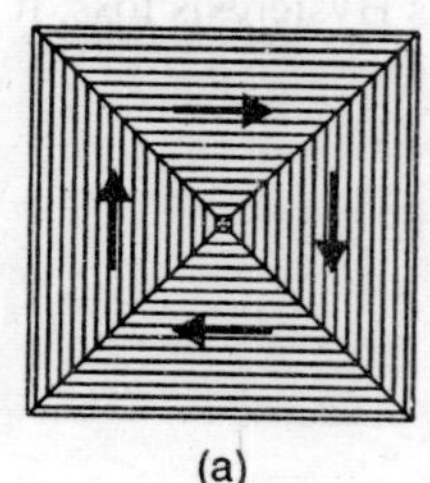
(a)

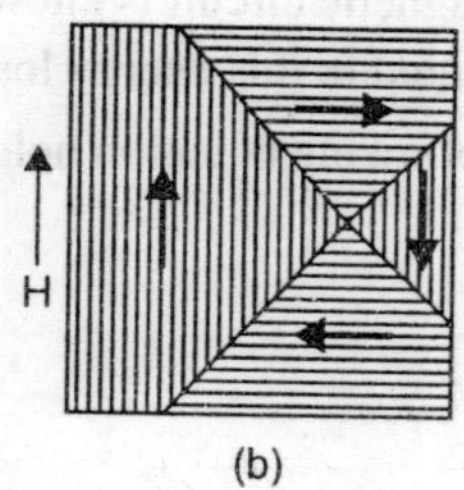

(b)

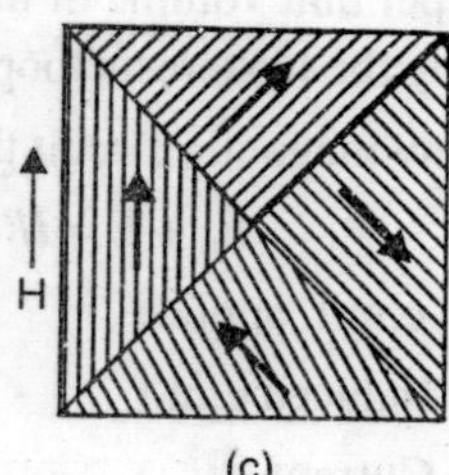

(c)

FIGURE 28.12 (*a*) Random orientation ; (*b*) Domain wall motion ; (*c*) Domain rotation.

Theory of ferrites. In the theory of ferromagnetism, Heisenberg showed that there was a change in the electrostatic energy of the material when the parallel spin alignment of atoms changes into antiparallel alignment. If the energy change favours the parallel spin alignment, the material is termed as ferromagnetic. If the energy change favours antiparallel spin alignment of atoms, it is still possible to obtain ordered spin structure with spins alternating from atom to atom as the crystal is traversed when the resulting magnetization vanishes, the material will be called antiferromagnet (see Fig. 28.13).

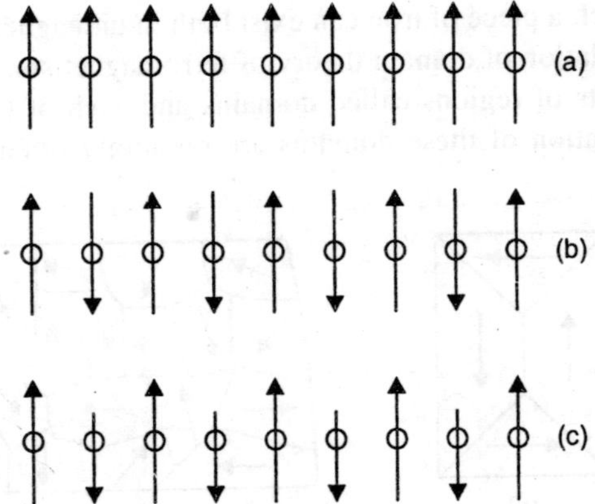

FIGURE 28.13 **Schematic representation of atomic spins in ordered spin structures : (*a*) Ferromagnetic ; (*b*) Antiferromagnetic ; (*c*) Ferrimagnetic.**

In certain crystals the resulting magnetization is non zero in the above mentioned ordered spin structure with atomic spins alternately parallel and antiparallel (Fig. 28.13*c*). Such materials are called ferromagnet or simply ferrite. These materials have chemical formula as $MoFe_2O_3$ with M some divalent metal such as CO, Ni, Cu, Mg, Cd or Fe. The mineral magnetite, Fe_3O_4, is a typical example of a ferrite which has been known since ancient times. These ferrites besides having saturation magnetization are poor conductor of electricity (ρ is 1 to 10^4 ohm-m) and thus have variety of applications.

28.14 HYSTERESIS LOSS

The plot of $\vec{B}$ and $\vec{H}$ when a ferromagnetic sample is subjected to a cycle of variation of $\vec{H}$ is a closed loop called Hysteresis loop when such a material is used in electric circuits there is always a loss of energy. This loss per cycle and per unit volume of the magnetic circuit is known as Hysteresis loss. It can be proved that the area enclosed by the hysteresis loop is equal to this energy loss.

Proof. In any magnetic circuit the following equation holds

$$E_0 + E = IR$$

where E_0 = e.m.f. applied

E = Induced e.m.f.

I = Current

and R = Ohmic resistance in the circuit.

Such a circuit could be a coil having turns and enclosing a piece of ferromagnetic material. Then the workdone by the external battery in moving a charge $dq = Idt$ around the circuit would be given by

$$dW = E_0 \times Idt = -EIdt + I^2Rdt$$

$$= +\frac{d\phi}{dt} Idt + I^2Rdt \qquad \left[\because \; E = -\frac{d\phi}{dt}\right]$$

where $\frac{d\phi}{dt}$ is the rate of change of flux.

Thus, $$dW = Id\phi + I^2Rdt$$

The second term on right hand side is Joule's heat where magnetic material plays no role. However, the first term concerns the magnetic flux and measures the work spent due to the presence of magnetic substance in the circuit. So, we call it magnetic energy loss and write

$dW_b = Id\phi$; this is +ve when change in ϕ is in same direction as the flux due to I, otherwise it is negative. If we have N turns in the coil each of A area of cross-section and denote $d\phi$ as the flux change in each turn, one gets

$$dW_b = NI\, d\phi$$

Now use Ampere's circuital law

$$\oint_c \vec{H}.\vec{dl} = NI$$

where c is a closed path as shown in Fig. 28.14.

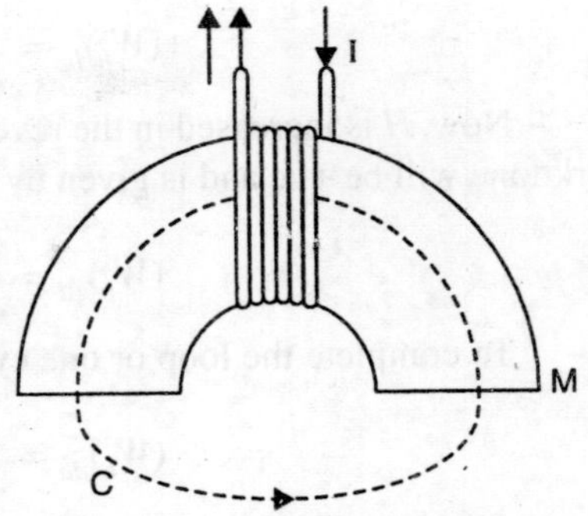

FIGURE 28.14 ***M*** **is a ferromagnetic piece enclosed by a coil of** ***N*** **turns ;** ***c*** **represents a typical closed flux path.**

Thus, $dW_b = \oint d\phi\, \vec{H}.\vec{dl} = \oint AdB\, \vec{H}.\vec{dl}$; dB = change in magnetic induction

As **B** is along **dl** on the flux lines, we may replace dB**dl** by $dl\, d$**B** to get

$$dW_b = \oint Adl\, \mathbf{H}.d\mathbf{B} = \int_v \mathbf{H}.d\mathbf{B}\, dv$$

where v is the volume of the magnetic circuit where flux is non zero. This gives us the amount of workdone per unit volume as

$$dW_b = \mathbf{H} \,.\, d\mathbf{B}$$

Now, if $\vec{H}$ is change over a complete cycle of variation by passing alternating current through the coil, we get total work per unit volume per unit cycle as

$$W_b = \oint \mathbf{H}.d\mathbf{B} = \oint HdB$$

when **H** and **B** are parallel.

H . d**B** will be positive if **H** and d**B** are both +ve or –ve, otherwise these element workdone on the system will be –ve. Let us next estimate W_b using a Hysteresis loop of the given magnetic specimen enclosed by the coil. Clearly, **H** . d**B** = HdB describes a strip between B-axis and the Hysteresis loop as shown in Fig. 28.15.

Suppose 'a' denotes the starting value of H which is subsequently increased to b. During this the amount of workdone on the system will be +ve and is given by

$$(W_b)_{ab} = \int_a^b \mathbf{H}.\,d\mathbf{B}$$

= Area between the loop ab and the B-axis, *i.e.* area $abeca$

Next, H is decreased upto zero to trace the loop bc. During this H remains positive while dB is negative. So, this workdone on the system would be negative and is given by

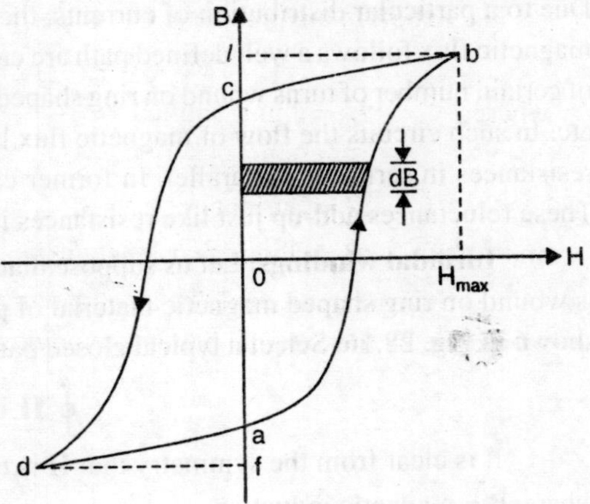

FIGURE 28.15 **Workdone per unit volume in cycling a ferromagnetic material.**

$$(W_b)_{bc} = \int_b^c \mathbf{H}.d\mathbf{B} = -\text{ Area } becb$$

Now, H is increased in the reverse direction from zero to $-H_{max.} = of$ to trace cd part of the loop. The workdone will be +ve and is given by

$$(W_b)_{cd} = \int_c^d \mathbf{H}.d\mathbf{B} = \text{Area } cdfac$$

To complete the loop or one cycle of variation of **H** is increased from $-H_{max.}$ to zero. So,

$$(W_b)_{da} = \int_d^a \mathbf{H}.d\mathbf{B} = -\text{ Area } dfad$$

On adding all these contribution we get the workdone on the material per unit volume per cycle of $\vec{H}$ as

$$W_b = \oint \mathbf{H}.d\mathbf{B}$$

= Area enclosed within the Hysteresis loop.

Thus, loss of energy is given by the size of the Hysteresis loop which depends on the nature of the magnetic material forming the core. Lastly, as $\mathbf{B} = \mu_0 (\mathbf{H} + \mathbf{M})$

$$\oint \mathbf{H}.d\mathbf{B} = \mu_0 \oint \mathbf{H}.d\mathbf{H} + \mu_0 \oint \mathbf{H}.d\mathbf{M}$$

$$= \mu_0 \oint \mathbf{H}.d\mathbf{M} \quad \text{since} \quad \oint \mathbf{H}.d\mathbf{H} = 0$$

$= \mu_0$ time the area of the M–H loop.

In C.G.S. it will be equal to $4\pi \oint \mathbf{H}.d\mathbf{M}$. This energy loss appears as heat.

28.15 MAGNETIC CIRCUITS

Due to a particular distribution of currents, the magnetic flux forms closed loops. The circuits in which the magnetic flux follows a well defined path are called magnetic circuits. Examples are toroidal windings, a coil of certain number of turns wound on ring shaped magnetic material, a circuit containing a permanent magnetic etc. In such circuits the flow of magnetic flux behaves like flow of currents in electric circuits consisting of resistances in series or in parallel. In former case the flow of flux encounters resistance called reluctance. These reluctances add-up just like resistances in series and in parallel.

Toroidal windings. Let us suppose that toroidal contains N turns each of area of cross-section A and is wound on ring shaped magnetic material of permeability μ. Let a current I be passed through the toroidal shown in Fig. 28.16. Select a typical closed path c for the flux and apply Ampere's circuital law,

$$\oint \mathbf{H}.d\mathbf{l} = NI$$

It is clear from the symmetry that **H** is tangent to c and hence is along **dl** all along c. Also $\mathbf{B} = \mu\mathbf{H}$.

where $\mathbf{B}$ = magnetic induction

such that $\quad \phi = BA$

Therefore, circuital law becomes

$$\oint \frac{\mathbf{B} \cdot \mathbf{dl}}{\mu} = NI$$

or

$$\oint \frac{\phi\, dl}{A\mu} = NI$$

ϕ being same for every dl, it can be taken out of the integral so that

$$\phi \oint \frac{dl}{A\mu} = NI$$

This relation agrees with

$$I \oint \frac{dl}{gA} = \xi, \text{ e.m.f.}$$

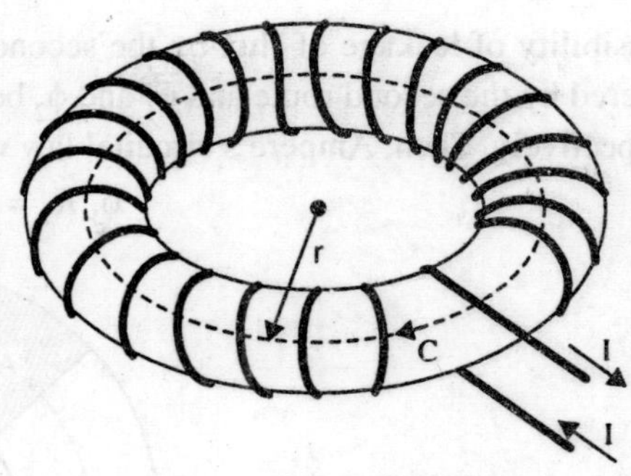

FIGURE 28.16 **A toroidal winding, C denotes a closed path.**

for an electric circuit (g denotes the conductance) in which $\oint \frac{dl}{gA}$ = resistance = R

Thus, if we call NI = magneto motive force (*m.m.f.*), ξ_m and $\oint \frac{dl}{A\mu}$ as reluctance R, we get

$$\phi R = NI = \xi_m$$

Series circuit. When the ring of the toroidal consists of parts of different materials with permeability μ_i, Area of connection A_i and lengths l_i,

$$\oint \frac{dl}{\mu A} = \sum_i \frac{l_i}{\mu_i A_i} = \sum_i R_i = R_A + R_B + R_C. \text{ (See Fig. 28.17)}$$

So, reluctance of different parts to the flow of flux, ϕ, adds up in the same way as the resistances are coupled in series. Infact, flux ϕ just as current I, satisfies the equation

$$\phi = \frac{\xi_m}{R}$$

This is the central equation of a magnetic circuit.

Parallel circuit. When the flux has two possible paths having reluctances R_h (high) and R_l (low), most of the flux will pass through the 2nd root of low reluctance just as current does in the electric circuits having resistances in parallel (*e.g.* Galvanometer and shunt arrangement). Also, the equivalent reluctance of the parallel magnetic circuit will be given by

$$R = \frac{R_h R_l}{R_h + R_l}$$

Such a circuit is being shown in Fig. 28.17.

The solenoid of N turns is wound on a ring shaped magnetic material of types A, B, C. Usually these magnetic materials have high values of μ, say μ_A, μ_B and μ_C as compared to air. Therefore, ring offers a low reluctance path to flow of flux, ϕ and remains confined to ring of the solenoid. However, if part C of the core is replaced by air gap, there will be a

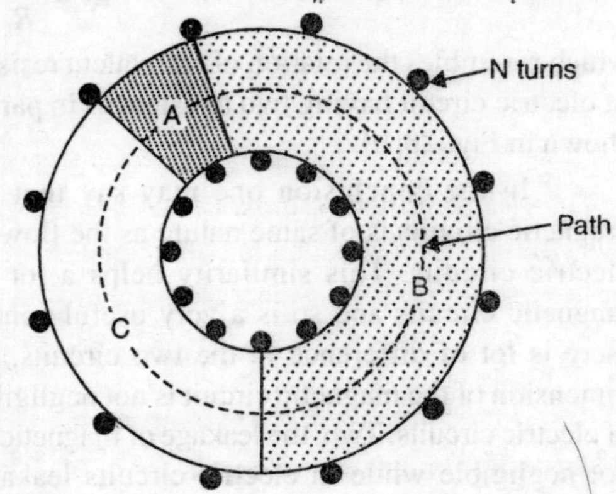

FIGURE 28.17 **A magnetic circuit consisting of different parts *A*, *B* and *C*.**

possibility of leakage of flux by the second route by air as shown in Fig. 28.18. Let R_2 be the reluctance offered by the second route and ϕ_1 and ϕ_2 be the flux through the two routes *i.e.* via the ring and via the air respectively. Then, Ampere's circuital law when applied for these two closed paths would give

$$\phi_1 R_1 = \phi_2 R_2 = NI = \xi_m$$

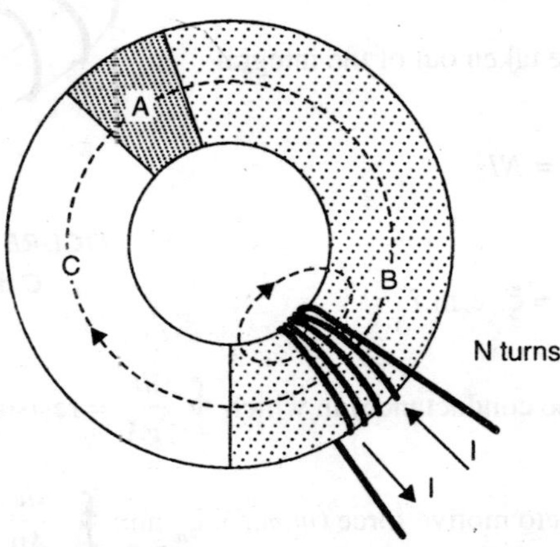

FIGURE 28.18 **This magnetic circuit is equivalent to the magnetic circuit of Fig. 28.17 if the permeabilities of *A*, *B*, and *C* are large.**

$$\therefore \quad \phi_1 = \xi_m / R_1$$

and

$$\phi_2 = \xi_m / R_2$$

Total flux in the circuit, ϕ, will be

$$\phi = \phi_1 + \phi_2 = \xi_m \left[\frac{1}{R_1} + \frac{1}{R_2}\right]$$

$$= \xi_m \frac{R_2 + R_1}{R_1 R_2} = \frac{\xi_m}{\left(\dfrac{R_1 R_2}{R_1 + R_2}\right)}$$

Thus, equivalent reluctance of the magnetic circuit mentioned above is

$$R = \frac{R_1 R_2}{R_1 + R_2}$$

which resembles the relation of equivalent resistance in the case of electric circuit having two resistances in parallel as has been shown in Fig. 28.19.

In the conclusion one may say that flow of flux in magnetic circuits is of same nature as the flow of current in an electric circuits. This similarity helps a lot in the study of magnetic circuits and so is a very useful concepts. However, there is lot of difference in the two circuits. One, the lateral dimension of the magnetic circuit is not negligible as in the case of electric circuits. Two, the leakage of magnetic flux is generally not negligible while in electric circuits leakage of current is almost nil.

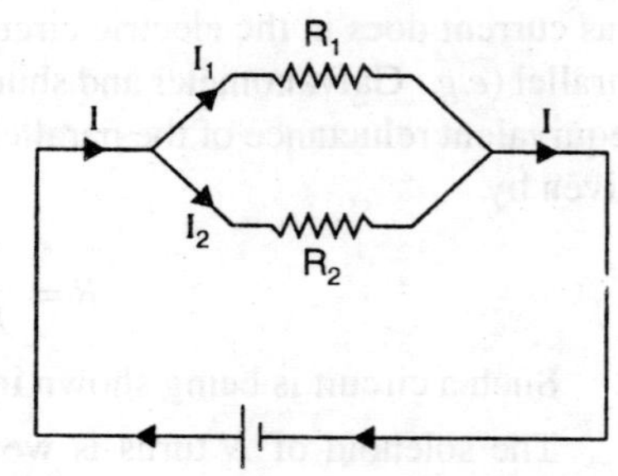

FIGURE 28.19 **Network of two resistances in parallel in an electric circuit.**

Magnetic circuit of permanent magnets. The magnetic circuit containing a permanent magnet is an other example worthy of discussion. Such a circuit is shown in Fig. 28. 20. Here *ab* is the permanent magnet (*P–M*) surrounded by soft iron pieces and the air gap between the soft iron poles completing the closed circuit for the magnetic flux.

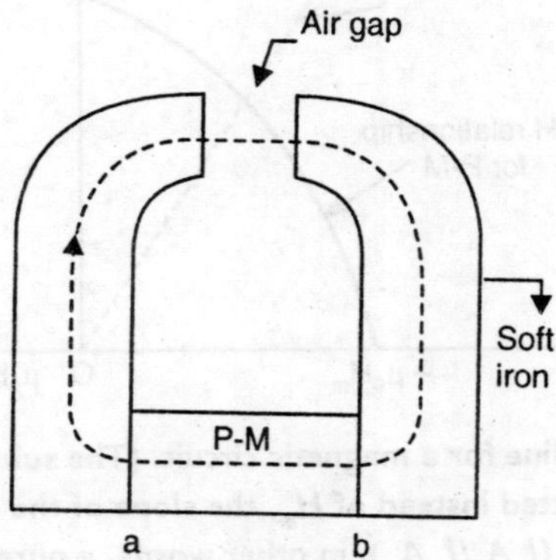

FIGURE 28.20 **A permanent magnet circuit. For the circuit shown, the magnet has a rather large demagnetizing field acting upon it ; the demagnetizing field can be reduced by increasing the length of *P-M* material, *e.g.* by placing additional magnets in the side arms of the circuit.**

Let us apply the Ampere's circuital law considering the dotted curve as the closed path of magnetic flux. We get

$$\oint H \,.\, dl = 0$$

because there is no current

or
$$\int_a^b H \,.\, dl = -\int_{b\,(P-M)}^{a} H_m . dl_m \;;$$

m denotes quantities in P-M.

Taking $H = \dfrac{B}{\mu} = \dfrac{\phi}{A\mu}$, we get

$$\phi\left[\frac{l_i}{\mu_i A_i} + \frac{l_g}{\mu_0 A_g}\right] = -H_m l_m \;; \; i \text{ denotes iron piece and } g \text{ the air gap.}$$

Since $\mu_i >> \mu_0$, we may write

$$\phi\left(\frac{l_g}{\mu_0 A_g}\right) = -H_m l_m.$$

Also,
$$\phi = A_m B_m,$$

$$B_m = -\left(\frac{l_m \mu_0 A_g}{l_g A_m}\right) H_m$$

$$= -C_m \mu_0 H_m \;; C_m = \frac{l_m A_g}{l_g A_m}$$

which is solved using the *B–H* curve for the permanent magnet material (see Fig. 28.21)

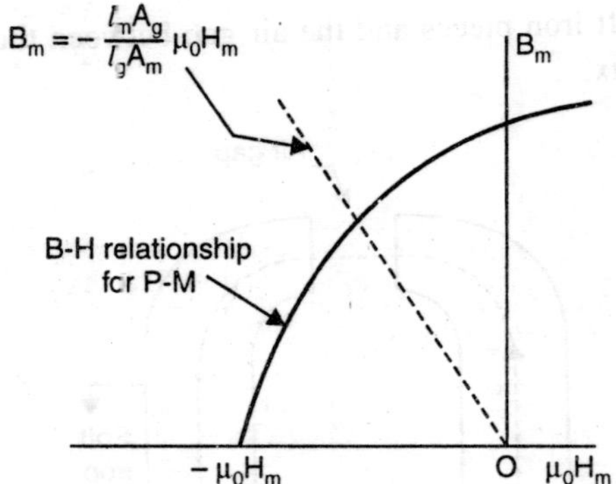

FIGURE 28.21 **Demagnetizing line for a magnetic circuit. (The subscript *m* stands for magnet.) Since $\mu_0 H_m$ is plotted instead of H_m, the slope of the demagnetizing line is just $-(l_m A_g / l_g A_m)$, in other words, a pure number.**

The point of intersection of the line.

$$B_m = -C_m \mu_0 H_m$$

with the $B_m - \mu_0 H_m$ curve gives the solution. From this value of B_m we calculate

$$\phi = B_m A_m$$

and

$$B_g = \frac{\phi}{A_g} = \frac{B_m A_m}{A_g}$$

which solves the problem. A_g denotes the area of cross-section of soft iron poles and A_m of the *P–M*.

APPENDIX

TABLE A.1 SI and CGS units of various magnetic quantities

Magnetic Quantity	*Units*		*Conversion*
	SI	*CGS*	
Flux(ϕ)	weber(w)	Maxwell	1CGS = 10^{-8} SI
Magnetic dipole moment (**m**)	Amp-m^2	e.m.u.-cm^2	1CGS = 10^{-3} SI
Magnetization (**M**)	Amp/m	e.m.u./cm	1CGS = 10^3 SI
Magnetic intensity (**H**)	Amp/m	e.m.u./cm	1CGS = 10^3 SI
Magnetic Induction (**B**)	n/Amp/m or w/m^2	dyne/e.m.u.-cm or Maxwell/cm^2(gauss)	1 gauss = 10^{-4} SI

Some useful vector identities

- $\vec{\nabla}.(\vec{\nabla} \times \vec{A}) = 0$
- Curlgrad $f = 0$; f = any scalar function
  i.e. Curlgrad = 0

- $\vec{\nabla} \times (\vec{A} f) = f \vec{\nabla} \times \vec{A} - \vec{A} \times \nabla f$
- $\vec{\nabla} \times (\vec{A} \times \vec{B}) = \vec{A} \vec{\nabla} . \vec{B} - \vec{B} \vec{\nabla} . \vec{A} + (\vec{B} . \vec{\nabla}) \vec{A} - (\vec{A} . \vec{\nabla}) \vec{B}$
- $\vec{\nabla} . (\vec{A} \times \vec{B}) = \vec{B} . \vec{\nabla} \times \vec{A} - \vec{A} . \vec{\nabla} \times \vec{B}$
- $\vec{\nabla} \times (\vec{\nabla} \times \vec{A}) = - \nabla^2 \vec{A} + \vec{\nabla} (\vec{\nabla} . \vec{A})$

EXERCISE

1. Explain what is meant by hysteresis. *(Jan. 1997. K.U.K.)*
2. Show that the area of a *B-H* curve denoted the energy dissipated per unit volume of material during each magnetizing cycle. *(June 2000, K.U.K.)*
3. What are the characteristics of para-, dia- and ferromagnetic substances ? Explain their behaviour with the examples. *(May 1999, June 1999, K.U.K.)*
4. Explain magnetic hysteresis. Show that the area of the *B-H* cycle represents the energy dissipated per cubic meter of the material per cycle of magnetization. *(Jan. 1999, K.U.K.)*
5. Explain the following terms :
   (*i*) Magnetization and magnetic intensity
   (*ii*) Magnetic permeability and susceptibility. *(May 1998, K.U.K.)*
6. Explain in details the magnetometer method for *B-H* curve. *(Dec. 2000)*
7. Describe *B-H* hysteresis loop and its significance. *(Dec. 2000, K.U.K.)*
8. A solenoid consisting of 500 turns and carrying 4 A current is 0.5 m long. Calculate (*i*) Magnetomotive force (*ii*) Total flux if area of cross-section is 0.0008 m^2. **[Ans. 251.3 Wb]**
9. A circular ring of iron is wound uniformly with 1000 turns of wire. Its mean diameter is 0.2 m and its cross-section is 4 cm^2. Find the current which will produce a flux of 2000 lines. Given permeability = 800. **[Ans. 31.25 mA]**
10. Give a detailed note on ferromagnetism. *(K. U. June, 2007)*

Chapter 29 SPECIAL TOPICS

29.1 GAUSS' LAW AND ITS APPLICATIONS

The Gauss' Law has lot of importance in electrostatic. It can be used to calculate the electric field due to symmetrical charge distributions such as a long line charge, infinite charged plane sheet etc. Since Gauss' law requires the knowledge of Coulomb law, we discuss this first

Coulomb Law

Towards the end of eighteenth century it was established through series of experiments that

(*a*) there are two kinds of charges ; positive and negative

(*b*) a force of attraction or repulsion acts between these depending on whether they are opposite or same types,

and (*c*) this force is given by

$$\vec{F} = \frac{1}{4\pi\varepsilon_0} \frac{q_1 q_2}{r_{12}^{\ 3}} \vec{r}_{12}$$

In SI system of units $\dfrac{1}{4\pi\varepsilon_0} = 9 \times 10^9\ n\ \text{m}^2/\text{C}^2$

This expression is known as Coulomb's law. Since this result is purely based on experimental work, this law is now known as expriemental law.

Now electric field vector $\vec{E}$ at a point is defined as the force which acts on a unit charge when placed at that point. Hence electric field at P due to a point charge q at O is given by

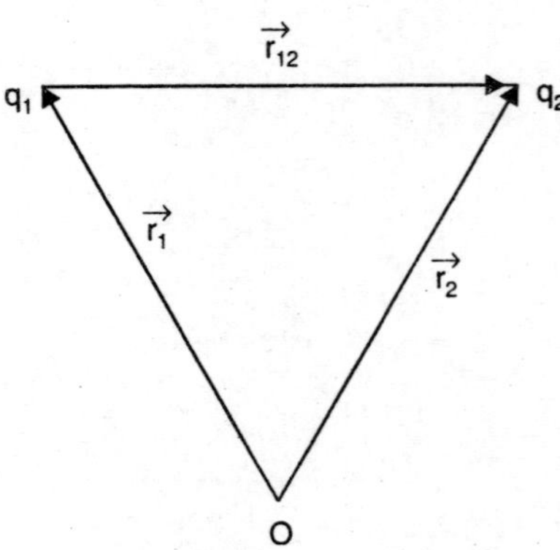

FIGURE 29.1 **q_1 and q_2 are two charges separated by a vector $\vec{r}_{12}$.**

$$\vec{E}(\vec{r}) = \frac{1}{4\pi\varepsilon_0} \frac{q}{r^3} \vec{r}$$

unit of $\vec{E}$ is volt/meter (= V/m)

Here, ε_0 represents the property of the free space and is called as permittivity of free space.

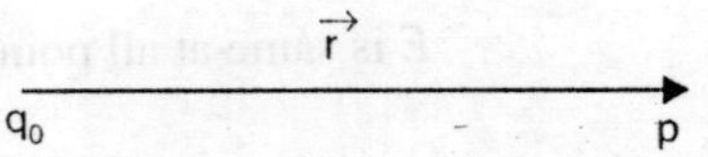

FIGURE 29.2 P is the point where field is $\vec{E}$.

Gauss' Law

Let us consider a closed surface S which encloses a point charge q at O (Fig. 29.3)

Now electric field $\vec{E}$ at P will be given by $\vec{E}(\vec{r}) = \frac{1}{4\pi\varepsilon_0} \frac{q\,\vec{r}}{r^3}$

Taking surface integral over the entire surface S we get

$$\oint_S \vec{E} \cdot d\vec{S} = \frac{q}{4\pi\varepsilon_0} \oint_S \frac{\vec{r} \cdot d\vec{S}}{r^3}$$

$$= \frac{q}{4\pi\varepsilon_0} \oint d\Omega$$

where $d\Omega = \frac{\vec{r} \cdot d\vec{S}}{r^3}$, the solid angle subtended by dS at O.

$$= \frac{q}{4\pi\varepsilon_0} \times 4\pi$$

$$= \frac{q}{\varepsilon_0}$$

Because total solid angle at any point within the enclosing surface is always equal to 4π.

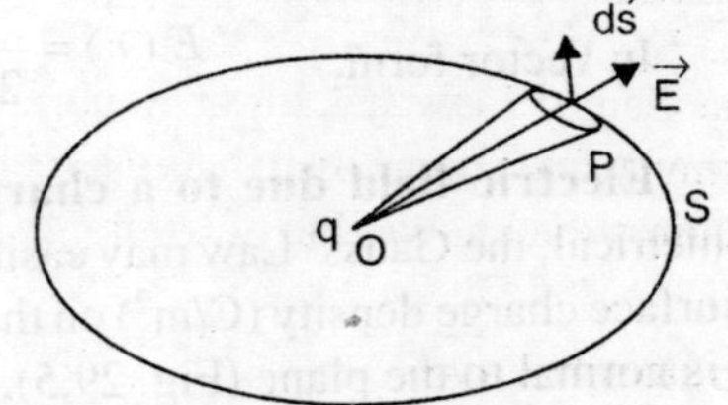

FIGURE 29.3 A point charge q at O and area $\vec{dS}$ located at P on the surface S. $\vec{E}$ is the electric field vector.

Hence, $\oint_S \vec{E} \cdot \vec{dS} = \frac{q}{\varepsilon_0}$ which is known as Gauss' Law.

Using divergence theorem and assuming q as uniformly distributed within S with ρ as the charge density of this distribution, we get

$$\int_V \vec{\nabla} \cdot \vec{E}\, dV = \frac{1}{\varepsilon_0} \int_V \rho\, dV, \text{ where } V \text{ is the total volume within } S.$$

or

$$\int \left(\vec{\nabla} \cdot \vec{E} - \frac{\rho}{\varepsilon_0} \right) dV = 0$$

Hence, $\vec{\nabla} \cdot \vec{E} - \frac{\rho}{\varepsilon_0} = 0$ or $\vec{\nabla} \cdot \vec{E} = \frac{\rho}{\varepsilon_0}$, known as the differential form of Gauss' Law.

Applications of Gauss' Law

Gauss' Law is very useful to find the electric field due to symmetrical charge distributions *e.g.*, a long line charge, a plane charged sheet of infinite size etc.

Electric field due to long line charge. Let λ be the charge on the unit length of AB and r, l represent respectively the radius and the length of the cylindrical surface S. Wire AB is the axis of this cylinder. Because of symmetry electric field $\vec{E}$ will be radial at *P*. Hence, $\vec{E}$ and $\vec{dS}$ will be parallel on the sides and perpendicular on the faces of the cylindrical surface.

Now, applying Gauss' Law we get

$$\oint \vec{E} \cdot \vec{dS} = \frac{q}{\varepsilon_0} = \frac{\lambda l}{\varepsilon_0}$$

or

$$\int_{sides} E\, dS = \frac{\lambda l}{\varepsilon_0} \qquad \because \text{ on the faces } \vec{E} \cdot \vec{dS} = 0$$

$$E \int_{sides} dS = \frac{\lambda l}{\varepsilon_0}$$

$\because$ *E* is same at all points on the sides of the cylinder.

$$E \cdot 2\pi r l = \frac{\lambda l}{\varepsilon_0}$$

Hence,

$$E = \frac{\lambda}{2\pi\varepsilon_0 r}$$

In vector form,

$$\vec{E}(\vec{r}) = \frac{\lambda}{2\pi\varepsilon_0} \left(\frac{\vec{r}}{r^2} \right)$$

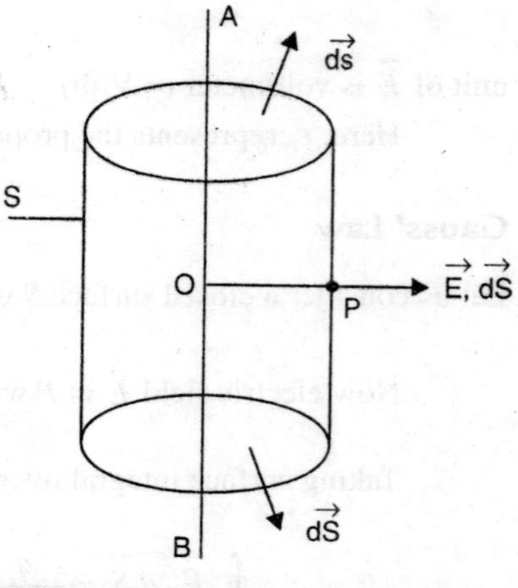

FIGURE 29.4** **AB is a part of long line charge. P is a point on the cylindrical surface S; OP = r, the radius.

Electric field due to a charged plane sheet. Since on plane sheet the charge distribution is symmetrical, the Gauss' Law may easily be used to calculate the electric field due to the charge on it. Let σ be the surface charge density (C/m^2) on the plane sheet. Construct a Gaussian surface of cylindrical shape whose axis is normal to the plane (Fig. 29.5).

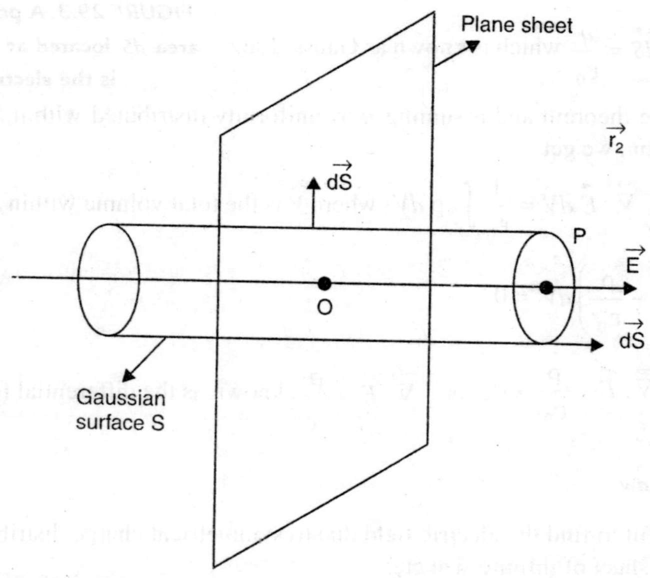

FIGURE 29.5** **P is a point on the face of the cylinder. ΔS is the area of each face and OP is normal to the plane.

Because of the symmetry $\vec{E}$ at P will be normal to the plane. Also, from the figure it is clear that elemental area $\vec{dS} \parallel \vec{E}$ on each face while $\vec{dS} \perp \vec{E}$ everywhere on the sides of the cylinder. Now, apply Gauss' Law over the cylindrical Gaussian surface as

$$\oint_{\text{cylinder}} \vec{E} \cdot \vec{dS} = \frac{q}{\varepsilon_0} = \frac{\sigma \Delta S}{\varepsilon_0}$$

because the total charge enclosed $= q = \sigma \Delta S$,

Hence, $\int_{\text{faces}} E\, dS = \frac{\sigma \Delta S}{\varepsilon_0}$ or $E \int_{\text{faces}} dS = \frac{\sigma \Delta S}{\varepsilon_0}$ $\because$ on the sides $\vec{E} \cdot \vec{dS} = 0$

$$E \times 2\,\Delta S = \frac{\sigma \Delta S}{\varepsilon_0}$$

Therefore, $E = \frac{\sigma}{2\varepsilon_0}$ and direction will be along OP.

29.2 INTERACTION OF NUCLEAR RADIATION WITH MATTER

The nuclear radiation consisting of α-particles, β-particles and gamma rays are detected with gas detectors as well as solid state detectors. α and β are charged particles which carry two and one unit of electronic charge respectively. Gamma rays, on the other hand, are electromagnetic radiation normally carrying energy up to a fraction of MeV. These nuclear radiations are therefore detected because of their interaction with the matter contained in the detector say gas or solid state. Thus, there is a need of studying the interaction of α, β and γ-rays with the matter in order to understand the working of nuclear detectors such as Ionization chamber, Geiger Muller Counter, solid state detectors etc. so, we now discuss interaction of γ- radiation with matter and then of charged particles.

Interaction of Gamma radiation

Essentially γ-radiation is electromagnetic in nature consisting of photons which interact with matter mainly through

1. Compton scattering,
2. Photo electric effect,

and 3. Pair production.

In every case energetic electrons are produced which would give rise to electric pulse for detection in the detectors.

Compton Scattering. In compton scattering the photon of the γ -radiation collide with the loosely bound electron of the atom and get scattered with reduced frequency. This collision is elastic and may be understood by classical approach treating the photon as a particle and using the conservation laws of total energy and momentum of colliding particles.

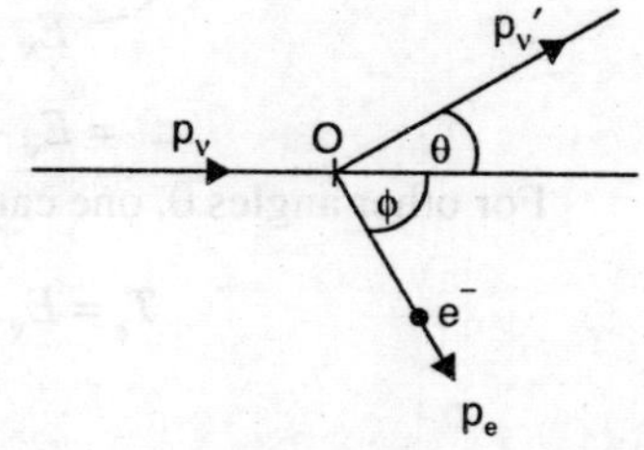

FIGURE 29.6 **The phenomenon of compton scattering ; e^- is the scattered electron.**

The incident photon with momentum $\vec{p}_\nu$ strikes the electron at rest at O. As a result of this collision photon scatters along θ with reduced momentum $\vec{p_\nu}'$ while electron gets recoiled along ϕ with momentum $\vec{p_e}$. The law of conservation of linear momentum leads to

$$p_\nu = p_\nu' \cos\theta + p_e \cos\phi,$$

and
$$0 = p_\nu' \sin\theta - p_e \sin\phi$$

Here, $p_\nu = h\nu/c$, $p_\nu' = h\nu'/c$ and $\vec{p_e} = \dfrac{m_0 \vec{v}}{\sqrt{1-\dfrac{v^2}{c^2}}}$, m_0 and c being the electron rest mass and velocity of light respectively. Next, the conservation law of total energy gives us

$$E_\nu + m_0c^2 = E_\nu' + E_e \quad \text{where}$$

$$E_\nu = h\nu,\ E_\nu' = h\nu' \quad \text{and} \quad E_e = \sqrt{p_e^2c^2 + m_0^2c^4} = m_0c^2 \Big/ \sqrt{1-\frac{v^2}{c^2}}$$

The simplification of these results finally gives

$$\lambda' - \lambda = \frac{h}{m_0c}(1-\cos\theta) = \frac{2h}{m_0c} \quad \text{at } \theta = 180°$$

where λ is the wavelength of the incident radiation and λ' that of scattered radiation ; h is the Planck's constant. The quantity $\dfrac{h}{m_0c}$ (= 2427 F) is called compton wavelength which is independent of γ-rays energy.

Further, the maximum reduction in wavelength of the scattered radiation occurs at $\theta = 180°$ (called head-on collision) thereby imparting max. recoil kinetic energy, T_e(max.), to the electron. This T_e (max.) is given by

$$T_e\,(\text{max.}) = h(\nu - \nu') \text{ at } \theta = 180°$$

$$= hc\left(\frac{1}{\lambda} - \frac{1}{\lambda'}\right)$$

$$= hc\left(\frac{1}{\lambda} - \frac{1}{\lambda + \dfrac{2h}{m_0c}}\right) \quad \text{by putting the value of } \lambda'$$

$$= \left(h\nu - \frac{h\nu}{1+\dfrac{2h\nu}{m_0c^2}}\right) = \left(E_\nu - \frac{E_\nu}{1+\dfrac{2E_\nu}{m_0c^2}}\right)$$

$$\approx E_\nu - \frac{m_0c^2}{2} \quad \text{when } E_\nu >> \frac{m_0c^2}{2} \quad \textit{i.e.}\ \text{for high energy radiation.}$$

$$= E_\nu - 0.255\ \text{MeV}.$$

For other angles θ, one can similarly obtain

$$T_e = E_\nu\left[1 - \frac{1}{1+\dfrac{E_\nu}{m_0c^2}(1-\cos\theta)}\right];\ T_e = E_e - m_0c^2$$

$$= E_\nu - \frac{m_0c^2}{1-\cos\theta} \quad \text{when } E_\nu >> m_0c^2.$$

The probability of compton scattering can only be calculated by quantum mechanical technique. It is found that the angular distribution of compton scattered gamma radiation of very small energy approaches the classical expression which is proportional to $1 - \cos^2 \theta$.

Also the cross-section of compton scattered electron as a function of recoil electron energy will drop to zero around its max. value

$$T_e(\text{max.}) = E_\gamma - 0.255 \text{ MeV}$$

The scattering of the radiation may also occur from bound electron in the atom. However, in this case the collision will be with the atom as a whole as the electron is a part of atom and the compton wavelength will be $\frac{h}{m_a c}$, with m_a as the mass of the atom instead of electron. This being small, the shift in the wavelength of the scattered radiation will be negligible. Such a collision plays no role in the nuclear detectors.

Photo Electric Effect. Gamma rays (photon) can be absorbed only by the bound electron of the atom. A loosely bound electron which is practically free can not do so because the conservation laws of momentum and energy can not be satisfied. For instance in a collision with a free electron these laws require that

$$p_\nu = \frac{h\nu}{c} = p_e \qquad \text{(Conservation of momentum)}$$

or

$$h\nu = p_e c$$

and

$$E_\nu = E_e = \sqrt{p_e^2 c^2 + m_0^2 c^4} \qquad \text{(Conservation of energy)}$$

or

$$h\nu = \sqrt{(h\nu)^2 + m_0^2 c^4} \quad \text{as } p_e c = h\nu,$$

These equations can not be satisfied as such. In other words it requires that p_e be less than photon momentum $= \frac{h\nu}{c}$. This can only happen when atomic electron is bound in which case the part of the photon momentum will be shared by the atom.

This phenomenon of emission of bound electron when an incident photon is absorbed is called photo electric effect. Conservation of momentum and energy would lead to

$$\vec{p}_\nu = \vec{p}_e + \vec{p}_a \quad ; \; \vec{p}_a = \text{recoil momentum of the parent atom.}$$

and $E_\nu = T_e + T_a + E_B$; T_e, T_a denote kinetic energies of electron (K.E.) and atom respectively.

where E_B is the binding energy of the electron, also called excitation energy of the atom while E_e and E_a denote respectively the kinetic energy of electron and the atom after collision. Usually, T_a is smaller than T_e by a factor $\frac{m_e}{m_a}$ ($\approx 10^{-4}$), it may be neglected so that

$$E_\nu = h\nu = T_e + E_B$$

or K.E. of photo electron, $T_e = h\nu - E_B$. The frequency of radiation for which K.E. of the electron is just zero is called threshold frequency ν_0 given by

$$0 = h\nu_0 - E_B$$

i.e., $\nu_0 = \frac{E_B}{h}$; h being the Planck's constant.

For γ-rays greater than $\frac{1}{2}$ MeV, the ejection of K-shell electrons is most probable because then resonance condition is satisfied. For low energy γ-rays angular distribution of photo electrons is symmetrical about $\theta = 90°$ as the electrons should be ejected in a direction parallel to $\vec{E}$ (electric field) of the incident radiation which is always normal to the incident direction. These distributions as a function of angle are shown in Fig. 29.7 for radiation of various energies.

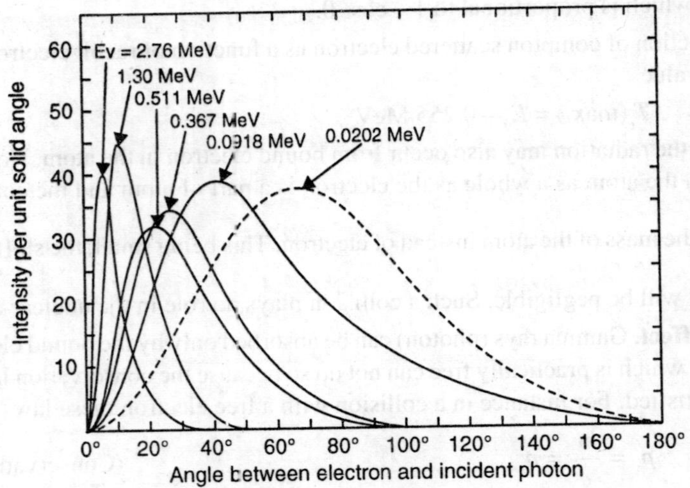

FIGURE 29.7 **Plot of intensity of photo-electrons as a function angle between these electrons and incident radiation.**

These ejected photo-electrons as well as the accompanied X-rays from excited atom cause electric pulses in the detectors particularly the scintillator detectors.

Pair Production. The third way in which the γ-rays disappears is the pair production *i.e.* pair of electron and positron. The positron has the same mass, m_0, as that of electron while it carries a positive charge equal to that of electron. Thus the minimum energy of photon required to create this pair would be equal to $2m_0c^2 = 2 \times 0.51 = 1.02$ MeV.

The existence of positron was first predicted by Dirac through the solution of relativistic Schrodinger equation called Dirac equation. That is the energy of the electron W moving in a field of potential V was to be given by

$$W = \pm (p_e^2c^2 + m_0^2c^4)^{1/2} + V$$

which for $V = 0$, becomes

$$W = \pm (p_e^2c^2 + m_0^2c^4)^{1/2}$$

Dirac developed his theory for relativistic electron using both signs for the energy of the electron. Accordingly, Dirac predicted both +ve as well as –ve energy states for the electron.

As the minimum value of p^2 is zero, so the +ve energy states will start from m_0c^2 upward while the –ve energy states from $-m_0c^2$ downwards as shown in Fig. 29.8.

So, there is an energy gap of $2m_0c^2$ between the two set of states : Dirac further assumed that in the absence of field or matter all +ve energy states are empty while all –ve energy states are filled. Further, according to Dirac the mass and momentum of electron

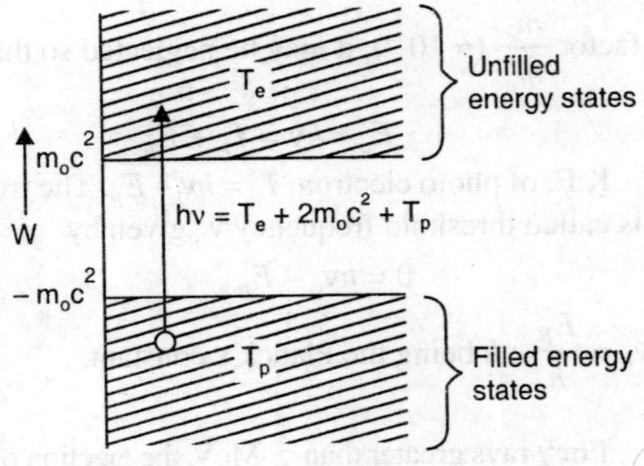

FIGURE 29.8 **Filled and unfilled energy states of electrons ; T_e and T_p denote kinetic energies of electron and positron.**

in the –ve energy states should be $-m_0$ and $-\vec{p}$ respectively. So, when a γ-ray *i.e.* photon of energy $h\nu$ is absorbed, the electron from –ve states jumps to the +ve states thus creating a hole in the –ve energy states. Because of this vacancy (hole) the system acquires a momentum $-(-\vec{p_e})$, mass $-(-m_0)$ and charge $-(-e)$ $= +e$. This hole in the system thus behaves like a particle of mass m_0, momentum $\vec{p_e}$ and charge $+e$.

This was called a positron. Such a prediction baffled people in the beginning till it was discovered by Anderson (1932) experimentally. The non relativistic kinetic energy of the hole (or positron) would be equal to $\frac{p^2}{2m_0}$. Further, it may be noted that when hole is created electron also appears. This phenomenon is known as Pair Production.

From the conservation of energy one gets

$$h\nu = T_e + T_p + 2m_0c^2$$

where T_e and T_p denote the kinetic energy of electron and positron respectively. Also conservation of linear momentum gives

$$\vec{p_\nu} = \vec{p_e} + \vec{p_p}$$

It is quite clear that the above two equations like in photoelectric effect can not be satisfied as such. There should be a third particle to share the momentum of the incident photon. It could be a nucleus to do this Job. However, the kinetic energy of the nucleus would be small and hence the above mentioned energy equation would remain valid.

Both these electron and positron cause ionisation while passing through the material of the detector and lose their energies. Positron when becomes slow annihilates the electron with the emission of twin γ-rays travelling in almost opposite directions. Energy of each γ-ray would be 0.51 MeV. Lastly, this pair may also form an atom called positronium with half life of 10^{-10} sec. or 10^{-7} sec. depending upon the relative spin alignment of the pair.

Absorption of Gamma Rays

A beam of γ-rays while passing through the material suffers attenuation due to the disappearance of photons from the beam through compton scattering, photoelectric effect and pair production. Let us suppose I_0 as the intensity of incident γ-rays and I_t its intensity after the beam comes out of the material (called absorber) of thickness t (see Fig. 29.9). We further suppose I to be the intensity of γ-rays at the penetration depth x and dI the amount absorbed when it further penetrates through a distance dx. Then the fraction of intensity of the beam lost will be proportional to dx as the three methods by which attenuation occurs are mutually independent. So, we may write

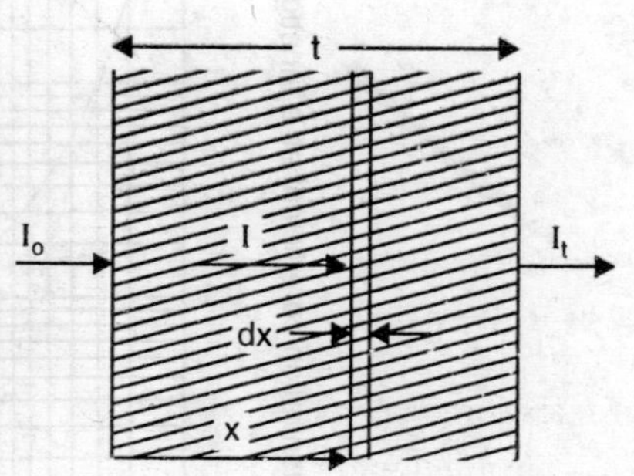

FIGURE 29.9 **The attenuation of the beam as it passes through the absorber of thickness *t*.**

$$\frac{dI}{I} = -\mu\, dx$$

where μ is the constant of proportionality and –ve sign is taken to indicate the loss of intensity. The constant μ is also called linear absorption co-efficient of the material or absorber. On integration this result we obtain

$$\int_{I_0}^{I_t} \frac{dI}{I} = -\mu \int_0^t dx \quad i.e. \quad \ln \frac{I_t}{I_0} = -\mu t \quad \text{or} \quad I_t = I_0 e^{-\mu t}$$

The half value thickness t_o will be given by

$$\frac{I_o}{2} = I_o e^{-\mu t_o} \quad i.e. \quad t_o = \frac{Ln(2)}{\mu} = \frac{0.693}{\mu}$$

As the three processes are independent, the corresponding absorption co-efficients may be related as

$$\mu = \mu_c + \mu_k + \mu_p$$

where μ_c = Compton absorption co-efficient

μ_k = Photoelectric absorption co-efficient

and μ_p = Co-efficient for pair production.

Since the attentuation of the beam depends on the number of atoms per unit volume we define mass absorption co-efficient by $\mu_m = \mu/\rho$ where ρ is the density of the absorber.

As a result $\qquad I_t = I_0 e^{-\mu_m \rho t}$

and μ_m will be independent of the state of the absorber.

In Fig. 29.10 we show a typical experimental plot of mass absorption co-efficient μ_m vs photon energy. The co-efficients for photoelectric effect ($= \mu_k/\rho$), compton scattering ($= \mu_c/\rho$) and pair production ($= \mu_p/\rho$) have been shown by dotted curves respectively. The energywise dominance of various co-efficients are obvious *e.g.* pair production occurs for energy > 1.02 MeV and compton scattering occurs at all energies.

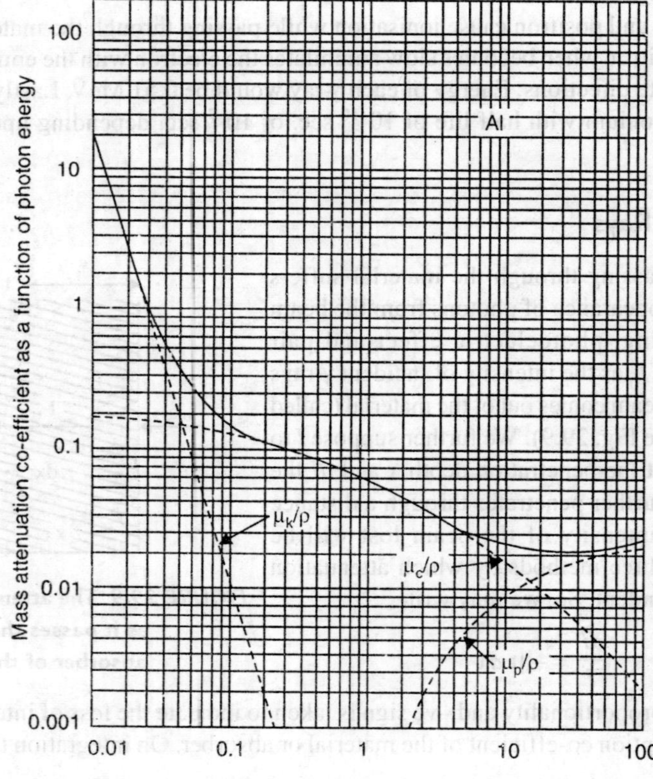

FIGURE 29.10 **Mass attenuation co-efficients as a function of photon energy.**

Interaction of Charged Particles with Matter

The detection of the charged particles particularly of nuclear radiations mainly depends on the extent of ionisation, atomic excitations and the deflection of these particles while passing through the matter. The loss of energy of the incident charged particles through ionisation and excitation of atoms encountered per unit path length is the greatest. These events of ionisation and excitation of atoms of the material of the detector give rise to detectable electric pulses and these could be counted subsequently leading to the identification and energy measurement of the incident charged particles such as α- and β-rays. The charged particles may loose energy through their interaction with the nucleus. However, these events are very rare and may be ignored. Further, at high energies comparable with the rest energy of the charged particles, the energy loss through emission of radiation due to their acceleration while passing near the field of the nucleus becomes increasing important. This is because the accelerated charge always emits electromagnetic radiation.

A rough estimate of the energy loss of the incident charged particle while it collides with the loosely bound atomic electron can be obtained through classical laws. One such encounters has been shown in Fig. 29.11. Here a charged particle of charge ze, mass M_0 and velocity v is shown as travelling along X-axis. The colliding electron of mass m_0, has been placed at a distance, b, along Y-axis called impact parameter. If the force on the electron is taken as

$$\vec{F} = \hat{i} F_x + \hat{j} F_y,$$

we have

$$\int_{\Delta t} F_x \, dt = p_x = 0;$$

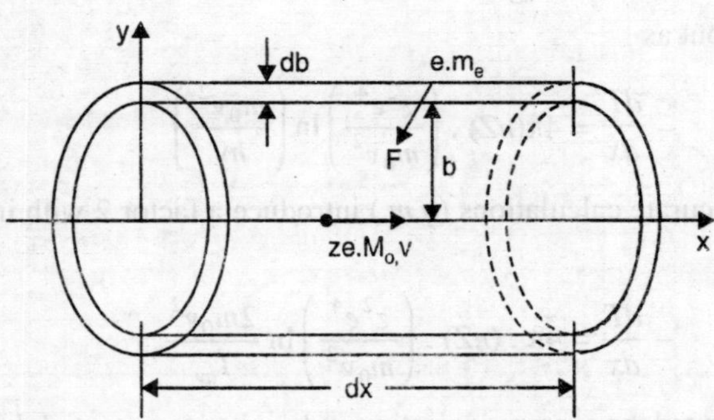

FIGURE 29.11 **Encounter between a heavy charged particle of mass M_0 and a free electron of mass m_0. The impact parameter b is indicated.**

being negligible and

$$\int_{\Delta t} F_y dt = p_y$$

where time of impact Δt may be taken as equal to $\frac{b}{v}$. The change in momentum, p_x, of the electron is assumed as negligible. Further, F_y may be taken as

$$F_y \approx \frac{ze^2}{b^2} \text{ (Coulomb law)}$$

thereby giving us

$$p_y \approx \frac{ze^2}{b^2} \times \frac{b}{v} = \frac{ze^2}{bv}$$

This gives the energy gained by the electron and hence loss of energy by incident charged particle during the encounter as

$$\frac{p_y^2}{2m_0} = \frac{z^2 e^4}{2m_0 b^2 v^2}$$

If there are n atoms of atomic number Z per unit volume, the total energy loss, will be

$$-\Delta T = \int_{b_{\min}}^{b_{\max}} nZ \,.\, 2\pi b \,.\, db \,.\, \Delta x \,.\left(\frac{z^2 e^4}{2m_0 b^2 v^2}\right)$$

Hence $$-\frac{dT}{dx} = (nZ)\,.\,2\pi\,.\left(\frac{z^2 e^4}{2m_0 v^2}\right)\int_{b_{\min.}}^{b_{\max.}} \frac{db}{b}$$

$$= \frac{\pi z^2 e^4}{m_0 v^2}\,.\,nZ\,.\ln\left(\frac{b_{\max.}}{b_{\min.}}\right)$$

However, a better estimate of energy loss may be obtained by using Gauss' theorem

$$\int \vec{E}\,.\,d\vec{s} = 4\pi q \quad \text{(C.G.S.)}$$

where various letters have usual meanings.

The result comes out as

$$-\frac{dT}{dx} = 4\pi(nZ)\,.\left(\frac{z^2 e^4}{m_0 v^2}\right)\ln\left(\frac{m_0 v^2}{h\nu}\right)$$

However, more accurate calculations (*q.m.*) introduce a factor 2 with in the natural logarithm. As a result, we get

$$-\frac{dT}{dx} = 4\pi\,.\,(nZ)\,.\left(\frac{z^2 e^4}{m_0 v^2}\right)\ln\frac{2m_0 v^2}{I_{av}}$$

where we have also substituted the average ionisation and excitation potential, I_{av}, of the atoms of the stopping material in place of $h\nu$.

If w is the energy lost to produce one ion pair and i is the no. of such ion pairs per unit path length,

$$-\frac{dT}{dx} = wi$$

This energy, w, is practically independent of the nature as well as the kinetic energy of the incident particle which is clear from Table 29.1.

TABLE 29.1 *w* in air

Energy	*w(eV)*
5 keV electrons	35.0
5.3 MeV alpha particles	35.2
340 MeV protons	33.3

Various quantities which appear in the expression $-\dfrac{dT}{dx}$ are :

n = No. of atoms per unit volume of the stopping material

Z = Atomic number of the constituent atoms

e = Electronic charge

m_0 = Rest mass of electron

ze = Charge of the incident particles

v = Velocity of the incident particles

and I_{av} = Average value of ionization and excitation potentials of atoms of the stopping material

It is worthnoting that

$$-\frac{dT}{dx} \propto \frac{1}{v^2}, \text{ provided the ln term does not very much with } v.$$

or

$$\propto \frac{1}{T}$$

Now, if N, A and ρ denote Avogadro's no., atomic wt. and density of the stopping material respectively, then

$$n = \frac{N\rho}{A}$$

As a result $-\dfrac{dT}{\rho\, dx}$ will become independent of the state of the stopping material. So, one prefers to plot $-\dfrac{dT}{\rho\, dx}$ as a function of the energy of the incident particles. These plots determined experimentally for protons at low energy as well as high enegy are shown in Fig. 29.12. It has been seen that theoretical predictions agrees quite well with the data except at very low energy (< 0.3 MeV) where the capture of electrons by slow protons occurs.

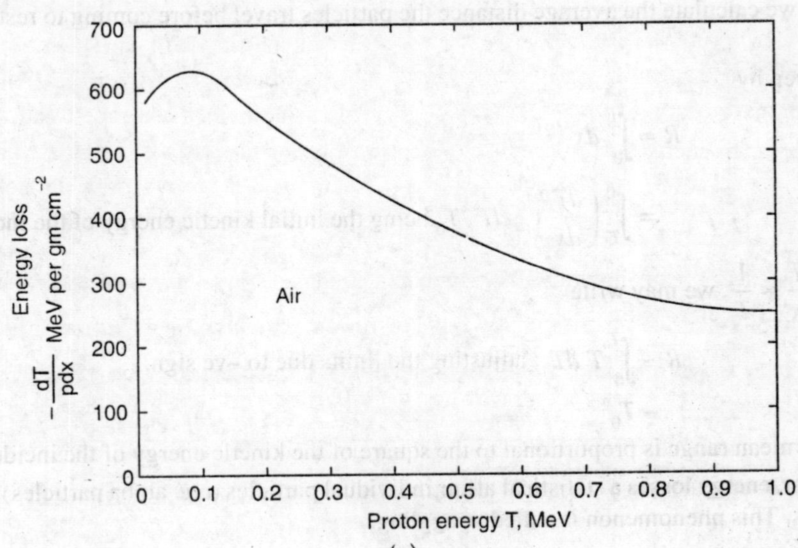

(*a*)

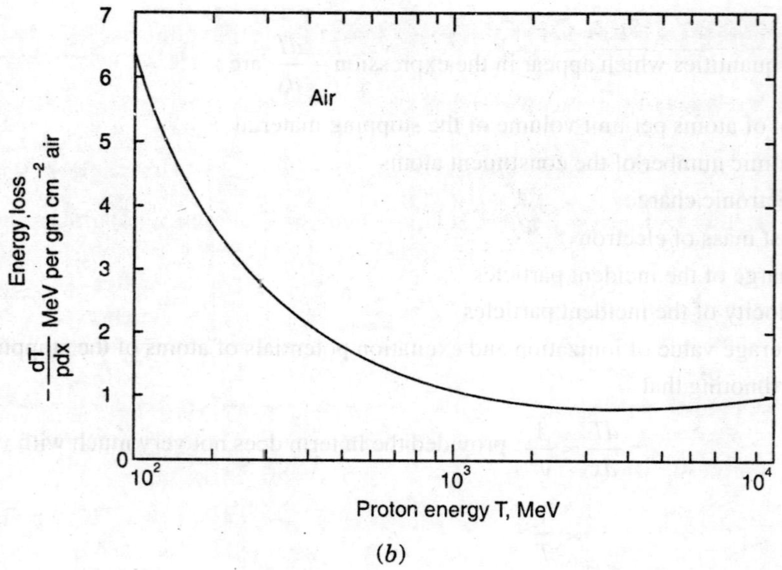

(*b*)

***FIGURE 29.12* Plots of energy loss per unit path length by protons in air at (*a*) low energy and at (*b*) high energy.**

Further, the above mentioned relation of $-\dfrac{dT}{dx}$ holds good even for electrons also. However, there are some noteworthy points of distinction.

(1) First, electrons cannot capture electron like protons or alpha particles. (2) Electrons scatter more in the material and has longer path length than it appears to be. That is electrons do not travel straight before coming to rest in the stopping material as heavy particles do. Thirdly, electrons at a particular energy has greater probability to lose energy through the emission of electro magnetic radiation than that of heavy particles at same energy. This is because electrons being lighter will have greater velocity and the emission of radiation depends upon the velocity but not on energy.

Lastly, we calculate the average distance the particles travel before coming to rest. This is known as Mean Range.

It is given by

$$\bar{R} = \int_0^R dx$$

$$= \int_{T_0}^{0} \left(\frac{dT}{dx}\right)^{-1} dT \text{, } T_0 \text{ being the initial kinetic energy of the incident particle.}$$

As $-\dfrac{dT}{dx} \propto \dfrac{1}{T}$, we may write

$$\bar{R} \sim \int_0^{T_0} T\, dT \text{ ; adjusting the limits due to –ve sign.}$$

$$\sim T_0^2$$

That is, mean range is proportional to the square of the kinetic energy of the incident particles.

However, energy loss is a statistical affair individual particles (*e.g.* alpha particles) may have slightly different ranges. This phenomenon is called straggling.

29.3 ELEMENTARY IDEA OF QUANTUM STATISTICS

In nature we have two types of particles, one having half integral spins and the other having integral spins. Particles like electrons, protons, neutrons which have $\frac{1}{2}$ spins fall in the first category while like photons, alpha particles, π-mesons which have integral spins fall in the second category. The dynamical properties of systems of such identical particles (non-interacting mutually) in thermal equilibrium are studied statistically. There are two types of quantum statistics, Fermi-Dirac and Bose-Einstein statistics. The former is applicable to the system of indentical particles having half integral spins while the latter to the system of identical particles having integral spins. For this reason particles of the first category mentioned above are called Fermions and particles of second category which have integral spins are called Bosons.

The quantum state of the system of fermions as has been verified is always anti symmetric under the interchange of pair of fermions and the quantum state of bosons is symmetric under this interchange. For instance in the case of two particles system the antisymmetric states is defined as

$\psi(1, 2) = -\psi(2, 1)$ where 1 and 2 denote the particle one and particle two. The symmetric state, on the other hand, is defined as

$$\psi(1, 2) = \psi(2, 1)$$

In other words, antisymmetric state changes sign under the interchange operation while symmetric state does not change sign.

Further, if ϕ_a and ϕ_b are single particle states we may write antisymmetric state as

$$\psi_A(1, 2) = \frac{1}{\sqrt{2}}[\phi_a(1)\phi_b(2) - \phi_b(1)\phi_a(2)]$$

and symmetric states as

$\phi_a(1)\phi_a(2)$; both particles in same state ϕ_a,

$\phi_b(1)\phi_b(2)$; both in the same state ϕ_b,

and $\frac{1}{\sqrt{2}}[\phi_a(1)\phi_b(2) + \phi_b(1)\phi_a(2)]$; representing particles in different states.

It is worth-noting from the expression of antisymmetric sate of two fermions that both cannot be placed in same state by nature. This is known as Pauli Exclusion principle. However, the other types of particles *i.e.,* bosons can occupy same state. It is also obvious that statistic properties of fermions are different from those of bosons. In fact, the probability of having both particles is same state is zero in case of fermions while for bosons the probability is $\frac{1}{3}$.

The quantum statistics are based on the assumption that similar particles can not be distinguished unlike in classical statistics (Maxwell's statistics) where one assumes that similar particles are distinguishable. In classical statistics states of pair of similar particles are

$$\phi_a(1)\phi_a(2),\ \phi_b(1)\phi_b(2),\ \phi_a(1)\phi_b(2) \quad \text{and} \quad \phi_b(1)\,\phi_a(2).$$

That is the probability of having pair of similar particles in same state is $\frac{1}{4}$.

The calculations have shown that the probability of occupancy of a state of energy E by a fermion in the assembly of fermion in thermal equilibrium is given by $\frac{1}{[\text{Exp}\,(\alpha + \beta E) + 1]}$ where α is constant and $\beta = \frac{1}{k_B T}$, k_B and T being Boltzmann constant and temperature respectively.

This function is called Fermi-Dirac statistical distribution. For the assembly of identical and non-interacting bosons this probability is given by $\frac{1}{[\text{Exp}\,(\alpha + \beta E) - 1]}$, known as Bose-Einstein statistical distributions function.

The Fermi-Dirac statistics has been applied to system of conduction electrons in solids and various properties of solid such as specific heat etc. explained successfully, Bose-Einstein distributions function was used for the assembly of photons of black body radiation and its properties like energy spectrum etc. studied quite successfully. (Note : In nature, particles are either fermions or bosons.). Lastly, we observe that at high energies both the probability distributions reduce to

$$\frac{1}{\text{Exp}\,(\alpha + \beta E)} = \text{Exp}\,[-(\alpha + \beta E)]$$

This distribution function represents classical statistics called Maxwell distribution function.

29.4 POINT DEFECTS IN CRYSTALS

The point imperfections in crystal due to chemical impurities, vacant lattice sites and extra atoms present at places other than regular lattice positions are known as Point Defects. There are also other types of imperfections called linear and planar imperfections.

Since the properties of the crystals significantly depend on these imperfections, the knowledge of these is essential. In the following we discuss only point imperfections called point defects. These are of mainly two types

1. Lattice Vacancy or Schottky defect

and 2. Frenkel defect.

Schottky Defect

The absence of positive ion, negative ion or pair of these are called Lattice vacancies or Schottky defect. They are shown by a square in the illustrations (See Fig. 29.13). One creates Schottky defect by transferring the ion from its site in the interior of crystal to the lattice site on the crystal surface.

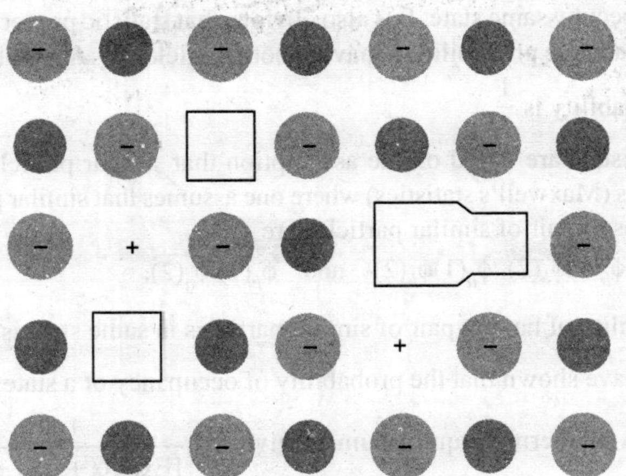

FIGURE 29.13 **A plane of a pure alkali halide crystal, showing a vacant positive ion site, a vacant negative ion site, and a coupled pair of vacant sites of opposite sign.**

In thermal equilibrium some lattice vacancies are always present in the crystal because of the principle of increase of entropy by the presence of disorder. The proportion of such lattice vacancies depends on the temperature and is of the order of 10^{-3} to 10^{-4} in metals. In metal carbides such as TiC it is as high as 50%.

The calculations have shown that for thermal equilibrium the proportions of lattice vacancies is given by

$$\frac{n}{N-n} = \text{Exp}\left[-\frac{E_v}{k_B T}\right]$$

where n = No. of lattice vacancies

N = Total no. of atoms

E_v = Energy required to cause Schottky vacancy

and k_B = Boltzmann constant

Obviously for $n << N$ it comes out as

$$\frac{n}{N} \approx \text{Exp}\left[-\frac{E_v}{k_B T}\right] \quad \text{called Boltzmann factor}$$

That is the proportion of Schottky defect is proportional to Boltzmann factor.

As an example when

$$E_v = 1 \text{ eV}$$

$$T = 1500 \text{ K}$$

$$\frac{n}{N} \approx e^{-7.73} = 4.4 \times 10^{-4}$$

Also, $\frac{n}{N}$ will decrease as T decreases and in practice actual value is found higher than this. Fig. 29.14 illustrates the creation of Schottky defect. It is very common in alkali halide crystals.*

Frenkel Defect

When an atom (or ion) is displaced from its regular site to an interstitial position we get a defect called Frenkel defect. It has been also illustrated in Fig. 29.14, the crystal being of alkali halides. In pure silver halide the most common defect is Frenkel defect.

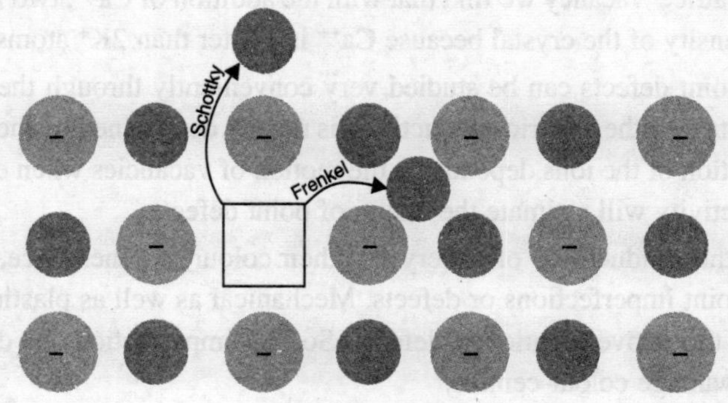

FIGURE 29.14 **Schottky and Frenkel defects in an ionic crystal. The arrows indicate the displacement of the ions.**

* In alkalihalides the creation of pair of positive and negative ions is more favourable. In which case $n/N \approx \text{Exp}\,[-E_p/2k_B T]$; E_p being the energy required to create the pair of vacancies and n is the concentration of pairs.

The proportion of Frenkel defect as calculations have shown, is given by

$$\frac{n}{(NN')^{1/2}} \approx \text{Exp}\left[-\frac{E_I}{2k_B T}\right]; \frac{n}{N} << 1$$

where n = No. of Frenkel defects

N = No. of lattice sites

N' = No. of interstitial sites

and E_I = Energy required to displace the atom from its lattice site to the interstitial position.

Lattice Vacancies due to Chemical Impurities

The point defects or lattice vacancies may also be created with the help of chemical impurities. If alkali halide (KCl) crystal is prepared under the controlled amount of $CaCl_2$, one obtains lattice vacancies in the crystal of KCl.

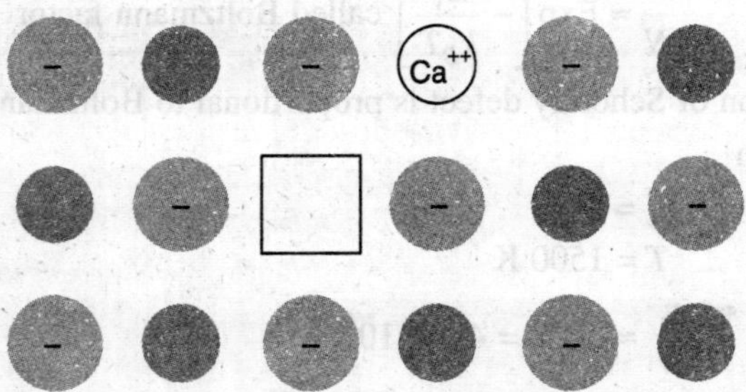

FIGURE 29.15 **Production of a lattice vacancy by the solution of $CaCl_2$ in KCl : to ensure electrical neutrality a positive ion vacancy is introduced into the lattice with each divalent cation Ca^{++}.**

In Fig. 29.15 such a point defect has been illustrated. The Ca^{++} enters at K^+ site while its Cl^- ions get positioned at the normal sites of Cl^- of KCl crystal. Now, in order to maintain the neutrality in the region, a vacancy at the K^+ site is also caused as Ca^{++} has two units of +ve charge. This situation was verified by measuring the density of the crystal. The observations show that the density decreases and by the above mechanism of causing lattice vacancy we find that with the addition of Ca^{++}, two K^+ atoms would disappear, thereby reducing the density of the crystal because Ca^{++} is lighter than $2K^+$ atoms.

The nature of point defects can be studied very conveniently through the measurements of electric conductivity of the crystal. As the electric conductivity is mainly determined by the motion of ions rather than electrons. Since the motion of the ions depends on the motion of vacancies when electric field is applied, the measurement of conductivity will estimate the nature of point defects.

Lastly, besides the conductivity of the crystals their colour, luminescence, and atomic diffusion may also be controlled by point imperfections or defects. Mechanical as well as plastic properties of the crystals are also influenced by the above mentioned defects. So, the imperfections or defects are usually created deliberately to ones advantage colour centres.

The role of point defects or imperfections in colouring the alkali halide crystals is most important. It led to a new field of research for the researchers. In fact, a lattice defect which absorbs visible light and gives colour to the crystal is defined as colour centres. These are deliberately created in the crystals for their investigations. Following types of colour centres are common in such studies.

- *F*-centre
- *M*-centre
- F_A-centre
- *R*-centre

***F*-centre**

In alkali halides when metal ions are deposited in excess we obtain *F*-centre. The loosely bound electron of excess metal ion gets trapped in the negative ion vacancy of the alkali halide crystal. This is because the negative ion vacancy in the crystal has attained equal positive charge and traps the loosely bound electron of the metal ion in excess. This trapped electron has excited states. As a result when visible radiation falls on the crystal, the system of this trapped electron absorbs the radiation through electric dipole excitation. So, we get absorption bands in the transmitted radiation thereby giving colour to the crystal. Such colour centres are called F-centres (See Fig. 29.16).

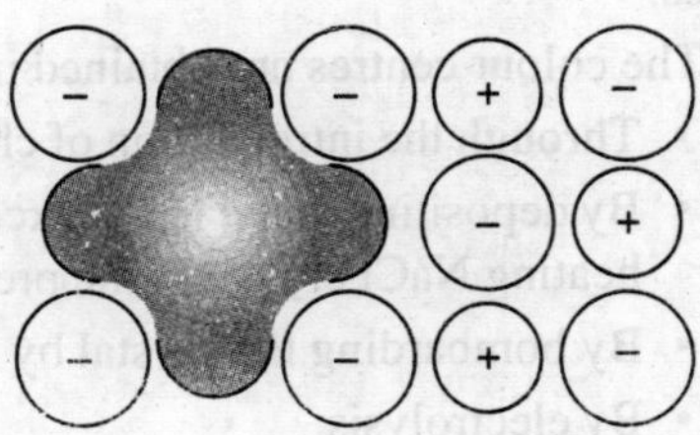

FIGURE 29.16 An *F* centre is a negative ion vacancy with one excess electron bound at the vacancy. The distribution of the excess electron is largely on the positive metal ions adjacent to the vacant lattice site.

F_A-centre

The negative ion vacancy in the alkali halide crystals always has six nearest neighbours of positive metal ions. When one of these is replaced by another metal ion *e.g.* K^+ in KCl is replaced by Na^+, we get F_A-centre. It also gives rise to colour to the crystal. Fig. 29.17 illustrates such F_A-centre.

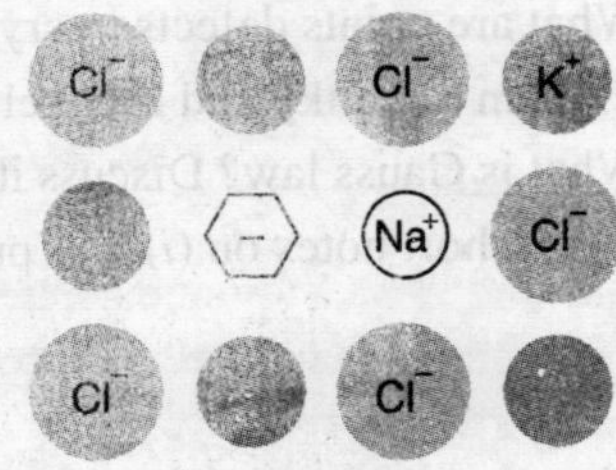

FIGURE 29.17 An F_A centre in KCl : one of the six K^+ ions which bind an *F* centre is replaced by another alkali ion, here Na^+.

***M*-centre**

Pair of adjacent *F*-centres constitutes what we call as *M*-centre. It has its characteristic absorption band when visible radiation is incident on the crystal. Fig. 29.18 displays system of two *F*-centres.

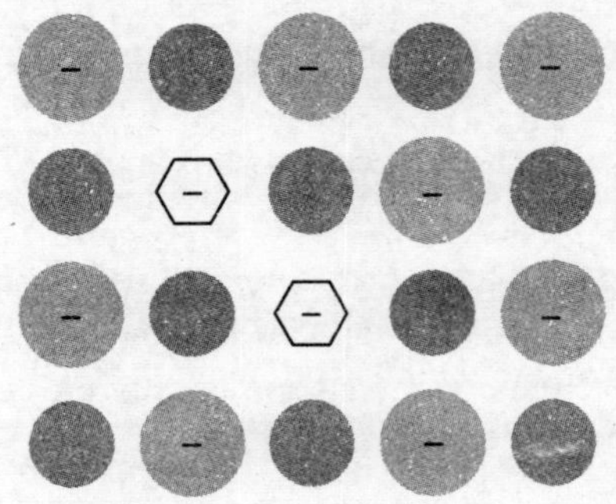

FIGURE 29.18 An *M*-centre consists of two adjacent *F* centres.

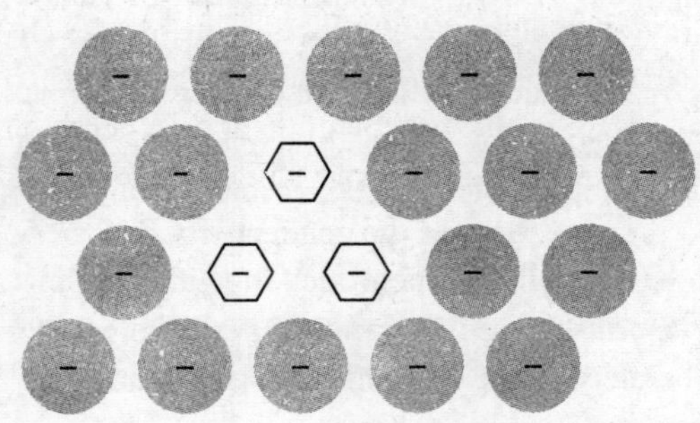

FIGURE 29.19 An *R*-centre consists of three adjacent *F* centres ; that is a group of three negative ion vacancies, with three associated electrons.

R-centres forming an M-centre

When there are three adjacent vacancies of negative ions *i.e.* *F*-centres in the alkali halide crystals (ionic crystal) we get a system called *R*-centre. This has been shown in Fig. 29.19. Here we have a system of three *F*-centres.

The colour centres are obtained in the following ways :

- Through the introduction of chemical impurities.
- By depositing metal ion in excess *e.g.* growing KCl crystal in the presence of potassium vapours or heating NaCl crystal in the presence of sodium vapours.
- By bombarding the crystal by X-rays, γ-rays, neutrons or electrons etc.
- By electrolysis.

EXERCISE

1. Explain the interaction of radiation with matter in detail.
2. Give elementary idea of quantum statistics.
3. What are points defects in crystals? Discuss these in detail.
4. Explain Schottky and Frenkel defect in solids.
5. What is Gauss law? Discuss its two applications.
6. Write short notes on (*i*) pair production (*ii*) Compton scattering (*iii*) photoelectric effect.

Chapter **30** *NANOSCIENCE*

30.1 INTRODUCTION

The concept of nanotechnology begins with the boldly speculative 1959 lecture '*There's Plenty of Rooms at the Bottom*' by Nobel Prize winning theoretical physicist Richard Feynman. In his lecture he said he was not afraid to consider the question whether 'ultimately—in the great future—we can arrange the atoms the way we want; the very atoms, all the way down!'. The word 'nanotechnology' was actually first coined by Japanese scientist Taniguchi Nori in 1974, but in the much narrower context of ultra fine machining.

Nanotechnology is receiving a lot of attention of late across the globe. The term nano originates from the Greek and it means "dwarf". The term indicates physical dimensions that are in range of one-billionth (10^{-9}) of a metre. One nanometre is approximately the length of two hydrogen atoms.

One nanometre (one billionth of a metre) is a magical point on the dimensional scale. Nanostructures are at the confluence of the smallest of human-made devices and the largest molecules of living things. Nanoscale science and engineering here refer to the fundamental understanding and resulting technological advances arising from the exploitation of new physical, chemical and biological properties of systems that are intermediate in size, between isolated atoms and molecules and bulk materials, where the transitional properties between the two limits can be controlled.

Why cannot we write the entire 24 volumes of the Encyclopaedia Brittanica on the head of a pin?

Let's see what would be involved. The head of a pin is a sixteenth of an inch across. If you magnify it by 25,000 diameters, the area of the head of the pin is then equal to the area of all the pages of the Encyclopaedia Brittanica. Therefore, all it is necessary to do is to reduce in size all the writing in the Encyclopaedia by 25,000 times. Is that possible? The resolving power of the eye is about 1/120 of an inch—that is roughly the diameter of one of the little dots on the fine half-tone reproductions in the Encyclopaedia. This, when you demagnify it by 25,000 times, is still 80 angstroms in diameter—32 atoms across, in an ordinary metal. In other words, one of those dots still would contain in its area 1,000 atoms. So, each dot can easily be adjusted in size as required by the photoengraving, and there is no question that there is enough room on the head of a pin to put all of the Encyclopaedia Brittanica.

30.2 CONCEPT OF NANOSYSTEMS

Although both fields deal with very small things, nanotechnology should not be confused with its sister field, which is even more of a mouthful-microelectromechanical systems (MEMS). MEMS scientists and engineers are interested in very small robots with manipulator arms that can do things like flow through the bloodstream, deliver drugs, and repair tissue. These tiny robots could also have a host of other applications including manufacturing's, assembling, and repairing larger systems. MEMS is already used in triggering mechanisms for automobile airbags and as well as other applications. But while MEMS does have some crossover with nanotechnology, they are by no means the same. For one thing, MEMS is concerned with structures between 1,000 and 1,000,000 nanometres, much bigger than the nanoscale. Further, nanoscience and nanotechnology are concerned with all properties of structures on the nanoscale, whether they are chemical, physical, quantum, or mechanical. It is more diverse and stretches into dozens of subfields. Nanotech is not nanobots.

A Different Kind of Small

Imagine something we would all like to have: a cube of gold that is 3 feet on each side. Now take the imaginary cube and slice it in half along its length, width, and height to produce eight little cubes, each 18 inches (50 centimetre) on a side. The properties (excepting cash value) of each of the eight smaller cubes will be exactly the same as the properties of the big one: each will still be gold, yellow, shiny, and heavy. Each will still be a soft, electrically conductive metal with the same melting point it had before you cut it. Aside from making your gold a bit easier to carry, you won't have accomplished much at all.

Now imagine taking one of the eight 18-inch (50-centimetre) cubes cutting it the same way. Each of the eight resulting cubes will now be 9 inches (25 centimetre) on a side and will have the same properties as the parent cube before we started cutting it. If we continue the gold in this way and proceed down in size from feet to inches, from inches to centimetres, from centimetres to millimetres, and from millimetres to microns, we will still notice no change in the properties of gold. Each time, the gold cubes will get smaller. Eventually we will not be able to see them with the naked eye and we'll start to need some fancy tools to keep cutting. Still, all the gold bricks' physical and chemical properties will be unchanged. This much is obvious from our real world experience—at the macro scale chemical and physical properties of materials are not size dependent. It doesn't matter whether the cubes are gold, iron, lead, plastic, ice, or brass.

When we reach the nanoscale, though, everything will change, including the gold's color, melting point, and chemical properties. The reason for this change has to do with the nature of interactions among the atoms that make up the gold, interactions that are averaged out of existence in the bulk material. Nano gold doesn't act like bulk gold.

The last few steps of the cutting required to get the gold cube down to the nanoscale represent a kind of nanofabrication, or nanoscale manufacturing. Starting with a suitcase-sized chunk of gold, our successive cutting has brought it down to the nanoscale. This particular kind of nanofabrication is sometimes called top-down nanofabrication because we started with a large structure and proceeded to make it smaller. Conversely, starting with individual atoms and building up to a nanostructure is called bottom-up nanofabrication. The tiny gold nanostructures that we prepared are sometimes called quantum dots or nanodots because they are roughly dot-shaped and have diameters at the nanoscale.

Gordon Moore, one of the founders of the Intel Corporation, came up with empirical law to describe the amazing advances in integrated circuit electronics. Moore's law says that the amount of space required to install a transistor on a chip shrinks by roughly half every 18 months. This means that the spot that could hold one transistor 15 years ago can hold 1,000 transistors today. Fig. 30.1 shows Moore's law in a graphical way. The line gives the size of a feature on a chip and shows how it has very rapidly got smaller with time.

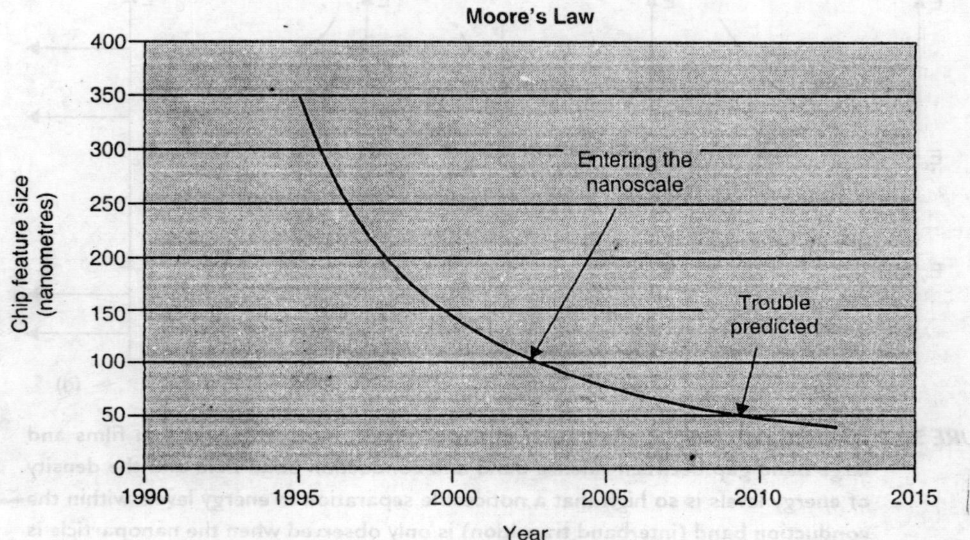

FIGURE 30.1 Moore's Law.

30.3 QUANTUM SIZE EFFECTS

Unique property of nanomaterials may also arise from another quantum size effect. When the size of nanocrystal (*i.e.*, a single crystal nanoparticle) is smaller than the de Broglie wavelength, electrons and holes are spatially confined and electric dipoles are formed, and discrete energy level would be formed in all materials. Similar to a particle in a box, the energy separation between the adjacent levels increases with decreasing dimensions. Fig. 30.2 schematically illustrates such discrete electronic configurations in nanocrystals, nanowires and thin films; the electronic configurations of nanomaterials are significantly different from that of their bulk counterpart. These changes arise through systematic transformations in the density of electronic energy levels as a functions of the size, and these changes result in strong variations in the optical and electric properties with size.

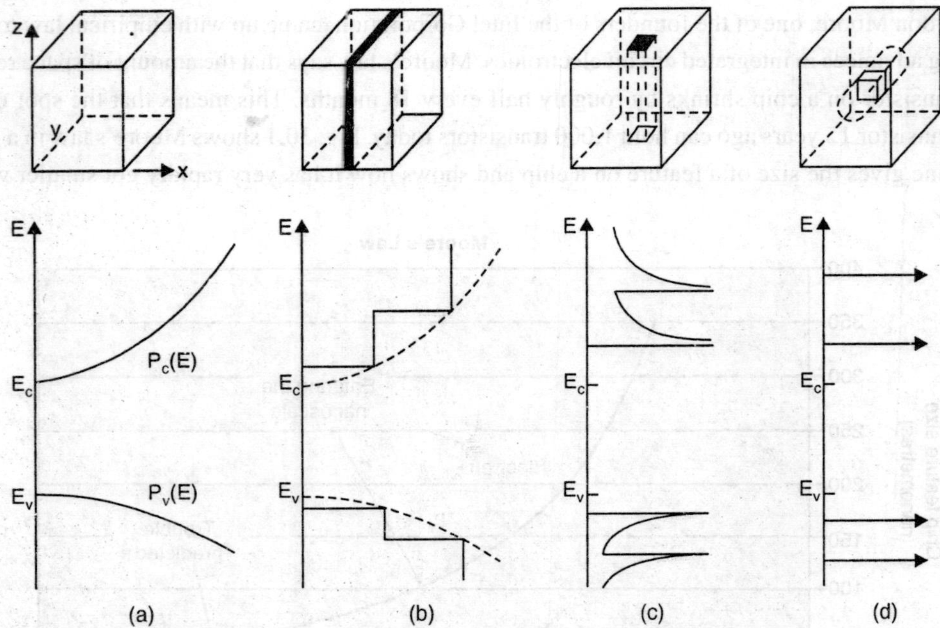

FIGURE 30.2 **Discrete electronic configuration in nanocrystals, nanowires and thin films and large band gap between valence band and conduction band field and the density of energy levels is so high that a noticeable separation in energy levels within the conduction band (interband transition) is only observed when the nanoparticle is made up of ~ 100 atoms.**

30.4 QUANTUM DOTS AND THEIR APPLICATIONS

A ***quantum dot*** is a semiconductor whose *excitons* are confined in all three spatial dimensions. As a result, they have properties that are between those of bulk semiconductors and those of discrete molecules. They were discovered by Louis E. Brus, who was then at Bell Labs. The term "Quantum Dot" was coined by Mark Reed.

Being zero dimensional, quantum dots have a sharper density of states than higher-dimensional structures. As a result, they have superior transport and optical properties, and are being researched for use in diode lasers, amplifiers, and biological sensors. Quantum dots may be excited within the locally enhanced electromagnetic field produced by the gold nanoparticles, which can then be observed from the surface Plasmon resonance in the photoluminescent excitation spectrum of (CdSe)ZnS nanocrystals. High-quality quantum dots are well suited for optical encoding and multiplexing applications due to their broad excitation profiles and narrow/symmetric emission spectra. The new generations of quantum dots have far-reaching potential for the study of intracellular processes at the single-molecule level, high-resolution cellular imaging, long-term in vivo observation of cell trafficking, tumor targetting, and diagnostics.

Quantum dots can best be described as false atoms. The primary material that a quantum dot is made out of is called a "hole", or a substance that is missing an electron from its valence band giving it a positive charge. The primary material is extremely small, that is why it is called a dot, and at that size, electrons start to orbit it. Since quantum dots do not have protons or neutrons in the centre, their mass is much smaller. Since the mass at the centre is smaller than that of an atom, quantum dots exert a smaller force on the orbiting electrons causing an orbit larger ... of a regular atom (Fig. 30.3) (K. Daneshvar, personal communication, Jul. 15, 2005). With a mass that ... scientists are able to precisely calculate and change the size of the band-gap of the quantum dot by adding ... electrons. The band-gap of a quantum dot is what determines which frequencies it will respond to, so ... to change the band-gap is what gives scientists more control and more flexibility when dealing with ...ions.

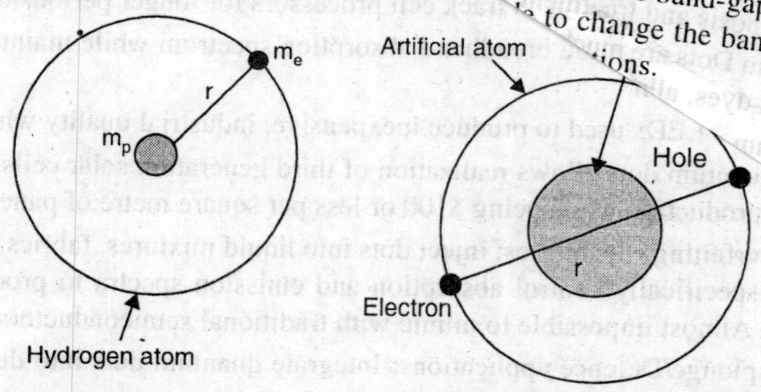

FIGURE 30.3 **The above image compares the orbit of a hydrogen atom to that of a quantum dot. Since the artificial atom has almost no mass compared to the hydrogen atom, the orbit is much larger.**

Applications

Microelectronic technology has focused on reducing the size of transistors to produce increasingly smaller; faster and more efficient computers (Shrinking Information Storage to the Molecular Level). However, this method is reaching its physical limit due to the restrictions placed by the laws of physics that do not allow these devices to operate below a certain size. With this advantageous feature of Quantum Dots, information storage can be brought down to the molecular level. Since no flow of electrons to transmit a signal is needed, electric current does not need to be produced and heat problems are avoided. Also, the quantum dot devices are sensitive enough to and can make a usage of the charges of single electrons. With improvements in quantum dot ordering and positioning, it is possible for us to hope in the near future to address and store information optically in a single quantum dot, thus opening the possibility of ultrahighdensity memory devices.

1. Quantum Dots have applications like quantum dot lasers which promises far more great advantages than quantum well lasers. Because Quantum Dot lasers are less temperature-dependent and less likely to degrade under elevated temperature, it allows more flexibility for lasers to operate more efficiently.
2. The Quantum Dots are used in optical switches would increase the rate at which data can be transferred. With regular switches, data can only travel as fast as the electrical current can, but optical switches can travel almost as fast as the speed of light.
3. The treasury could engineer quantum dots to be responsive to a specific frequency of light and suspend them in ink that they would print onto money. Shining light with the same frequency that the ink solution has been engineered for would reveal whether or not the money is real or counterfeit. This idea can be used for just about any substance that could be illegally duplicated.
4. Though Quantum Dots are still under research for other possible applications and need more technological advancement in order to be put into use, the features introduced will grant far better optical communication, significant change in electronic devices, and even detection of antigens in the body tissues.

cancerous cells and providing a means to

5. Quantum Dots can be useful tool for mon[...]e armed with tumor-fighting toxic therapies to provide understand its evolution. [...]er.
6. In the future, Quantum Dots resistant to degradation than other optical imaging probes such as the diagnosis and trealem to track cell processors for longer periods of time.
7. Quantum Dot[...]er a wide broadband absorption spectrum while maintaining a distinct, static organic [...]avelength.
8. [...]um Dots LEDs used to produce inexpensive, industrial quality white light.
[...]tilizing quantum dots allows realization of third generation solar cells at ~ 60% efficiency in electricity production while being $100 or less per square metre of paneling necessary.
11. Anti-counterfeiting capabilities: inject dots into liquid mixtures, fabrics, polymer matrices, etc. Ability to specifically control absorption and emission spectra to produce unique validation signatures. Almost impossible to mimic with traditional semiconductors.
12. Counter-espionge/Defence applications: Integrate quantum dots into dust that tracks enemies. Protection against friendly-fire events.
13. Research Continues: The possibilities seem endless...

EXERCISE

1. Explain the concept of quantum size effect.
2. What are quantum dots? Write down in detail the applications of quantum dots.
3. What do you mean by Nanosystems? Explain in detail.
4. Define 'Nanotechnology'. Give the applications of Nanotechnology.

APPENDICES

APPENDIX I

YOUNG'S DOUBLE SLIT EXPERIMENT

The earliest experiment to demonstrate the interference was Young's double slit experiment, which is described below :

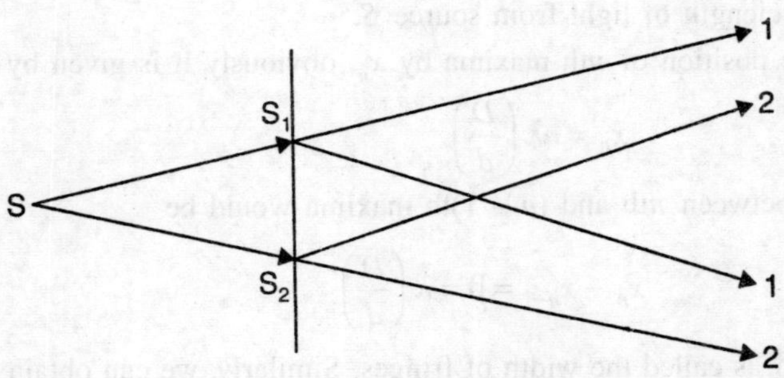

FIGURE I(a) **S is the monochromatic source while S_1 and S_2 are the pin holes. ray 1 and ray 2 represent light from S_1 and S_2 respectively.**

Obviously, S_1 and S_2 being illuminated by light from S, act as two coherent sources regulated from the division of the wavefront falling on the screen A. The interference will occur in the region, where light from both S_1 and S_2 overlap and can be observed on the photographic plate, B. In order to determine the position of maxima and minima we consider Fig. I(b).

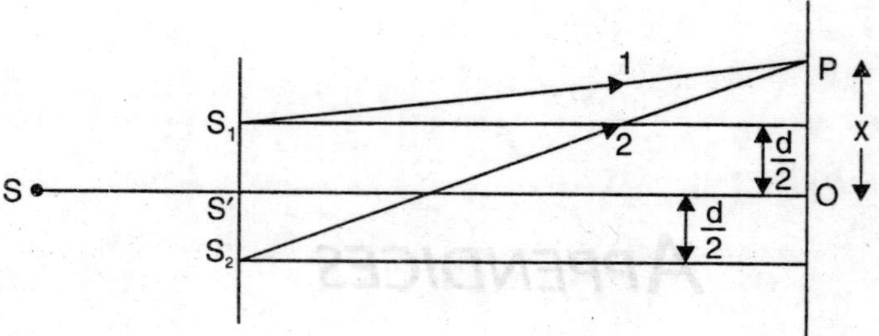

FIGURE I(b) **Illustrating interference from two point sources S_1 and S_2, $S_1S_2 = d$, $OP = x$, and $S'O = D$.**

Since S_1 and S_2 are equidistance from the primary source S, the path difference between two rays 1 and ray 2 will be given by

$$S_2P - S_1P = \sqrt{\left(x + \frac{d}{2}\right)^2 + D^2} - \sqrt{\left(x - \frac{d}{2}\right)^2 + D^2}$$

$$= D\left[1 + \frac{\left(x + \frac{d}{2}\right)^2}{2D^2} - 1 - \frac{\left(x - \frac{d}{2}\right)^2}{2D^2}\right] \qquad \left(\text{Using binomial expansion when } \frac{x}{d} << 1\right)$$

$$= \frac{xd}{D}$$

Thus, P will be nth maxima if $\qquad \frac{xd}{D} = n\lambda\,; \qquad n = 0, 1, 2, \ldots\ldots \qquad$ ...(I-1)

While it will be one of the minima if $\qquad \frac{xd}{D} = (2n-1)\frac{\lambda}{2}\,; \qquad n = 1, 2, \ldots\ldots \qquad$ ...(I-2)

where l denotes the wavelength of light from source S.

Let us denote the position of nth maxima by x_n, obviously it is given by

$$x_n = n\lambda\left(\frac{D}{d}\right) \qquad \text{[From eq. (I-1)]}$$

Then separation between nth and $(n - 1)$th maxima would be

$$x_n - x_{n-1} = \beta = \lambda\left(\frac{D}{d}\right) \qquad \text{...(I-3)}$$

This separation, β, is called the width of fringes. Similarly, we can obtain the separation between two consecutive minima(dark fringes) as

$$= \frac{D}{d} \cdot \frac{\lambda}{2}[(2n-1) - \{2(n-1) - 1\}]$$

$$= \frac{\lambda D}{d} \quad \text{same as between two consecutive maxima (bright fringes)}$$

Thus, we obtain bright and dark interference fringes of equal widths.

Other arrangement of getting pair of coherent sources are also being shown below *i.e.* Fresnel's double mirrors [Fig. I(*c*)], Lloyd's mirror [Fig. I(*d*)] and Fresnel's biprism [Fig. I(*e*)].

The interference fringes which have fixed location are called localised fringes and those which do not have fixed location are called non localised fringes. Interference fringes obtained in Michelson's interferometer, Newton's rings experiments are localised fringes while those of Fresnel's biprism are Non-localised.

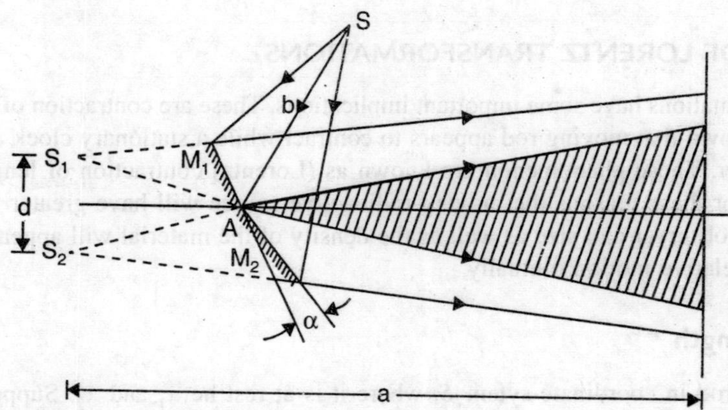

FIGURE I(c) **Arrangement of Fresnel's double mirrors M_1 and M_2 S_1 and S_2 form a pair of coherent sources.**

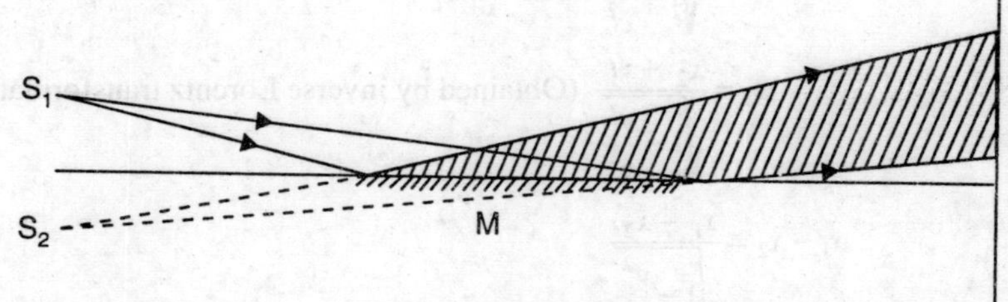

FIGURE I(d) **Lloyd's mirror *M*. S_2 is the virtual image of primary source S_1. S_1 and S_2 form a pair of coherent sources.**

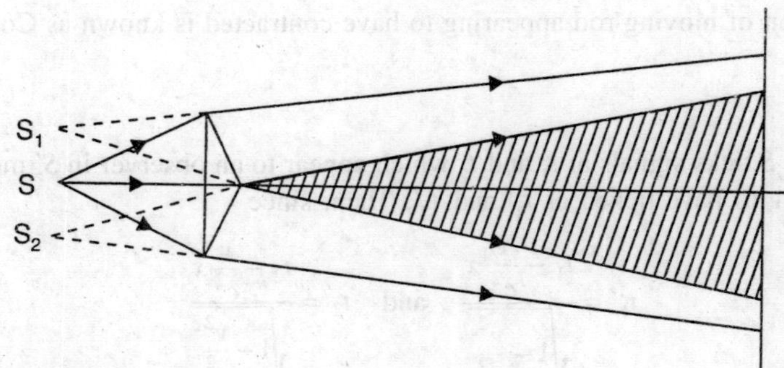

FIGURE I(e) **Fresnel's biprism producing a pair of coherent sources from the primary source *S*.**

APPENDIX II

IMPLICATIONS OF LORENTZ TRANSFORMATIONS

The Lorentz transformations have some important implications. These are contraction of length and Dilation of time. It can be shown that moving rod appears to contract while a stationary clock appears to run slow to a moving observer. These phenomenon are known as (Lorentz) contraction of length and Dilation of time respectively. For these reasons the moving radioactive source will have greater half-life than when it is at rest w.r.t. the observer and size as well as the density of the material will appear differently to two observers when in relative motion mutually.

Contraction of Length

Let the ends of the rod in coordinate sytem S, where it is at rest be x_1 and x_2. Suppose the observer in S' which is moving with respect to S along common X-axis with velocity v observes that these ends are at x_1' and x_2' at the same time t such that

$$x_1 = \frac{x_1' + vt'}{\sqrt{1 - \frac{v^2}{c^2}}}$$

$$x_2 = \frac{x_2' + vt'}{\sqrt{1 - \frac{v^2}{c^2}}} \quad \text{(Obtained by inverse Lorentz transformations)}$$

Then $$x_1 - x_2 = \frac{x_1' - x_2'}{\sqrt{1 - \frac{v^2}{c^2}}}$$

That is $$l' = l\sqrt{1 - \frac{v^2}{c^2}} \quad \text{or} \quad l' < l \quad \left(\text{If } l = 1 \text{ m}, \frac{v}{c} = \frac{1}{2}\,;\ l' = 0.866 \text{ m}\right)$$

This phenomenon of moving rod appearing to have contracted is known as Contraction of length.

Time Dilation

Let a clock be at rest in S ticks signals at t_1 and t_2 which appear to an observer in S' moving w.r.t. S along X-axis with a velocity v to have ticked at t_1' and t_2'. Then, since

$$t_1' = \frac{t_1 - \frac{v}{c^2}x}{\sqrt{1 - \frac{v^2}{c^2}}} \quad \text{and} \quad t_2' = \frac{t_2 - \frac{v}{c^2}x}{\sqrt{1 - \frac{v^2}{c^2}}}$$

$$t_2' - t_1' = \Delta t' = \frac{t_2 - t_1}{\sqrt{1 - \frac{v^2}{c^2}}} = \frac{\Delta t}{\sqrt{1 - \frac{v^2}{c^2}}} \quad \left(\text{If } \Delta t = 1 \text{ hr}, \frac{v}{c} = \frac{1}{2}\,;\ \Delta t' = 1.155 \text{ hr}\right)$$

So, $$\Delta t' > \Delta t$$

Hence the clock at rest in S appears to be slow to the observer in relative motion. This phenomenon is known as dilation of time.

APPENDIX III

VALUES OF PHYSICAL CONSTANTS

(The values listed below are taken from the book, *The Fundamental Constants of Physics,* by E. R. Cohen, K.M. Crowe, and J. W. M. DuMond. New York : Interscience Publishers, Inc., 1957.)

Constant	*Value*
Avogadro's number	$N_0 = (6.02486 \pm 0.00016) \times 10^{23}$ per gram mole
Velocity of light	$c = (2.997930 \pm 0.000003) \times 10^{10}$ cm/sec
Standard volume of a perfect gas	$V_0 = 22420.7 \pm 0.6$ cm^3/mole
Gas constant per mole	$R_0 = (8.31696 \pm 0.00034) \times 10^7$ erg/mole°C
Boltzmann's constant	$k = R_0/N_0$
	$= (1.38044 \pm 0.00007) \times 10^{-16}$ erg/°K
	$= (8.6167 \pm 0.0004) \times 10^{-5}$ ev/°K
Charge on the electron	$e = (4.80286 \pm 0.00009) \times 10^{-10}$ esu
	$= (1.60206 \pm 0.00003) \times 10^{-20}$ emu
Faraday constant (physical scale)	$F = N_0 e = (2.89366 \pm 0.00003) \times 10^{14}$ esu/(gm mole)
	$= (9652.19 \pm 0.11)$ emu/(gm mole)
Planck's constant	$h = (6.62517 \pm 0.00023) \times 10^{-27}$ erg-sec
	$h/2\pi = (1.05443 \pm 0.00004) \times 10^{-27}$ erg-sec
Rydberg constants	$R_\infty = 109737.309 \pm 0.012$ em^{-1}
	$R_H = 109677.576 \pm 0.012$ cm^{-1}
	$R_D = 109707.419 \pm 0.012$ cm^{-1}
	$R_{He} = 109722.269 \pm 0.012$ cm^{-1}
Rest masses	neutron : 1.008982 ± 0.000003 amu
	$= (1.67470 \pm 0.00004) \times 10^{-24}$ gm
	proton : 1.007593 ± 0.000003 amu
	$= (1.67239 \pm 0.00004) \times 10^{-24}$ gm
	hydrogen atom : 1.008142 ± 0.000003 amu
	ratio : $\frac{\text{mass of hydrogen atom}}{\text{mass of proton}}$
	$= 1.000544613 \pm 0.000000006$
	electron : $(5.48763 \pm 0.00006) \times 10^{-4}$ amu
	$= (9.1083 \pm 0.0003) \times 10^{-28}$ gm
	ratio : $\frac{\text{proton mass}}{\text{electron mass}}$
	$= 1836.12 \pm 0.02$
	deuterium atom : 2.014735 ± 0.000006 amu

(*Continued*)

VALUES OF PHYSICAL CONSTANTS—(*Continued*)

Constant	*Value*
Mass-energy conversion factors	1 amu = 931.141 ± 0.010 Mev
	1 proton mass = 938.211 ± 0.010 Mev
	1 neutron mass = 939.505 ± 0.010 Mev
	1 electron mass = 0.510976 ± 0.000007 Mev
	1 gm = $(5.61000 \pm 0.00011) \times 10^{26}$ Mev
Energy conversion factor	1 ev = $(1.60206 \pm 0.00003) \times 10^{-12}$ erg
Energy of a 2200 m/sec neutron	E_{2200} = (0.0252973 ± 0.0000003 ev
	T_{2200} = 293.585 ± 0.012°K
	= 20.435 ± 0.012°C.
First Bohr radius	$a_0 = h^2/4\pi^2 me^2$
	$= (5.29172 \pm 0.00002) \times 10^{-9}$ cm
Classical electron radius	$r_0 = e^2/mc^2$
	$= (2.81785 \pm 0.00004) \times 10^{-13}$ cm
Thomson cross section	$\frac{8}{3}\pi r_0^2 = (6.65205 \pm 0.00018) \times 10^{-25}$ cm²
Compton wavelength	electron : $\lambda_{ce} = (24.2626 \pm 0.0002) \times 10^{-11}$ cm
	proton : $\lambda_{cp} = (13.2141 \pm 0.0002) \times 10^{-14}$ cm
	neutron : $\lambda_{cn} = (13.1959 \pm 0.0002) \times 10^{-14}$ cm

APPENDIX IV

ALPHABETIC LIST OF THE ELEMENTS

Element	Symbol	Atomic number Z	Element	Symbol	Atomic number Z
Actinium	Ac	89	Helium	He	2
Aluminum	Al	13	Holmium	Ho	67
Americium	Am	95	Hydrogen	H	1
Antimony	Sb	51	Indium	In	49
Argon	A	18	Iodine	I	53
Arsenic	As	33	Iridium	Ir	77
Astatine	At	85	Iron	Fe	26
Barium	Ba	56	Krypton	Kr	36
Berkelium	Bk	97	Lanthanum	La	57
Beryllium	Be	4	Lead	Pb	82
Bismuth	Bi	83	Lithium	Li	3
Boron	B	5	Lutetium	Lu	71
Bromine	Br	35	Magnesium	Mg	12
Cadmium	Cd	48	Manganese	Mn	25
Calcium	Ca	20	Mendelevium	Md	101
Californium	Cf	98	Mercury	Hg	80
Carbon	C	6	Molybdenum	Mo	42
Cerium	Ce	58	Neodymium	Nd	60
Cesium	Cs	55	Neon	Ne	10
Chlorine	Cl	17	Neptunium	Np	93
Chromium	Cr	24	Nickel	Ni	28
Cobalt	Co	27	Niobium	Nb	41
Copper	Cu	29	Nitrogen	N	7
Curium	Cm	96	Nobelium	No	102
Dysprosium	Dy	66	Osmium	Os	76
Einsteinium	E	99	Oxygen	O	8
Erbium	Er	68	Palladium	Pd	46
Europium	Eu	63	Phosphorus	P	15
Fermium	Fm	100	Platinum	Pt	78
Fluorine	F	9	Plutonium	Pu	94
Francium	Fr	87	Polonium	Po	84
Gadolinium	Gd	64	Potassium	K	19
Gallium	Ga	31	Praseodymium	Pr	59
Germanium	Ge	32	Promethium	Pm	61
Gold	Au	79	Protactinium	Pa	91
Hafnium	Hf	72	Radium	Ra	88

Element	*Symbol*	*Atomic number Z*	*Element*	*Symbol*	*Atomic number Z*
Radon	Rn	86	Tellurium	Te	52
Rhenium	Re	75	Terbium	Tb	65
Rhodium	Rh	45	Thallium	Tl	81
Rubidium	Rb	37	Thorium	Th	90
Ruthenium	Ru	44	Thulium	Tm	69
Samarium	Sm	62	Tin	Sn	50
Scandium	Sc	21	Titanium	Ti	22
Selenium	Se	34	Tungsten	W	74
Silicon	Si	14	Uranium	U	92
Silver	Ag	47	Vanadium	V	23
Sodium	Na	11	Xenon	Xe	54
Strontium	Sr	38	Ytterbium	Yb	70
Sulfur	S	16	Yttrium	Y	39
Tantalum	Ta	73	Zinc	Zn	30
Technetium	Tc	43	Zirconium	Zr	40

APPENDIX V

WAVE GUIDES

Wave guides are the devices to allow the propagation of electromagnetic radiation up to an appropriate wavelength. Such a device could be cylindrical or rectangular in shape with walls made of conducting material.

In order to understand the principle of such wave guides we consider first a device of parallel conducting plates for the propagation of electromagnetic radiation [see Fig. 1].

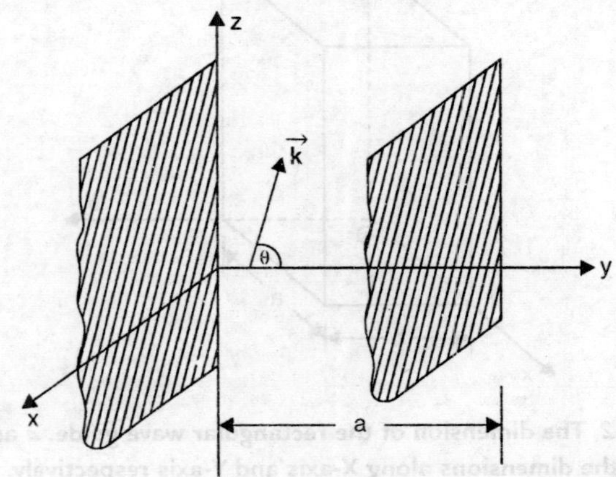

FIGURE 1 **The device of parallel conducting plates '*a*' distance apart. $\vec{k}$ is the direction of propagation.**

The pair of plates are parallel to *X-Z* plane while the wave vector $\vec{k}$ is in *Y-Z* plane making an angle θ with *Y*-axis. The flow of radiation would be either transverse electric (*TE*) or Transverse magnetic (*TM*). It can be shown that the electric field vector for *TE* will be

$$\mathbf{E} = \mathbf{i}E' \sin (ky \cos \theta) \text{ Exp. } [i(kz \sin \theta - \omega t)]$$

where $E' = 2iE_0$ with E_0 as constant.

From the form of **E** it is obvious that wavelength along *Y*-axis is given by $\lambda_c = 2\pi/k \cos \theta$ and wavelength along *Z*-direction as

$$\lambda_g = 2\pi/k \sin \theta. \qquad k = |\mathbf{k}| = 2\pi/\lambda_0 = \omega/c.$$

These results further give

$$k^2 = (2\pi/\lambda_c)^2 + (2\pi/\lambda_g)^2$$

and hence $$(2\pi/\lambda_0)^2 = (2\pi/\lambda_c)^2 + (2\pi/\lambda_g)^2$$

or $$1/\lambda_g^{\,2} = 1/\lambda_0^{\,2} - 1/\lambda_c^{\,2}\ ; \quad \lambda_0 = 2\pi c/\omega$$

The boundary conditions for the electromagnetic radiation between conducting plates also require that

$$ka \cos \theta = n\pi \text{ ; } n = +\text{ve integer ; } a \text{ being the distance between plates}$$

That is
$$\lambda_c = 2\pi a/n\pi = 2a/n$$
$$= 2a \text{ for } n = 1$$
$$= a \text{ for } n = 2 \text{ and so on.}$$

Here, *n* determines the mode of *TE* radiation.

For a given λ_c or mode of radiation, as λ_0 increases in $1/\lambda_g^{\,2} = 1/\lambda_0^{\,2} - 1/\lambda_c^{\,2}$ a stage will reach when, $\lambda_g^{\,2}$ becomes negative and λ_g an imaginary quantity. As a result the exponential factor in **E** becomes exponentially decreasing.

So, there will be no radiation for

$$1/\lambda_0^2 \le 1/\lambda_c^2 \quad \text{or} \quad \lambda_0 \ge \lambda_c = 2a/n.$$

Thus, this explains how the frequency $\omega(= 2\pi c/\lambda_0)$ of radiation gets restricted when it propagates between parallel plates. Further the group velocity is found as given by $v_g = c\lambda_0/\lambda_g < c$ as $\lambda_0/\lambda_g < 1$. (= sin θ)

Wave guides are usually cylindrical or rectangular in shape. If it is rectangular [see Fig. 2]

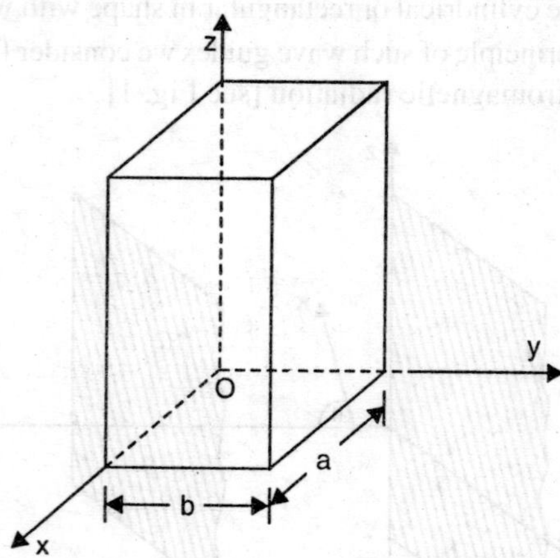

FIGURE 2 **The dimension of the rectangular wave guide. *a* and *b* are the dimensions along X-axis and Y-axis respectively.**

Let us suppose that electromagnetic radiation is to propagate along Z-axis. Then variation of electric as well as of magnetic fields along Z-axis will be of the form

$$\text{Exp}\left[i\left(\frac{2\pi}{\lambda_g}\right)z\right]$$

where λ_g is the wavelength along *Z*-axis. If we consider *TE* type of radiation, the *Z*-component of **E** would be zero.

That is, $E_z = 0$

with this situation only five field components are left to be determined.

These are E_x, E_y, H_x, H_y, and H_z.

We also consider the variation of these components with time as of the form Exp $(-i\omega t)$, (monochromatic).

Further, it can be shown with the help of Maxwell's equations that E_x, E_y and H_x, H_y all four can be expressed in terms of H_z. Hence we need to determine only H_z from the corresponding wave equation

$$\nabla^2 H_z - \varepsilon_0\mu_0\, \partial^2 H_z/\partial t^2 = 0 \text{ in the free space.}$$

Since we know the variation of H_z w.r.t. time,

we get
$$\nabla^2 H_z + \varepsilon_0\mu_0\, \omega^2 H_z = 0$$

or
$$\partial^2 H_z/\partial x^2 + \partial^2 H_z/\partial y^2 + \partial^2 H_z/\partial z^2 + \varepsilon_0\mu_0\, \omega^2 H_z = 0$$

here ε_0, μ_0 are constants for the vacuum such that $\varepsilon_0\mu_0 = 1/c^2$, *c* being the velocity of light.

Next, as variation of H_z w.r.t. *z* is governed by Exp $[i(2\pi/\lambda_g z)]$, the third term may be simplified to get

$$\partial^2 H_z/\partial x^2 + \partial^2 H_z/\partial y^2 + (\varepsilon_0\mu_0\, \omega^2 - (2\pi/\lambda_g)^2)H_z = 0$$

which equation can be solved by the technique of separation of variables. As a result, the variation of H_z w.r.t. x, y and z would be

$$H_z(x, y, z) = [\mathrm{A} \cos(k_x x) \cos(k_y y) + \mathrm{B} \cos(k_x x) \sin(k_y y) + \mathrm{C} \sin(k_x x) \cos(k_y y) + \mathrm{D} \sin(k_x x) \sin(k_y y)] \times \mathrm{Exp}\,(i\, 2\pi/\lambda_g z)$$

and $$H_z(x, y, z, t) = H_z(x, y, z)\ \mathrm{Exp}\,(-i\omega t)$$

alongwith $$-(k_x^2 + k_y^2) + (\varepsilon_0 \mu_0\, \omega^2 - (2\pi/\lambda_g)^2) = 0$$

Now, if we apply the boundary conditions such as E_x, E_y to vanish at $x = 0$ and a and at $y = 0$ and b, we find that

$$H_z(r, t) = \mathrm{A} \cos(k_x x) \cos(k_y y)\ \mathrm{Exp}\,[i(2\pi/\lambda_g z - \omega t)]$$

and $$k_x a = m\pi \text{ as well as } k_y b = n\pi$$

and that $$-((m\pi/a)^2 + (n\pi/b)^2) + \varepsilon_0 \mu_0\, \omega^2 - (2\pi/\lambda_g)^2 = 0$$

The last condition further simplifies to $(2\pi/\lambda_g)^2 = (2\pi/\lambda_0)^2 - (2\pi/\lambda_c)^2$

where $(2\pi/\lambda_0)^2 = \varepsilon_0 \mu_0\, \omega^2 = (\omega/c)^2$ and $(2\pi/\lambda_c)^2 = (m\pi/a)^2 + (n\pi/b)^2$

The above result shows that the transverse electric radiation, *TE* derived from

$$H_z(x, y, z, t) = A \cos(k_x x) \cos(k_y y)\ \mathrm{Exp}\,[i(2\pi/\lambda_g z - \omega t)]$$

where $k_x = m\pi/a$ as well as $k_y = n\pi/b$

has to satisfy $$(2\pi/\lambda_g)^2 = (2\pi/\lambda_0)^2 - (2\pi/\lambda_c)^2\ ;\ 2\pi/\lambda_c = [(m\pi/a)^2 + (n\pi/b)^2]^{1/2}$$

So, for a given $2\pi/\lambda_0 = \omega/c$, wave guide wavelength λ_g as well as the velocity $v_g = c\lambda_0/\lambda_g$ depend on m, n integral values called the mode of the radiation. It is denoted as TE_{mn}. Here 'a' and 'b' are the dimensions of the rectangular wave guide.

Further for $\lambda_0 > \lambda_c$, λ_g^2 becomes negative and λ_g a purely imaginary. This will make Exp $[i(2\pi/\lambda_g)z]$, an exponentially decreasing factor. So, there is no propagation for $\lambda_0 > \lambda_c$.

Its maximum value is

$$\lambda_0(\text{max.}) = \lambda_c = 2/[(m/a)^2 + (n/b)^2]^{1/2}$$

Usually, wave guide is so designed that the microwave power propagates only through TE_{10}. For this, cut off wavelength is given

$$\lambda_0(\text{max.}) = \lambda_c = 2/[(1/a)^2 + 0]^{1/2} = 2a$$
$$= 4.56 \text{ cm for } a = 2.28 \text{ cm}$$

Similarly, the upper limit of wavelength for the transmission of TE_{20} will be $\lambda_0(\text{max.}) = \lambda_c$

$$= 2/[(2/a)^2 + 0]^{1/2} = a = 2.28 \text{ cm}$$

The above discussion, thus, explains the transmission of microwave power through rectangular wave guides. The cylindrical wave guides also act in the same way (see Fig. 3).

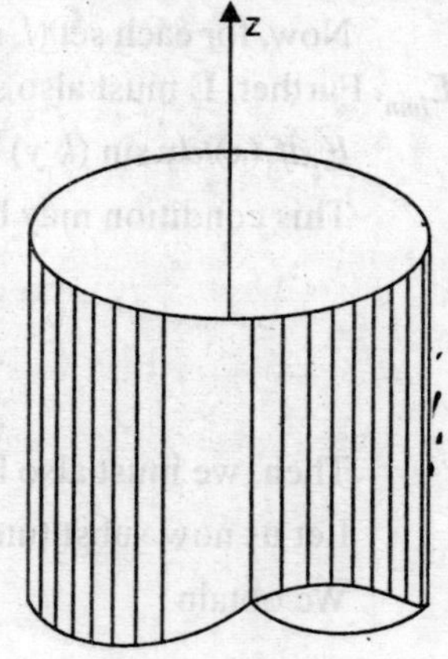

FIGURE 3 **Cylindrical wave guide.**

CAVITY RESONATOR

Cavity resonator is another device like a wave guide. Like *L-C* resonating circuits cavity resonators store electromagnetic radiations but of higher frequencies of the order of Megacycles. Another point of difference from *L-C* resonator is that loss of energy per cycle in the case of cavity resonators is very small, smaller by factor 1/20 of what a *L-C* circuit does.

Usually cavity resonator is a rectangular parallelopiped having perfectly conducting walls [see Fig. 4]. Thus, the appropriate boundary conditions would be that the normal component of **B** and the tangential components of **E** to be zero on the walls.

The wave equation for the electric field is

$$\nabla^2\mathbf{E} - \varepsilon_0\mu_0\partial^2\mathbf{E}/\partial t^2 = 0 \quad \text{(SI)}$$

Taking the variation of **E** w.r.t. time as $e^{-i\omega t}$, we get

$$\nabla^2\mathbf{E} + \varepsilon_0\mu_0\omega^2\mathbf{E} = 0$$

As $\varepsilon_0\mu_0 = 1/c^2$, this equation becomes

$$\nabla^2\mathbf{E} + \omega^2/c^2\ \mathbf{E} = 0$$

which for x-component is

$$\nabla^2 E_x + \omega^2/c^2\ E_x = 0$$

Its solution is to satisfy that

$E_x = 0$ at $y = 0$ and b as well as at $z = 0$ and d. The obvious solution is

$$E_x = E_1 f_1(x) \sin(k_y y) \sin(k_z z)\, e^{-i\omega t}$$

FIGURE 4 **Rectangular parallelopiped with sides *OA* = *a*, *OB* = *b* and *OC* = *d*.**

so that $k_y b = m\pi$

and $k_z d = n\pi$

where m and n are positive integers. Here

$$k_x^2 + k_y^2 + k_z^2 = k^2 = (\omega/c)^2$$

If we proceed in the similar fashion for E_y as well as E_z, we get $E_y = E_2 f_2(y) \sin(k_x x) \sin(k_z z)\, e^{-i\omega t}$

with $k_x a = l\pi$; l = +ve integer

and $E_z = E_3 f_3(z) \sin(k_x x) \sin(k_y y) e^{-i\omega t}$

Here, $f_1(x), f_2(y), f_3(z)$ are the arbitrary functions.

Now, for each set (l, m, n) we have a particular mode of TE radiation (Transverse electric) denoted as TE_{lmn}. Further, **E** must also satisfy the Maxwell equation $\nabla \cdot \mathbf{E} = 0$ which requires that

$$E_1 df_1(x)/dx \sin(k_y y) \sin(k_z z) + E_2\, df_2(y)/dy \sin(k_x x) \sin(k_z z) + E_3\, df_3(z)/dz \sin(k_x x) \sin(k_y y) = 0$$

This condition may be satisfied if we choose

$$f_1 = \cos(k_x x)$$

$$f_2 = \cos(k_y y)$$

$$f_3 = \cos(k_z z)$$

Then, we must also have $k_x E_1 + k_y E_2 + k_z E_3 = 0$

Let us now substitute E_x, E_y or E_z in the wave equation $\nabla^2\mathbf{E} - \varepsilon_0\mu_0\,\partial^2\mathbf{E}/\partial t^2 = 0$

We obtain

$$-(k_x^2 + k_y^2 + k_z^2) + \omega^2/c^2 = 0$$

or $\quad (l\pi/a)^2 + (m\pi/b)^2 + (n\pi/d)^2 - (2\pi f/c)^2 = 0$; $\omega = 2\pi f$, f being the frequency.

Hence, $\quad (l/a)^2 + (m/b)^2 + (n/d)^2 - (2f/c)^2 = 0$

For a typical wave guide 0.4" × 0.9" in size the frequency of TE_{102} will be given by

$$(1/a)^2 + (2/d)^2 - (2f/c)^2 = 0 \text{ ; } a = 0.4"$$

That is, the frequency f will be determined by the z-dimension of the cavity resonators *i.e., d*.

Cavity resonators in right cylindrical shape are also used to store electromagnetic radiation of high frequency. The calculations of f may be carried out in the similar manner.

TABLE Conversin of SI units into Gaussian units

Quantity	*SI*	*Gaussian*
Charge q	1 coul	3×10^9 esu
Current I	1 amp	3×10^9 esu or $\frac{1}{10}$ emu
Potential V	1 Volt	$\frac{1}{300}$ esu
Electric Field $\vec{E}$	1 Volt/m	$\frac{1}{3} \times 10^{-4}$ esu/cm
Magnetic Induction $\vec{B}$	1 n/Amp-m	10^4 gauss
Magnetic Intensity $\vec{H}$	1 Amp/m	$4\pi \times 10^{-3}$ oersted
Electric Displacement $\vec{D}$	1 coul/m^2	$12\pi \times 10^5$ esu
Flux ϕ	1 weber	10^8 Maxwells

INDEX

A

B

C

D

E

F

G

H

I

Q

R

S

T

U

V

W

X

Z